The
Herpesviruses

Volume 4

IMMUNOBIOLOGY AND PROPHYLAXIS OF HUMAN HERPESVIRUS INFECTIONS

THE VIRUSES

Series Editors
HEINZ FRAENKEL-CONRAT, *University of California*
Berkeley, California

ROBERT R. WAGNER, *University of Virginia School of Medicine*
Charlottesville, Virginia

THE HERPESVIRUSES
Volumes 1–3 • Edited by Bernard Roizman
Volume 4 • Edited by Bernard Roizman and Carlos Lopez

THE REOVIRIDAE
Edited by Wolfgang K. Joklik

THE PARVOVIRUSES
Edited by Kenneth I. Berns

THE ADENOVIRUSES
Edited by Harold S. Ginsberg

THE VIRUSES: Catalogue, Characterization, and Classification
Heinz Fraenkel-Conrat

The Herpesviruses

Volume 4

IMMUNOBIOLOGY AND
PROPHYLAXIS OF HUMAN
HERPESVIRUS INFECTIONS

Edited by

BERNARD ROIZMAN

University of Chicago
Chicago, Illinois

and

CARLOS LOPEZ

Sloan-Kettering Institute for Cancer Research
New York, New York

PLENUM PRESS • NEW YORK AND LONDON

Library of Congress Cataloging in Publication Data

Main entry under title:

The Herpesviruses.

The Viruses.
Includes bibliographies and indexes.
1. Herpesvirus diseases. 2. Herpesviruses. I. Roizman, Bernard, 1929–
[DNLM: 1. Herpesviridae. QW 165.5.H3 H5637]
RC147.H6H57   1982   616.9′25   82-15034
ISBN-13: 978-1-4615-8023-2   e-ISBN-13: 978-1-4615-8021-8
DOI: 10.1007/978-1-4615-8021-8

© 1985 Plenum Press, New York
Softcover reprint of the hardcover 1st edition 1985
A Division of Plenum Publishing Corporation
233 Spring Street, New York, N.Y. 10013

Contributors

Lawrence Corey, Department of Laboratory Medicine and Microbiology and Virology Division, University of Washington, Seattle, Washington 98195

M. A. Epstein, Department of Pathology, University of Bristol Medical School, Bristol BS8 1TD, England

Anne A. Gershon, New York University School of Medicine, New York, New York 10016

Martin S. Hirsch, Department of Medicine, Massachusetts General Hospital, and Harvard Medical School, Boston, Massachusetts 02114

Eva Klein, Department of Tumor Biology, Karolinska Institutet, S 104 01 Stockholm, Sweden

François Lebel, Department of Medicine, Massachusetts General Hospital and New England Deaconess Hospitals, and Harvard Medical School, Boston, Massachusetts 02114

K.-N. Leung, Department of Pathology, University of Cambridge, Cambridge CB2 1QP, England

Carlos Lopez, Sloan–Kettering Institute for Cancer Research, New York, New York 10021

Bernard Meignier, Institut Merieux, Marcy l'Etoile, Charbonnieres les Bains, France

Joel D. Meyers, Fred Hutchinson Cancer Research Center and the University of Washington School of Medicine, Seattle, Washington 98104

A. A. Nash, Department of Pathology, University of Cambridge, Cambridge CB2 1QP, England

Bodil Norrild, Institute of Medical Microbiology, University of Copenhagen, DK-2100 Copenhagen Ø, Denmark

Gary R. Pearson, Section of Microbiology, Mayo Foundation, Rochester, Minnesota 55905

Stanley A. Plotkin, The Children's Hospital of Philadelphia, The University of Pennsylvania, and the Wistar Institute, Philadelphia, Pennsylvania 19104

Gerald V. Quinnan, Jr., National Center for Drugs and Biologics, Division of Virology, Office of Biologics, Food and Drug Administration, Bethesda, Maryland 20205

Harvey Rabin, DuPont Biomedical Products, North Billerica, Massachusetts 01862.

Barry T. Rouse, Department of Microbiology, College of Veterinary Medicine, University of Tennessee, Knoxville, Tennessee 37996

John L. Sullivan, Department of Pediatrics, University of Massachusetts Medical School, Worcester, Massachusetts 01605

Jean-Louis Virelizier, Groupe d'Immunologie et de Rhumatologie Pediariques (INSERM U 132) Hopital Necker-Enfants Malades, 75743 Paris Cedex 15, France

Georges H. Werner, Rhône-Poulenc Santé, Centre de Recherches de Vitry, 94407 Vitry-sur-Seine, France

Richard J. Whitley, Department of Pediatrics and Microbiology, The University of Alabama, Birmingham, Alabama 45294

P. Wildy, Department of Pathology, University of Cambridge, Cambridge CB2 1QP, England

Aurelio Zerial, Rhône-Poulenc Santé, Centre de Recherches de Vitry, 94407 Vitry-sur-Seine, France

Preface

What makes herpesviruses unique? It is certainly not the size of their genomes or the individual features of their reproductive cycle, although *in toto* striking features that are exclusive to the herpesviruses abound. Unquestionably, the pre-eminent feature is the relationship of herpesviruses with their natural hosts. As described in preceding volumes, all herpesviruses seem to be able to colonize and to remain in a latent, nonproductive form for life of their hosts. Once established in the host, the relationship is best described as that of an armed truce. What happens when this truce breaks down or when the host encounters the virus for the first time is the subject of this volume. We have focused primarily on the five human herpesviruses [herpes simplex virus 1 (HSV-1), herpes simplex virus 2 (HSV-2), cytomegalovirus (CMV), varicella zoster virus (VZV) and Epstein–Barr virus (EBV)] because much more is known about them than about any other herpesviruses, and because it is of interest to compare both the diversity of manifestations of infections with human herpesviruses and the spectrum of human responses to these viruses.

This volume summarizes the current knowledge of the pathogenesis and immunobiology of herpesvirus infections in man and describes new and developing approaches to prophylaxis and treatment. It contains contributions from distinguished research scientists presently engaged at the forefront of these scientific investigations. Reactivated HSV-2 infections occur in a large number of otherwise normal individuals, resulting in recrudescent genital disease. The first chapter of this book presents the natural history of genital HSV infections, and thus, defines one of the more common disease problems associated with a human herpesvirus. The next three chapters describe the natural and adaptive host defense mechanisms thought to be required for protection against HSV infections, while chapter 5 deals with the damage caused by the same immune response induced by the virus infection. Chapter 6 defines the cell-mediated immune responses to CMV and Chapters 7 through 9 are concerned with the immunobiology of EBV infections in primates and man. Infections with any of the herpesviruses can be devastating in immunosuppressed patients and Chapter 10 describes those that are most difficult to deal with: CMV infections in organ transplant recipients. Chapters 11 and 12 describe patients with inherited host defense deficiencies, which make them unusually susceptible to severe disease caused by herpesviruses. Chapters 13 through 16 describe the current status of vaccine development for the herpesviruses. Finally, Chapters 17 through 19 focus on, antiviral chemotherapy, interferon therapy, and immunopotentiators.

It is the understandable wish of editors of volumes such as this that their compilations serve a useful and lasting reference to the current state of knowledge. We are not immune to such thoughts, but the best way we can thank the contributors for their efforts and especially for the timeliness of their contributions is to wish this volume a precocious obsolescence through their efforts.

Bernard Roizman
Carlos Lopez

Chicago, Illinois
New York, New York

Contents

Chapter 1

The Natural History of Genital Herpes Simplex Virus: Perspectives on an Increasing Problem

Lawrence Corey

I. Introduction	1
II. History	1
III. Epidemiology of Genital Herpes	2
A. Prevalence of Genital HSV Infections	2
B. Evidence for an Increasing Prevalence of Infection	3
C. Prevalence of HSV Antibody	4
D. Incidence of Genital Herpes	6
E. Comment	6
IV. Clinical Manifestations of Genital Herpes	7
A. First Episodes of Genital Herpes	7
B. Primary Genital Herpes	8
C. Complications of Genital Herpes	13
D. Summary of the Clinical Course of First Episodes of Genital Herpes	17
E. Future Directions for Research on the Clinical Manifestations of Genital Herpes	18
F. Recurrent Genital Herpes	19
G. Other Clinical Syndromes Associated with Genital HSV Infection	20
V. Chronicity of Genital HSV Infections	23
VI. Sexual Transmission of Genital HSV	27
VII. Conclusions	28
References	29

Chapter 2

Natural Resistance Mechanisms in Herpes Simplex Virus Infections

Carlos Lopez

I.	Introduction	37
II.	Genetic Resistance to Lethal HSV-1 in the Mouse	40
III.	Genetic Resistance to Lethal HSV-2 in the Mouse	43
IV.	Genetic Resistance to HSV-2-Induced Hepatitis	43
V.	Macrophages	43
	A. Role of Macrophages in Resistance to HSV-1 in the Mouse	44
	B. Role of Macrophages in Resistance to HSV-2 in the Mouse	46
	C. Role of Macrophages in Resistance to HSV-2-Induced Hepatitis	47
	D. Role of Macrophages in Resistance to HSV-1 and HSV-2 in Man	48
VI.	Natural Killer Cells	49
	A. Heterogeneity of NK Cells	50
	B. Cell Lineage of NK(HSV-Fs) Effectors	52
	C. Role of IFN in NK(HSV-Fs) Response	53
	D. NK Cells Limit HSV-1 Replication *in Vitro*	53
	E. Role of NK Cells in Resistance to HSV-1 in the Mouse	54
	F. Low NK(HSV-Fs) Function Associated with Susceptibility to Herpesvirus Infections in Man	55
VII.	Interferon	57
	A. Role of IFN in Genetic Resistance to HSV-1 in the Mouse	57
	B. IFN-α-Generating Cells in Man	58
	C. Role of IFN in Resistance to Herpesvirus Infections in Man	59
VIII.	Concluding Remarks	60
	References	60

Chapter 3

Humoral Response to Herpes Simplex Virus Infections

Bodil Norrild

I.	Introduction	69
II.	HSV-Specific Antibodies in Human Convalescent Sera	70
III.	Measurement of HSV-Specific Neutralizing Antibodies	71

   IV. Measurement of HSV Antibodies by Immunofluorescence . 72
    V. Measurement of HSV-Cytolytic Antibodies 72
   VI. Measurement of HSV-Precipitating Antibodies 72
  VII. Measurement of HSV Antibodies with ELISA and RIA . . . 73
 VIII. Measurement of HSV Antibodies by Immunoblotting 74
   IX. Immunogenicity of the HSV-1 and HSV-2 Proteins 74
    X. Measurement of HSV-Specific Antibodies Belonging to the
       IgM and IgA Classes 77
   XI. Function of HSV-Specific Antibodies *in Vivo* 79
  XII. Conclusion 81
References 82

Chapter 4

**The T-Cell-Mediated Immune Response of Mice to Herpes
Simplex Virus**

A. A. Nash, K.-N. Leung, and P. Wildy

    I. Introduction 87
   II. Evidence for Thymus-Dependent Resistance to Herpes
      Infections 88
      A. Use of Athymic Mice and Immunosuppression 88
      B. Use of B-Cell-Suppressed Mice 89
  III. T-Cell Subsets Induced in Herpes Infections 89
      A. Cytotoxic T Lymphocytes 89
      B. T Cells Mediating Delayed Hypersensitivity 91
      C. T-Helper Lymphocyte Response 92
      D. T-Suppressor Cells 92
   IV. T Cells Involved in Antiherpes Responses *in Vivo* 93
    V. Regulation of Immune Responses to Herpes—A Role in
      Prevention of Immunopathology? 97
   VI. T-Cell Memory 98
  VII. Conclusions 99
 VIII. Appendix: Immunological Terms Used 99
References 100

Chapter 5

Immunopathology of Herpesvirus Infections

Barry T. Rouse

  I. Introduction 103
 II. IgE-Mediated Immunopathology 105
III. Immune Complex Immunopathology 106

    IV. Immunopathology Mediated by T Lymphocytes 109
     V. Management of Immunopathological Reactions 114
References 116

Chapter 6

Cell-Mediated Immunity in Cytomegalovirus Infections

Gerald V. Quinnan, Jr.

      I. Introduction 121
     II. Characteristics of Effector Cell Functions Relevant in
         CMV Infections 122
         A. NK Cells and Other Large Granular Lymphocytes 122
         B. Antibody-Dependent Cell-Mediated Cytotoxicity 123
         C. Cytotoxic T Cells 124
    III. Virus-Specific Cytotoxic Lymphocyte Responses during
         CMV Infections 126
     IV. Evidence That Virus-Specific Cytotoxic Lymphocytes
         Determine the Outcome of CMV Infections 131
      V. Evidence That NK Cells and Other Large Granular
         Lymphocytes Are Important in CMV Infections 132
     VI. Causes of Depressed Cytotoxic T-Cell Responses and NK
         Cell Activity in High-Risk Populations 134
    VII. Delayed-Type Hypersensitivity Responses to CMV 139
   VIII. CMV-Induced Immune Suppression 140
     IX. CMV-Induced *in Vitro* Lymphocyte Blastogenesis 141
      X. Cellular Immune Responses to CMV Vaccine 141
     XI. Conclusion 142
References 142

Chapter 7

***In Vivo* Studies of Epstein–Barr Virus and Other Lymphotropic Herpesviruses of Primates**

Harvey Rabin

      I. Introduction 147
     II. EBV in New World Primates 149
         A. Host Range for Disease Induction 149
         B. Clonality of Tumors in Cotton-Topped Marmosets ... 150
         C. Potential Use of Cotton-Topped Marmosets in EBV
            Vaccine Development 152
    III. *Herpesvirus papio* and Lymphoproliferative Disease in
         Baboons 152
         A. Background 152

B. Virus Isolates 153
C. Antibody Patterns to *H. papio*-Associated Antigens ... 153
D. Presence of Viral DNA in Tumor Tissue and Spleen ... 154
E. Experimental Disease in Cotton-Topped Marmosets ... 155
F. Summary of Features of the Disease and Possible
Similarity to Other Diseases of Captive Monkeys 155
IV. Other Lymphotropic Herpesviruses of Old World
Primates and Their Possible Relationship to Disease 156
V. T-Cell-Tropic Herpesviruses 157
A. Pathogenicity and Host Range 157
B. Lymphocyte Tropism and Antigen Modulation 158
C. *H. saimiri* Tumor Cell Lines 159
D. *In Vitro* Transformation 161
VI. *Tupaia* Herpesviruses 162
VII. Summary and Conclusions 163
References 165

Chapter 8

Cell-Mediated Immunity to EBV

Eva Klein

I. Introduction 171
II. Inhibition of the *in Vitro* Growth of B Cells 173
III. EBV-Induced Cell Surface Antigens Detected by
Lymphocyte Cytotoxicity 175
IV. Immunological Memory to EBV-Determined Antigens
Detected by Lymphokine Production 176
V. EBV-Related Immune Parameters in Pathological
Conditions 178
References 179

Chapter 9

Immunobiology of EBV-Associated Cancers

Gary R. Pearson

I. Introduction 183
II. Diagnostic Value of Antibodies to EBV Antigens 184
III. Prognostic Value of EBV Serology 187
VI. Identification and Purification of EBV Proteins 190
V. Cellular Immunity 193
VI. Conclusions 194
References 196

Chapter 10

Cytomegalovirus Infection after Organ Allografting: Prospects for Immunoprophylaxis

Joel D. Meyers

I. Introduction 201
II. Pathogenesis of CMV Disease 202
III. Epidemiologic Risk Factors 205
  A. Infection Due to Blood Products in Seronegative Patients 206
  B. Transmission of Virus in Transplanted Tissue 208
  C. Reactivation of Latent Virus 208
IV. Prevention of Primary Infection 210
  A. Donor Selection 210
  B. Control of Blood Products 211
  C. Passive Immunoprophylaxis with Plasma or Globulin . 212
  D. CMV Vaccine 215
V. Prevention of Virus Reactivation 216
  A. Interferon 217
  B. Antiviral Agents 219
VI. Immunosuppressive Regimen 219
VII. Treatment of CMV Infection 220
VIII. Conclusions 220
References 222

Chapter 11

Epstein–Barr Virus and the X-Linked Lymphoproliferative Syndrome

John L. Sullivan

I. Introduction 229
II. Immunopathogenesis of EBV Infection in the Normal Host 230
III. X-Linked Lymphoproliferative Syndrome 231
  A. General Characteristics 231
  B. Immunologic and Virologic Studies in Males before and during Fatal EBV Infection 232
  C. Immunologic Studies in Surviving Males 239
  D. Immunological and Virological Studies in Carrier Females 240
IV. Pathogenesis of the X-Linked Lymphoproliferative Syndrome 242
V. Treatment of Fulminant EBV Infections 246
References 247

Chapter 12

Abnormal Responses to EBV Infection in Patients with Impairment of the Interferon System

Jean-Louis Virelizier

    I. Introduction 251
   II. Immunological Control of EBV Infection 252
  III. The IFN System and Its Place in Antiviral Immunity 254
  IV. Evidence for a Role of IFN in the Control of EBV Infection
       in Vitro 255
   V. Evidence for a Role of IFN in the Control of EBV
       Infection in Patients 257
       A. IFN Administration in Immunosuppressed Patients ... 257
       B. Selective Defects of IFN Secretion 257
  VI. Chronic Neutropenia Triggered by EBV Infection and
       Associated with Impaired Activation of NK Cells by IFN .. 259
  VII. Conclusion 261
References 261

Chapter 13

Vaccination against Herpes Simplex Virus Infections

Bernard Meignier

    I. Introduction 265
   II. The Use of Wild-Type Virus 266
  III. The Use of Inactivated (Killed) Virus 267
       A. The Placebo Effect 267
       B. Studies without a Placebo Group 268
       C. Studies Including a Placebo Group 271
  IV. Subunit Vaccines 273
       A. Preparation 274
       B. Experience in Animals 274
       C. Experience in Man 276
       D. Future Trends 278
   V. Live Vaccines 279
       A. HSV Mutants 280
       B. Heterologous Herpesviruses 281
       C. Antigens Expressed by Non-HSV Viral Vectors 281
       D. Genetically Engineered HSV 281
  VI. Conclusion 289
References 290

Chapter 14

CMV Vaccines

Stanley A. Plotkin

    I. Introduction 297
   II. Donor Selection 298
  III. Passive CMV Antibody 298
  IV. Vaccination 300
    V. Subunit Vaccines 307
  VI. Interferon 309
 VII. Summary 310
References 310

Chapter 15

Live Attenuated Varicella Vaccine

Anne A. Gershon

    I. Historical Perspective 313
   II. Development of Varicella Vaccine 315
      A. First Studies in Japan 315
      B. Subsequent Studies in Japan 316
  III. Studies in the United States on Varicella Vaccine 317
      A. Studies in Normal Children 317
      B. Studies in Children with an Underlying Malignancy .. 318
  IV. Problems Requiring Further Resolution 320
      A. Persistence of Immunity after Vaccination 320
      B. Dose of Vaccine Virus Necessary to Induce Immunity . 321
      C. Safety 322
    V. Logistics of Future Vaccine Use 323
      A. Potential Vaccine Candidates 323
      B. Future Planning 324
References 324

Chapter 16

A Subunit Vaccine against Epstein–Barr Virus

M. A. Epstein

    I. Introduction 327
      A. Immunological Control of EBV Infection 327
      B. EBV and Human Tumors 328
      C. Comment 329

   II. Reasons for an Antiviral Vaccine 329
  III. EBV Membrane Antigen 330
     A. Quantification of MA gp340 330
     B. A Preparation Method for MA gp340 330
     C. Immunization with MA gp340 Using Novel Adjuvants . 331
     D. Structure of MA gp340 332
  IV. Discussion 332
References 334

Chapter 17

A Perspective on the Therapy of Human Herpesvirus Infections

Richard J. Whitley

    I. Introduction 339
     A. Development of Antivirals for Clinical Evaluation . . . 340
     B. Antiviral Drugs 343
   II. The Road to Controlled Trials 345
     A. Vidarabine 346
     B. Acyclovir 347
  III. Clinical Trials 350
     A. Herpes Simplex Keratoconjunctivitis 350
     B. Cutaneous HSV Infections 351
     C. Herpes Simplex Encephalitis 354
     D. Neonatal HSV Infections 358
     E. VZV Infections 361
  IV. Conclusion 364
References 365

Chapter 18

The Role of Interferon in Immunity and Prophylaxis

François Lebel and Martin S. Hirsch

    I. Introduction 371
   II. The Interferons 371
  III. Antiviral Mechanisms of Interferon 373
  IV. Immunolomodulatory Effects of Interferon 374
     A. Immunoenhancement 374
     B. Immunosuppression 375
    V. Interferon and the Herpesviruses 375
     A. Cytomegalovirus 375
     B. Varicella–Zoster Virus 377
     C. Esptein–Barr Virus 377
     D. Herpes Simplex Virus 379

VI. Prophylaxis and Therapy with Interferon 380
    A. Animal Studies 381
    B. Human Studies 381
VII. Future Directions 385
References 385

Chapter 19

**Effects of Immunopotentiating and Immunomodulating Agents
on Experimental and Clinical Herpesvirus Infections**

Georges H. Werner and Aurelio Zerial

I. Introduction 395
II. Microbial Agents and Substances of Microbial Origin 397
III. Transfer Factor 401
IV. Thymic Hormones (Factors) 401
V. Synthetic Interferon Inducers 403
VI. Levamisole 405
VII. Inosiplex 407
VIII. Miscellaneous Immunomodulating Agents 410
IX. Conclusion 411
References 412

Index 417

CHAPTER 1

The Natural History of Genital Herpes Simplex Virus
Perspectives on an Increasing Problem

LAWRENCE COREY

I. INTRODUCTION

Genital herpes simplex virus (HSV) infection has emerged as a disease of public health importance. The morbidity of the illness, its frequent recurrence, its complications such as aseptic meningitis and neonatal transmission, and its epidemiologic association with cervical carcinoma have made this entity of great concern to patients and health care providers. In the last three decades major advances in the understanding of the natural history and epidemiology of this entity have occurred. However, major gaps in our knowledge of the epidemiology, clinical course, and immunology of this infection still remain. Much is yet to be learned before we can expect to develop and test effective methods to control recurrences and prevent the transmission of disease.

II. HISTORY

Genital herpes was first described by the French physician John Astruc in 1736, and the first English translation of his *Treatice of Venereal Disease* appeared in 1754. In 1893, genital herpes was diagnosed in 9.1% of 846 prostitutes visiting an infirmary (Hutfield, 1966). In 1886, Diday and Doyon published the monograph *Les Herpes Genitaux* in which they observed that genital herpes often appeared after a venereal infection such

LAWRENCE COREY • Department of Laboratory Medicine and Microbiology and Virology Division, University of Washington, Seattle, Washington 98195.

1

as syphilis, chancroid, or gonorrhea. They also described cases of recurrent genital herpes. They proposed that the eruptions were related to nervous "trigger mechanisms" acting by way of the sacral plexus. Menstruation was felt to be the most important trigger mechanism in women (Diday and Doyon, 1886).

Fluid from orolabial infection was shown to be infectious to other humans in the late 19th century. The disease was successfully transferred to rabbits in 1920, and HSV was grown *in vitro* in 1925 (Baum, 1920; Cruter, 1924; Parker and Nye, 1925). In the 1920s Lipschutz inoculated material from genital herpetic lesions into the skin of humans, eliciting clinical infection within 48–72 hr in six persons and 24 days in one case. In other experiments, he observed that protection in rabbits from corneal infection with specimens from genital herpes occurred only with strains originating from the genital sites and not from orolabial sites (Lipschutz, 1921). From these and other experiments, he surmised that there were epidemiologic and clinical differences between oral and genital herpes. However, most workers felt that the viruses of genital and labial herpes were identical. In the early 1960s, Dowdle and Nahmias reported that HSV could be divided by neutralization tests into two antigenic types, and that there was an association between the antigenic type and the site of viral recovery (Dowdle *et al.,* 1967; Nahmias and Dowdle, 1968). These observations led to the benchmark studies on the epidemiology of genital herpes in the late 1960s and early 1970s.

III. EPIDEMIOLOGY OF GENITAL HERPES

A. Prevalence of Genital HSV Infections

The reported prevalence of genital herpes appears to depend upon the demographic and clinical characteristics of the patient population studied and whether clinical and/or laboratory techniques are used for diagnoses. In the United Kingdom in 1979, genital HSV infections accounted for approximately 2.2% of all visits to genitourinary medicine clinics (Catterall, 1981). Nationwide reporting of statistics for genital herpes is not available in the United States. Genital HSV infection was diagnosed in 5.6% of all persons attending the sexually transmitted disease (STD) clinics in King County, Washington in 1982 (Table I).

HSV has been isolated from 0.3–5.4% of males and 1.6–8% of females attending STD clinics (Wentworth *et al.,* 1973; Jeansson and Molin, 1970, 1974). In non-VD clinic patient populations, HSV has been isolated from the genital tract in 0.25–4.0% of patients (Rauh *et al.,* 1977; Taintivanich and Tharavawij, 1980; Knox *et al.,* 1979; Vesterinen *et al.,* 1977). Many of these patients are asymptomatic. Asymptomatic shedding of virus at or near term in women attending OB-GYN clinics has ranged from 0.24%

TABLE I. Prevalence of Genital HSV Infections in Sexually Transmitted
Disease Clinics of King County, Washington, 1976–1982

	1976	1977	1978	1979	1980	1981	1982
No. cases of HSV	1,014	1,011	1,146	1,251	1,525	1,686	1,922
First episode	—[a]	—	—	—	—	—	942
Recurrent episode	—[a]	—	—	—	—	—	980
% HSV cases Caucasian	94%	92%	89%	87%	88%	87%	—[a]
No. cases of *N. gonorrhoeae*	4,834	4,501	4,664	4,266	3,789	2,883	2,526
Total No. STD clinic visits	39,803	38,722	39,440	35,890	35,207	33,313	34,196
% STD visits due to HSV	2.5%	2.6%	2.9%	3.5%	4.3%	5.1%	5.6%

[a] Not available.

to 4% of attendees (Vontver *et al.*, 1982; Bolognese *et al.*, 1976; Harger *et al.*, 1983; Tejani *et al.*, 1979; Scher *et al.*, 1982).

The prevalence of clinically diagnosed genital herpes is greater in Caucasian than non-Caucasian populations (Table I). In STD clinics seeing a high proportion of non-Caucasians, genital herpes is reported only 1/10th as frequently as *Neisseria gonorrhoeae* infection (STD Fact Sheet). However, student health centers seeing middle- and upper-class young adults, in whom the prevalence of gonococcal infection is very low, report that genital HSV infections are 7–10 times more common than gonorrhea (Sumaya *et al.*, 1980). In addition, genital HSV infections appear to be more frequently diagnosed in heterosexual than homosexual men (Judson *et al.*, 1980).

B. Evidence for an Increasing Prevalence of Infection

In many populations, genital HSV infections appear to have increased in prevalence in the last decade. Consultations for genital herpes to private practitioners increased in the U.S. from 3.4 per 100,000 patient consultations in 1966 to 29.2 per 100,000 consultations in 1979 (Anonymous, 1982). In Rochester, Minnesota, a population-based epidemiologic study indicated that the incidence of genital HSV increased from 12.5 cases per 100,000 population in the years 1965 through 1970 to 48.1 in the years 1970 thorugh 1975 and then to 82.3 cases per 100,000 population in the years 1975 through 1980 (Chuang *et al.*, 1983). During the years 1976 through 1982, the frequency of reported genital herpes in STD clinics in King County, Washington, increased by 47% (Table I). In this population, the ratio of genital HSV cases to *N. gonorrhoeae* cases rose from 1:4.8 in 1976 to 1:1.3 in 1982, a result of both a decrease in *N. gonorrhoeae* and an increase in genital HSV infections. In a middle-class student population in Ann Arbor, Michigan, the prevalence of HSV in cytologic smears increased 10-fold over the years 1965–1975 (Britto et al., 1976). Between 1968 and 1973, the reported incidence of genital HSV infection nearly doubled each year in Auckland, New Zealand (MacDougall, 1975). In the years 1970 to 1976, the incidence of HSV infections, as detected by the routine cytologic screening of Papanicolaou smears obtained from cervical scraping, from women attending a gynecology clinic in Bangkok, Thailand, increased from 0.2 case per 1000 patients to 6.9 cases per 1000 patients (Srinannaboon, 1979). Recently, Sullivan-Bolyai et al. (1983) reported that the incidence of neonatal HSV infection in King County, Washington, increased from 2 cases per 100,000 live births in the years 1966–1969 to 12.8 cases per 100,000 live births in the years 1978–1981.

Thus, in many populations throughout the world, the prevalence of genital HSV infections has markedly increased over the last 10 years. However, in other population groups, no evidence of an increasing prevalence of genital HSV infection has been noted. For example, the high prevalence of cytologically detected genital HSV infections noted in the predominantly black, lower socioeconomic populations studied in Atlanta, Georgia, in the 1960s, has remained relatively constant over the decade of the 1970s (Nahmias et al., 1966, 1969a; Rawls et al., 1971; Josey et al., 1972; Nahmias, 1983). These data suggest that genital herpes may be increasing in frequency in some population groups, especially those with a previously low prevalence of genital herpes, e.g., middle-class Caucasions. Whether these apparent increases reflect a true increase in the prevalence of genital herpes or an increasing recognition and use of diagnostic facilities to confirm the diagnosis is uncertain. The increased incidence of neonatal HSV infection recently demonstrated in King County, Washington, indicates that at least part of the observed increase is probably related to a real increase in the incidence of infection.

C. Prevalence of HSV Antibody

The prevalence of antibody to HSV increases with age and varies with socioeconomic status (Nahmias and Roizman, 1973; Nahmias et al., 1970; Wentworth and Alexander, 1971; Holzel et al., 1953). Studies of the seroepidemiology of HSV-2 infections are affected by the difficulty of distinguishing the human immune response between HSV-1 and HSV-2 infections. Some serologic tests such as the standard complement fixation antibody assay do not distinguish the antibody response between the two HSV types (Rawls, 1979). Several serologic tests (neutralization, indirect immunofluorescence, passive hemagglutination, indirect hemagglutination, ELISA, and RIA) have been developed that distinguish antibody to HSV-1 and to HSV-2 (Rawls et al., 1970; Plummer et al., 1970; Prakash and Seth, 1979; Lerner et al., 1974). Nevertheless, serologic cross-reactions between the two HSV types are common. For example, in persons with prior HSV-1 infections, much of the antibody response to a new HSV-2 infection appears to be directed at common rather than type-specific antigenic determinants (Yeo et al., 1981). In addition, it may be difficult to detect the presence of HSV-2 antibody in the presence of high titers of HSV-1 antibody (McClung et al., 1976). For example, in 78 patients with confirmed documented HSV-2 infections, we have found HSV-1 neutralizing antibody specificity in the sera of 9% of these patients. Because of this, many of the currently available serologic techniques may underestimate the true prevalence of HSV-2 infection. The development of new diagnostic reagents using type-specific antigens and/or antibody systems will hopefully improve the specificity and sensitivity of sero-epidemiologic surveys of HSV-2 infection (Eberle and Courtney, 1981; Pereira et al., 1982).

Serologic studies of Western industrialized populations in the post-World War II era found that 80–100% of middle-aged adults of lower socioeconomic status possessed antibodies to HSV, as compared to 30–50% of adults of higher socioeconomic groups (Nahmias et al., 1970). In the U.S. in the early 1960s, HSV antibody was detected in 40% of those aged 8–14 seen in a private pediatric group compared to 80% of those of a similar age attending outpatient clinics at an urban public hospital (Porter et al., 1969). The higher prevalence of antibody to HSV in persons of lower socioeconomic class was due to an increased frequency of both HSV-1 and HSV-2 infections (Nahmias et al., 1970; Nahmias and Roizman, 1973).

Antibodies to HSV-2 are not routinely detected in sera until puberty, and antibody prevalence rates correlate with past sexual activity. HSV-2 antibodies were detected in 80% of female prostitutes, up to 60% of adults of lower socioeconomic status, 10% of higher socioeconomic groups, and 3% of nuns (Duenas et al., 1972). In the 1970s, the anti-HSV-2 prevalence rate was 7–20% in male and female voluntary blood donors in England compared to 60% in voluntary blood donors from prison populations (Roome et al., 1975). In Ibadan, Nigeria, HSV-2 antibodies were detected in 20% of voluntary blood donors and in 27% of women attending family planning clinics (Montefiore et al., 1980). A small population of young children in tropical areas have HSV-2 antibodies, suggesting that nonsexual transmission may occasionally occur in this population and climate (Montefiore et al., 1980; Johnson et al., 1981). In the late 1970s, using an ELISA technique, HSV-2 antibodies were detected in 50% of men and women reporting to an outpatient medical dispensary in Scandinavia (Vestergaard and Rune, 1980; Grauballe and Vestergaard, 1977). Recently, a randomized survey of middle-class Caucasian adults in metropolitan Toronto indicated a prevalence of HSV-2 antibody in 17.8% of women and 12.1% of men; only 20% of men and 13% of women gave a history of having symptomatic genital lesions (Stavraky et al., 1983).

Few serosurveys are avilable in similar population groups comparing antibody prevalence to HSV-1 and HSV-2 over time. However, separate studies of the prevalence of HSV antibody in England indicated that in 1953, 85% of Caucasian children between 3 and 15 years of age possessed CF antibody to HSV-1 as compared to only 41% in 1965. In the 1965 survey, antibody to HSV was detected in 35% of 10-14 year olds and 69% of 15–19 years olds (Smith and Pleutherer, 1967).

D. Incidence of Genital Herpes

Little data are available regarding the serologic or clinical incidence of genital HSV infections. In one study of genital HSV infections in college students in the U.S., the mean average incidence of clinically apparent first episodes of genital HSV presenting to a student health service was 5.2 cases per 10,000 students per month, or approximately 0.62% of students per year developed clinically symptomatic first episodes of disease (Sumaya et al., 1980). A major problem with these data is that many students may present to other types of health care facilities especially for a sexually acquired disease.

Little is known about the incidence of serologic infection or the infection-to-disease ratio. In a recent study we prospectively followed a cohort of homosexually active men. Antibody prevalence rates to HSV-1 and HSV-2 were high; 44% had HSV-2-neutralizing specificity, 23% had HSV-1-neutralizing antibody, and only 13% were seronegative to HSV. The incidence of seroconversion among the seronegative men was 0.46 cases per person year. Of interest was that of the eight seroconversions and/or serologic rises to HSV-1 or HSV-2 in this cohort of men, only two were associated with symptoms that brought the patients to clinical attention (Mann et al., 1984).

E. Comment

Several important epidemiologic questions remain to be answered before one can adequately assess the prevalence and incidence of genital HSV infections. There appears to be a substantial difference between the prevalence of HSV antibody and the prevalence of disease. Whereas the prevalence of HSV-2 antibody is greatest in persons of lower socioeconomic status, the prevalence of recognized genital herpes even in STD clinics is greater in middle-class Caucasians. Recent serologic and epidemiologic data suggest that the age-specific prevalence of HSV-1 infection may be falling in some populations, especially middle-class populations in Western industrialized nations, whereas the prevalence of sexually acquired HSV infections may be increasing. In contrast, data exist that other population groups may have experienced no recent change in the prevalence and incidence of asymptomatic HSV infection.

Several interesting questions require definition: What is the serologic prevalence of genital herpes in a variety of social groups and locales? What are the most important epidemiologic correlates of disease? Do age of acquisition, crowding, racial factors affect the proportion of persons who acquire symptomatic versus asymptomatic infection? What is the effect of prior HSV-1 infection on the subsequent development of antibody or disease? Randomized point prevalence serosurveys of several different socioeconomic groups should be performed to assess the prevalence of HSV-2 antibody and excretion of HSV-2 from the genital tract. Data collected should allow one to assess what effect cofactors (e.g., socioeconomic status, social crowding, contraceptive practices, age, and past sexual activity) have on the prevalence of antibody and clinical disease. The frequency of asymptomatic acquisition of disease also requires clarification. While retrospective surveys suggest only one-third of patients with antibody have genital lesions, these surveys, because of the time lapse between acquisition of disease and questioning of symptoms or signs, do not assess how many have ignored symptomatic versus truly asymptomatic infection. Prospective studies defining the frequency of symptomatic to asymptomatic acquisition of disease are needed. Does prior HSV-1 antibody influence the frequency of persons with asymptomatic disease?

IV. CLINICAL MANIFESTATIONS OF GENITAL HERPES

A. First Episodes of Genital Herpes

The clinical manifestations and recurrence rate of genital herpes are influenced both by host factors (past exposure to HSV-1, previous episodes of genital herpes, gender) and by viral type (Corey et al., 1983a). First episodes of genital herpes are often associated with systemic symptoms, involve multiple genital and extragenital sites, and have a prolonged duration of viral shedding and lesions (Adams et al., 1976; Vontver et al., 1979, Kaufman et al., 1973). Patients with first episodes of genital herpes who have clinical or serologic evidence of prior HSV infection tend to have a milder illness than those experiencing true primary infection, i.e., experiencing their first infection with either HSV-1 and HSV-2 (Corey et al., 1981). About 60% of persons who attend our genital HSV clinic with their first episode of symptomatic genital herpes had primary infection with either HSV-1 or HSV-2. Most persons with nonprimary first episodes of genital HSV infection have serologic evidence of past HSV-1 infection. However, 10–30% of persons with first episodes of disease have serologic evidence of past HSV-2 in acute-phase sera, indicative of past asymptomatic acquisition of HSV-2 (Corey et al., 1978a, 1982).

Past orolabial HSV-1 infection may decrease the acquisition of genital HSV-1 infections (Reeves et al., 1981). Genital HSV-1 infections have been reported with increasing frequency in some populations, especially in some persons experiencing first episodes of infection (Kalinyak et al.,

1977; Kawana *et al.*, 1976; Barton *et al.*, 1982). In Seattle, in the years 1975–1980, HSV-1 was isolated from genital lesions in 21 (7%) of 286 persons with first episodes of genital herpes. Between the years 1980–1983, 14% of first episodes of genital herpes have been due to HSV-1. Ninety-five percent of the patients with first-episode HSV-1 infection have lacked anti-HSV in acute-phase sera, i.e., had true primary infection, suggesting that prior HSV-1 infection protects against the acquisition of genital HSV-1 disease.

Prior HSV-1 infection also ameliorates the severity of first episodes of genital herpes. For example, only 16% of persons with first-episode nonprimary genital HSV-2 infection demonstrate systemic symptoms during the course of illness as compared to 62 and 68% of patients with primary HSV-2 or primary genital HSV-1 infections. The severity of local pain (8.7 days), duration of viral shedding from lesions (6.8 days), and duration of lesions (15.5 days) are also less than in patients with primary HSV-2 disease (11.8, 11.4, and 18.6 days, respectively) (Fig. 1). The duration of symptoms and lesions are similar between patients with primary HSV-1 and primary HSV-2 infections (Reeves *et al.*, 1981; Corey *et al.*, 1983a).

B. Primary Genital Herpes

1. Symptoms

Primary genital HSV infection is characterized by a high frequency and prolonged duration of systemic and local symptoms. Fever, headache, malaise, and myalgias are reported in nearly 40% of men and 70% of women with primary HSV-2 disease ($p < 0.05$) (Table II). Systemic symptoms appear early in the course of the disease, usually reaching a peak within 3–4 days after onset of lesions, and gradually recede over the subsequent 3–4 days.

Pain, itching, dysuria, vaginal or urethral discharge, and tender inguinal adenopathy are the predominant local symptoms of disease. Painful lesions are reported in 95% of men (mean duration 10.9 days) and 99% of women (mean duration 12.2 days) with primary HSV infection. Dysuria, both external and internal, appears more frequently in women (83%) than men (44%). HSV has been isolated from the urethra and urine of both men and women with primary genital herpes, suggesting that HSV urethritis and/or cystitis in addition to external dysuria resulting from urine touching active genital HSV lesions may account for the high frequency and long duration of dysuria in women (Corey *et al.*, 1983a,b; Person *et al.*, 1973).

Urethral discharge and dysuria are noted in about one-third of men with primary HSV-2 infection of the external genitalia. HSV can be isolated from a urethral swab or first voiding urine of these men. The urethral

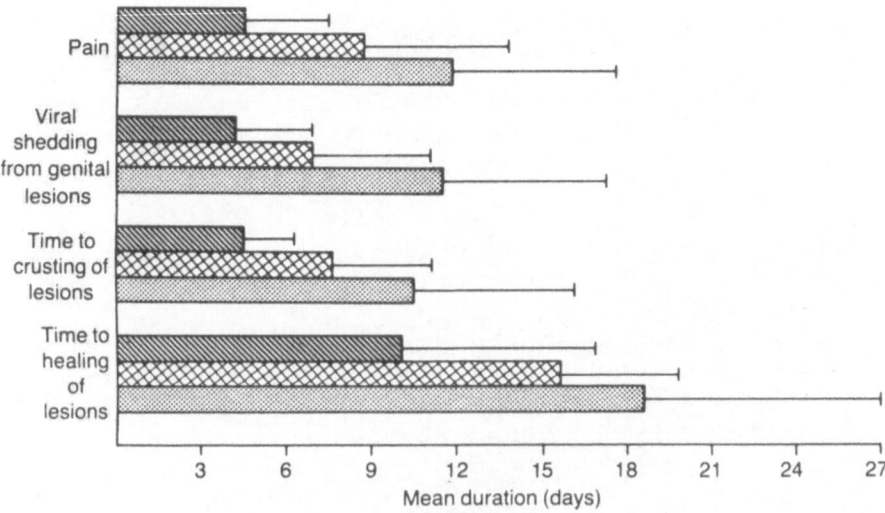

FIGURE 1. Clinical course of untreated genital HSV infection. Comparison of the duration of local pain, viral shedding, time to crusting of genital lesions, and time to healing of genital lesions in patients with first-episode primary (▨), first episode nonprimary (▨), and recurrent genital herpes (▨). The mean difference in the duration of each of the listed signs between each of these three stages of disease is significant ($p < 0.01$ for each comparison). —— = standard deviation.

discharge is usually clear and mucoid and the severity of dysuria is often "out of proportion" to the amount of urethral discharge elicited on genital exams. Gram stain of the urethral discharge usually reveals between 5 and 15 polymorphonuclear leukocytes per oil immersion field. Occasionally a mononuclear cell response is seen.

The clinical symptoms of pain and irritation from lesions gradually increase over the first 6–7 days of illness, reaching their maximum intensity between days 7 and 11, and gradually recede over the second week of illness. Tender inguinal adenopathy usually appears during the second and third weeks of disease and often is the last symptom to resolve.

TABLE II. Clinical Symptoms and Signs of Primary Genital HSV-2

	Men	Women
% with constitutional symptoms[a]	39%	68%
Mean duration (days) local pain	10.9	11.9
% with dysuria[a]	44%	83%
Mean duration (days) dysuria[a]	7.2	11.9
Mean area (mm^2) of lesions	427	550
Mean duration (days) viral shedding from lesion	10.5	11.8
Mean duration (days) lesions	16.5	19.7

[a] $p < 0.05$ by χ^2.

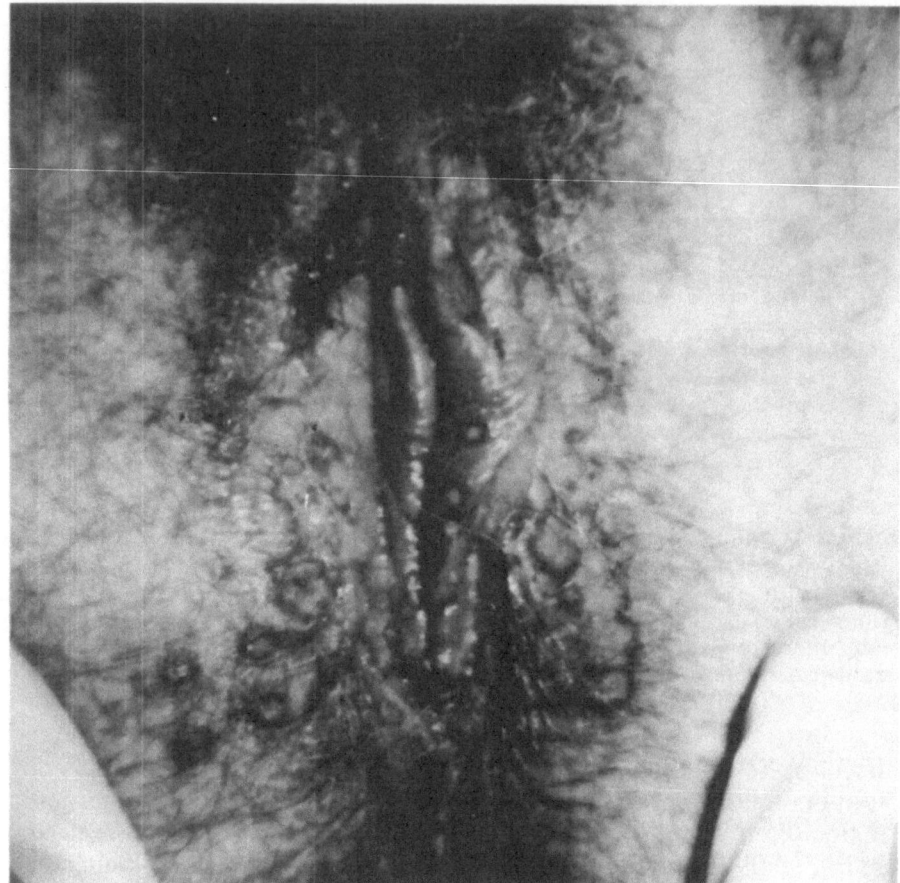

FIGURE 2. Ulcerative genital lesions in a woman with primary genital HSV-2 infection.

Inguinal and femoral lymph nodes are generally firm, nonfluctuant, and tender when palpated. Suppurative lymphadenopathy is a very uncommon manifestation of genital herpes.

2. Signs

In both men and women with primary genital HSV infection, widely spaced bilateral pustular or ulcerative lesions on the external genitalia are the most frequent presenting sign. Lesions are characteristically described as starting as papules or vesicles that rapidly spread over the genital area. At the time of the first clinic visit, multiple small pustular lesions that coalesce into large areas of ulcerations are usually present (Fig. 2). The size and shape of the ulcerative lesions vary greatly between patients. These ulcerative lesions persist between 4 and 15 days until crusting and/or reepithelialization occurs. In general, lesions in the penile and mons areas crust over before complete reepithelialization ensues. Crusting does not occur on mucosal surfaces. Residual scarring from lesions is uncommon. New lesion formation (the development of new areas of vesiculation or ulceration during the course of infection) occurs in over 75% of patients with primary genital herpes. New lesions usually form between days 4 and 10 of disease.

The median duration of viral shedding as defined from onset of lesions to the last positive culture is 12 days. The mean time (about 10.5 days) from the onset of vesicles to the appearance of the crust stage correlates well with the duration of viral shedding (Fig. 1). However, because there is considerable overlap between the duration of viral shedding and the duration of crusting and because mucosal lesions do not crust, patients should be advised not to resume sexual activity until lesions have completely reepithelialized. The mean time from the onset of lesions to complete reepithelialization of all lesions appears to be slightly longer in women (19.5 days) than men (16.5 days).

3. Concomitant Vulvar and Cervical HSV Infection

Ninety percent of women with primary genital HSV-2 infection, 70% of women with primary genital HSV-1 infection, and 70% of women with nonprimary genital HSV-2 infection have concomitant HSV cervicitis (Adams et al., 1976; Corey et al., 1983a; Barton et al., 1981). The high rate of isolation of HSV from the cervix in first episodes contrasts sharply with the 12–20% isolation rate in women who present with recurrent external genital lesions (Adams et al., 1976; Guinan et al., 1981; Corey et al., 1983a). Primary genital HSV cervicitis may be symptomatic (purulent vaginal discharge) or asymptomatic (Josey et al., 1966). In most cases of primary HSV cervicitis, the cervix appears abnormal to inspection. The appearance may include areas of diffuse or focal friability and redness, extensive ulcerative lesions of the exocervix, or severe necrotic cervicitis. HSV infection of the cervix usually involves the squamous epithelium of the exocervix in contrast to the mucopurulent cervicitis of C. trachomatis and N. gonorrhoeae infection. However, clinical differentiation is difficult. Recent studies of women with mucopurulent cervicitis without any external genital lesions suggest that HSV may cause 5–20% of cases (J. Pavonnen and L. Corey, unpublished results).

In women with primary genital herpes, the mean duration of viral shedding from the cervix (11.4 days) is similar to that from lesions of the external genitalia, and there is a close correlation between the duration of cervical viral excretions and the duration of viral shedding from external genital lesions (Fig. 3).

4. Pharyngeal Infection

HSV of the pharynx is commonly seen in association with primary genital herpes and may be the presenting complaint. Both HSV-1 and

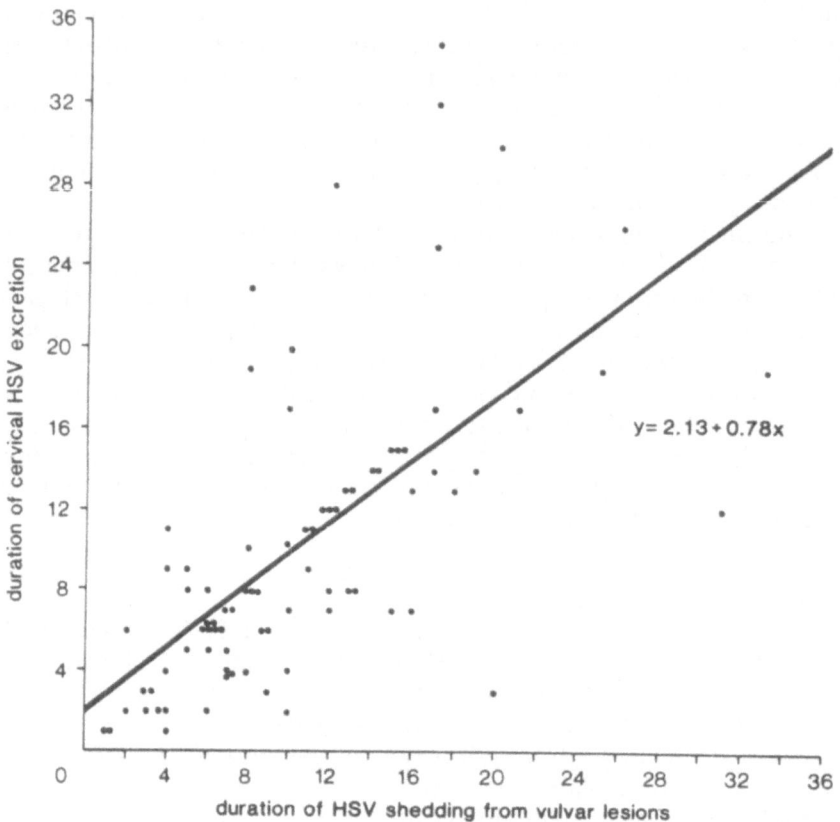

FIGURE 3. Comparison of the duration of viral shedding from the vulva (x-axis) and the cervix (y-axis) in women with untreated first episodes of genital herpes (N = 85). Includes only those women from whom HSV was isolated from both the external genitalia and cervix.

HSV-2 may cause pharyngitis. Both HSV types may be associated with orogenital exposure to the source contact (Embil et al., 1981; Corey et al., 1983a). We have isolated HSV from the pharynx of 11% of patients with primary HSV-2 infection, 20% of patients with primary HSV-1 infection, 1% of patients with nonprimary initial genital herpes, and 1% of persons with recurrent genital herpes. Among persons with primary genital herpes who complained of sore throat during the acute episode of disease, HSV was isolated from the pharynx in 70%. Viral cultures of the pharynx were obtained on 20 patients with primary HSV-2 disease who did not complain of sore throat and HSV was not isolated from the pharynx in any of these patients, indicating that HSV pharyngitis is usually symptomatic. Clinical signs of HSV pharyngitis may vary from mild erythema to a diffuse ulcerative pharyngitis (Evans and Dick, 1964; Glezen *et al.*, 1975). The inflammatory response to these large areas of ulceration may produce a whitish exudate; when wiped away, the extensive ulceration may be visualized. In rare cases, severe swelling of the posterior pharynx resulting in obstruction of the airway may occur (Tustin and Kaiser, 1979). Extension of the ulcerative posterior pharyngeal lesions into the anterior gingival area may occur. Most patients with HSV pharyngitis have tender cervical nodes, and constitutional symptoms, such as fever, malaise, myalgia, and headache, are common. Many are misdiagnosed as having streptococcal pharyngitis.

Previous studies of HSV pharyngitis have shown the entity to be a common form of infection in college-age students and almost invariably due to primary infection with HSV-1. The more frequent isolation of HSV-2 in patients with primary than with nonprimary or recurrent genital disease suggests that HSV-2 pharyngitis is also a manifestation of primary rather than recurrent genital herpes.

C. Complications of Genital Herpes

The complications of first episodes of genital herpes are related to both local extension and spread of virus to extragenital sites. The most frequent complication is the development of lesions at extragenital sites (Table III). CNS involvement and fungal superinfection are also frequently encountered. Complications of primary genital herpes occur more frequently in women than in men (Corey *et al.*, 1983a).

1. CNS Complications

CNS involvement may occur in several forms, including aseptic meningitis, transverse myelitis, or what has been called the sacral radiculopathy syndrome (Ross and Stevenson, 1961; Klastensky *et al.*, 1972; Caplan *et al.*, 1977). In patients with primary genital HSV-2 infection seen at the University of Washington Genital Herpes Clinic, meningeal irritation (defined as the presence of stiff neck, headache, and photophobia on two consecutive examinations) was reported in 36% of women and 13% of men ($p < 0.001$). Hospitalization was required for clinically overt aseptic meningitis in 6.4% of women and 1.6% of men with primary HSV-2 infections. All of the hospitalized patients were febrile, had Kernig's and Brudzinski's signs on physical examination, and had a CSF pleocytosis. Routine lumbar punctures were not performed on all nonhospitalized patients. However, a study of primary genital herpes in the early 1900s found a high frequency of CSF pleocytosis in patients without overt clinical evidence of meningeal irritation, suggesting that meningeal involvement may be a frequent occurrence with primary genital herpes (Ravaut and Darre, 1904).

Both HSV-1 and HSV-2 have been isolated from CSF. HSV has been isolated from 0.5 to 3.0% of patients presenting to hospital with aseptic

TABLE III. Complications of Primary Genital HSV-2 Infection

	Approximate frequency
CNS complications	
% with stiff neck, headache, and photophobia[a]	28%
% hospitalized with aseptic meningitis	5%
Sacral autonomic nervous system radiculopathy[b]	1%
Transverse myelitis	rare
HSV pharyngitis	10%
Development of extragenital lesions	21%
Lip	3%
Buttock, groin	9%
Breast	2%
Finger	6%
Eye	1%
Disseminated cutaneous infection	rare
Direct extension of disease	
Pelvic inflammatory disease syndrome (including endometritis)	1.8%
Pelvic cellulitis	rare
Suppurative lymphadenitis	<0.5%
Fungal superinfection	
Yeast vaginitis	14%

[a] All three symptoms present on two consecutive examinations.
[b] Constipation, urinary retention, and sacral anesthesia.

meningitis (Skoldenberg *et al.*, 1975; Bayer and Gear, 1955). HSV aseptic meningitis appears to be more frequently associated with genital than orolabial infection (Wolontis and Jeansson, 1977; Craig and Nahmias, 1973). Fever, headache, vomiting, photophobia, and nuchal rigidity are the predominant symptoms of HSV aseptic meningitis. Meningeal symptoms usually start from 3 to 12 days after onset of genital lesions. Symptoms generally reach a maximum 2–4 days into the illness and gradually recede over 36–72 hr. The CSF in HSV aseptic meningitis is usually clear or slightly cloudy, and the opening pressures may be elevated. White blood cell counts in CSF may range from 10 to over 1000 per mm^3 (mean 550). The pleocytosis is predominantly lymphocytic in adults, although early in the course of the disease and in neonates a predominantly polymorphonuclear response may be seen. The CSF glucose is usually more than 50% of the blood glucose, although hypoglycorrhachia has been reported; the CSF protein is slightly elevated (Brenton, 1980). If cultures are obtained within 24 hr of onset of headache and photophobia, HSV may be grown from the CSF. The differential diagnosis of HSV aseptic meningitis includes diseases that result in neurologic involvement and genital ulcerations: sacral herpes zoster, Behcet's syndrome, collagen vascular disease, inflammatory bowel disease, porphyria, and benign recurrent aseptic meningitis (Mollaret's syndrome).

Craig and Nahmias (1973) reported the isolation of HSV-2 from buffy coat collected in adults with HSV-2 meningitis. These investigators hypothesized hematogenous spread of virus into the CNS. In animal models, HSV appears to travel from involved mucocutaneous sites through peripheral nerves into the CNS (Overall *et al.*, 1975; Renis *et al.*, 1976; Kristensson *et al.*, 1978). This form of neurotropic spread is often associated with the development of paralysis and/or death due to encephalitis, which rarely occur in human cases of aseptic meningitis associated with genital herpes. The pathogenesis of HSV meningitis is unclear. We have been unable to isolate HSV from buffy coat in patients with primary genital herpes (Corey *et al.*, 1983b). Whether hematogenous, neurotropic, or a combination of both, means by which virus reaches the CNS requires further study.

Aseptic meningitis associated with genital herpes appears to be a benign albeit uncomfortable disease in immunocompetent persons. Signs and symptoms of encephalitis are unusual, and neurologic sequelae are rare. Risk factors associated with the development of aseptic meningitis are not well understood. Whether antiviral chemotherapy will shorten the course or prevent the development of HSV-2 aseptic meningitis is currently unknown.

Both transverse myelitis and autonomic nervous system dysfunction have been described in association with genital HSV infection (Klastensky *et al.*, 1972; Caplan *et al.*, 1977; Craig and Nahmias, 1973). Symptoms of sacral nervous system dysfunction include hyperesthesia or anesthesia of the perineal, lower back, and anal regions; difficulty in urinating; and constipation. Physical examination reveals a large bladder, decreased sacral sensation, and poor rectal and perineal sphincter tone. In men, a history of impotence and absent bulbocavernous reflexes may be present. CSF pleocytosis may be present in some cases. Electromyography usually reveals slowed nerve conduction velocities and fibrillation potentials in the affected area, and urinary cystometric exam shows a large atonic bladder. Most cases gradually resolve over 4–8 weeks. In cases of transverse myelitis, decreased deep tendon reflexes and muscle strength in the lower extremities, as well as the above-described sacral neurologic dysfunction are present. In one reported case, significant residual dysfunction was present years later (Craig and Nahmias, 1973).

Whether these neurologic abnormalities result from viral invasion of the CNS or an unusual immunologic response to infection is unknown. Paresthesias, urinary retention, and/or impotence have been noted in approximately 50% of men who present with HSV proctitis (Goodell *et al.*, 1983). We have observed these symptoms in only 2 of 123 heterosexual women and 0 of 63 heterosexual men who presented with HSV infection of the penile or scrotal area. Of interest was the fact that both of the women who presented with sacral neurologic dysfunction also had anal HSV lesions. The mechanism for the more frequent association between HSV proctitis and these symptoms is unclear. Involvement of the dorsal nerve root ganglia (ganglionitis) has been postulated as one potential mechanism. Patients with these neurologic symptoms have been noted to have CSF pleocytosis. However, pleocytosis is common in primary genital herpes and is not specific for ganglionitis. To date, our group has studied 6 women and 2 men with these symptoms. In 4 women, sacral nerve conduction studies revealed slowed nerve conduction and the presence of bladder dysfunction: usually motor and/or mixed sensory/motor involvement. In both men, studies revealed no evidence of sacral nerve dysfunction, but both had severe perineal pain associated with inability to fully relax the perineal musculature. Thus, the pathogenesis of these complications may be varied. Controlled clinical trials of the use of antivirals or anti-inflammatory medications in these sacral neurologic syndromes have not been performed.

2. Extragenital Lesions

Extragenital lesions appearing during the course of infection was present in 16% of persons with primary first episodes, 8% with nonprimary first episodes, and 4% with recurrent genital herpes. Among patients with primary genital HSV-2 infection, extragenital lesions developed in 26% of females compared to 8% of males ($p < 0.05$). These lesions were most frequently located in the buttock, groin, or thigh area, although finger and eye sites were also involved. Characteristically, the extragenital lesions developed after the onset of genital lesions, often during the second week of disease. The distribution of lesions on the extremities and/or areas near the genital lesions and their occurrence well into the course of disease suggest that the majority of extragenital lesions develop by autoinoculation of virus rather than viremic spread. Anatomic differences, especially contact with infected cervical–vaginal secretions, are the most likely explanation for the apparent increased risk of extragenital lesions in women.

3. Disseminated Infection

Blood-borne dissemination evidenced by the appearance of multiple vesicles over widespread areas of the thorax and extremities occurs rarely ($< 0.5\%$) in persons with primary mucocutaneous herpes (Nahmias, 1979; Ruchman and Dodd, 1950). Cutaneous dissemination usually occurs early in the disease and is often associated with aseptic meningitis, hepatitis, pneumonitis, or arthritis. Other complications of primary genital HSV-2 infection include monoarticular arthritis (Friedman et al., 1980), hepatitis (Joseph and Vogt, 1974; Flewett et al., 1969), thrombocytopenia (Whittaker and Hardson, 1978), and myoglobinuria (Schlesinger et al., 1978). Pregnancy may predispose to severe visceral dissemination of primary genital HSV disease (Koberman et al., 1980; Peacock and Sarubb., 1983; Young et al., 1976). Severe mucocutaneous and occasionally visceral dissemination of disease may occur in patients with atopic eczema (Wheeler and Abele, 1966). In immunosuppressed patients, especially those with impaired cellular immune responses, reactivation of genital HSV infection can be associated with viremic spread of virus to multiple organs (Linnemann *et al.*, 1976; Sutton *et al.*, 1974; Ramsey *et al.*, 1982; Meyers *et al.*, 1980). These patients may develop interstitial pneumonia, hepatitis, and pneumonitis, similar to the manifestations of disseminated infection of the neonate (Nahmias *et al.*, 1969b). Disseminated visceral infection in the immunosuppressed and pregnant patient is a disease of potentially high mortality. Systemic antiviral chemotherapy should be considered.

4. Local Extension of Disease

Extension of HSV infection into the uterine cavity is apparently uncommon. It is uncertain whether genital HSV causes pelvic inflammatory disease (Abramham, 1978). This author has occasionally seen patients with primary cervical HSV infection who also have lower abdominal pain and adnexal uterine tenderness. Whether this represents dual infection with other sexually transmitted pathogens such as *N. gonorrhoeae* and *C. trachomatis* or extension of HSV infection into the uterine cavity is unknown. Laparoscopic examination of one woman revealed vesicular lesions on the fimbrionated end of the fallopian tube. However, although HSV was isolated from vulvar and cervical lesions, virus was not isolated from a sample of the fluid taken at laparoscopy.

5. Superinfection

Bacterial superinfection of genital herpes in nonimmunosuppresed patients is uncommon. In rare cases, pelvic cellulitis presenting as an advancing erythema and swelling of the perineal area is encountered. In this instance, systemic antimicrobial therapy should be administered.

Fungal superinfection is frequently encountered during the course of initial genital herpes. We have noted the development of monilial vaginitis in 14% of women with first episodes of genital disease. Characteristically, monilial infection developed during the second week of disease, was associated with a change in character of the vaginal discharge and reemergence of local symptoms such as vulvar itching and irritation. Typical hyphal yeast forms can be demonstrated on KOH examination of vaginal secretions.

D. Summary of the Clinical Course of First Episodes of Genital Herpes

First-episode genital HSV infection is a disease of both systemic and local manifestations. Over one-half of the patients with primary genital herpes suffer from constitutional complaints, and one-third complain of headache, stiff neck, and mild photophobia during the first week of disease. Most patients have involvement of multiple anatomic sites. Herpes progenitalis is clearly a misnomer for first-episode genital HSV infection.

Patients with serologic evidence of prior HSV-1 infection are less apt to have constitutional symptoms, have a lower rate of complications and a shorter duration of disease than persons with true primary genital herpes. Neutralizing antibody to HSV has been shown to inactivate extracellular virus and interrupt the spread of HSV infection (Notkins, 1974). In addition, a cellular immune response to HSV antigens appears earlier in persons with nonprimary genital HSV than in persons with true primary infection (Corey et al., 1978a). It is likely that both of these immune mechanisms account for the clinical differences between primary and nonprimary first episodes of genital infection.

Although the duration of viral shedding from external genital lesions is similar in men and women with primary genital herpes, the duration of symptoms and the frequency of complications are greater in women. One potential explanation may be the high frequency of cervical and urethral involvement in women; the greater surface area of involvement may account for the increased severity of clinical disease in women.

E. Future Directions for Research on the Clinical Manifestations of Genital Herpes

While recent studies of the clinical manifestations of genital herpes have gone a long way toward defining the severity, diverse manifestation and complications of infection, little research emphasis has been placed upon studies that define the epidemiologic, anatomic, and/or host factors associated with disease acquisition and/or expression. For example: are there gender, racial, or age-related factors that influence the frequency of asymptomatic versus symptomatic infection? Is primary infection in the adolescent different in severity from the older adult? While surface area of involvement may help explain the higher rate of complications in women, are there hormonal factors that also affect disease severity? While HSV-type does not appear to affect the severity of first-episode disease, are there strain virulence factors between patients with asymptomatic primary versus symptomatic primary infection? Are there neurotropic strains associated with HSV meningitis or sacral neuropathy?

Animal models of HSV suggest that host immunologic factors influence the severity of disease (Lopez et al., 1980). Are there identifiable genetic factors associated with the expression of disease; or likelihood of complications? Kindreds with severe HSV infections have been described (Kapadia et al., 1979). In humans, several host factors have been shown to affect some of the clinical manifestations of primary genital herpes. Merriman et al. (1984) have demonstrated an inverse correlation between the development of local IgA antibody to HSV in cervicovaginal secretions and the duration of cervical viral shedding. Women who developed high titers of anti-HSV secretory IgA titer in cervicovaginal secretions had a shorter duration of viral shedding than those whose local immune response was slower to develop. The duration of viral shedding of primary first-episode genital herpes has also been correlated with the *in vitro* cellular immune response to HSV antigens (Corey *et al.*, 1978a). These data indicate that the role local, systemic, and cellular immune responses play in severity of first-episode disease should be studied in more detail. Newer techniques to dissect out the functional and morphologic cell types associated with the cellular immune response to HSV-specific antigens should be pursued. Are the local, systemic and cellular responses integrated? Studies to determine which one of these factors is correlated best with lesion or viral shedding, for example, could be approached by investigating enough patients with each assay to allow more detailed dissection of what are the determinant factors in disease expression.

F. Recurrent Genital Herpes

In contrast to first episodes of genital infection, the symptoms, signs, and anatomic sites of infection of recurrent genital herpes are localized to the genital region (Adams *et al.*, 1976; Guinan *et al.*, 1981; Corey *et al.*, 1983a). Local symptoms such as pain and itching are mild compared to first episodes of genital infection and the duration of the episode usually ranges from 8 to 12 days (Fig. 1). Approximately 50% of persons with recurrent genital herpes develop symptoms in the prodromal phase of illness, i.e., prior to the appearance of lesions. Prodromal symptoms vary from a mild tingling sensation, occurring $\frac{1}{2}$ to 48 hr prior to the eruption, to shooting pains in the buttocks, legs, or hips 1–5 days prior to the episode. In many patients these symptoms of sacral neuralgia are often the most bothersome part of the episode.

Symptoms of recurrent genital herpes tend to be more severe in women. In the 362 patients with untreated recurrent genital herpes followed at the Genital Herpes Clinic at the University of Washington, pain associated with lesions was a complaint of 88% of women (mean duration 5.9 days), as compared to 67% of men (mean duration 3.9 days). In addition, pain was more severe in women. Dysuria was reported in only 27% of women with recurrent disease. Most reported only external dysuria, and isolation of HSV from the urethra was uncommon in both sexes (3–9%).

Lesions of recurrent genital HSV are usually confined to one side, with an area of involvement approximately 1/10th that of primary genital infection (Figs. 4 and 5). The average duration of viral shedding from the onset of lesions was about 4 days; and the mean time from the onset of lesions to crusting of lesions averaged between 4 and 5 days for both men

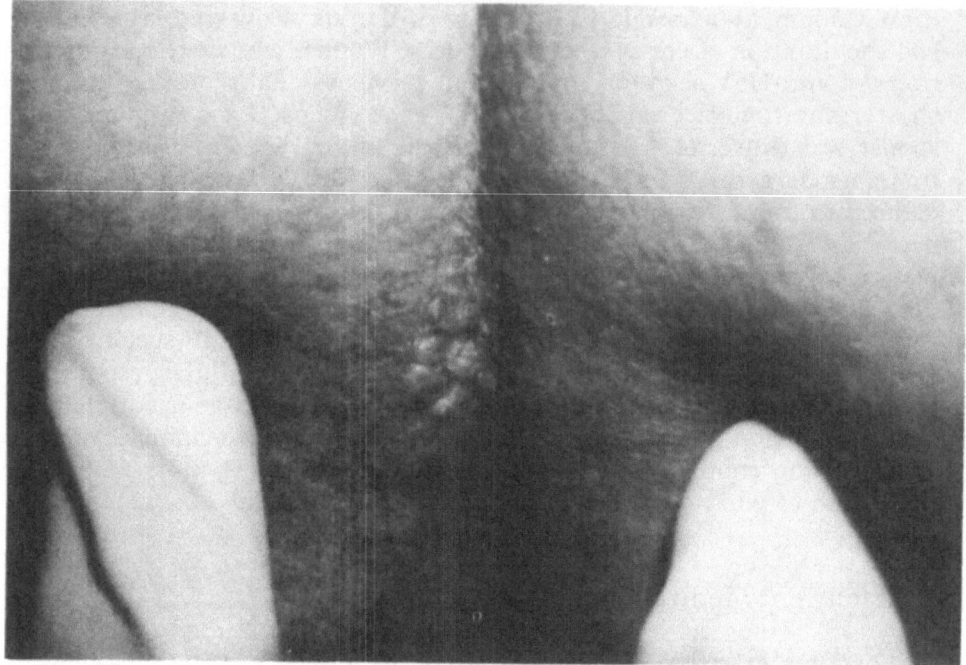

FIGURE 4. Recurrent buttock lesion.

and women. The mean time from onset of vesicles to complete reepithelialization of lesions was about 10 days. Although symptoms of recurrent genital disease tend to be more severe in females, objective signs of disease are similar in the two sexes.

Considerable overlap in the severity and duration of disease exists between patients (Fig. 1—note standard deviation for lesions, pain). In addition, the severity of any one individual episode of disease in any one patient may vary greatly during the course of disease. Some recurrences have only 1–2 lesions lasting 6–7 days, while others may be associated with 15–20 lesions lasting 12–16 days. Factors that are related to these variations in clinical expression have been unexplored. For example, does the time between recurrences or the severity of the prior recurrence influence the duration and time to the next recurrence? Data such as this would be useful in evaluating whether host immune factors are likely to play an important role in reactivation of infection.

G. Other Clinical Syndromes Associated with Genital HSV Infection

1. HSV Cervicitis

HSV may involve the cervix alone, without involvement of the external genitalia (Josey et al., 1966). Cervical HSV infection may be asymp-

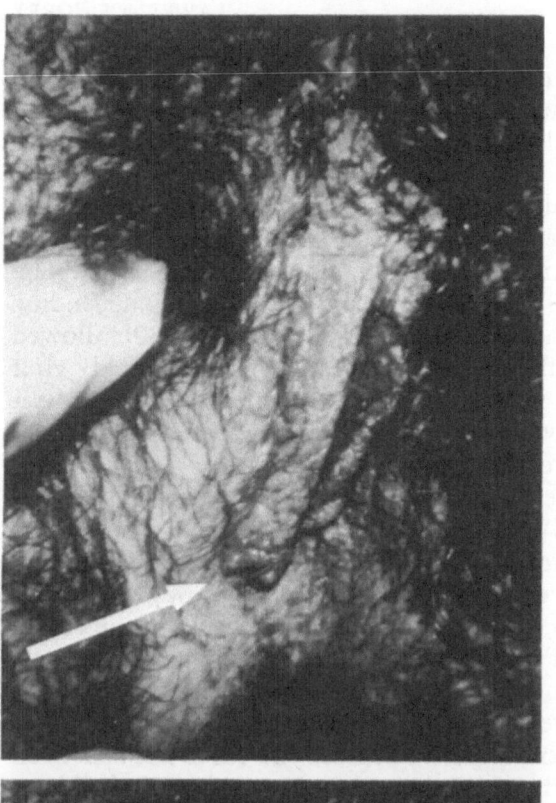

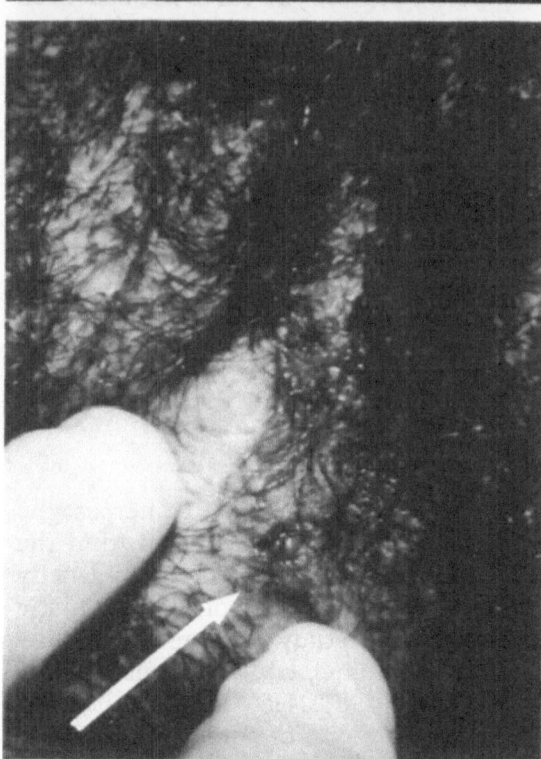

FIGURE 5. (A) Group of unilateral vesicular lesions in a woman with recurrent genital herpes. (B) Same lesions 24 hr later; small vesicles have formed coalescent painful ulcer.

tomatic or may present as a mucopurulent cervicitis. It is currently unknown what percentage of women who contract primary genital HSV
infection develop cervicitis alone compared with those with both vulvar
and cervical disease.

Asymptomatic viral shedding from the cervix, both during episodes
of vulvar lesions as well as between episodes of recurrent vulvar disease,
has been demonstrated (Rattray *et al.*, 1980; Vontver *et al.*, 1982). The
frequency of detection of HSV infection of the cervix varies depending
on the frequency that patients are sampled. Rattray *et al.* (1980) followed
six women with recurrent vulvar genital herpes with twice-weekly viral
cultures of the cervical and vulvar area. Twenty-three clinical and virologic recurrences of HSV were recorded in these women during 190
weeks of follow-up. Three recurrences were associated with asymptomatic excretion of virus: one with a small vulvar lesion noted only by
the examiner, from which HSV was isolated; one with no external lesions,
although HSV was isolated from a vulvar skin culture; and one in which
HSV was isolated from the cervix. Thus, in this small group of women
with recurrent genital herpes, asymptomatic HSV infection of the cervix
accounted for 1 of 23 clinical and virologic recurrences of genital herpes.
Adam *et al.* (1979) also demonstrated transient asymptomatic excretion
of HSV from the cervix in three women who were sampled regularly over
3 months. Asymptomatic excretion of virus from the vulva has also been
described and may be even more frequent than cervical viral shedding.
Vontver *et al.* (1982) described asymptomatic vulvar shedding in 0.75%
and asymptomatic cervical excretion in 0.66% of cultures in a cohort of
pregnant women with known genital herpes.

HSV antigen has been demonstrated in cervicovaginal secretions
when HSV culture was negative (Adam *et al.*, 1979; Moseley *et al.*, 1981;
Goldstein *et al.*, 1983). It is currently unknown whether this is a reflection
of short but frequent periods of viral shedding in which infectious virus
is quickly inactivated by local immune mechanisms, or defective viral
replication. These studies do, however, indicate that the pattern of viral
shedding from the cervix is intermittent, similar to the intermittent appearance of external genital lesions (August *et al.*, 1979).

2. HSV Urethritis

As discussed earlier, among patients with primary genital herpes who
present with external genital lesions, HSV has been isolated from the
urethra in 28% of men and 82% of women. This urethritis is usually
symptomatic. HSV urethritis may also occur as the sole symptomatic
manifestation of genital herpes. In a recent study of women with the
urethral syndrome (dysuria-frequency syndrome), HSV was isolated from
the urethra or cervix in 5% (Stamm *et al.*, 1980). HSV has been isolated
from urine of both men and women with dysuria or hematuria, and cys-
toscopy has in some of these cases revealed mucosal ulcerations. It is

likely that HSV cystitis occasionally occurs as a result of ascending infection from the urethra into the bladder.

3. Herpes Simplex Proctitis

HSV has been isolated from rectal mucosal and rectal biopsies in men and women with symptoms of rectal pain and discharge (Goddell *et al.*, 1983; Quinn *et al.*, 1983; Waugh, 1976). In a prospective study of 100 consecutive homosexual men who presented to an STD clinic with symptoms of rectal discharge and pain, HSV was isolated from rectal swabs and/or rectal biopsies in 23% and was the most frequent cause of non-gonococcal proctitis in homosexual men (Goodell *et al.*, 1983). Patients with HSV proctitis usually present with the acute onset of rectal pain, discharge, tenesmus, constipation, and blood and/or mucosy rectal discharge. Fever, malaise, and myalgia are common, and urinary retention, dysesthesia of the perineal region, and impotence may be reported. External perianal lesions are seen in about one-half the cases. Anoscopy and/or sigmoidoscopy generally reveal a diffuse, friable rectal mucosa, although occasionally discrete ulcerations of the rectal mucosa may be present. In most cases the pathology is limited to the lower 10 cm of the rectum. Rectal biopsies of involved mucosa generally reveal diffuse ulcerations and lymphocytic infiltration. If multiple histologic sections are performed, intranuclear inclusions may be demonstrated in rectal biopsies in about 50% of cases.

Both HSV-1 and HSV-2 have been isolated from patients with HSV proctitis (Levine and Saeed, 1979; Goodell *et al.*, 1983). Recurrences of this disease have been described and may be mild and/or asymptomatic. Controlled clinical evaluation of the use of systemic antivirals on this entity is currently unavailable.

4. Genital Ulcerations

HSV infection causes 40–60% of genital ulcerations in patients presenting to gynecologic practices or STD clinics in Western industrialized countries (Chapel *et al.*, 1978; Kinghorn *et al.*, 1982). In underdeveloped nations, however, genital HSV infection is an infrequent cause of genital ulcerations (Plummer *et al.*, 1983). The varied size, symptoms, and appearance of genital HSV lesions may make clinical diagnosis of genital ulcerations, especially single ulcerations, difficult. HSV may be isolated from many lesions attributed by patients to "trauma" or irritation. In addition, clinical differentiation of genital ulceration due to HSV, *Treponema pallidum*, and *Hemophilus ducreyi* is often uncertain.

V. CHRONICITY OF GENITAL HSV INFECTIONS

There are still large gaps in our knowledge concerning the natural history and chronicity of genital herpes. Long-term prospective studies

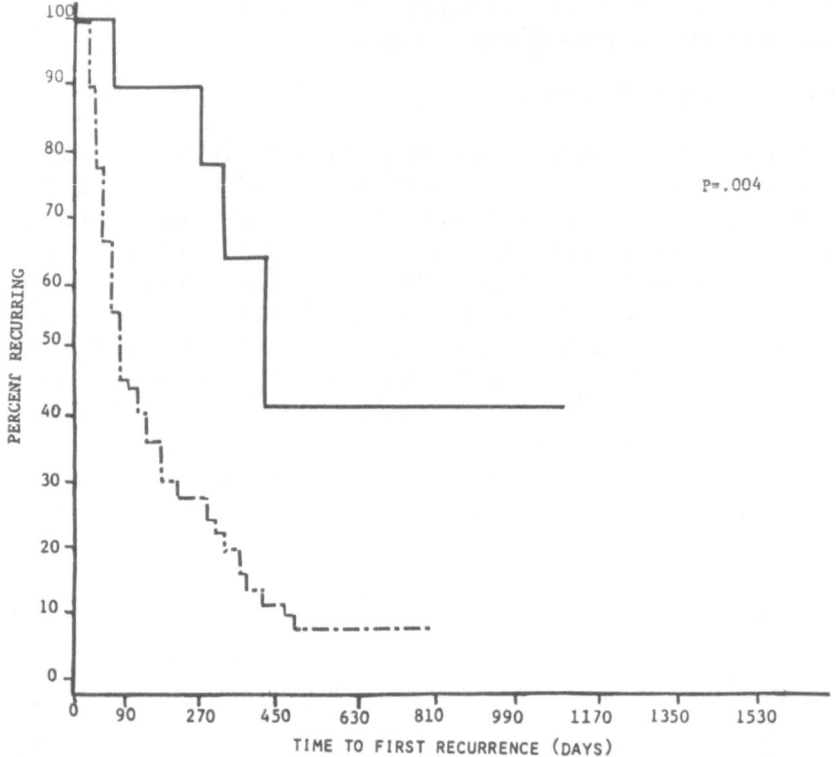

FIGURE 6. Survivorship analysis of the time to first clinical recurrence among untreated patients with (—) primary genital HSV-1 ($N = 18$) and (---) primary genital HSV-2 infection ($N = 157$). ($p < 0.001$, Mantel–Cox test for comparison between genital HSV-1 and genital HSV-2 patients.)

of the subsequent rate of recurrence of persons who acquire symptomatic primary infection are unavailable. Many basic questions remain unanswered. Does the frequency of disease change over time? What trigger factors influence recurrences of infection? How frequent is reinfection by the same subtype of HSV or by another subtype?

To date, prospective studies of patients who present with symptomatic primary genital herpes infection have shown that after 1 year of follow-up, 55% of patients with HSV-1 genital infections will develop recurrences; compared with 88% of patients with primary HSV-2 infection (Fig. 6). The median time to first recurrence was 42.5 days in patients with primary HSV-1 compared to 115 days in patients with primary HSV-2 infections. The mean rate of recurrences over time was also significantly less in patients with HSV-1 as compared to patients with primary HSV-2 infections, 0.10 recurrence per month in patients with primary HSV-1 infections versus 0.32 in patients with genital HSV-2 infection ($p < 0.001$). Of interest is that the subsequent recurrence rate of disease was similar in patients with primary and nonprimary first-episode HSV-2 infections (0.35 and 0.31 recurrence per month, respectively).

Vaginal inoculation of mice with HSV-1 strains, but not HSV-2, has been shown to result in latent infection of sacral nerve root ganglia (Richards *et al.*, 1981). This suggests that HSV-1 is much less likely to establish latency in sacral nerve root ganglia than HSV-2 infection. This biologic difference between the two viruses may account for their difference in recurrence rates, and the well-known anatomic predilection for HSV-2 infections to be "below the waist."

The severity of the initial episode as well as the host immune response to disease also appear to influence the subsequent recurrence rate of disease. Patients with primary genital HSV-2 infection who developed high titers of HSV-2 complement-independent neutralizing antibody in convalescent sera were more likely to develop recurrences than those who did not develop anti-HSV neutralizing antibody (Reeves *et al.*, 1981). More severe disease as measured by prolonged viral shedding and lesions was also associated with the development of higher titers of neutralizing antibodies as well as antibodies to HSV-specified glycoproteins (Zweerink and Corey, 1982). Development of high titers of complement-independent neutralizing antibody in convalescent phase sera after primary infection may reflect a high degree of antigenic exposure and/or a large number of latently infected cells in sacral nerve root ganglia. In a mouse model of mucocutaneous HSV infections, high levels of neutralizing antibody in convalescent sera were associated with increased numbers of latently infected ganglionic cells (Klein *et al.*, 1979a,b).

A past history of symptomatic genital HSV infection also appears to affect the subsequent time to next recurrence of disease. The median time to the next recurrence was 120 days after primary genital HSV-2 infection, compared to a median of 42 days between episodes in women with recurrent genital HSV-2 disease (Reeves *et al.*, 1981). As the duration of lesions and viral shedding is significantly longer in first versus recurrent genital herpes, these data suggest that the greater amount of antigenic exposure may affect host factors that influence the time to the subsequent recurrence of disease.

A large nationwide questionnaire survey of over 6000 persons with symptomatic recurrent genital herpes indicated that the median number of yearly clinical recurrences of genital herpes was between five and eight (Knox *et al.*, 1982). Recurrences of genital herpes do not follow a uniform pattern. The time of onset of disease and the number of recurrences over time that an individual experiences may vary greatly between patients and over time in any individual patient. For example, we have recently followed a cohort of patients with frequently recurring genital herpes over an 8-month period. In this population, which averaged 1.17 recurrences per month, only 28 of 92 patients exhibited little change (< 30%) in their clinical recurrence rate in the first as compared to the second 4 months of observation. Twenty-three of the ninety-two patients had at least a

TABLE IV. Recurrence Rates of Genital Herpes in Male
Patients With Symptomatic Genital Herpes ($N = 69$)

Duration of past disease	No. recurrences/month of follow-up
3 months–2 years	0.73
2 years–4 years	0.51
>4 years	0.42

50% increase and 25 of 92 a 50% decrease in their recurrence rates in the first as compared to subsequent 4-month periods. These observations are of interest in that they indicate that anecdotal observations regarding recurrence rates are unlikely to have much validity and that studies involving therapy of genital herpes must account for the varying natural history of untreated infection.

Little is also known regarding the long-term natural history of genital herpes. Does the disease change in frequency over time? Unfortunately, no prospective cohort of patients with first-episode disease followed over time has yet been studied. We have prospectively followed a cohort of men with recurrent genital herpes over a year's time interval and analyzed their recurrence rate of disease as a function of how long they previously had genital herpes. As shown in Table IV, the mean rate of recurrence of men with disease of greater than 4 years' duration was 42% less than those with disease of 3 months' to 2 years' duration. However, frequent recurrences of disease are often seen in persons who have had genital herpes for greater than 10 years' duration. The nationwide survey by Knox et al. (1982) indicated no significant difference in the reported recurrence rate over time. However, this latter study was a self-selected population of persons with severe genital herpes.

One of the major unanswered questions is what factors affect recurrence rates. "Trigger" mechanisms associated with recurrent genital herpes are largely anecdotal and unproven. Guinan et al. (1981) found a clustering of genital HSV episodes during the menstrual period. Others have shown no relationship between onset of lesions and menstrual cycle or frequency of sexual activity (Vontver et al., 1979; Rattray et al., 1978). Anecdotal observations have suggested that such diverse phenomena as emotional stress, heat, moisture, climate change, pregnancy, oral contraceptive use, anesthesia, and trauma might be "trigger" factors associated with recrudescences of HSV; however, objective data are lacking. Have patients with frequently recurrent disease for greater than 10 years had recent reinfections with new strains of HSV-2?

Reinfection versus Recrudescence

Using the technique of restriction enzyme analysis, Buchman et al. (1979) demonstrated that reinfection with different strains of HSV-2 can occur. Fife *et al.* (1983) have documented dual HSV-1 and HSV-2 infection. Since we have started to routinely type our clinical HSV isolates with monoclonal antibodies, we have documented dual HSV-1 and HSV-2 infection in 0.5% of genital HSV infections. The question of how frequently reinfection with different strains of HSV-2 accounts for recurrent episodes of clinical disease is largely unanswered. We have recently evaluated widely spaced genital HSV-2 isolates from 40 patients with genital HSV-2 infection for the possibility of reinfection with different viral strains. Patients selected for analysis were those who reported new sexual partners during the follow-up period. The first isolates obtained from the acute episode of 20 patients with primary herpes and an isolate obtained a mean of 14.9 months later were subjected to restriction enzyme analysis using five enzymes (*HpA*I, *Bam*HI, *Kpn*I, *Sal*I or *Pvu*I, *Bgl*II). No reinfections with different strains were noted. Similarly, a cohort of 21 patients with recurrent genital herpes, also had multiple genital isolates analyzed in a similar manner. The isolates were obtained a mean of 22.5 months apart. The mean number of recurrences between isolates was 13.5 and the mean number of new sexual partners was 2.9 (range 1–9) Again, no infections with new strains of HSV-2 were noted. Thus, in heterosexual patients with genital HSV, reinfection with new strains of HSV-2 appears uncommon (Schmidt 1984). Whether reinfection with multiple strains is a more common occurrence in more promiscuous populations will require further study.

VI. SEXUAL TRANSMISSION OF GENITAL HSV

As in the area of chronicity of infection, little is known about the sexual transmission of genital herpes. The risk of acquiring disease from a sex partner with infection is unknown. Does the risk differ between patients with symptomatic versus asymptomatic infection? What is the relative risk of acquiring disease from sexual contact with infected partners during the time of lesions? Mertz *et al.* (1984) evaluated the source contacts of 63 patients who presented with first episodes of genital herpes. The median time from first sexual exposure to the source contact and development of disease in the index patient was 4 months. Two-thirds of the source contacts had an obvious clinical history of genital HSV infection. However, only one-third of the source contacts were aware of the diagnosis at the time of presumed transmission. These data suggest that educating patients regarding the high transmissibility of infection during periods when mildly symptomatic lesions are present might decrease transmission of HSV disease.

Sexual contact with asymptomatic excreters of virus appears to be an important route of infection. Asymptomatic excretion of HSV has been demonstrated in saliva, cervical and seminal secretions. Douglas and Couch (1970) demonstrated asymptomatic excretion of HSV in saliva in

2% of adults, and there are reports of transmission of genital HSV-1 infection from orogenital sex during periods of asymptomatic salivary excretion of virus (Embil *et al.*, 1981). Asymptomatic cervical viral shedding has been shown to result in transmission to sexual contacts or infants (Mertz, *et al.*, 1984; Whitely *et al.*, 1980a). In the few studies in which virus has been titered from cervical secretions during asymptomatic episodes, the titer of HSV is 2–3 logs less than that recovered from lesions (Guinan *et al.*, 1980; Merriman *et al.*, 1984; Brown *et al.*, 1979). Whether the frequency of transmission from asymptomatic shedding is less than symptomatic vulvar lesions is unknown and a question of great clinical importance.

The source of asymptomatic shedding of HSV into the male genital tract is also not well clarified. Centifanto *et al.* (1982) isolated HSV from seminal secretions from nearly 20% of men undergoing routine vasectomy, but Corey *et al.* (1983a) failed to isolate HSV from any of 86 urethral and 41 prostatic secretion cultures from 47 men with recurrent genital herpes during intercurrent episodes. HSV has been isolated infrequently from prostatic secretions or biopsies (Deture *et al.*, 1976). Further studies are needed of the frequency and site(s) of asymptomatic excretion of HSV in the male genitourinary tract.

VII. CONCLUSIONS

Genital HSV infection is a disease of major public health importance. In the last 10 years, genital herpes has increased in prevalence in some population groups, especially white middle-class Caucasian men and women between the ages of 15 and 35 years. First episodes of genital HSV infection involve multiple anatomic sites, last 3–4 weeks, and have a high rate of complications. In contrast, episodes of recurrent genital disease are of much milder intensity and duration. The major morbidity of recurrent genital herpes is its frequency of recurrence, its chronicity, and its effects on the patients' personal relationships and sexuality. A possibility of increased risk of subsequent cervical carcinoma and the potential transmission of the disease to the neonate are major concerns to women. Studies in the last few years with new antiviral compounds such as acyclovir suggest that these substances may reduce the severity and duration of acute episodes of genital herpes. The effects of these substances on reducing the transmission of disease and subsequent recurrence rate of disease are, however, unknown. The development of effective forms of immunoprophylaxis of HSV infection offers several new potential approaches in the management of HSV infections especially in reducing the acquisition of disease. Further investigations on the mechanism of recurrence and risk factors associated with recurrence will hopefully provide the understanding to design effective forms of immunoprophylaxis. As animal models of recurrent genital HSV infection are not analogous to human infection, well-controlled clinical evaluations of humans are essential if we are to further our understanding of the pathogenesis of human HSV infections. Fortunately, genital HSV patients are some of the most cooperative and rewarding patient populations in which to conduct long-term studies of the natural history of disease. These controlled studies of the natural history of disease are critical in order to direct laboratory research into new diagnostic and therapeutic uses, as well as to generate the information necessary to design appropriate clinical trials evaluating new therapies and/or vaccines. In the meantime, knowledge of the natural history of the disease is also of direct vital importance to the physician in providing the patient with the information necessary to understand this complex entity, and to identify risk factors that will decrease the transmission of disease to sexual partners and neonates.

REFERENCES

Abraham, A. A., 1978, Herpesvirus hominis endometritis in a young woman wearing an intrauterine device, *Am. J. Obstet. Gynecol.* **131**:340–343.

Adam, E., Kaufman, R. H., Mirkovic, R. R., and Melnick, J. L., 1979, Persistence of virus shedding in asymptomatic women atter recovery from herpes genitalis, *Obstet. Gynecol.* **54**:171–173.

Adams, H. G., Benson, E. A., Alexander, E. R., Vontver, L. A., Remington, M. A., and Holmes, K. K., 1976, Genital herpetic infection in men and women: Clinical course and effect of topical application of adenine arabinoside, *J. Infect. Dis.* **133**:A151–A159.

Anonymous, 1982, Estimated rate of patients consultations with personal physicians for genital herpes: United States 1966–1979, *Morbidity and Mortality Weekly Report* **31**:138–139.

Astruc, J., 1736, *De Morbis Venereis Libri Sex*, Paris.

August, M. J., Nordlund, J. J., and Hsiung, G. D., 1979, Persistence of herpes simplex virus types 1 and 2 in infected individuals, *Arch. Dermatol* **115**:309–310.

Barton, I. G., Kinghorn, G. R., Walker, M. J., Al-Omar, L. S., Potter, C. W., and Gunner, E. B., 1981, Association of HSV-1 with cervical infection, *Lancet* **2**:1108.

Barton, I. G., Kinghorn, G. R., Najem, S., Al-Omar, L. S., and Potter, C. W., 1982, Incidence of herpes simplex virus types 1 and 2 isolated in patients with herpes genitalis in Sheffield, *Br. J. Vener. Dis.* **58**:44–47.

Baum, O., 1920, Uber die ubentragbarkeit des herpes simplex auf die kaninchen hourhaut, *Dermatol. Wochenschr.* **70**:105.

Bayer, P., and Gear, J., 1955, Virus meningo-encephalitis in South Africa, *S. Afr. J. Lab. Clin. Med.* **1**:22.

Bierman, S. M., 1978, Double-blind crossover study of levamisole as immunoprophylaxis for recurrent herpes progenitalis, *Cutis* **21**:352–354.

Bolognese, R. J., Corson, S. L. Fuccillo, D. A., Traub, R., Moder, F., and Sever, J. L., 1976, Herpesvirus hominis type II infections in asymptomatic pregnant women, *Obstet. Gynecol.* **48**:507–510.

Brenton, D. W., 1980, Hypoglycorrhachia in herpes simplex type 2 meningitis, *Arch. Neurol.* **37**:317.

Britto, E., Dikshit, S. S., and Naylor, B., 1976, Herpes virus simplex infection of the female genital tract, *Univ. Mich. Med. Bull.* **42**:152–154.

Brown, Z. A., Kern, E. R., Spruance, S. L., and Overall, J. C., Jr., 1979, Clinical and virologic course of herpes simplex genitalis (Medical Progress), *West. J. Med.* **130**:414.

Buchman, T. G., Roizman, B., and Nahmias, A. J., 1979, Demonstration of exogenous genital reinfection with herpes simplex virus type 2 by restriction endonuclease fingerprinting of viral DNA, *J. Infect. Dis.* **140:**295–304.

Caplan, L. R., Kleman, F. J., and Berg, S., 1977, Urinary retention probably secondary to herpes genitalis, *N. Engl. J. Med.* **297:**920–921.

Catterall, D., 1981, Biological effects of sexual freedom, *Lancet* **1:**315–319.

Centifanto, Y. M., Drylie, D. M., Deardourff, S. L., and Kaufman, H. 1972, Herpesvirus type 2 in the male genitourinary tract, *Science* **178:**318–319.

Chapel, T., Brown, W. J., Jeffres, C., and Stewart, J. A., 1978, Microbiological flora of penile ulcerations, *J. Infect. Dis.* **137:**50–57.

Chuang, T.-Y., Daniel, W. P. Su, W. P., Perry, H. O., Ilstrup, D. M., and Kurland, L. T., 1983, Incidence and trend of herpes progenitalis, a 15-year population study, *Mayo Clin. Proc.* **58:**436–441.

Corey, L., and Holmes, K. K., 1983, Genital herpes simplex virus infection: Current concepts in diagnosis, therapy and prevention, *Ann. Intern. Med.* **98:**973–978.

Corey, L., Reeves, W. C., and Holmes, K. K., 1978a, Cellular immune response in genital herpes simplex virus infection, *N. Engl. J. Med.* **299:**986–991.

Corey, L., Reeves, W. C., Chiang, W. T., Vontver, L. A., Remington, M., Winter, C., and Holmes, K. K., 1978b, Ineffectiveness of topical ether for the treatment of genital herpes simplex virus infection, *N. Engl. J. Med.* **299:**237–239.

Corey, L., Holmes, K. K., Benedetti, J., and Critchlow, C., 1981, Clinical course of genital herpes: Implications for therapeutic trials, in: *The Human Herpes Viruses* (A. J. Nahmias, W. R. Dowdle, and R. F. Schinazi, eds.), pp. 496–502, Elsevier, Amsterdam.

Corey, L., Nahmias, A. J., Guinan, M. E., Benedetti, J. K., Critchlow, C. W., and Holmes, K. K., 1982, A trial of topical acyclovir in genital herpes simplex virus infections, *N. Engl. J. Med.* **306:**1313–1319.

Corey, L., Adams, H. G., Brown, Z. A., and Holmes, K. K., 1983a, Genital herpes simplex virus infection: Clinical manifestations, course and complications, *Ann. Intern. Med.* **98:**958–972.

Corey, L., Fife, K. H., Benedetti, J. K., Winter, C., Fahnlander, A., Connor, J. D., Hintz, M. A., and Holmes, K. K., 1983b, Intravenous acyclovir for the treatment of primary genital herpes, *Ann. Intern. Med.* **98:**914–921.

Craig, C., and Nahmias, A., 1973, Different patterns of neurologic involvement with herpes simplex virus types 1 and 2: Isolation of herpes simplex virus from the buffy coat of two adults with meningitis, *J. Infect. Dis.* **127:**365–372.

Cruter, W., 1924, Das herpesvirus seine aetiologische und klinische bedeutung, *Muench. Med. Wochenschr.* **71:**1058–1060.

Deture, F. A., Drylie, D. M., Kaufman, H. E., and Centifanto, Y. M., 1976, Herpesvirus type 2 isolation from seminal vesicle and testes, *Urology* **7:**541.

Diday, P., and Doyon, A., 1886, *Les Herpes Genitaux*, Masson, Paris.

Douglas, R. G., Jr., and Couch, R. B., 1970, A prospective study of chronic herpes simplex virus infection and recurrent herpes labialis in humans, *J. Immunol.* **104:**289–295.

Dowdle, W. R., Nahmias, A. J., Harwell, R. W. and Pauls, F. P. 1967, Association of antigenic type of herpesvirus hominis with site of viral recovery, *J. Immunol.* **99:**974–980.

Duenas, A., Adam, E., Melnick, J. L., and Rawls, W. E., 1972, Herpesvirus type 2 in a prostitute population, *Am. J. Epidemiol.* **95:**483.

Eberle, R., and Courtney, R. J., 1981, Assay of type-specific and type-common antibodies to herpes simplex virus types 1 and 2 in human sera, *Infect. Immun.* **31:**1062–1070.

Embil, J. A., Manuel, F. R., and McFarlane, S., 1981, Concurrent oral and genital infection with an identical strain of herpes simplex virus type 1, *Sex. Transm. Dis.* **8:**70–73.

Evans, A. S., and Dick, E. C., 1964, Acute pharyngitis and tonsilitis in University of Wisconsin students, *J. Am. Med. Assoc.* **190:**699–708.

Fife, K. H., Schmidt, O., Remington, M., and Corey, L., 1983, Primary and recurrent concomitant genital infection with herpes simplex virus types 1 & 2, *J. Infect. Dis.* **147:**163.

Flewett, T. H., Parker, R. G. F., and Philip, W. M., 1969, Acute hepatitis due to herpes simplex in an adult, *J. Clin. Pathol.* **22**:60–66.

Friedman, H. M., Pincus, T., Gibilisco, P., 1980, Acute monoarticular arthritis caused by herpes simplex virus and cytomegalovirus, *Am. J. Med.* **69**:241–247.

Glezen, W. P., Fernald, G. W., and Lohr, J. A., 1975, Acute respiratory disease of university students with special reference to the etiologic role of herpesvirus hominis, *Am. J. Epidemiol.* **101**:111–120.

Goldstein, L. C., Corey, L., McDougall, J., Tollentino, E., Spear, P., and Nowinski, R. C., 1983, Monoclonal antibodies to herpes simplex viruses: Use in antigenic typing and rapid diagnosis, *J. Infect. Dis.* **147**:829–837.

Goodell, S. E., Quinn, T. C., Mkritchian, E. E., Schuffler, M. D., Corey, L., and Holmes, K. K., 1983, Herpes simplex virus: An important cause of acute proctitis in homosexual men, *N. Engl. J. Med.* **308**:868–871.

Grauballe, P. C., and Vestergaard, B. F., 1977, ELISA for herpes simplex virus type 2 antibodies, *Lancet* **2**:1038–1039.

Guinan, M. E., MacCalman, J., Kern, E. R., Overall, J. C., and Spruance, S. L., 1980, Topical ether and herpes simplex labialis, *J. Am. Med. Assoc.* **243**:1059–1061.

Guinan, M. E., MacCalman, J., Kern, E. R., Overall, J. C., and Spruance, S. L., 1981, Course of an untreated episode of recurrent genital herpes simplex infection in 27 women, *N. Engl. J. Med.* **304**:759–763.

Harger, J. H., Pazin, G. J., Armstrong, J. A., Breinig, M. C., and Ho, M., 1983, Characteristics and management of pregnancy in women with genital herpes simplex virus infection, *Am. J. Obstet. Gynecol.* **145**:784.

Hirsch, M. S., Zisman, B., and Allison, A. C., 1970, Macrophages and age-dependent resistance to herpes simplex virus in mice, *J. Immunol.* **104**:1140–1165.

Holzel, A., Feldman, G. V., Tobin, J. O., and Harper, J., 1953, Herpes simplex: A study of complement fixing antibodies at different ages, *Acta Paediatr. Scand.* **42**:206–214.

Hutfield, D. C., 1966, History of herpes genitalis. *Br. J. Vener. Dis.* **42**:263–268.

Jeansson, S., and Molin, L., 1970, Genital herpesvirus hominis infection: A venereal disease, *Lancet* **1**:1064.

Jeansson, S., and Molin, L., 1974, On the occurrence of genital herpes simplex virus infection, *Acta Derm. Venereol.* **54**:479–485.

Johnson, A. O., Salimonu, L. S., and Osunkoya, B. O., 1981, Antibodies to herpesvirus hominis types 1 and 2 in malnourished Nigerian children, *Arch. Dis. Child.* **56**:45–48.

Joseph, T. J., and Vogt, R. J., 1974, Disseminated herpes with hepatoadrenal necrosis in an adult, *Am. J. Med.* **56**:735–739.

Josey, W. E., Nahmias, A. J., Naib, Z. M., Utley, D. M., McKenzie, W. J., and Coleman, M. T., 1966, Genital herpes simplex infection in the female, *Am. J. Obstet. Gynecol.* **96**:493–501.

Josey, W., Nahmias, A., and Naib, Z., 1972, The epidemiology of type 2 (genital) herpes simplex virus infection, *Obstet. Gynecol. Surv.* **27**:295–302.

Judson, F. N., Penley, K. A., Robinson, M. E., and Smith, J. K., 1980, Comparative prevalence rates of sexually transmitted diseases in heterosexual and homosexual men, *Am. J. Epidemiol.* **112**:836–843.

Kalinyak, J. E., Fleagle, G., and Docherty, J. J., 1977, Incidence and distribution of herpes simplex virus types 1 and 2 from genital lesions in college women, *J. Med. Virol.* **1**:175–181.

Kapadia, A., Gupta, S., Good, R. A., and Day, N. K., 1979, Familial herpes simplex infections associated with activation of the complement system, *Am. J. Med.* **67**:122–126.

Kaufman, R. H., Gardner, H. L., Rawls, W. E., Dixon, R. E., and Young, R. L., 1973, Clinical features of herpes genitalis, *Cancer Res.* **33**:1446–1451.

Kawana, T., Kawaguchi, T., and Sakamoto, S, 1976, Clinical and virological studies on genital herpes [letter], *Lancet* **2**:964.

Kern, E. R., Glasgow, L. A., Klein, R. J., and Friedman-Kien, A. E., 1982, Failure of 2-deoxy-D-glucose in treatment of experimental cutaneous and genital herpes simplex virus (HSV) infections, *J. Infect. Dis.* **146**:159–166.

Kinghorn, G. R., Hafiz, S., and McEntegart, M. G., 1982, Pathogenic microbial flora of genital ulcers in Sheffield with particular reference to herpes simplex virus and *Haemophilus ducreyi, Br. J. Vener. Dis.* **58**:377–380.

Klastensky, J., Cappel, R., Snoeck, J. M., Flament, J., and Thiry, L., 1972, Ascending myelitis in association with herpes simplex virus, *N. Engl. J. Med.* **287**:182–184.

Klein, P. J., Friedman-Kien, A. E., and DeStefano, E., 1979a, Latent herpes simplex virus infections in sensory ganglia of hairless mice prevented by acycloguanosine, *Antimicrob. Agents Chemother.* **15**:723–729.

Klein, R. J., Friedman-Kien, A. E., and Yellin, P. B., 1979b, Orofacial herpes simplex virus infection in hairless mice: Latent virus in trigeminal ganglia after topical antiviral treatment, *Infect. Immun.* **20**:130–135.

Knox, G. E., Pass, R. F., Reynolds, D. W., Stagno, S., and Alford, C. A., 1979, Comparative prevalence of subclinical cytomegalovirus and herpes simplex virus in the genital and urinary tracts of low-income urban women, *J. Infect. Dis.* **140**:419–422.

Knox, S. R., Corey, L., Blough H. A., and Lerner, A. M., 1982, Historical findings in subjects from a high socioeconomic group who have genital infections with herpes simplex virus, *Sex. Transm. Dis.* **9**:15–20.

Koberman, T., Clark, L., and Griffin, W. T., 1980, Maternal death secondary to disseminated herpesvirus hominis *Am. J. Obstet. Gynecol.* **137**:742–743.

Kristensson, K., Vahlne, A., Person, L. A., and Lycke, E., 1978, Neural spread of herpes simplex virus type 1 and 2 in mice after corneal or subcutaneous (footpad) inoculation, *J. Neurol.* **35**:331–340.

Lerner, A. M., Shippey, M. J., and Crane, L. R., 1974, Serological responses to herpes simplex virus in rabbits: Complement requiring neutralizing, conventional neutralizing, and passive hemagglutinating antibodies, *J. Infect. Dis.* **129**:623–636.

Levine, J. B., and Saeed, M., 1979, Herpesvirus hominis (type 1) proctitis, *J. Clin. Gastroenterol.* **1**:225–227.

Linnemann, C. C., First, M. R., Alvira, M. M., Alexander, J. W., and Schiff, G. M., 1976, Herpesvirus hominis type 2 meningoencephalitis following renal transplantation, *Am. J. Med.* **61**:703–708.

Lipschutz, B., 1921, Untersuchungen uber die aetiologic der krankheiten der Herpesgruppe (Herpes zoster, Herpes genetalis, Herpes februlis), *Arch. Dermatol. Syph.* **136**:428–482.

Lopez, C., Ryshke, R., and Bennett, M., 1980, Marrow-dependent cells depleted by ^{89}Sr mediate resistance to herpes simplex virus type 1 infection in mice, *Infect. Immun.* **28**:1028–1032.

McClung, H., Seth, P., and Rawls, W. E., 1976, Relative concentration in human sera of antibodies to cross reacting and specific antigens of HSV virus types 1 and 2, *Am. J. Epidemiol.* **104**:192–201.

MacDougall, M. L., 1975, Genital herpes simplex in the female 1968 to 1973, *N. Z. Med. J.* **82**:333–335.

Mann, S. L., Meyers, J. D., Holmes, K. K., and Corey, L., 1984, Prevalence and incidence of herpesvirus infections among homosexually active men, *J. Infect. Dis.* **149(6)**:1026–1027.

Merriman, H. G., Woods, S., Winter, C., Fahnlander, A., and Corey, L., 1984, Secretory IgA antibody in cervicovaginal secretions in women with genital herpes simplex virus infection, *J. Infect. Dis.* **149(4)**:505–510.

Mertz, G. J., Schmidt, O., Jourden, J. L., Guinan, M. E., Remington, M. L., Fahnlander, A., Winter, C., Holmes, K. K., and Corey, L., 1984, Frequency of acquisition of first episode genital herpes simplex virus infection from symptomatic and asymptomatic source contacts, *Sex. Trans. Dis.* (in press).

Meyers, J. D., Flournoy, N., and Thomas, E. D., 1980, Infection with herpes simplex virus and cell-mediated immunity after marrow transplant, *J. Infect. Dis.* **142**:338–346.

Montefiore, D., Sogbetun, A. O., and Anong, C. N., 1980, Herpesvirus hominis type 2 infection in Ibadan: Problem of non-venereal transmission, *Br. J. Vener. Dis.* **56**:49–53.

Moseley, R., Corey, L., Winter, C., and Benjamin, D., 1981, Comparison of the indirect immunoperoxidase and direct immunofluoresence techniques with viral isolation for the diagnosis of genital herpes simplex virus infection, *J. Clin. Microbiol.* **13**:913–918.

Nahmias, A. J., 1970, Disseminated herpes simplex virus infections, *N. Engl. J. Med.* **282**:684–685.

Nahmias, A. J., and Dowdle, W. R., 1968, Antigenic and biologic differences in herpesvirus hominis, *Prog. Med. Virol.* **10**:110–159.

Nahmias, A. J., and Roizman, B., 1973, Infection with herpes simplex virus 1 and 2, *N. Engl. J. Med.* **299**:667–674, 719–725, 781–789.

Nahmias, A. J., Naib, Z. M., Josey, W. E., and Clipper, A. C., 1966, Genital herpes simplex infection: Virologic and cytologic studies, *Obstet. Gynecol.* **29**:395–400.

Nahmias, A. J., Dowdle, W. R., Naib, Z. M., Josey, W. E., McLane, D., and Domecsik, G., 1969a, Genital infections with type 2 herpes virus hominis: a commonly occuring venereal disease, *Brit. J. Vener. Dis.* **45**:294–298.

Nahmias, A. J., Dowdle, W. R., Josey, W. E., Naib, Z. M., Pawler, L. M., and Luce, C., 1969b, Newborn infection with herpesvirus hominis types 1 and 2, *J. Pediatr.* **75**:1194–1203.

Nahmias, A. J., Josey, W. E., Naib, Z. M., Luce, C. F., and Duffey, A., 1970, Antibodies to herpesvirus hominis types 1 and 2 in humans, *Am. J. Epidemiol.* **91**:539–546.

Nahmias, A. J., Keyserling, H. L., Kerrick, G. M., 1983, Herpes Simplex Virus: in *Infectious Diseases of the Fetus and Newborn Infant* (J. S. Remington and J. O. Klein, eds.), pp. 642–644, W. B. Saunders, Phililedphia.

Notkins, A. L., 1974, Immune mechanisms by which the spread of viral infections is stopped, *Cell. Immunol.* **11**:478–483.

Overall, J. C., Jr., Kern, E. R., Schlitzer, R. L., Friedman, S. B., and Glasgow, L. A. 1975, Genital herpesvirus hominis infection in mice. I. Development of an experimental model, *Infect. Immun.* **11**:476–480.

Parker, F., and Nye, R. W., 1925, Studies on filterable viruses. II. Cultivation of herpes virus, *Am. J. Pathol.* **1**:337–340.

Peacock, J. E., and Sarubbi, F. A., 1983, Disseminated herpes simplex virus infection during pregnancy, *Obstet. Gynecol.* **61**:135–137.

Pereira, L., Dondero, D. V., Gallo, P., Devlin, V., and Woodie, J. D., 1982, Serological analysis of herpes simplex virus types 1 and 2 with monoclonal antibodies, *Infect. Immun.* **35**:363–367.

Person, D. A., Kaufman, R. H., Gardner, H. L., and Rawls, W. E., 1973, Herpesvirus type 2 in genitourinary tract infection, *Am. J. Obstet. Gynecol.* **116**:993–995.

Plummer, F. A., Nsanze, H., Costa, L. J., Karasira, P., Maclean, I. W., Ellison, R. H., and Ronald, A. R., 1983, Single-dose therapy of chancroid with trimethoprim-sulfametrole, *N. Engl. J. Med.* **309**:67–71.

Plummer, G., Waner, J. L., Phuangsab, A., and Goodheart, C. R. 1970, Type 1 and 2 herpes simplex viruses: Serological and biological differences, *J. Virol.* **5**:51–59.

Porter, D. D., Wimberly, I., and Benyesh-Melnick, M., 1969, Prevalence of antibodies to EB virus and other herpesviruses, *J. Am. Med. Assoc.* **208**:1675–1679.

Prakash, S. S., and Seth, P., 1979. Evaluation of indirect hemagglutination and its inhibition in the differentiation between antibodies to herpes simplex virus types 1 and 2 for seroepidemiologic studies: Use of a II/I index threshold of 85 and an assay of type-specific antibodies, *J. Infect. Dis.* **139**:524–528.

Quinn, T. C., Stamm, W. E., Goodell, S. E., Mkritchian, E., Benedetti, J., Corey, L., Schuffler, M. D., and Holmes, K. K., 1983, The polymicrobial origin of intestinal infections in homosexual men, *N. Engl. J. Med.* **309**:576–582.

Ramsey, P. G., Fife, K., Hackman, R., Meyers, J. D., and Corey, L., 1982, Herpes simplex virus pneumonia: Clinical presentation and pathogenesis, *Ann. Intern. Med.* **97**:813–820.

Rattray, M. C., Corey, L., Reeves, W. C., Vontver, L. A., and Holmes, K. K., 1978, Recurrent genital herpes among women: Symptomatic versus asymptomatic viral shedding, *Bri. J. Vener. Dis.* **54**:262–265.

Rattray, M. C., Peterman, G. M., Altman, L. C., Corey, L., and Holmes, K. K., 1980, Lymphocyte-derived chemotactic factor synthesis in initial genital herpesvirus infections: Correlation with lymphocyte transformation, *Infect. Immun.* **30**:110–116.

Rauh, J. L., Brookman, R. R., and Schiff, G. M., 1977, Genital surveillance among sexually active adolescent girls, *J. Pediatr.* **90:**844.

Ravaut, P., and Darre, M., 1904, Les reactions nerveuses au cours des herpes genitaux, *Ann. Dermatol. Syphiligr.* **5:**481–496.

Rawls, W. E., 1979, Herpes simplex viruses type 1 and 2, in: *Diagnostic Procedures for Viral, Rickettsial and Chlamydial Infections*, 5th ed. (E. H. Lennette and N. J. Schmidt, eds.), pp. 309–354, American Public Health Association, Washington, D. C.

Rawls, W. E., Iwamoto, K., Adam, E., and Melnick, J. L., 1970, Measurement of antibodies to herpesvirus types 1 and 2 in human sera, *J. Immunol.* **104:**599–606.

Rawls, W. E., Gardner, H. L., Flunders, R. W., Lawry, S. P., Kaufman, R. H., and Melnick, J. L., 1971, Genital herpes in 2 social groups, *Am. J. Obstet. Gynecol.* **110:**682–689.

Reeves, W. C., Corey, L., Adams, H. G., Vontver, L. A., and Holmes, K. K., 1981, Risk of recurrence after first episodes of genital herpes: Relation to HSV type and antibody response, *N. Engl. J. Med.* **305:**315–319.

Renis, H. E., Eidson, E. E., Mathews, J., and Gray, J. E., 1976, Pathogenesis of herpes simplex virus types 1 and 2 in mice after various routes of inoculation, *Infect. Immun.* **14:**571–578.

Richards, J. T., Kern, E. R., Overall, J. C., and Glasgow, L., 1981, Differences in neurovirulence of herpes simplex virus type 1 and type 2 isolates in four experimental infections of mice, *J. Infect. Dis.* **144:**464–471.

Roome, APCH, Montefiore, D., and Waller, D., 1975, Incidence of herpesvirus hominis antibodies among blood donor populations, *Br. J. Vener. Dis.* **51:**324–328.

Ross, CAC, and Stevenson, J., 1961, Herpes simplex meningoencephalitis, *Lancet* **2:**682–685.

Ruchman, I., and Dodd, K., 1950, Recovery of herpes simplex virus from the blood of a patient with herpetic rhinitis, *J. Lab. Clin. Med.* **35:**434–439.

Scher, J., Bottone, E., Desmond, E., and Simons, W., 1982, The incidence and outcome of asymptomatic herpes simplex genitalis in an obstetric population, *Am. J. Obstet. Gynecol.* **144:**906–907.

Schlesinger, J. J., Gandara, D., and Bensch, K. G., 1978, Myoglobinuria associated with herpes-group viral infections, *Arch. Intern. Med.* **138:**422–424.

Schmidt, O. W., Fife, K. H., and Corey, L., 1984, Reinfection is an uncommon occurence in patients with symptomatic recurrent genital herpes, *J. Inf. Dis.*, **149(4):**645–646.

Shore, S. L., Milgrom, H., Wood, P. A., and Nahmias, A. J., 1977, Antibody dependent cellular cytotoxicity to target cells infected with herpes simplex virus: Functional adequacy in the neonate, *Pediatrics* **59:**22–28.

Skoldenberg, B., Jeansson, S., and Wolontis, S., 1975, Herpes simplex virus 2 and acute aseptic meningitis, *Scand. J. Infect. Dis.* **7:**227–232.

Smith, I. W., and Pleutherer, J. F., 1967, The incidence of herpesvirus hominis antibody in the population, *J. Hyg.* **65:**395–408.

Srinannaboon, S., 1979, Cytologic study of herpes simplex infection and dysplasia in the female genital tract, *J. Med. Assoc. Thailand* **62:**201–207.

Stamm, W. E., Wagner, K. F., Amsel, R., Alexander, E. R., Turck, M., Counts, G. W., and Holmes, K. K., 1980, Causes of the acute urethral syndrome in women, *N. Engl. J. Med.* **303:**409–415.

Stavraky, K. M., Rawls, W. E., Chiavetta, J., Donner, A. P., and Wanklin, J. M., 1983, Sexual and socioeconomic factors affecting the risk of past infections with herpes simplex virus type 2, *Am. J. Epidemiol.* **118:**109–121.

STD Fact Sheet, Edition 35, pp. 5–12, Centers for Disease Control, Atlanta.

Sullivan-Bolyai, J., Hull, H. F., Wilson, C., and Corey, L., 1983, Neonatal herpes simplex virus infection in King County, Washington: Increasing incidence and epidemiologic correlates, *J. Am. Med. Assoc.* **250:**3059–3062.

Sumaya, C. V., Marx, J., and Ullis, K., 1980, Genital infections with herpes simplex virus in university student populations, *Sex. Transm. Dis.* **7:**16–20.

Sutton, A. L., Smithwick, E. M., Seligman, S. J., and Kim, D. S., 1974, Fatal disseminated herpesvirus hominis type 2 infection in an adult with associated thymic dysplasia, *Am. J. Med.* **56**:545–553.

Taintivanich, S., and Tharavawij, V., 1980, Prevalence of genital herpes virus infection in Thai women, *Southeast Asian J. Trop. Med. Public Health* **11**:127–130.

Tejani, N., Klein, S. W., and Kaplan, M., 1979, Subclinical herpes simplex genitalis infections in the perinatal period, *Am. J. Obstet. Gynecol.* **135**:547.

Tustin, A. W., and Kaiser, A. B., 1979, Life threatening pharyngitis caused by herpes simplex virus type 2, *Sex. Transm. Dis.* **6**:23–24.

Vestergaard, B. F., and Rune, S. J., 1980, Type-specific herpes simplex virus antibodies in patients with recurrent duodenal ulcer, *Lancet* **1**:1273–1274.

Vesterinen, E., Purola, E., Sadsela, E., and Lenikkin, P., 1977, Clinical and virological findings in patients with cytologically diagnosed gynecologic herpes simplex infection, *Acta Cytol* **21**:199–205.

Vontver, L. A., Reeves, W. C., Rattray, M., Corey, L., Remington, M. A., Tolentino, E., Schwed, A., and Holmes, K. K., 1979, Clinical course and diagnosis of genital herpes simplex virus infection and evaluation of topical surfactant therapy, *Am. J. Obstet. Gynecol.* **133**:548–554.

Vontver, L. A., Hickok, D. E., Brown, Z., Reid, L., and Corey, L., 1982, Recurrent genital herpes simplex virus infection in pregnancy: Infant outcome and frequency of asymptomatic recurrences, *Am. J. Obstet. Gynecol.* **142**:75.

Waugh, M. A., 1976, Anorectal herpesvirus hominis infection in man, *J. Am. Vener. Dis. Assoc.* **3**:68.

Wentworth, B. B., and Alexander, E. R., 1971, Seroepidemiology of infections due to member of the herpesvirus group, *Am. J. Epidemiol.* **94**:496–507.

Wentworth, B. B. Bonin, P., Holmes, K. K., and Alexander, E. R. 1973, Isolation of viruses, bacteria and other organisms from venereal disease clinic patients: Methodology and problems associated with multiple isolations, *Health Lab. Sci.* **10**:75–81.

Wheeler, C. E., Jr., and Abele, D. C., 1966, Eczema herpeticum, primary and recurrent, *Arch. Dermatol.* **93**:162–173.

Whitley, R. J., Nahmias, A. J., Visintine, A. M., Fleming, C. L., and Alford, C. A., 1980, The natural history of herpes simplex virus infection of mother and newborn, *Pediatrics* **66**:489–494.

Whittaker, J. A., and Hardson, M. D., 1978, Severe thrombocytopenia after generalized HSV-2 infection, *South. Med. J.* **71**:864–866.

Wolontis, S., and Jeansson, S., 1977, Correlations of herpes simplex virus types 1 and 2 with clinical features of infection, *J. Infect. Dis.* **135**:28–33.

Yeo, J., Killington, R. A., Watson, D. H., and Powell, K. L., 1981, Studies on cross reactive antigens in herpes viruses, *Virology* **108**:256–266.

Young, E. J., Killam, A. P., and Greene, J. F., 1976, Disseminated herpesvirus infection association with primary genital herpes in pregnancy, *J. Am. Med. Assoc.* **235**:2731–2733.

Zweerink, H. J., and Corey, L., 1982, Virus-specific antibodies in sera from patients with genital herpes simplex virus infection, *Infect. Immun.* **37**:413–421.

CHAPTER 2

Natural Resistance Mechanisms in Herpes Simplex Virus Infections

CARLOS LOPEZ

I. INTRODUCTION

The normal host has a series of defense mechanisms that participate in sequestering and clearing invading viral pathogens (Allison, 1974). The defense systems involved fall into two major categories: Natural resistance mechanisms are immediately available to respond to the microorganism and require no prior exposure to foreign antigen in order to be operative; adaptive immune mechanisms require presensitization and are not active until several days after primary virus infection. The adaptive immune response against herpesvirus infections consists of the antibody response (discussed in Chapter 3 by Norrild) and the cell-mediated immune response (discussed in Chapter 4 by Nash, Leung, and Wildy). This chapter will focus on the role of natural resistance mechanisms in defense against herpes simplex virus (HSV) infections. The specific systems to be discussed include macrophages, natural killer (NK) cells, and interferon (IFN).

The pathogenesis of HSV infections is complex and host defense mechanisms may interact with infection at several different points during the disease process. Primary infections may be subclinical or may be associated with significant morbidity (Nahmias and Roizman, 1973; Corey, this volume). Latent infections appear to develop readily after primary infections and may be reactivated at some later date to cause

CARLOS LOPEZ • Sloan–Kettering Institute for Cancer Research, New York, New York 10021.

subclinical infection or recrudescent disease (Douglas and Couch, 1970; Docherty and Chopan, 1974; Corey et al., 1983). Natural resistance mechanisms are thought to be active during the earliest phases of an infection and to play an important role in sequestering and preventing dissemination of the infection. These systems may also have a direct or indirect effect on the development of latent infections. Thus, the same mechanisms that act to control primary infections may act locally to restrict the development of latently infected neurons. For example, controlling the primary virus infection may indirectly reduce the number of latently infected ganglia or the incidence of latently infected animals. In turn, this might lead to fewer recurrences because fewer latent infections are available for reactivation. As patients with severe primary herpes genitalis demonstrated recurrent disease more often than did patients with subclinical primary infections (Allen and Rapp, 1982; Corey, this volume), either the host defense restricts both the primary infection and reactivated (latent) infections or, by restricting the primary infection, these mechanisms may reduce the load of latent infection that is available for reactivation at a later date. The first possibility seems more likely because antiviral drugs, which significantly reduce primary infection, appear to have little effect on recurrent disease (Corey, this volume).

Animal models of recurrent herpesvirus infections are poor examples of human disease and offer little opportunity to study the role of defense mechanisms required for control of reactivated infections. Therefore, this chapter will focus on the role of natural resistance mechanisms in defense against primary infection and will only discuss latent infections when they are relevant to the mechanism under consideration.

When one thinks of both natural resistance and adaptive immunity, it is usually with the thought that these systems are required for protecting the normal host from a plethora of infectious pathogens. However, it is also possible that these mechanisms can cause great damage to the infected host. An immunopathological response has been demonstrated in HSV infections and is discussed in Chapter 5 by Rouse.

Although HSV types 1 and 2 (HSV-1, HSV-2) are similar in many ways, including 48% homology of viral DNA, colinearity of DNAs, and exchangeability of genetic materials (Kieff et al., 1972; Nahmias and Roizman, 1973; Cassai et al., 1975; Morse et al., 1977), there are important differences in the natural resistance systems involved in host defense. In mice, for example, HSV-2 causes a focal, necrotic hepatitis whereas HSV-1 failed to do so in immunologically competent mice (Mogensen et al., 1974). Also, when inoculated intraperitoneally (i.p.), HSV-2 is more neurovirulent than HSV-1 in the mouse (Ruchmann, 1954; McKendall, 1980; Richards et al., 1981), whereas in man, HSV-1, but not HSV-2, causes encephalitis and thus appears to be more neurovirulent (Leider et al., 1965; Graig and Nahmias, 1973). These differences may be dependent, at least in part, on the natural resistance systems and particular attention will be given to those differences that highlight a role for these mechanisms.

Three approaches have been used for the study of defense mechanisms in HSV infections: (1) patients with primary or secondary immunodeficiencies have been studied to determine whether a specific deficiency is associated with susceptibility to severe infections; (2) animals have been treated with agents to suppress or augment one or another aspect of host defense to determine whether that was necessary for protection against challenge with virus; and (3) more recent studies have utilized animal models of genetic resistance to virus infections in order to determine the mechanisms deficient in the genetically susceptible. The studies of immunodeficient humans and immunosuppressed animals have, for example, clearly shown that the capacity to generate an antibody response can be normal in the face of a severe virus infection (Meyers *et al.*, 1980; Verhoef *et al.*, 1980; Bloomfield and Lopez, 1980; Lopez *et al.*, 1974). These observations have been taken as an indication that cell-mediated immunity, which is often found suppressed in these hosts, plays a major role in defense against infection. Unfortunately, however, the immunosuppressive regimens used to define the specific systems involved usually affect more than the one mechanism being studied. Consequently, these studies fail to define the specific deficiencies responsible for susceptibility to severe infections. In fact, some studies can be very misleading. For example, the studies of Schlabach *et al.* (1979) determined that antithymocyte antisera made mice much more susceptible to HSV-1 infection only if the mice were challenged immediately after treatment because the antisera used had a relatively short-term suppressive effect on macrophage function and thus made the mice more susceptible. When the mice were challenged later and T cells were still deficient but macrophage function had returned to normal, treated mice were no more susceptible than the controls. Although one would normally have expected that the effect of this antiserum on T-cell-mediated immunity was responsible for the increased susceptibility to herpesvirus infections, in fact it was its effect on macrophages that resulted in the lethal infections of the mice. Thus, immunosuppression, especially with agents that have effects on more than one subpopulation of effector cells, has been less than definitive when used to determine the mechanisms required for resistance to herpesvirus infections.

In recent years, models of genetic resistance to HSV-1 and HSV-2 have been defined in the mouse (Lopez, 1975; Mogensen *et al.*, 1974; Kirchner *et al.*, 1978b; Armerding *et al.*, 1981a,b). These models have the advantage over the other animal models that one can use genetic tools to define the systems required for resistance. Thus, *in vitro* or *in vivo* correlates of resistance can be evaluated in crosses and backcrosses of resistant and susceptible mice to determine whether the functions in question segregate with resistance or independently. Another advantage is that these models use genetically homogeneous mice, which can be used for cell transfer experiments. In contrast to the obvious advantages of these genetic models, certain possible disadvantages must be considered. First, HSV-1 and HSV-2 are viruses that are indigenous to man and not the mouse. As herpesviruses have coevolved with animals and many species have been shown to have herpesviruses that are found only in that species (Nahmias, 1972), it is possible that the resistance systems found to be operative for HSV-1 in the mouse may not be the same as those that are operative in man. The genetic model may, in fact, be an artifact. Whether this turns out to be the case will depend on studies in man using the information generated in the study of the mouse. In fact, preliminary evidence (to be discussed below) suggests that study of the mouse model of genetic resistance to HSV-1 has prompted a number of studies the results of which indicate that similar mechanisms probably play an important role in resistance to HSV-1 infections in man.

II. GENETIC RESISTANCE TO LETHAL HSV-1 IN THE MOUSE

The genetic model of resistance to HSV-1 (Lopez, 1975) has been used by several investigators in attempts to define mechanisms required for protection of the host. This model and models of genetic resistance to HSV-2 will be described prior to discussion of the various mechanisms that might be required for genetic resistance.

Adult, inbred strains of mice have been challenged with 10-fold dilutions of HSV-1, strain 2931, a virulent strain of virus (Lopez, 1975). These mice were found to be either resistant, moderately susceptible, or very susceptible. C57BL/6 mice are the prototype resistant strain of mice and survived challenge with 10^6 PFU of HSV-1. Moderately susceptible mice (BALB/c is the prototype) had LD_{50}s of 10^3-10^4 PFU, and very susceptible strains of mice (A/J is the prototype) had LD_{50}s of 10^1-10^2 PFU. Male and female mice have been inoculated with no indication that resistance is sex-linked. Many of these observations have been confirmed by Kirchner et al. (1978b).

The studies of Kirchner et al. (1978a) also showed that two closely related strains of mice differed significantly in their resistance to HSV-1. In these studies, C3H/HeJ mice were found to be more resistant to challenge than C3H/FeJ and to be resistant to all known effects of bacterial endotoxin, as well. Because endotoxin was found to convert lymphoid cells from nonpermissive to cells capable of replicating HSV-1 in vitro, these data have been taken to suggest an important role for endotoxin-stimulated cells in susceptibility to infection.

As noted above, our studies were carried out with HSV-1 strain 2931 while the studies of Kirchner et al. (1978a,b) were conducted with HSV-1 strain WAL. In addition, we have evaluated 12 other strains of HSV-1, including strains that had been passaged only two or three times in culture and strains that had been maintained as laboratory strains for many years and had been passaged hundreds of times (Lopez, 1975, and unpublished). The same pattern of resistance was found with all strains of virus tested: C57BL/6 mice were resistant to 10^6 PFU of each of the strains tested, and A/J mice had LD_{50}s that were lower than those of BALB/c mice. To date, strain 2931 has been the most virulent strain that we have tested and has been used for most of the other studies.

Genetic studies indicated that resistance to lethal HSV-1 infections was a dominant trait, governed by two independently segregating, non-*H-2* loci (Lopez, 1980). A challenge dose of 10^6 PFU of HSV-1 (strain 2931) was used because it appeared to clearly distinguish resistant from susceptible strains of mice. F_1 crosses between resistant and susceptible and between resistant and moderately susceptible mice were found to be resistant to challenge with virus, indicating that resistance was a dominant trait. In backcross experiments, 25% of the progeny of crosses between $F_1 \times$ susceptible A/J mice or $F_1 \times$ moderately susceptible BALB/c mice were found to be resistant to virus challenge, indicating that resistance was governed by two loci, both of which were required for resistance. Studies with F_2 generations also indicated that two independently segregating, genetic loci governed resistance. An indication that resistance was not *H-2* linked was obtained in earlier studies where mice of the H-2^k phenotype were found to be both resistant and susceptible (Lopez, 1975). Also, congenic strains of mice were challenged with HSV-1 and the results showed that the background (non-*H-2*) loci were important in determining resistance and susceptibility whereas the *H-2* genes were not (Lopez, 1980).

The results of several studies suggest that the host's hemopoietic defense system is responsible for genetic resistance to HSV-1 (Lopez, 1981). Resistant mice survived challenge with 10^6 PFU of HSV-1 when inoculated i.p. or intravenously (i.v.) and yet were found to be as susceptible as the susceptible strains of mice to an intracerebral challenge of 10^1 PFU of virus. Thus, resistance appeared to depend on the ability of the host to inhibit virus from reaching the CNS, which is the target organ for HSV-1. Most significantly, transplantation of bone marrow stem cells from resistant (F_1) mice into lethally irradiated susceptible mice resulted in chimeras with greatly enhanced resistance (Lopez, 1981). Thus, replacement of the hemopoietic system of susceptible mice with one from resistant mice rectified the deficiency that had been responsible for their lack of resistance.

An indication of the mechanisms that might be involved in resistance derived from the striking similarity between genetic resistance to HSV-1 and genetic resistance to allogeneic bone marrow grafts (Lopez, 1981). The latter was defined earlier by Cudkowicz and Bennett (1971) as the ability of lethally irradiated mice to resist engraftment with allogenic bone marrow cells. As earlier studies by Bennett (1973) showed that ^{89}Sr treatment of mice abrogated allogeneic resistance without significantly affecting their capacity to reject skin grafts, studies were carried out to determine whether similar treatment of mice would abrogate genetic resistance to HSV-1 (Lopez et al., 1980). Challenge of ^{89}Sr-treated mice with HSV-1 indicated that this treatment made genetically resistant mice as susceptible to HSV-1 as the genetically most susceptible strains and that this treatment reduced resistance more effectively than any other treatment except for high-dose γ-irradiation. The spleen is forced to take over hemopoiesis and antibody responses and skin graft rejection remain normal. In ^{89}Sr-treated mice, virus persisted in the visceral tissues and spread to the spinal cord whereas untreated mice cleared virus from viscera by 3 days postinoculation and virus could not be detected in the spinal cord.

In contrast to the marked suppression of genetic resistance resulting from ^{89}Sr treatment, a deficiency of T-cell function and cell-mediated immunity was not associated with increased susceptibility to i.p. infection with HSV-1 (Lopez, 1978; Zawatzky et al., 1979; Schlabach et al., 1979). Studies with athymic nude mice and thymectomized and anti-thymocyte serum-treated mice showed that such mice were as resistant to an i.p. challenge with HSV-1 as were the normal (euthymic) or untreated littermates. When challenged by the subcutaneous or intraocular routes, athymic nude mice were found to be more susceptible to HSV-1 infection than the normal littermates (Nagafuchi et al., 1979; Metcalf et al., 1979). These results suggest that route of inoculation is an important determinant of the defense mechanisms induced during a virus infection and, therefore, the mechanisms required for resistance.

Most of the studies described above were carried out using i.p. inoculation of virus into adult mice. Although not extensive, other studies of genetic resistance to HSV-1 have used i.v. (Lopez, 1981), intravaginal (Schneweis and Saftig, 1981), and intraocular (Price and Schmitz, 1978) infection with results that are, in general, similar to those for i.p. infection. Price and Schmitz (1978) showed that C57BL/6 mice inoculated intraocularly required about 100 times more virus inoculum to establish a latent infection than did BALB/c mice. It is thus possible that the mechanisms (or closely related ones) responsible for genetic resistance to i.p. infection are also responsible for resistance to the development of latent virus infections. Studies by Schneweis and Saftig (1981) showed that inbred strains of mice resistant to i.p. challenge were also resistant to intravaginal infection with HSV-1 and that more virus was found in the sacral ganglia of susceptible than of resistant mice after infection. In addition, strain distribution was found to be similar to that found with i.p. infected mice and resistance to inoculation by both routes was diminished by X-irradiation, cyclophosphamide, and [89]Sr. Thus, the same or closely related host defense mechanisms may play roles in resistance to HSV-1 infections by i.p. and intravaginal routes of inoculation.

III. GENETIC RESISTANCE TO LETHAL HSV-2 IN THE MOUSE

In a series of studies similar to those reported for HSV-1, Armerding *et al* (1981b) found that inbred strains of mice could be classified as resistant, moderately susceptible, or susceptible to a lethal infection with HSV-2. The strain distribution for these categories was similar to that presented for HSV-1 and no significant differences were noted with respect to the other parameters evaluated.

IV. GENETIC RESISTANCE TO HSV-2-INDUCED HEPATITIS

Mogensen *et al.* (1974) showed that inoculation of mice with various strains of HSV-2 resulted in a focal, necrotic hepatitis 4 or 5 days postinoculation. In contrast, infection with strains of HSV-1 caused very few, if any, necrotic lesions (Mogensen, 1976). The hepatitis caused by HSV-2 began to resolve spontaneously after 4 days and was not the cause of death in inoculated mice. If mice died, it was due to an ascending myelitis or encephalitis at about 6–10 days after infection. The liver lesions caused by HSV-2 were easily visible to the naked eye whereas those caused by HSV-1 were not. Histologically, the HSV-2 induced lesions were characterized by areas of degenerating liver cells that were infiltrated with polymorphonuclear leukocytes causing central necrosis.

Resistance to HSV-2-induced hepatitis was found to be age dependent, 3-week-old mice being more susceptible than 8-week-old mice (Mogensen, 1978). Using mature mice, Mogensen (1977b) showed that resistance to hepatitis varied in the inbred strains of mice tested; GR mice were resistant whereas BALB/c mice were susceptible to inoculation with 10^6 PFU of HSV-2. In addition, genetic studies were used to show that resistance to HSV-2-induced hepatitis was governed by one X-linked, dominant gene.

V. MACROPHAGES

Because of their very nature, macrophages are prime candidates for the first line of defense against herpesvirus infections. These cells are relatively large with abundant cytoplasm and a reniform nucleus (Hirsch and Fedorko, 1970; von Furth *et al.*, 1972). The precursors for this lineage of cells are found in the bone marrow and are recognized as monocytes and promonocytes. Precursors are distributed by the blood to the visceral organs and other tissues of the body where they mature to macrophages and become an active part of the reticuloendothelial system. Mature cells are found in the liver, lungs, pleural and peritoneal cavities, lymph nodes, and spleen and are often among the first cells encountered by an invading virus. Macrophages are highly pinocytotic and phagocytic cells that adhere firmly to glass and plastic surfaces. In addition, these cells contain lysosomal enzymes, which might be used to degrade phagocytosed material. The role of macrophages in defense against HSV infections has been evaluated by several different approaches. In the following paragraphs, I will summarize the data generated and discuss the strengths and weaknesses of each approach.

A. Role of Macrophages in Resistance to HSV-1 in the Mouse

Over 50 years ago, Andervont (1927) showed that newborn mice are much more susceptible to lethal infection with HSV than are adult mice. Much later, Johnson (1964a,b) showed that the age-dependent susceptibility of newborn mice was due to the inability of peritoneal macrophages from these animals to restrict replication of the virus in culture. In infectious center assays, macrophages from newborns readily transmitted virus infection to nearby, susceptible cells whereas adult cells did so to a much lower degree. More recently, studies of Stevens and Cook (1971) showed that exposure of adult macrophages to HSV-1 resulted in an abortive infection. Viral antigens were expressed but infectious progeny virus was not generated by cells from adult mice.

Although these results seem to strongly indicate a dominant role for macrophages and their inability to sequester virus replication as an important determinant of susceptibility in the newborn, it is probably not that simple. Newborn animals have been shown to have a broad array of cellular deficiencies (Stiehm, 1980), any or all of which could contribute to their susceptibility. For example, a deficiency of NK cell function (Ching and Lopez, 1979; Lopez et al., 1983a; Kohl et al., 1981), of the capacity to generate IFN-γ (Bryson et al., 1980), or of neutrophil function (Miller, 1979) might be equally responsible for susceptibility to HSV-1 in the newborn.

In other studies aimed at determining whether the macrophage plays an important role in resistance to HSV-1, experiments were undertaken to evaluate the effect of treating animals with agents that selectively inactivate macrophages. This is accomplished by taking advantage of the fact that macrophages are highly phagocytic and pinocytotic cells. Macrophage "poisons" such as silica particles (Allison et al., 1966) or solutions of κ-carrageenan (Schwartz and Leskowitz, 1969) are readily taken in by macrophages but not by most other cells. These agents eventually result in the death of macrophages and in the marked diminution of resistance to i.p.-inoculated HSV-1 (Zisman et al., 1970; du Buy, 1975). Interpretation of these results depends on the selectivity of the macrophage "poisons" being used in the study. Although suspensions of small (1–10 μm in diameter) silica particles were thought to be relatively selective for macrophages (as contrasted with T cells), more recent data indicate that this agent also depresses NK cell function (Herberman and Holden, 1978) and thus may reduce resistance by this mode, as well.

Several immunopotentiators have been shown to enhance macrophage function and to augment resistance to HSV-1 concomitantly (Halpern *et al.*, 1973). Agents such as *C. parvum* have been shown to enhance clearance of carbon particles, immunologic response to an antigenic stimulus, and resistance to challenge with bacteria, protozoans, or tumors. Mice treated with *C. parvum* have also been shown to be much more resistant to i.p. infection with HSV-1 than were untreated mice (Kirchner *et al.*, 1977). Unfortunately, agents that enhance macrophage function are no more selective than are the macrophage "poisons" in that other cells also demonstrate augmented function. NK cells, for example, are also much more active after treatment of mice with *C. parvum* than are the cells from untreated mice (Herberman and Holden, 1978). Thus, this evidence for the role of macrophages in resistance to HSV-1 infections is also not conclusive.

Another approach to the study of the role of macrophages in resistance to HSV-1 has been to attempt to transfer resistance with relatively pure populations of these effectors. Because immature macrophages appeared to cause the susceptibility of the newborn, Hirsch *et al.* (1970) transferred adherent cells from adult mice to newborns and showed that this resulted in increased resistance to an i.p. infection. Although this appears to be a straightforward approach to determining whether macrophages mediate the resistance found in adult mice as compared to newborns, there are difficulties that must be considered. For example, if even a small percentage of the transferred cells were NK cells, then the contaminating cells may be responsible for resistance. Also, separation procedures and other manipulations might activate cells and lead to the same observations but, in fact, these results might not reflect the role of resident macrophages. There is also the possible interaction of transferred and resident cells, which might result in resistance. Thus, even cell transfer experiments may not clearly define a role for macrophages in resistance to HSV-1.

Studies have also been undertaken to investigate the role of macrophages in genetic resistance to HSV-1 in the mouse. In earlier studies, macrophage poisons were shown to suppress genetic resistance to virus infection in strains of mice normally resistant to infection (Lopez 1978). Because susceptible adult mice might be similar to newborn mice and not be able to sequester virus replication, experiments were carried out to determine whether macrophages from susceptible mice replicated HSV-1 better than cells from resistant mice. Adherent peritoneal cells from all strains of mice tested restricted replication of HSV-1 when infected immediately after formation of the monolayers but replicated the virus to high titer when inoculated after 3–7 days of culture (Lopez and Dudas, 1979). Although macrophages from a genetically resistant strain of mice restricted virus replication better than cells from a susceptible strain, this function did not segregate with genetic resistance, for macrophages from resistant F_1 mice failed to restrict HSV-1 replication. Comparable studies have recently been carried out by H. Kirchner (personal communication) with only slightly different results. They found that utilizing a different medium allowed virus replication without prior culturing of the adherent peritoneal cells. Using cultured cells in order to enhance the purity of the macrophages, they showed that cells from resistant mice failed to replicate virus to high titers whereas cells from susceptible mice generated significantly more virus. Further, they found that macrophage monolayers from resistant mice generated more IFN than did cells from susceptible mice and that macrophages were exceedingly sensitive to the antiviral activity of IFN.

Other studies suggested that restriction of HSV-1 replication by macrophages that had not been cultured is dependent on the induction of IFN by high concentrations of virus (Linnavuori and Hovi, 1981). Infection with a much lower dose of virus resulted in replication of HSV-1. These investigators also showed that culturing cells for 3–7 days greatly diminished their capacity to generate IFN in response to HSV-1.

B. Role of Macrophages in Resistance to HSV-2 in the Mouse

Many of the studies presented above with HSV-1 have also been carried out with HSV-2 with similar results (Starr *et al.*, 1976; Morahan *et al.*, 1977; McGeorge and Morahan, 1978). The evidence supporting an important role for macrophages in resistance against HSV-2 inoculated i.p. is as strong as that presented for HSV-1 and the same reservations pertain. These studies will not be discussed here as they add little to our understanding of the role of macrophages in defense against these infections.

Two observations on HSV-2-infected mice appear to be important. Most studies of resistance to HSV infections have been carried out with i.p.-infected mice because of the ease of carrying out such studies. This is not a natural mode of infection and more recent studies have attempted to use natural routes of inoculation such as into the lip or intravaginally. In one such study by McGeorge and Morahan (1978), macrophage poisons increased susceptibility to an i.v. infection with HSV-2 and yet failed to augment susceptibility to an intravaginal infection. As the latter is the more natural route of inoculation for HSV-2, it appears that macrophages are not required for defense against such an infection. The second observation was made by Armerding *et al.* (1981a) using a genetic model of resistance to HSV-2 infections in the mouse. These investigators noted that i.p. infection with HSV-2 was an excellent inducer of macrophage phagocytic activity in mice. Thus, activated macrophages, possibly induced by locally generated IFN (Zawatzky *et al.*, 1981) or by the virus infection alone (Casali *et al.*, 1981), might be required for resistance to an i.p. infection.

C. Role of Macrophages in Resistance to HSV-2-Induced Hepatitis

Studies have been carried out to determine the mechanisms responsible for genetic resistance to hepatitis caused by HSV-2. Treatment of mice to inhibit macrophage function has been found to markedly diminish genetic resistance as well as the age-dependent resistance of resistant strains of mice (Mogensen, 1977b, 1978). Silica treatment was also used to show that the virus-strain-dependent (HSV-1 vs. HSV-2) differences were also due to macrophage function, as such treatment abrogated those differences. Finally, Mogensen and Andersen (1978) had earlier noted that athymic, nude mice were actually more resistant to hepatitis than their normal littermates. This enhanced resistance to HSV-2-induced hepatitis in nude mice was due to activated macrophages found in these mice.

Mogensen (1977a,b, 1978) has carried out a number of *in vitro* studies to determine the mechanisms involved in resistance to HSV-2-induced hepatitis. For example, the relative differences between young and mature mice were reflected in the capacity of macrophages from these mice to inhibit spread of virus in an infectious center assay. Similarly, the genetic resistance of mature mice could be ascribed to the capacity of the macrophages from these mice to inhibit replication of HSV-2 *in vitro*. In these studies, genetic resistance to HSV-2-induced hepatitis and inhibition of viral replication segregated together in backcrossed mice. Lastly, the differences found between the capacities of HSV-1 and HSV-2 to induce hepatitis were reflected in the infectious center assay. Thus, HSV-2 but not HSV-1 readily replicated in peritoneal macrophages, causing viral plaques to form in the surrounding susceptible cells.

Although these data strongly suggest that resistance to hepatitis caused by HSV-2 is dependent on the macrophage, there are some important questions left unanswered. For example, is there a deficiency of the macrophage itself or could there be a deficit of some humoral factor (such as IFN or macrophage-activating factor) that might be responsible for the lack of macrophage function? If the deficiency is in the macrophage, is there a defect in its capacity to respond to a humoral factor, as is seen with the *Mx* gene and genetic resistance to influenza in the mouse? And, finally, if there is a defect in a humoral factor-producing cell, is this a lacunar deficiency with respect to HSV-2 or does it affect other infectious microorganisms, as well? In a more recent study, Mogensen and Andersen (1981) showed that sensitized splenic cells could adoptively transfer resistance to HSV-2-induced hepatitis. These cells could be producing factors that interact with macrophages and make them sequester virus infection.

HSV-2 very rarely causes hepatitis in man so that one must ask whether this animal model is a laboratory artifact. Wiskott—Aldrich syndrome is, however, a primary immunodeficiency disorder of man that, like genetic susceptibility to HSV-2-induced hepatitis, is a sex-linked recessive disorder (Aldrich *et al.*, 1954). Patients with this disorder are known to be highly susceptible to herpesvirus infections and one of the first reports describing this phenomenon noted the finding of focal, necrotic lesions in the liver of a patient at autopsy (St. Geme *et al.*, 1965).

D. Role of Macrophages in Resistance to HSV-1 and HSV-2 in Man

Very few studies have been carried out to evaluate the role of macrophages in resistance against herpesvirus infections in man. Human monocyte/macrophage populations may be active in spontaneous cytotoxicity of HSV-1-infected fibroblast targets (Stanwick *et al.*, 1980) as well as in antibody-dependent cell-mediated cytotoxicity (ADCC) (Kohl *et al.*, 1977). ADCC has been studied extensively and has been found to be mediated by killer (K) cells (Shore *et al.*, 1976), monocyte/macrophage cells (Stanwick *et al.*, 1980), and neutrophils (Oleski *et al.*, 1977). A deficiency of monocyte-mediated ADCC has been found in patients with Wiskott—Aldrich syndrome (Blaese *et al.*, 1975). As these patients often suffer from severe herpesvirus infections, this monocyte deficiency was thought to be associated with susceptibility to infection.

Because of the earlier studies showing that the age-dependent resistance of mice to HSV-1 appeared to be reflected in the capacity of peritoneal macrophages to replicate virus, similar studies were carried out with adherent mononuclear cells derived from human peripheral blood (Daniels *et al.*, 1978; Trofatter *et al.*, 1979; Linnavuori and Hovi, 1981; Grogan *et al.*, 1981). As with adherent peritoneal cells from mice, adherent cells (mostly monocytes) from human mononuclear cell fractions failed to replicate HSV-1 unless they were first cultured for 4–7 days prior to infection. In contrast, similar studies, carried out with adherent cells from cord blood, showed that the latter was capable of replicating HSV-2 (Daniels *et al.*, 1978). As the human newborn, like the newborn mouse, is prone to develop severe disease as a consequence of HSV-2 infection, these results suggest that a deficiency of macrophage function in the human newborn may account, at least in part, for the severity of the disease in this host. However, the newborn has been found to have a number of other deficiencies (Stiehm, 1980) that might also account for their susceptibility and further studies are required to determine whether the human newborn lacks fully developed monocytes and what role such a deficiency might have in susceptibility to HSV-2 infections. Future studies should be carried out with cells derived from newborns and not cord blood to avoid any possible role of maternal cells and with purified cell populations in order to define the abnormality more specifically.

Although the experience is not extensive, one other approach has been used that might indicate a role for macrophages in resistance to HSV in man. Thus, attempts have been made to augment macrophage function in patients with recurrent herpesvirus infections and determine whether such treatment diminishes severity of infections or incidence of recurrences. BCG was used to treat patients with recurrent herpes genitalis but failed to demonstrate efficacy (Bierman, 1976). However, it is important to remember that macrophage function was found to be critical in mice inoculated i.p. or i.v. and that diminution of macrophage function with silica failed to augment susceptibility to intravaginal infection, suggesting that macrophages may not play an important role if the infection is strictly localized (Armerding et al., 1981a; McGeorge and Morahan, 1978).

VI. NATURAL KILLER CELLS

NK cells were first described as effector cells from normal individuals that gave the "background" lysis in studies of specific T-cell-mediated cytotoxicity of tumor targets [see Herberman (1980) for historical perspective]. Although NK cells were first thought to be a nuisance, more recent studies suggest that they may be important effectors in surveillance against malignant adaptation (Herberman and Ortaldo, 1981), as a first line of defense against virus infections (Welsh, 1978b; Lopez, 1981), as mediators of bone marrow graft rejection (Kiessling et al., 1977) and inducers of graft-versus-host reactions in bone marrow transplant recipients (Lopez et al., 1979), and as mediators of hemopoietic homeostasis (Kiessling and Haller, 1978). NK cells have been described in all mammals tested to date and in some invertebrates. Many of the early studies of NK cell function in the mouse utilized Yac-1 lymphoma cells as targets and most of the human studies have used K562 erythroleukemia tumor cells as targets.

The studies of Haller et al. (1977) showed that NK cells were derived from bone marrow stem cell precursors. Thus, transplantation of lethally irradiated, low-responder mice with bone marrow from high-responder mice resulted in chimeras with the high-responder phenotype. Conversely, chimeras developed by transfer of marrow from low responders to high responders resulted in low-responder mice. In addition, Haller and Wigzell (1977) showed that NK cells are marrow dependent in that ablation of bone marrow by treatment of mice with ^{89}Sr also markedly diminished NK cell function. Estradiol also causes a severe aplasia in the marrow and this is associated with diminished NK cell function (Seaman et al., 1979). In addition, mice with congenital osteopetrosis similarly have very low NK responses associated with a lack of normal marrow cavities (Seaman et al., 1979). In the latter, the marrow cavities are obliterated by bone matrix due to a failure of normal osteoclasts to resorb bone.

IFN (α, β, and γ) has been shown to augment NK cell function (Gidlund et al., 1978; Djeu et al., 1979; Trinchieri and Santoli, 1978; Ortaldo et al., 1981). IFN functions both by increasing the lytic efficiency of mature NK cells (Ullberg and Jondal, 1981) and by recruiting immature cells to functional, mature cells (Saksela et al., 1978; Minato et al., 1980). Many research groups have shown that virus infections of mice rapidly augment the NK cell activity of the effector cells from these mice (Welsh, 1978a,b). Most viruses tested, including both HSV-1 (Engler et al., 1981) and HSV-2 (Armerding et al., 1981a), augment NK activity 1–3 days after infection with virus. Welsh (1978a) showed that the augmented NK cell function followed closely the generation of IFN in the virus-infected mice and was probably dependent on this cytokine.

A number of different laboratories, using both RNA and DNA viruses, have found that virus-infected cells are much better targets for NK effectors than are the uninfected cells (Diamond et al., 1977; Anderson, 1978; Santoli et al., 1978a,b; Ching and Lopez, 1979; Minato et al., 1979; Welsh and Hallenbeck, 1980). As with lysis of the tumor targets, NK lysis of virus-infected cells did not require previous exposure of the donor to the virus. In addition, the effector cells responsible for this lytic activity do not have the cell surface characteristics of mature T cells, B cells, or macrophages.

As our interest has been in the role of NK cells in resistance to herpesvirus infections, we developed an assay using HSV-1-infected fibroblasts as targets [NK(HSV-Fs)] (Ching and Lopez, 1979). As with NK cells that lyse the commonly used K562 erythroleukemia targets [NK(K562)], the effectors of NK(HSV-Fs) are large granular lymphocytes (Timonen and Saksela, 1980; Timonen et al., 1981; Fitzgerald et al., 1983) and require no prior sensitization to be operative. In addition, we have shown that NK(HSV-Fs) effectors are derived from bone marrow stem cells, as transplantation of patients with low NK activity resulted in chimeras with normal levels of NK(HSV-Fs) (Lopez et al., 1979). Lastly, we have also shown that these effectors are marrow dependent, as patients with aplastic anemia without evidence of marrow precursors and patients with congenital osteopetrosis whose marrow has been obliterated by bone matrix, demonstrated greatly diminished NK(HSV-Fs) responses (Lopez et al., 1982).

A. Heterogeneity of NK Cells

Studies of murine cells first provided evidence that NK cells are heterogeneous. Effector cells can be differentiated by their in vivo susceptibility to [89]Sr (Kumar et al., 1979; Lust et al., 1981), their cell surface markers (Ault and Springer, 1981), and cold target inhibition studies or adsorption studies (Stutman *et al.*, 1978, 1981). The use of different target cells in NK assays can reflect one or more of the subpopulations of effector cells.

Heterogeneity of human NK effector cells has also been demonstrated. Thus, adsorption of peripheral blood cells onto monolayers of one tumor target cell was found to remove the cytolytic activity against that target but not against another (Phillips *et al.*, 1980; Tai *et al.*, 1982). In addition, certain monoclonal antibodies have been shown to recognize a portion (but not all) of the effector cells that lyse K562 erythroleukemia targets, suggesting that there is heterogeneity even within the cells capable of lysing this one target cell (Zarling *et al.*, 1981; Kay and Horowitz, 1980).

In our studies of the characteristics of NK(HSV-Fs) effectors, we compared them to the effector cells that lysed K562 erythroleukemia tumor targets [NK(K562)] and have developed evidence suggesting that, although they are both found in Percoll gradient fractions enriched for large granular lymphocytes and require no presensitization, the effector cells for these two targets are clearly different (Fitzgerald *et al.*, 1983). We have shown that NK(HSV-Fs) effectors differ from NK(K562) cells using monoclonal antibodies to cell surface markers of human mononuclear cells. A portion of the NK(K562) effectors were positive for Leu-1 and Leu-4 (pan T-cell) surface markers while the NK(HSV-Fs) cells were negative. Most of the NK(K562) cells were positive for Lyt-3 (sheep red blood cell receptor) while again the NK(HSV-Fs) were negative. In contrast, a portion of the NK(HSV-Fs) effectors were found to express Ia (framework) while the NK(K562) effectors were negative for this marker. Monoclonal antibodies have also been used to show that these effectors share other characteristics. Thus, both NK cells were found to express Leu-11, which detects an Fc receptor on the surfaces of effector cells.

Cold target inhibition experiments have also been used to support the concept of heterogeneity (Fitzgerald *et al.*, 1983). In these studies, unlabeled K562 cells effectively reduced the lysis of labeled K562 targets but reduced lysis of HSV-Fs to a much lower extent. Similarly, infected and uninfected fibroblasts reduced lysis of HSV-Fs much better than did K562 cells. These results have been interpreted to suggest that different structures are recognized on these different target cells.

Probably the best evidence for heterogeneity of human effector cells has been the observation that certain patient groups have effector cells capable of lysing one target cell but not another and vice versa (Fitzgerald *et al.*, 1983). Thus, a group of five adult patients with aplastic anemia were found to have effector cells that repeatedly lysed K562 targets normally while lysis of HSV-1-infected targets was found to be markedly reduced (greater than 3 S.D. below the normal mean). In contrast, we have found patients whose effector cells consistently lysed HSV-Fs targets but not K562 targets. As significant deficiencies have been found in both directions, these results provide strong evidence that at least two subpopulations of NK effector cells mediate lysis of these targets.

B. Cell Lineage of NK(HSV-Fs) Effectors

There is much debate over whether NK cells belong to the T-cell lineage (Fast *et al.*, 1981; Grossman and Herberman, 1982), the monocyte/macrophage lineage (Kay and Horowitz, 1980; Ault and Springer, 1981), or an independent lineage. The results of studies of cell surface markers have been confusing with regard to this issue. Thus, NK(K562) effectors have been shown to have markers characteristic of T cells and macrophages. In an attempt to approach this question from another perspective, we have carried out studies of "experiments of nature," i.e., on patients with defects of the lymphoid or monocyte/macrophage cell lineages, which might shed some light on the cell lineage of NK(HSV-Fs) effectors (Lopez *et al.*, 1982).

Our studies on patients with primary immunodeficiency disorders indicate that the NK(HSV-Fs) effectors probably do not belong to the lymphoid lineage. Thus, patients with severe deficiencies of B cells (X-linked hypogammaglobulinemia) usually had normal responses (Lopez *et al.*, 1982). In addition, all of the patients tested with the classical form of severe combined immunodeficiency disease (SCID) (defined as 0% B cells and no real T cells of their own) had normal NK(HSV-Fs) responses. Most of these patients were evaluated for the presence of pre-T cells in their marrow (defined as cells that would differentiate to T-cell markers or function upon culturing them with thymic humoral factors). There was no correlation between NK(HSV-Fs) function and the presence of pre-T cells in the marrows of these patients, indicating that these effector cells were probably not pre-T cells either. These data suggest that the NK(HSV-Fs) effectors do not belong to the lymphoid lineage. Studies of patients with deficiencies of the monocyte/macrophage lineage were less conclusive. Some patients with abnormalities also demonstrated NK(HSV-Fs) deficiencies although others did not. In sum, NK(HSV-Fs) effectors may be related to the precursors of the monocyte/granulocyte series but may also belong to an independent hemopoietic cell lineage.

Recently, Peter *et al.*, (1983) evaluated NK(K562) activity of eight patients with SCID and found that two patients with no B cells had normal activity while five of six patients with some B cells expressed low levels of activity. These results are in agreement with those of Lipinski *et al.* (1980) and R. Buckley *et al.* (personal communication). Although Peter *et al.* (1983) obtained results similar to ours using different targets, these authors concluded that NK(K562) effectors either constitute an independent lineage of cells or may be closely related to early differentiation steps of the T-cell lineage.

C. Role of IFN in the NK(HSV-Fs) Response

As noted earlier, IFN is capable of augmenting NK cell function *in vitro* and *in vivo*. Trinchieri and Santoli (1978) showed that many of the target cells that were lysed efficiently by human NK cells also induced the generation of IFN and that the levels were great enough to account for the lysis observed. Also, these investigators noted that overnight incubation of mononuclear cells resulted in the loss of NK function as well as the ability of these cells to generate IFN (Trinchieri *et al.*, 1978). Further, in kinetic studies, they found that NK lytic activity failed to increase until after IFN generation was detected *in vitro*. These results were taken to indicate that NK cell lysis of virus-infected targets depended on the generation of IFN (caused by the virus infection) that, in turn, induced lysis of whatever target cells might be available. Thus, Trinchieri *et al.* (1978) proposed that selective lysis of virus-infected versus uninfected cells was dependent on the IFN generated during the assay.

We have carried out a number of studies indicating that the generation of IFN is a function independent of the NK(HSV-Fs) lytic function (Fitzgerald *et al.*, 1982). Although IFN-α was generated during normal NK(HSV-Fs) assays, there was no indication that the amount of IFN produced during the assay correlated with the level of ^{51}Cr release. In addition, anti-IFN antibodies were added to the NK(HSV-Fs) assay and, even though all detectable IFN was neutralized, this treatment did not significantly diminish lysis of the target cells. More recently, we have found that the effector cells that make IFN in response to HSV-Fs are similar to the lytic effectors in that they are enriched in Percoll gradient fractions containing large granular lymphocytes but they can be differentiated from NK(HSV-Fs) effectors by monoclonal antibodies to cell surface markers (Fitzgerald *et al.*, in preparation). Most importantly, we have now found many patients with normal levels of NK(HSV-Fs) activity even though less than 10 IU of IFN was generated during the assay (Lopez *et al.*, 1983b). Conversely, other patients were found whose mononuclear cell generated high levels of IFN whereas the same effector cells were incapable of lysing these target cells (P. Fitzgerald *et al.*, unpublished results). Taken together, these results indicate that although IFN is often generated in response to the HSV-Fs, the production of this cytokine is independent of the lytic function of NK(HSV-Fs) and IFN is not necessary for normal target cell lysis *in vitro*. It is possible, however, that IFN generated in response to virus infection *in vivo* might be an important part of the natural defense of the host (Minato *et al.*, 1979).

D. NK Cells Limit HSV-1 Replication *in Vitro*

If NK cells are to play an important role in defense against HSV-1 infections, these effectors must reduce the amount of virus progeny produced by infected cells. In an attempt to determine whether NK cells limit HSV-1 replication during the *in vitro* assay, we measured the amount of HSV-1 secreted into supernatants of cultures containing effector cells and HSV-1-infected fibroblasts at the termination of the NK assay and compared these to HSV-1 produced by the infected fibroblasts in the absence of effectors (Fitzgerald *et al.*, in preparation). Cytotoxic cells reduced the virus yield by 90% or more at an effector:target (E:T) ratio of 200:1 and lower reductions were found at lower E:T ratios. Peripheral blood mononuclear cells were fractionated on Percoll gradients in order to characterize the effector cells responsible. Virus replication was significantly lower in wells containing effector cells from fractions enriched for NK(HSV-Fs) effectors (and large granular lymphocytes) as compared to the other fractions. As IFN-α is also generated during this assay and might be responsible for the reduced virus titers, experiments were carried out in the presence of sufficient anti-IFN-α antibody to neutralize all the antiviral activity generated during the assay. Although the IFN was neutralized, there was no difference in virus production of wells with the antibody as compared to those without it. Further, most of the IFN-α-generating cells can be separated from the peak NK(HSV-Fs) effector cells on Percoll gradients and the reduction of virus titer was associated with the cytotoxic activity and not the peak IFN production. These results indicated that the cytotoxic NK effectors themselves were able to directly inhibit virus replication, most likely by lysing target cells prior to the production of infectious new progeny. *In vivo* these effectors may significantly reduce the virus load that the infected host must deal with.

E. Role of NK Cells in Resistance to HSV-1 in the Mouse

Studies have shown that many of the treatments that reduce or abrogate NK cell function in mice also reduce genetic resistance to HSV-1. Thus, ^{89}Sr treatment of mice completely abrogated NK cell function as well as genetic resistance to HSV-1 (Lopez *et al.*, 1980). In a less dramatic manner, chronic estradiol treatment resulted in reduced NK cell function and reduced resistance to HSV-1 (C. Lopez *et al.*, unpublished results). In other studies, mice carrying the bg/bg phenotype were found to be slightly more susceptible to HSV-1 than were their normal littermates. As they also has lower NK responses, these results again showed a correlation between low NK and susceptibility to infection. Genetically resistant mice were found to have good NK responses whereas most susceptible strains of mice lacked significant lysis of HSV-1-infected target cells (C. Lopez *et al.*, unpublished results). Of interest was the observation that, although CBA mice had high NK activity, they were moderately susceptible to HSV-1 infection. As CBA mice lack only one of the two resistance genes, NK may be necessary but not sufficient for resistance.

The studies of Engler *et al.* (1981) suggest that the production of early IFN rather than NK cell function plays an important role in resistance to HSV-1 in the mouse. These investigators cite SJL mice as relatively resistant mice that have low NK activity against Yac targets but high production of early IFN, results that they say indicate that the early IFN is more important than NK in conferring resistance to HSV-1 (H. Kirchner, personal communication). However, as noted above, there is heterogeneity of NK effector cells and these investigators may not have evaluated the correct subpopulation of effectors.

The studies suggesting an important role for NK cells in resistance to HSV-1 in mouse are interesting but are only preliminary and require substantiation. In particular, it will be important to determine whether NK segregates with genetic resistance in backcrossed mice. Additional studies must also take into account the heterogeneity of NK effector cells.

F. Low NK(HSV-Fs) Function Associated with Susceptibility to Herpesvirus Infections in Man

Because of the studies suggesting an important role for NK cells in genetic resistance to HSV-1 in the mouse, studies have been carried out to determine whether individuals susceptible to severe herpesvirus infections have deficiencies of NK(HSV-Fs) function (Lopez *et al.*, 1983a). We have evaluated two groups of patients known to be susceptible to severe herpesvirus infections. Cord blood lymphocytes were studied because of the newborns' known susceptibility to disseminated infections. Only 30% of the cord blood lymphocytes tested demonstrated normal responses. Similar observations have been made by Kohls *et al.* (1981). Peripheral blood cells from nine premature infants were also tested and all had responses greater than 3 S.D. below the normal mean (Lopez *et al.*, 1983a, and unpublished). This discrepancy between the results with newborns and cord blood might be real and reflect the significantly greater risk of the premature newborn infant to severe herpesvirus infection (Visintine *et al.*, 1981). However, this difference might also reflect differences found between cord blood and peripheral blood of newborn infants.

Wiskott–Aldrich syndrome is an X-linked, primary immunodeficiency disorder characterized by the triad of eczema, thrombocytopenia, and recurrent infections (Aldrich *et al.*, 1954). Most patients with this syndrome have also been shown to be susceptible to severe herpesvirus infections (St. Geme *et al.*, 1965). We have evaluated the NK(HSV-Fs) responses of six patients with Wiskott–Aldrich syndrome and found that only one had a normal capacity (Lopez *et al.*, 1983a, and unpublished). The one patient with a normal response has survived much longer than usual and has had little trouble with infections. Low NK(HSV-Fs) responses in the other individuals with this syndrome correlated with a history of severe infections including some with herpesviruses. In addition to these five patients, we have recently evaluated two other patients with this disease. As noted with the other patients, these two patients had low NK(HSV-Fs) responses that correlated with their susceptibility to herpesvirus infections. We found, however, that these patients had normal responses when evaluated for NK(K562) activity, indicating that a deficiency of NK function might have been missed had we only tested these patients with K562 tumor targets.

Newborns and patients with Wiskott–Aldrich syndrome have been shown to have deficiencies in other aspects of their defense systems that might also contribute to their susceptibility to infection (Stiehm, 1980; Blaese et al., 1975). We therefore evaluated the NK capacities of nine individuals susceptible to herpesvirus infection as evidenced by the unusually severe infection that each had when studied. The mean NK(HSV-Fs) response of this group was significantly below the normal mean and below the mean of normal individuals with an occasional virus infection. Two of these individuals were evaluated between bouts of infection and each demonstrated equally depressed responses, suggesting that low NK(HSV-Fs) capacity might predispose such patients to unusually severe herpesvirus infections. As with the Wiskott–Aldrich patients, four of these patients were also evaluated for NK(K562) activity; two were found to have low responses and two had normal responses (C. Lopez et al., unpublished results). Again, NK(K562) did not invariably correlate with susceptibility to herpesvirus infection whereas NK(HSV-Fs) responses distinguished those patients susceptible to infection, suggesting that NK cell heterogeneity is an important consideration when attempting to correlate biological function with NK activity. As these patients had also been evaluated for primary cellular immunodeficiencies and were found to have normal T-cell numbers and function, the deficiency of NK(HSV-Fs) function was the best correlate of susceptibility to infection.

Other studies also suggest a role for NK effectors in resistance to virus infections. Sullivan et al. (1980) demonstrated low NK(K562) capacity in patients with X-linked lymphoproliferative syndrome. This disorder is associated with susceptibility to Epstein–Barr virus-induced lymphoproliferative disease as well as to other infections. More recently, these investigators have determined that suppressed NK(K562) responses may be a result of infections in these patients and not the cause of the infections (Sullivan et al., 1983). Studies by Virelizier and his colleagues (Virelizier, 1981; Virelizier et al., 1979,1981) have also demonstrated low NK(K562) capacity in a group of patients unusually susceptible to a number of infectious agents including herpesviruses. More recently, Stein et al., (1983) described strikingly low NK(K562) activity in badly burned patients, who are unusually susceptible to virus infections. Further, these investigators found that other immunological abnormalities found in these patients resolved during treatment but that NK(K562) function remained low for up to 80 days after the injury. Similar studies by Kohl and Ericsson (1982) utilizing another target cell for their NK assay, failed to demonstrate that low function correlated with susceptibility to virus infection in burn patients. This, however, may be due to the target cells selected for the study (see description of heterogeneity of NK above).

VII. INTERFERON

IFN was first described as a substance, made during a virus infection, that was capable of converting susceptible, uninfected cells into cells resistant to virus infection (Isaacs and Lindenmann, 1956). These early studies also showed that the effect of IFN was on the cell and not on the extracellular virus. Over the years, many biological effects have been ascribed to IFN and each appears to be induced by the interaction of IFN with the cell membrane (Gresser, 1977).

The nomenclature for IFN was recently changed by an expert committee (Stewart et al., 1977). The three major types of IFN, previously known as leukocyte, fibroblast, and immune, are now referred to as IFN-α, IFN-β, and IFN-γ. In addition, a new IFN with properties of both IFN-α and IFN-γ has been described and is referred to as acid-labile IFN-α (Preble et al., 1982). Most studies of IFNs and their actions were conducted with relatively impure preparations. More recently, however, the modern tools of molecular biology have been applied to the study of IFN with spectacular results. Many of the 12 or more IFN-α genes have been cloned from human cells and are being mass-produced in bacterial cultures (Stiehm et al., 1982). The gene products from these various clones differ by 15–17% in their amino acid sequence and appear to differ in certain biological activities as well. IFN-β and IFN-γ genes are also being cloned and pure preparations are now, or will soon be available for study of their biological properties.

There are at least three mechanisms by which IFNs might play a role in the natural resistance to herpesvirus infections. IFN has been shown to inhibit the replication of herpesviruses and could act by limiting the quantities of infectious virus produced at the focus of infection (Rasmussen and Farley, 1975). Also, IFN might act by augmenting the efficiency of cytolytic NK cells or by recruiting pre-NK cells to differentiate into mature, functional effector cells (Ullberg and Jondal, 1981; Saksela et al., 1978). In addition, IFN could function by activating macrophages so that they take up and sequester the replication of HSV-1 (Linnavuori and Hovi, 1981) or lyse HSV-1-infected target cells (Stanwick et al., 1980). IFN might also act by "priming" the production of more IFN (Stewart, 1979). Although evidence will be presented suggesting that the rapid production of IFN is required for host defense against herpevirus infections, the specific mechanism involved has not been determined.

A. Role of IFN in Genetic Resistance to HSV-1 in the Mouse

Studies by Kirchner's group (Zawatzky et al., 1979; Engler et al., 1981) have shown that strains of genetically resistant mice made IFN rapidly after infection whereas susceptible mice failed to generate this response. The IFN generated in resistant mice was found to be α/β, and more recent experiments suggest that this IFN works by activating macrophages that in turn sequester the replication of virus (H. Kirchner, personal communication). The IFN is generated in athymic, nude mice, which also demonstrate resistance to i.p. challenge with HSV-1 (Zawatzky et al., 1979). In preliminary studies, we also have found that most resistant mice rapidly make IFN in response to HSV-1 infection (C. Lopez et al., unpublished results). One exception has been the CBA strain mouse, which makes high levels of IFN but is moderately susceptible to HSV-1 infection. As noted for NK cell responses, definition of the mechanisms responsible for genetic resistance awaits backcross experiments with genetically susceptible and resistant mice.

The studies of Gresser et al. (1976) have provided some of the strongest evidence suggesting a role for IFN in resistance to HSV-1. These investigators produced a potent, heterologous anti-mouse IFN serum and then used it to make IFN-deficient mice to determine the effect on the pathogenesis of a number of virus infections including HSV-1. This IFN deficiency markedly increased the susceptibility of the mice to HSV-1. These investigators used outbred mice for these studies so that it is difficult to determine whether genetic resistance was being abolished. In preliminary studies, we have found that a heterologous anti-mouse IFN serum dramatically reduced resistance of C57BL/6 mice to challenge with HSV-1, suggesting that IFN generation is probably an integral part of genetic resistance against this virus (C. Colmenares et al., unpublished results).

B. IFN-α-Generating Cells in Man

The effector cells that produce IFN-α in man are similar to NK cytolytic cells in that they are null cells (Peter et al., 1980; Kirchner et al., 1979) found in Percoll gradient fractions enriched for large granular lymphocytes (Djeu et al., 1982; Fitzgerald et al., in preparation) and require no presensitization in order to be operative. More recently, however, we have found that the IFN-α-generating cells can be distinguished from the lytic effectors by cell surface markers as well as by studies of certain patient populations. Thus, the cells that make IFN-α lack Lyt-3, OKM1, and Leu-11, markers found on all or most NK lytic effector cells (Fitzgerald et al., in preparation). Conversely, the IFN-α-generating cells express Ia on their cell surface, a marker also expressed on some NK(HSV-Fs) but not on NK(K562) effectors. Furthermore, patient studies indicate that NK IFN-α generation are independently regulated functions. Thus, we have found patients with normal NK(HSV-Fs) lytic capacity but with a marked deficiency of IFN-generating capacity and vice versa. These studies indicate that the IFN-α-generating cells are probably a subpopulation of large granular lymphocytes different from NK lytic effectors and are independently regulated.

C. Role of IFN in Resistance to Herpesvirus Infections in Man

A number of studies have associated a deficiency of IFN generation with increased susceptibility to herpesvirus infections. Cord blood cells have been shown to make normal levels of IFN-α and IFN-β in response to virus infections *in vitro* but to be deficient in IFN-γ production in response to mitogens (Cantell *et al.*, 1968; Carter *et al.*, 1971). The latter appeared to be due to immature macrophage function required, in general, for lymphokine production (Taylor and Bryson, 1981). Patients with deficiencies of IFN-γ generation in response to stimulus with a B-cell line have been described to be susceptible to severe infection with a number of pathogens including herpesviruses (Virelizier, 1981; Virelizier *et al.*, 1978, 1979). Some of these patients only had deficiencies of IFN generation while other had other deficits as well.

Patients with AIDS are susceptible to a variety of intracellular pathogens, such as *P. carinii* pneumonia, which usually cause disease only in patients with primary or secondary immunodeficiency disorders or AIDS patients develop infections that are much more severe than infections in nonimmunocompromised individuals, such as ulcerative HSV infections rather than the self-limited infections usually found in otherwise normal individuals (Siegal *et al.*, 1981; Gottlieb *et al.*, 1981; Masur *et al.*, 1981; Curran, 1983). Patients at risk of developing AIDS include male homosexuals (71% of all current patients), men and women who are i.v. drug abusers (17%), Haitians (5.2%), heterosexual sexual partners of AIDS patients (1%), hemophiliacs (1%), blood transfusion recipients (1%), and patients where risk factors were not documented (Curran, 1983). Patients with AIDS and opportunistic infections have been found to have deficiencies of T-cell numbers and function as well as of B-cell function. However, many of these deficiencies are also found in homosexual controls who do not develop AIDS and in non-AIDS patients with acute virus infections. Analysis of data generated in the study of over 200 patients indicates that the best correlate with susceptibility to opportunistic infections was the deficiency of IFN-α generation by mononuclear cells challenged with HSV-1-infected fibroblasts. This deficiency appeared to be selective in that it was rarely found repeatedly in individuals who did not develop an infection and was found in 99 of 101 AIDS patients who developed opportunistic infections. When compared to the other immunological functions evaluated, the deficiency in ability to produce IFN-α was the best correlate of susceptibility to infection (Lopez *et al.*, 1983b).

Patients undergoing surgical procedures that result in manipulation of the trigeminal root often demonstrate reactivated HSV-1 infections following surgery. Pazin *et al.* (1979) showed that IFN treatment of such patients resulted in a significantly reduced reactivation rate. Thus, exogenous IFN was sufficient to inhibit recurrent HSV-1 infections. Although low levels of IFN were sufficient, the mechanisms involved remain undetermined.

VIII. CONCLUDING REMARKS

Natural resistance mechanisms probably play a decisive role during early stages of primary herpesvirus infections. Unlike antibody and cell-mediated immunity, these mechanisms do not require prior exposure to viral antigens to be operative and a significant response can be detected as early as 2–4 hr after infection. Failure of the natural defense systems theoretically would result in widely disseminated infection long before an adaptive immune response might be induced and, therefore, capable of clearing the infection. The fact that natural resistance mechanisms probably play a decisive role in primary infection does not rule out a role for these mechanisms in defense against recrudescent HSV disease. Animal models for the study of recurrent disease are poor models of human disease and may not offer the experimental capabilities needed to determine whether natural resistance systems play a role in maintaining latent infections or controlling reactivated disease. These questions may, in the future, be addressed by models of genetic resistance to the development of latent infections.

Deficiencies of natural resistance mechanisms have been associated with unusual susceptibility to herpesvirus infections. An understanding of the basic biology of the effector cells and the interactions needed for a normal response is the necessary first step toward developing new modalities of treatment that either augment these defense mechanisms or replace necessary humoral factors in order to circumvent a deficiency. The heterogeneity of effector cells of natural resistance must be defined in order to develop an understanding of the biological roles played by these cells. Furthermore, humoral factors that act to differentiate, suppress, or augment effectors of natural resistance must be defined in order to develop the means for controlling these cells.

Natural resistance mechanisms constitute a formidable barrier to a herpesvirus infection. The proper functioning of these newly described mechanisms is required by the host in order to survive our numerous encounters with these pathogens. With a thorough understanding of the effector cells involved and their interactions, it should be possible to develop ways of manipulating them to the advantage of the host.

REFERENCES

Aldrich, R. A., Steinberg, A. G., and Campbell, D. C., 1954, Pedigree demonstrating sex-linked recessive condition characterized by draining ears, eczematoid dermatitis, and bloody diarrhea, *Pediatrics* 13:133.

Allen, W. P., and Rapp, F., 1982, Concept review of genital herpes vaccines, *J. Infect. Dis.* **145**:413.

Allison, A. C., 1974, Interactions of antibodies, complement components and various cell types in immunity against virus and pyogenic bacteria, *Transplant. Rev.* **19**:3.

Allison, A. C., Harington, J. S., and Birbeck, M., 1966, An examination of the cytotoxic effects of silica on macrophages, *J. Exp. Med.* **124**:141.

Anderson, M. J., 1978, Innate cytotoxicity of CBA mouse spleen cells to Sendai-virus infected L-cells, *Infect. Immun.* **20**:608.

Andervont, H. B., 1927, Activity of herpetic virus in mice, *Am. J. Hyg.* **14**:383.

Armerding, D., Mayer, P., Scriba, M., Hren, A., and Rossiter, H., 1981a, In vivo modulation of macrophage functions by herpes simplex virus type 2 in resistant and sensitive inbred mouse strains, *Immunobiology* **160**:217.

Armerding, D., Simon, M., Hammerling, U., Hammerling, G. J., and Rossiter, H., 1981b, Function, target cell preference and cell surface characteristics of herpes simplex virus type 2 induced non antigen specific killer cells, *Immunobiology* **158**:347.

Ault, K. A., and Springer, T. A., 1981, Cross-reaction of a rat-anti-mouse phagocyte-specific monoclonal antibody (anti-Mac-1) with human monocytes and natural killer cells, *J. Immunol.* **126**:359.

Bennett, M., 1973, Prevention of marrow allograft rejection with radioactive strontium: Evidence for marrow-dependent effector cells, *J. Immunol.* **110**:510.

Bierman, S. M., 1976, BCG, immunoprophylaxis of recurrent herpes progenitalis, *Arch. Dermatol.* **112**:1410.

Blaese, M., Strober, W., and Waldman, T. A., 1975, Immunodeficiency in the Wiscott–Aldrich syndrome, in: *Immunodeficiency in Man and Animals* (D. Bergsma, R. A. Good, and J. Finstad, eds.), p. 250, Sinauer, Sunderland, Md.

Bloomfield, S. E., and Lopez, C., 1980, Herpes infections in the immunosuppressed host, *Am. Acad. Ophthalmol.* **87**:1226.

Bryson. Y. J., Winter, H. S., Gard, S. E., Fischer, T. J., and Stiehm, E. R., 1980, Deficiency of immune interferon production by leukocytes of normal newborns, *Cell. Immunol.* **55**:191.

Cantell, K., Strander, J., Saxen, L., and Meyer, B., 1968, Interferon response of human leukocytes during intrauterine and postnatal life, *J. Immunol.* **100**:1304.

Carter, W. A., Hande, K. R., Essian, B., Prochownik, E., and Kaback, M. M., 1971, Comparative production of interferon by human fetal and neonatal and maternal cells, *Infect. Immun.* **3**:671.

Casali, P., Sissons, J. G. P., Buchmeier, M. J., and Oldstone, M. B. A., 1981, In vitro generation of human cytotoxic lymphocytes by virus: Viral glycoproteins induce nonspecific cell-mediated cytotoxicity without release of interferon, *J. Exp. Med.* **154**:840.

Cassai, E. N., Sarmiento, M., and Spear, P. G., 1975, Comparisons of the virion proteins specified by herpes simplex virus types 1 and 2, *J. Virol.* **16**:1327.

Ching, C., and Lopez, C., 1979, Natural killing of herpes simplex virus type 1-infected target cells: normal human responses and influence of antiviral antibody, *Infect. and Immun.* **26**:49.

Corey, L., Adams, H. G., Brown, Z. A., and Holmes, K. K., 1983, Genital herpes simplex virus infection: Clinical manifestations, course, and complications, *Ann. Intern. Med.* **98**:958.

Cudkowicz, G., and Bennett, M., 1971, Peculiar immunobiology of bone marrow allografts. II. Rejection of parental grafts by resistant F_1 hybrid mice, *J. Exp. Med.* **134**:1513.

Curran, J. W., 1983, AIDS—Two years later, *N. Engl. J. Med.* **309**:609.

Daniels, C. A., Kleinerman, E. S., and Snyderman, R., 1978, Abortive and productive infections of human mononuclear phagocytes by type 1 herpes simplex virus, *Am. J. Pathol.* **91**:119.

Diamond, R. D., Keller, R., Lee, G., and Finkel, D., 1977, Lysis of cytomegalovirus-infected human fibroblasts and transformed human cells by peripheral blood lymphoid cells from normal human donors (39650), *Proc. Soc. Exp. Biol. Med.* **154**:259.

Djeu, J. D., Heinbaugh, J. A., Holden, H. T., and Herberman, R. B., 1979, Augmentation of mouse natural killer cell activity by interferon and interferon inducers, *J. Immunol.* **122:**175.

Djeu, J. D., Stocks, N., and Zoom, K., 1982, Positive self regulation of cytotoxicity upon exposure to influenza and herpes viruses, *J. Exp. Med.* **156:**1222.

Docherty, J. J., and Chopan, M., 1974, The latent herpes simplex virus, *Bacteriol. Rev.* **38:**337.

Douglas, R. G., Jr., and Couch, R. B., 1970, A prospective study of chronic herpes simplex virus infection and recurrent herpes labialis in humans, *J. Immunol.* **104:**289.

du Buy, H., 1975, Effect of silica on virus infections in mice and mouse tissue culture, *Infect. Immun.* **11:**996.

Engler, H., Zawatzky, R., Goldbach, A., Schroder, C. H., Weyand, C., Hammerling, G. J., and Kirchner, H., 1981, Experimental infection of inbred mice with herpes simplex virus. II. Interferon production and activation of natural killer cells in the peritoneal exudate, *J. Gen. Virol.* **55:**25.

Fast, L. D., Hansen, J. A., and Newman, W., 1981, Evidence for T cell nature and heterogeneity within natural killer and antibody-dependent cellular cytotoxicity (ADCC) effectors: A comparison with cytolytic T lymphocytes (CTL), *J. Clin. Immunol.* **1:**51.

Fitzgerald, P. A., von Wussow, P., and Lopez, C., 1982, Role of interferon in natural kill of HSV-1 infected fibroblasts, *J. Immunol.* **129:**819.

Fitzgerald, P. A., Evans, R., Kirkpatrick, D., and Lopez, C., 1983, Heterogeneity of human NK cells: Comparison of effectors that lyse HSV-1-infected fibroblasts and K562 erythroleukemia targets, *J. Immunol.* **130:**1663.

Gidlund, M., Orn, A., Wigzell, H., Senik, A., and Gresser, I., 1978, Enhanced NK cell activity in mice injected with interferon and interferon inducers, *Nature* **273:**759.

Gottlieb, M. S., Schroff, R., Schanker, H. M., Weisman, J. D., Pan, P. I., Wolf, R. A., and Saxon, A., 1981, Pneumocystis carinii pneumonia and mucosal candidiasis in previously healthy homosexual men: Evidence of a new acquired cellular immunodeficiency, *N. Engl. J. Med.* **305:**1425.

Graig, C. P., and Nahmias. A. J., 1973, Different patterns of neurologic involvement with herpes simplex virus types 1 and 2: Isolation of herpes simplex virus type 2 from the buffy coat of two adults with meningitis, *J. Infect. Dis.* **127:**365.

Gresser, I., 1977, Commentary: On the varied biologic effects of interferon, *Cell. Immunol.* **34:**406.

Gresser, I., Tovey, M. G., Maury, C., and Bandu, M.-T., 1976, Role of interferon in the pathogenesis of virus diseases in mice as demonstrated by the use of anti-interferon serum. II. Studies with herpes simplex, Maloney sarcoma, vesicular stomatitis, Newcastle disease, and influenza viruses, *J. Exp. Med.* **144:**1316.

Grogan, E., Miller, G., Moore, T., Robinson, J., and Wright, J., 1981, Resistance of neonatal human lymphoid cells to infection by herpes simplex virus overcome by aging cells in culture, *J. Infect. Dis.* **144:**6.

Grossman, Z., and Herberman, R. B., 1982, Hypothesis on the development of natural killer cells and their relationship to T cells, in: *NK Cells and Other Natural Effector-Cells* (R. B. Herberman, ed.), p. 229, Academic Press, New York.

Haller, O., and Wigzell, H., 1977, Suppression of natural killer cell activity, with radioactive strontium: Effector cells are marrow dependent, *J. Immunol.* **119:**1503.

Haller, O., Kiessling, R., Orn, A., and Wigzell, H., 1977, Generation of natural killer cells: An autonomous function of the bone marrow, *J. Exp. Med.* **145:**1411.

Halpern, B., Fray, A., Crepin, O., Platica, O., Lorinet, A. M., Rabourdin, A., Sparros, L., and Isac, R., 1973, *Corynebacterium parvum*, a potent immunostimulant in experimental infections and in malignancies, in: *Immunopotentiation, Vol. 18, Ciba Foundation Symposium*, p. 217, Associated Scientific Publishers, New York.

Herberman, R. B., 1980, in: *Natural Cell-Mediated Immunity against Tumors* (R. B. Herberman, ed.), Academic Press, New York.

Herberman, R. B., and Holden, H. T., 1978, Natural cell-mediated immunity, *Adv. Cancer Res.* **27**:305.

Herberman, R. B., and Ortaldo, J. R., 1981, Natural killer cells: Their role in defenses against disease, *Science* **214**:24.

Hirsch, J. G., and Fedorko, M. E., 1970, Morphology of mouse mononuclear phagocytes, in: *Mononuclear Phagocytes* (R. von Furth, ed.), p. 7, Blackwell, Oxford.

Hirsch, M. S., Zisman, B., and Allison, A. C., 1970, Macrophages and age-dependent resistance to herpes simplex virus in mice, *J. Immunol.* **104**:1160.

Isaacs, A., and Lindenmann, J., 1956, Virus interference. I. The interferons, *Proc. R. Soc. London Ser. B* **147**:258.

Johnson, R. T., 1964a, The pathogenesis of herpes virus encephalitis. I. Virus pathways to the nervous system of suckling mice demonstrated by fluorescent antibody staining, *J. Exp. Med.* **119**:343.

Johnson, R. T., 1964b, The pathogenesis of herpes virus encephalitis. II. A cellular basis for the development of resistance with age, *J. Exp. Med.* **120**:359.

Kay, H. D., and Horowitz, D. A., 1980, Evidence by reactivity with hybridoma antibodies for a probable myeloid origin of peripheral blood cells active in natural cytotoxicity and antibody-dependent cell-mediated cytotoxicity, *J. Clin. Invest.* **66**:847.

Kieff, E., Hoyer, B., Bachenheimer, S., and Roizman, B., 1972, Genetic relatedness of type 1 and type 2 herpes simplex viruses, *J. Virol.* **9**:738.

Kiessling, R., and Haller, O., 1978, Natural killer cells in the mouse, an alternative immune surveillance mechanism, *Contemp. Top. Immunobiol.* **8**:171.

Kiessling, R., Hochman, P. S., Haller, O., Shearer, G. H., Wigzell, H., and Cudkowicz, G., 1977, Evidence for a similar or common mechanism for natural killer cell activity and resistance to hemopoietic grafts, *Eur. J. Immunol.* **7**:655.

Kirchner, H., Hirt, H. M., and Munk, K., 1977, Protection against herpes simplex virus infection in mice by *Corynebacterium parvum*, *Infect. Immun.* **16**:9.

Kirchner, H., Hirt, H. M., Rosenstreich, D. L., and Mergenhagen, S. E., 1978a, Resistance of C3H/HeJ mice to lethal challenge with herpes simplex virus, *Proc. Soc. Exp. Biol. Med.* **157**:29.

Kirchner, H., Hochen, M., Hirt, H. M., and Munk, K., 1978b, Immunological studies of HSV infections of resistant and susceptible inbred strains of mice, *Z. Immunitaetsforsch. Immunobiol. [Suppl]* **154**:147.

Kirchner, H., Peter, H. H., Hirt, H. M., Zawatzky, R., Dalugge, H., and Bradstreet, P., 1979, Studies of the producer cell of interferon on human lymphocyte cultures, *Immunobiology* **156**:6575.

Kohl, S., and Ericsson, C. D., 1982, Cellular cytotoxicity to herpes simplex virus-infected cells of leukocytes from patients with serious burns, *Clin. Immunol. Immunopathol.* **24**:171.

Kohl, S., Frazier, J. P., Greenberg, S. B., Pickering, L. K., and Loo, L. S., 1981, Interferon induction of natural killer cytotoxicity in human neonates, *J. Pediatr.* **98**:379.

Kohl, S., Starr, S. E., Oleske, J. M., Shore, S. L., Ashman, R. B., and Nahmias, A. J., 1977, Human monocyte-macrophage-mediated antibody-dependent cytotoxicity to herpes simplex virus-infected cells, *J. Immunol.* **118**:729.

Kumar, V., Ben-Ezra, J., Bennett, M., and Sonnenfeld, G., 1979, Natural killer cells in mice treated with [89]Sr: Normal target binding cell numbers but inability to kill even after interferon administration, *J. Immunol.* **123**:1832.

Leider, W., Magoffin, R. L., Lenette, E. H., and Leonards, L. N. R., 1965, Herpes simplex-virus encephalitis: Its possible association with reactivated latent infection, *N. Engl. J. Med.* **273**:341.

Linnavuori, K., and Hovi, T., 1981 Herpes simplex virus infection in human monocyte cultures: Dose-dependent inhibition of monocyte differentiation resulting in abortive infection, *J. Gen. Virol.* **52**:381.

Lipinski, M., Virelizier, J.-L., Tursz, T., and Griscelli, C., 1980, Natural killer cell activities in patients with primary immunodeficiencies or defects in immune interferon production, *Eur. J. Immunol.* **10**:246.

Lopez, C., 1975, Genetics of natural resistance to herpesvirus infections in mice, *Nature* **258**:152.

Lopez, C., 1978, Immunological nature of genetic resistance of mice to herpes simplex-virus-type 1 infection, in: *Oncogenesis and Herpesviruses III* (G. de Thé, W. Henle, and F. Rapp, eds.), p. 775, IARC, Lyon, France.

Lopez, C., 1980, Resistance to HSV-1 in the mouse is governed by two major, independently segregating non-H-2 loci, *Immunogenetics* **11**:87.

Lopez, C., 1981, Resistance to herpes simplex virus-type 1 (HSV-1), *Curr. Top. Microbiol. Immunol.* **92**:15.

Lopez, C., and Dudas, G., 1979, Replication of herpes simplex virus type 1 in macrophages from resistant and susceptible mice, *Infect. Immun.* **23**:432.

Lopez, C., Simmons, R. L., Park, B. H., Najarian, J. S., and Good, R. A., 1974, Cell-mediated and humoral immune responses of renal transplant recipients with cytomegalovirus infections, *Clin. Exp. Immunol.* **16**:565.

Lopez, C., Kirkpatrick, D., Sorell, M., O'Reilly, R. J., and Ching, C., 1979, Association between pre-transplant natural kill and graft-versus-host disease after stem-cell transplantation, *Lancet* **2**:1103.

Lopez, C., Ryshke, R., and Bennett, M., 1980, Marrow dependent cells depleted by ^{89}Sr mediate genetic resistance to herpes simplex virus 1 infection, *Infect. Immun.* **28**:1028.

Lopez, C., Kirkpatrick, D., Fitzgerald, P. A., Ching, C. Y., Pahwa, R. N., Good, R. A., and Smithwick, E. M., 1982, Studies of the cell lineage of the effector cells which spontaneously lyse HSV-1 infected fibroblasts NK(HSV-1), *J. Immunol.* **129**:824.

Lopez, C., Kirkpatrick, D., Read, S., Fitzgerald, P. A., Pitt, J., Pahwa, S., Ching, C. Y., and Smithwick, E. M., 1983a, Correlation between low natural kill of HSV-1 infected fibroblasts, NK(HSV-1) and susceptibility to herpesvirus infections, *J. Infect. Dis.* **147**:1030.

Lopez, C., Fitzgerald, P. A., and Siegal, F. P., 1983b, Severe acquired immunodeficiency syndrome in male homosexuals: Diminished capacity to make interferon *in vitro* is associated with susceptibility to opportunistic infections, *J. Infect. Dis.* **148**:962.

Lust, J. A., Kumar, V., Burton, R. C., Bartlett, S. P., and Bennett, M., 1981, Heterogeneity of natural killer cells in the mouse, *J. Exp. Med.* **154**:306.

McGeorge, M. B., and Morahan, P. S., 1978, Comparison of various macrophage-inhibitory agents on vaginal and systemic herpes simplex virus type 2 infections, *Infect. Immun.* **22**:623.

McKendall, R. R., 1980, Comparative neurovirulence and latency of HSV-1 and HSV-2 following footpad inoculation in mice, *J. Med. Virol.* **5**:25.

Masur, H., Michelis, M. A., Green, J. B., Onorato, I., Vande Stouwe, R. A., Holzman, R. S., Wormser, G., Brettman, L., Lange, M., Murray, H. W., and Cunningham-Rundles, S., 1981, An outbreak of community-acquired *Pneumocystis carinii* pneumonia: Initial manifestation of cellular immune dysfunction, *N. Engl. J. Med.* **305**:1431.

Metcalf, J. F., Hamilton, D. S., and Reichert, R. W., 1979, Herpetic keratitis in athymic (nude) mice, *Infect. Immun.* **26**:1164.

Meyers, J. D., Fluornoy, N., and Thomas, E. D., 1980, Infection with herpes simplex virus and cell-mediated immunity after marrow transplant, *J. Infect. Dis.* **142**:338.

Miller, M., 1979, Phagocyte function in the neonate: Selected aspects, *Pediatrics* **64**:709.

Minato, N., Bloom, B. R., Jones, C., Holland, J., and Reid, L. M., 1979, Mechanism of rejection of virus persistently infected tumor cells by athymic nude mice, *J. Exp. Med.* **149**:1117.

Minato, N., Reid, L., Cantor, H., Lengyel, P., and Bloom, B. R., 1980, Mode of regulation of natural killer cell activity by interferon, *J. Exp. Med.* **152**:124.

Mogensen, S. C., 1976, Biological conditions influencing the focal necrotic hepatitis test for differentiation between herpes simplex virus type 1 and 2, *Acta Pathol. Microbiol. Scand. Sect. B* **84**:154.

Mogensen, S., 1977a, Role of macrophages in hepatitis induced by herpes simplex virus types 1 and 2, *Infect. Immun.* **15**:686.

Mogensen, S. C., 1977b, Genetics of macrophage-controlled resistance to hepatitis induced by herpes simplex virus type 2 in mice, Infect. Immun. 17:268.

Mogensen, S. C., 1978, Macrophages and age-dependent resistance to hepatitis induced by herpes simplex virus type 2 in mice, Infect. Immun. 19:46.

Mogensen, S. C., and Andersen, H. K., 1978, Role of activated macrophages in resistance of congenitally athymic nude mice to hepatitis induced by herpes simplex virus type 2, Infect. Immun. 19:792.

Mogensen, S. C., and Andersen, H. K., 1981, Recovery of mice from herpes simplex type 2 hepatitis: Adoptive transfer of recovery with immune spleen cells, Infect. Immun. 33:743.

Mogensen, S. C., Teisner, B., and Andersen, H. K., 1974, Focal necrotic hepatitis in mice as a biological marker for differentiation of herpesvirus hominis type 1 and type 2, J. Gen. Virol. 25:151.

Morahan, P. S., Kern, E. R., and Glasgow, L. A., 1977, Immunomodulator-induced resistance against herpes simplex virus, Proc. Soc. Exp. Biol. Med. 154:615.

Morse, L. S., Buchman, T. G., Roizman, B., and Schaffer, P. A., 1977, Anatomy of herpes simplex virus DNA. IX. Apparent exclusion of some parental DNA arrangements in the generation of intertypic (HSV-1 × HSV-2) recombinants, J. Virol. 24:231.

Nagafuchi, S., Oda, H., Mori, R., and Taniguchi, T., 1979, Mechanism of acquired resistance to herpes simplex virus infection as studied in nude mice, J. Gen. Virol. 44:715.

Nahmias, A. J., 1972, Herpesviruses from fish to man—A search for pathobiologic unity, Pathobiol. Annu. 2:143.

Nahmias, A. J., and Roizman, B., 1973, Infection with herpes-simplex viruses 1 and 2, N. Engl. J. Med. 289:667, 719, 781.

Oleski, J. M., Ashman, R. B., Kohl, S., Shore, S. S., Starr, S. E., Wood, P., and Nahmias, A. J., 1977, Human polymorphonuclear leukocytes as mediators of antibody-dependent cellular cytotoxicity to herpes simplex virus-infected cells, Clin. Exp. Immunol. 27:446.

Ortaldo, J. R., Lang, N. P., Timonen, T., and Herberman, R. B., 1981, Augmentation of human natural killer cell activity by interferon: Conditions required for boosting and characteristics of the effector cells, J. Interferon Res. 1:253.

Pazin, G. J., Armstrong, J. A., Lam, M. T., Tarr, G. C., Jannetta, P. J., and Ho, M., 1979, Prevention of reactivated herpes simplex infection by human leukocyte interferon after operation on the trigeminal root, N. Engl. J. Med. 301:225.

Peter, H. H., Dallugge, H., Zawatzky, R., Euler, S., Leibold, W., and Kirchner, H., 1980, Human peripheral null lymphocytes. II. Producers of type-1 interferon upon stimulation with tumor cells, herpes simplex virus and Corynebacterium parvum, Eur. J. Immunol. 10:547.

Peter, H. H., Friedrich, W., Dopfer, R., Muller, W., Kortmann, C., Werner, J. P., Heinz, F., and Rieger, C. H. L., 1983, NK cell function in severe combined immunodeficiency (SCID): Evidence of a common T and NK cell defect in some but not all SCID patients, J. Immunol. 131:2332.

Phillips, W. H., Ortaldo, J. R., and Herberman, R. B., 1980, Selective depletion of human natural killer cells on monolayers of target cells, J. Immunol. 125:2322.

Preble, O. T., Block, R. J., Friedman, R. M., Klippel, J. H., and Vilcek, J., 1982, Systemic lupus erythematosis: Presence in human serum of an unusual acid-labile leukocyte interferon, Science 216:429.

Price, R. W., and Schmitz, J., 1978, Reactivation of latent herpes simplex virus infection of the autonomic nervous system by postganglion neurectomy, Infect. Immun. 19:523.

Rasmussen, L., and Farley, L. B., 1975, Inhibition of herpesvirus hominis replication by human interferon, Infect. Immun. 12:104.

Richards, J. T., Kern, E. R., Overall, J. C., and Glasgow, L. A., 1981, Differences in neurovirulence among isolates of herpes simplex virus types 1 and 2 in mice using four routes of infection, J. Infect. Dis. 144:464.

Ruchmann, I., 1954, Virulence of strains of herpes simplex virus for mice, *Proc. Soc. Exp. Biol. Med.* **86**:649.

St. Geme, J. W., Prince, J. T., Burke, B. A., Good, R. A., and Krivitt, W., 1965, Impaired cellular resistance to herpes-simplex virus in Wiskott–Aldrich syndrome, *N. Engl. J. Med.* **273**:229.

Saksela, E., Timonen, T., and Cantell, K., 1978, Human natural killer cell activity is augmented by interferon via recruitment of "pre-NK" cells, *Scand. J. Immunol.* **10**:257.

Santoli, D., Trinchieri, G., and Lief, F. S., 1978a, Cell-mediated cytotoxicity against virus-infected target cells in humans. I. Characterization of the effector lymphocyte, *J. Immunol.* **121**:526.

Santoli, D., Trinchieri, G., and Koprowski, H., 1978b, Cell-mediated cytotoxicity against virus-infected target cells in humans, *J. Immunol.* **121**:538.

Schlabach, A. J., Martinez, D., Field, A. K., and Tytell, A. A., 1979, Resistance of C57 mice to primary systemic herpes simplex virus infection, macrophage dependence and T-cell independence, *Infect. Immun.* **26**:615.

Schneweis, K. E., and Saftig, V., 1981, The vaginal herpes simplex virus infection of resistant (C57 BL) mice, *Int. Herpesvirus Workshop*, p. 144.

Schwartz, H. J., and Leskowitz, C., 1969, The effect of carrageenan on delayed hypersensitivity reactions, *J. Immunol.* **103**:87.

Seaman, W. E., Gindhart, T. D., Greenspan, J. S., Blackman, M. A., and Talal, N., 1979, Natural killer cells, bone, and the bone marrow: Studies in estrogen-treated mice and in congenitally osteopetrotic (*mi/mi*) mice, *J. Immunol.* **122**:1226.

Shore, S. L., Black, C. M., Melewick, F. M., Wood, P. A., and Nahmias, A. J., 1976, Antibody-dependent cell-mediated cytotoxicity to target cells infected with type 1 and type 2 herpes simplex virus, *J. Immunol.* **116**:194.

Siegal, F. P., Lopez, C., Hammer, G. S., Brown, A. E., Kornfeld, S. J., Gold, J., Hassett, J., Hirschman, S. Z., Cunningham-Rundles, C., Adelsberg, B. R., Parham, D. M., Siegal, M., Cunningham-Rundles, S., and Armstrong, D., 1981, Severe acquired immunodeficiency in male homosexuals, manifested by chronic perianal ulcerative herpes simplex lesions, *N. Engl. J. Med.* **305**:1439.

Stanwick, T. L., Campbell, D. E., and Nahmias, A. J., 1980, Spontaneous cytotoxicity mediated by human monocyte-macrophages against human fibroblasts infected with herpes simplex virus–Augmentation by interferon, *Cell. Immunol.* **53**:413.

Starr, S. E., Visintine, A. M., Tomeh, M. O., and Nahmias, A. J., 1976, Effects of immunostimulants on resistance of newborn mice to herpes simplex virus type 2 infection, *Proc. Soc. Exp. Biol. Med.* **152**:57.

Stein, M. D., Klimpel, G. R., and Herndon, D. N., 1983, Defective natural killer activity in thermal injury, *Fed. Proc.* **42**:5425.

Stevens, J. G., and Cook, M. L., 1971, Restriction of herpes simplex virus by macrophages: An analysis of the cell–virus interaction, *J. Exp. Med.* **133**:19.

Stewart, W. E., II, 1979, *The Interferon System*, 2nd ed., Springer-Verlag, Berlin.

Stewart, W. E., II, Blalock, J. E., Burke, D. C., Chang, C., Dunnick, J. K., Falcoff, E., Friedman, R. M., Galasso, G. J., Joklik, W. K., Vilcek, J. T., Youngner, J. S., and Zoon, K. C., 1977, Interferon nomenclature, *Nature* **268**:110.

Stiehm, R. E., 1980, The human neonate as an immunocompromised host, in: *Infections in the Immunocompromised Host—Pathogenesis, Prevention and Therapy, Vol. 11,* (J. Verhoef, P. K. Peterson, and P. G. Quie, eds.), p. 77, Elsevier/North-Holland, New York.

Stiehm, E. R., Kronenberg, L. H., Rosenblatt, H. M., Bryson, Y., and Merigan, T. C., 1982, Interferon: Immunobiology and clinical significance, *Ann. Intern. Med.* **96**:80.

Stutman, O., Paige, C. J., and Figarella, E., 1978, Natural cytotoxic cells against solid tumors in mice. I. Strain and age distribution and target cell susceptibility, *J. Immunol.* **121**:1819.

Stutman, O., Lattime, E. C., and Figarella, E. F., 1981, Natural cytotoxic cells against solid tumors in mice: A comparison with natural killer cells, *Fed. Proc.* **40**:2699.

Sullivan, J. L., Byron, K. S., Brewster, F. E., and Purtilo, D. T., 1980, Deficient natural killer cell activity in X-linked lymphoproliferative syndrome, *Science* **210**:543.

Sullivan, J. L., Byron, K. S., Brewster, F. E., Baker, S. M., and Ochs, H. D., 1983, X-linked lymphoproliferative syndrome, *J. Clin. Invest.* **71**:1765.

Tai, A. S., Safilian, B., and Warner, N. L., 1982, Identification of distinct target-specific subsets of Nk cells in peripheral blood of normal donors, *Hum. Immunol.* **4**:123.

Taylor, S., and Bryson, Y. J., 1981, Impaired production of immune (PHA-induced) interferon in newborns is due to a functionally immature macrophage, *Pediatr. Res.* **15**:604.

Timonen, T., Saksela, E., 1980, Isolation of human NK cells by density gradient centrifugation, *J. Immunol. Meth.* **36**:285.

Timonen, T., Ortaldo, R. R., and Herberman, R. B., 1981, Characteristics of human granular lymphocytes and relationship to natural killer cells, *Fed. Proc.* **40**:2705.

Trinchieri, G., and Santoli, D., 1978, Antiviral activity induced by culturing lymphocytes with tumor-derived or virus-transformed cells: Enhancement of human natural killer cell activity by interferon and antagonistic inhibition of susceptibility of target cells to lysis, *J. Exp. Med.* **147**:1299.

Trinchieri, G., Santoli, D., and Koprowski, H., 1978, Spontaneous cell-mediated cytotoxicity in humans: Role of interferon and immunoglobulins, *J. Immunol.* **120**:1849.

Trofatter, K. F., Jr., Daniels, C. A., Williams, R. J., Jr., and Gall, S. A., 1979, Growth of type 2 herpes simplex virus in newborn and adult mononuclear leukocytes, *Intervirology* **11**:117.

Ullberg, M., and Jondal, M., 1981, Recycling and target-binding capacity of human natural killer cells, *J. Exp. Med.* **153**:651.

Virelizier, J.-L., 1981, Viral infections in patients with selective disorders of the interferon system, *Fifth International Congress of Virology*, p. 152.

Virelizier, J.-L., Lenoir, G., and Griscelli, C., 1978, Persistent Epstein–Barr virus infection in a child with hypergammaglobulinemia and immunoblastic proliferation associated with a selective defect in immune interferon secretion, *Lancet* **2**:231.

Virelizier, J.-L., Lipinski, M., Tursz, T., and Griscelli, C., 1979, Defects of immune interferon secretion and natural killer activity in patients with immunological disorders, *Lancet* **2**:696.

Visintine, A., Nahmias, A., Whitley, R., and Alford, C., 1981, The natural history and epidemiology of neonatal herpes simplex virus infection, in: *The Human Herpesviruses* (A. J. Nahmias, W. R. Dowdle, and R. F. Schinazi, eds.), p. 599, Elsevier/North-Holland, Amsterdam.

von Furth, R., Cohn, Z. A., Hirsch, J. G., Humphrey, J. H., Spector, W. G., and Langevoort, H. L., 1972, The mononuclear phagocyte system: A new classification of macrophages, monocytes, and their precursor cells, *Bull. WHO* **46**:845.

Welsh, R. M., 1978a, Cytotoxic cells induced during lymphocytic choriomeningitis virus infection of mice. I. Characterization of natural killer cell induction, *J. Exp. Med.* **148**:163.

Welsh, R. M. 1978b, Mouse natural killer cells: Induction specificity and function, *J. Immunol.* **121**:1631.

Welsh, R. M., and Hallenbeck, L. A., 1980, Effect of virus infections on target cell susceptibility to natural killer cell mediated lysis, *J. Immunol.* **124**:2491.

Zarling, J. M., Clouse, K. A., Biddison, W. E., and Kung, P. C., 1981, Phenotypes of human natural killer cell populations detected with monoclonal antibodies, *J. Immunol.* **127**:2275.

Zawatzky, R., Hilfenhaus, J., and Kirchner, H., 1979, Resistance of nude mice to herpes simplex virus and correlation with *in vitro* production in interferon, *Cell. Immunol.* **47**:424.

Zawatzky, R., Hilfenhaus, J., Marucci, F., and Kirchner, H., 1981, Experimental infection of inbred mice with herpes simplex virus type 1. I. Investigation of humoral and cellular immunology and of interferon induction, *J. Gen. Virol.* **43:**31.

Zisman, B., Hirsch, M. S., and Allison, A. C., 1970, Selective effects of antimacrophage serum, silica and anti-lymphocyte serum on pathogenesis of herpes virus infection of young adult mice, *J. Immunol.* **104:**1155.

CHAPTER 3

Humoral Response to Herpes Simplex Virus Infections

BODIL NORRILD

I. INTRODUCTION

Herpes simplex virus types 1 and 2 (HSV-1, HSV-2) give rise to a variety of human diseases of different severity and morbidity. It is well known that herpes virus remains latent in the human body after the first exposure to the virus (the primary infection) and colonizes either the trigeminal or the sacral ganglia, from where the virus is occasionally reactivated (for review see Wildy *et al.*, 1982). The result of reactivation is either a clinical or a subclinical recurrent infection, and viral cellular proteins synthesized during a primary as well as during a recurrent infection interact with the immune system of the infected host organisms and give rise to the formation of both circulating HSV-specific antibodies and activated lymphocytes that have the potential to eliminate infectious virus from the infected host.

The interaction between virus and antibodies was studied by use of both human convalescent serum and rabbit hyperimmune serum, and it was recognized early that various virus isolates reacted most strongly with antibodies raised to the homologous virus (Roizman *et al.*, 1973; Schneweis, 1962), a finding that allowed the identification of the two HSV strains HSV-1 and HSV-2. The serological differentiation was later extended with biochemical and genetic analyses of HSV-1 and HSV-2 DNA (Kieff *et al.*, 1971, 1972; Ludwig *et al.*, 1972) and proteins (Cassai *et al.*, 1975; Heine *et al.*, 1974; Morse *et al.*, 1978; Pereira *et al.*, 1976; Powell *et al.*, 1977; Spear and Roizman, 1972).

BODIL NORRILD • Institute of Medical Microbiology, University of Copenhagen, DK-2100 Copenhagen Ø, Denmark.

The present review will focus on the ability of HSV to induce specifically circulating antibodies in the human host, and the following aspects will be discussed:

1. The categories of antibodies detected in various immunological tests
2. The immunogenicity of individual viral proteins
3. The significance of the induction of antibodies belonging to different immunoglobulin classes during HSV infection
4. The possible protective effect of specific HSV antibodies, with special emphasis on the monitoring of a human vaccine by measurement of the induction of specific HSV antibodies

II. HSV-SPECIFIC ANTIBODIES IN HUMAN CONVALESCENT SERA

The identification of the two strains of HSV and the observation that especially HSV-2 might be of significance for the development of cervical cancer (Nahmias and Norrild, 1980; Nahmias et al., 1974; Rawls et al., 1977) made it desirable to develop serological tests that allowed both qualitative and quantitative measurements of HSV-1 and HSV-2 antibodies in human convalescent sera. In early studies the neutralization tests and the complement-binding test were most commonly used (Pauls and Dowdle, 1967; Schneweis, 1962) but these methods were both time-consuming and tedious and for seroepidemiological purposes methods allowing a rapid screening of large numbers of sera were needed. Many tests have been developed over the years for the analysis of antibodies reactive to both HSV type-common and HSV type-specific antigens. Among the tests to be mentioned, the inhibition–passive hemagglutination test (Schneweis and Nahmias, 1971), the immunofluorescence tests (Geder and Skinner, 1971; Nahmias et al., 1971a,b), the immunocytolytic test (Rager-Zisman and Bloom, 1974; Rawls and Tompkins, 1975; Shore et al., 1976; Subramanian and Rawls, 1977), and the immunoprecipitation tests (Tokumaru, 1965; Vestergaard, 1980) were all useful mainly for measurement of HSV-specific antibodies of the IgG class. The recent development of the ELISA and RIA tests has been a great improvement for both seroepidemiological and diagnostic work, and these methods also allow the measurement of antibodies belonging to the IgM and IgA immunoglobulin subclasses more easily than by the immunofluorescence test used in previous studies (Nahmias et al., 1971b). (For review see Plummer, 1973.) The methods to be discussed in this review are listed in Table I.

Only the major findings obtained by the seroepidemiological studies and of relevance for the understanding of the humoral immune response to HSV will be discussed in the following.

TABLE I. Methods in Use for the Analysis of Human Sera for HSV-Specific Antibodies

Method	Reactive proteins	Immunoglobulin class detectable
Virus neutralization	Membrane glycoproteins	IgG, IgM
Fluorescence microscopy	Intracellular proteins and/or membrane glycoproteins	IgG, IgM, IgA
Antibody-dependent immunocytolytic tests		
Complement-mediated	Membrane glycoproteins	IgG, IgM
K-cell-mediated	Membrane glycoproteins	IgG
Immunoprecipitation	Soluble infected cell proteins	IgG
ELISA/RIA	Total or purified proteins	IgG, IgM, IgA
Immunoblotting	Soluble and insoluble infected cell proteins	IgG

III. MEASUREMENT OF HSV-SPECIFIC NEUTRALIZING ANTIBODIES

Neutralizing antibodies that are produced in response to HSV antigens present on the surface of the virus envelope appear after primary HSV infections 2 weeks after onset of clinical lesions, and the titer increases until 3–4 weeks after clinical symptoms (Kohl *et al.*, 1982; Mann and Hilty, 1982; Rawls, 1973; Zweerink and Corey, 1982).

The IgG antibody titers to HSV usually remain fairly stable and even after clinical recurrent infections fluctuations in neutralizing antibody titers occur only occasionally (Cesario *et al.*, 1969; Douglas and Couch, 1970; reviewed by Doerr *et al.*, 1976). Measurement of neutralizing antibodies is a very sensitive test but it has been reported that a certain subset of the antibodies is dependent on the presence of complement, according to studies first made in a rabbit model system. The complement-dependent fraction of antibodies was found especially in early hyperimmune sera (Wallis and Melnick, 1971; Yoshino and Taniguchi, 1965; Yoshino *et al.*, 1977) but also late sera contained a fraction of complement-requiring antibodies (Hamper *et al.*, 1968; Yoshino *et al.*, 1979). Complement was shown to influence mainly antibodies of the IgM class but also a subfraction of the IgG antibodies (Ohashi and Ozaki, 1981). Complement also enhanced the neutralizing capacity of human sera, an effect that was found in both early and late convalescent sera (Zweerink and Corey, 1982).

The percentage of seropositive adults as measured by the presence of neutralizing antibodies is dependent on the population tested and is related to a variety of factors, including socioeconomic parameters, number of sex partners, age, etc. (Nahmias and Roizman, 1973a–c; Wentworth and Alexander, 1971). The percentage of HSV-1-seropositive women in Scandinavia was as high as 87% and for HSV-2 47% (Vestergaard *et al.*, 1972) whereas figures as low as 7–9% were found in Israel and Hungary (for review see Rawls *et al.*, 1980).

IV. MEASUREMENT OF HSV ANTIBODIES BY IMMUNOFLUORESCENCE

The immunofluorescence test made it possible to measure both antibody binding to the surface of infected cells and binding to internal HSV proteins (Nahmias *et al.*, 1971a,b). However, the major contributions obtained by immunofluorescence studies are the measurements of HSV-specific antibodies belonging to the different immunoglobulin classes (described in detail later in this review).

V. MEASUREMENT OF HSV-CYTOLYTIC ANTIBODIES

Complement-dependent immunocytolysis of HSV-infected cells in the presence of specific antibodies was recognized in early studies where HSV-1-infected cells were injured in the presence of hyperimmune rabbit sera and complement (Notkins, 1971; Oldstone, 1977; Rawls and Tompkins, 1975; Roizman and Roane, 1961; Smith *et al.*, 1972). The application of the test to the analysis of human sera showed that convalescent sera added even in high dilutions were able to lyse both HSV-1- and HSV-2-infected cells. The sensitivity of the immunocytolysis was increased in the antibody-dependent cell-mediated cytotoxicity (ADCC) tests, but the method allowed only the measurement of antibodies belonging to the IgG class reactive with the HSV surface antigens (Rager-Zisman and Bloom, 1974; Shore *et al.*, 1977). (For review see Norrild *et al.*, 1984). Sera that were originally scored as negative for HSV antibodies by use of the complement-dependent immunocytolytic test were positive in the ADCC test, and out of 100 sera of which 60 were negative in the complement-dependent test, only 27 were negative in the ADCC test (Subramanian and Rawls, 1977). The ADCC technique also made it possible to follow some fluctuations in the antibody titer during various stages of malignant disease, and it was found that low antibody titer correlated with severe disease (Christenson, 1978). The antibody population measured in the immunocytolytic test is likely to be the same subset of antibodies as measured in neutralization reactions.

VI. MEASUREMENT OF HSV-PRECIPITATING ANTIBODIES

Immunoprecipitation of HSV proteins with human sera was first done in gels (Tokumaru, 1965). Five to seven precipitin lines were found, most in convalescent sera. Vestergaard (1979) applied the crossed immunoelectrophoresis for the analysis of precipitating antibodies present in human sera. The human sera were applied in an intermediate gel and their interaction with the HSV immunoprecipitates found by precipitation of HSV-infected cell extracts into antibody-containing gel was measured. Antibodies to the glycoproteins gB and gD were found to be dominant in the 100 human sera tested.

Analysis of HSV-specific precipitating antibodies present in human sera was also done in suspensions by the binding of the sera to radiolabeled HSV proteins extracted from infected cells. The immune complexes formed were bound to protein A and the specificity of the reactive antibodies was visualized as the autoradiographic image of the bound HSV proteins after their electrophoretic separation on SDS-polyacrylamide gels (Eberle and Courtney, 1981; Gilman et al., 1981; Mann and Hilty, 1982). The method is especially valuable because antibodies to both membrane and intracellular—but soluble—viral proteins are measured and at the same time this kind of study identifies the subset of viral proteins that are immunogenic during an infection, a topic that will be dealt with in detail in a later section of this review. Immunoprecipitation studies are possible only if the antibody titer is reasonably high, and although the population of antibodies measured is different from the one measured by neutralization, there seems to be a correlation between the titer of the serum as measured by neutralization and its ability to immunoprecipitate (Eberle and Courtney, 1981; Mann and Hilty, 1982). It should be noted that in sequential sera from a patient with a primary infection, antibodies to the known glycoproteins of HSV-1, gB, gC, gD, and gE, were present from day 14 after onset of the illness, and the antibody profile remained constant over a period of 2 years. Sequential sera from a patient with recurrent HSV-1 showed a similar antibody pattern and no major changes up to or during a recurrence (Mann and Hilty, 1982).

Another approach where immunoprecipitation was shown to be useful was for typing of human sera for their content of antibodies to either HSV-1 or HSV-2 (Eberle and Courtney, 1981). The investigators took advantage of the existence of an HSV-1-specific glycoprotein designated gC, and sera that precipitated gC of HSV-1 contained HSV-1-specific antibodies whereas sera that did not precipitate gC of HSV-1 apparently precipitated an HSV-2 glycoprotein with an apparent molecular weight similar to that of gC of HSV-1. This approach for typing human sera might not be specific enough as it is well known that strains of HSV-1 lacking gC may occur (Hoggan and Roizman, 1959).

VII. MEASUREMENT OF HSV ANTIBODIES WITH ELISA AND RIA

Development of the ELISA and RIA tests allowed a quick screening of a large number of human sera for the content of antibodies to HSV proteins. The methods may be used for quantitative analysis of both HSV-specific IgG antibodies and for the analysis of the IgA and IgM subclasses as described in a later section (Jordan and Rytel, 1981; Kalimo et al., 1977; Vestergaard and Grauballe, 1979).

VIII. MEASUREMENT OF HSV ANTIBODIES BY IMMUNOBLOTTING

The immunoblotting test has been used by several investigators for the analysis of the immunological specificity of hybridoma antibodies made to HSV proteins. Protein extracts made from infected cells were separated electrophoretically in SDS-polyacrylamide gels and then transferred to nitrocellulose paper. The immobilized proteins were incubated with antibodies followed by a horseradish peroxidase-coupled conjugate that by reaction with a substrate for the enzyme stained the bound antibodies (Braun et al., 1983; Fujinami et al., 1983).

Application of the technique for the analysis of the HSV-specific antibody repertoire present in hyperimmune human sera made it possible to detect antibodies reactive with both soluble and insoluble HSV proteins (Stubbe Teglbjaerg and Norrid, in press; Vass-Sørensen et al., 1984). As many as 23 HSV-1 proteins bound specific antibodies when reacted with hyperimmune human sera, whereas 10–15 HSV-2 proteins at the most bound antibodies from the same sera.

In order to understand the humoral immune response induced by HSV, it is necessary to sum up the information obtained by the various tests. The antibodies made to the antigen exposed on the surface of infected cells and on the virion envelope have been considered the ones most important for the limitation of the spread of virus and for the elimination of infected virus from the infected host. Our present knowledge about the spectrum of antibodies induced by both structural and intracellular proteins leaves us with speculations about the biological significance of these antibodies. The most likely hypothesis is that various antibodies form immune complexes with the corresponding proteins, complexes that are then eliminated from the host by uptake into macrophages.

IX. IMMUNOGENICITY OF THE HSV-1 AND HSV-2 PROTEINS

During an infectious cycle, HSV-1 and HSV-2 specify the synthesis of more than 50 proteins. These have been identified in infected cell extracts (Heine et al., 1974; Morse et al., 1978), and among the cellular HSV-1 and HSV-2 proteins, five glycoproteins, designated gB, gC, gD, gE, and gG, are inserted in both the plasma membrane of infected cells and

the virus envelope. It should be noted that gC of HSV-2 in the old nomenclature was named gF and that gG is a newly discovered glycoprotein
specific for HSV-2 (Balachandran *et al.*, 1981; Bauke and Spear, 1979;
Cohen *et al.*, 1978; Heine *et al.*, 1972; Norrild *et al.*, 1980; Powell *et al.*,
1974; Roizman *et al.*, 1984; Spear, 1976; Vestergaard and Norrild, 1978).
Among the virus-specified proteins identified in infected cells, approximately half are structural proteins that comprise the virus particle (Cassai
et al., 1975; Heine *et al.*, 1972).

Several of the HSV proteins are strong immunogens, and especially
the glycoproteins that are exposed on the surface of infected cells and
virions have been studied extensively with respect to their ability to induce the production of both specific antibodies and activated cytolytic
lymphocytes (Eberle and Courtney, 1981; Kapoor *et al.*, 1982a; Mann and
Hilty, 1982; Norrild, 1980; Rouse and Lawman, 1980; Zweerink and
Corey, 1982). As the immunogenicity of the glycoproteins is reviewed
elsewhere (Norrild, 1980), only the following features shall be mentioned:
each of the glycoproteins carries a specific set of antigen-determinant
sites, which are not present on any of the other glycoproteins from the
homologous virus. All HSV-1 glycoproteins have immunologically cross-
reactive counterparts in HSV-2 virus, and only gG of HSV-2 has no counterpart in HSV-1 (Roizman *et al.*, 1984b).

Among the nonglycosylated proteins, several structural and nonstructural proteins are strong immunogens in the human host. By immunoprecipitation studies it has been reported that the HSV-1 proteins
with apparent molecular weights of 159, 152, 132, 125, 96, 67, and 63K
could be precipitated by antibodies produced 2 weeks after a primary
encephalitis. The proteins represented both the major capsid protein
(159K in this study) and the glycoproteins gB (125K) and gC (132K). The
subset of proteins precipitated with sera from patients with a recurrent
herpes labialis was very much the same, but instead of the 67 and 63K,
a protein of 48K was demonstrated (Mann and Hilty, 1982). In another
study where 231 sera were tested by immunoprecipitation as controls for
a cancer study, 31 HSV-1 and 27 HSV-2 proteins were immunoprecipit-
able. However, the individual sera reacted with 12–14 proteins each, and
the patterns obtained with the various sera differed. The protein that was
precipitated most frequently from HSV-1 was a glycoprotein of 133K,
most likely gC, and from HSV-2 a glycoprotein of 131K (Gilman *et al.*,
1981).

Studies done in my laboratory confirm the data obtained by Gilman
et al. (1981). Twelve to fourteen proteins from either HSV-1 or HSV-2 are
precipitated by human sera, but the profiles obtained are very consistent
from serum to serum in the series of 50 sera analyzed (Fig. 1). In primary
infections the precipitating antibodies were weak in acute serum but the
protein profile obtained with convalescent sera was similar to the one
obtained with hyperimmune sera from persons without clinical recurrence (Figs. 1 and 2). The HSV-1 and HSV-2 proteins precipitated by the

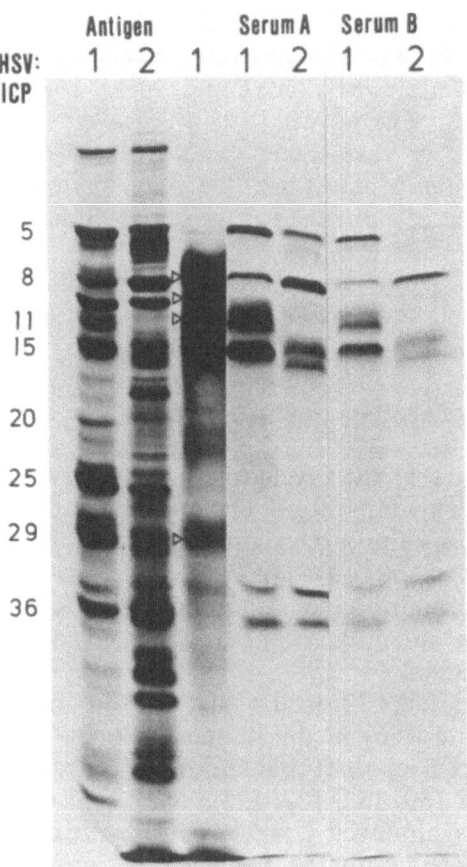

Antigen   Serum A   Serum B
HSV: 1 2   1   1 2   1   2
ICP

5
8
11
15
20
25
29
36

FIGURE 1. SDS-polyacrylamide gel electrophoresis of HSV-1 and HSV-2 proteins immunoprecipitated with human hyperimmune serum. HSV-1 and HSV-2 infected Vero cells were labeled with 2 μCi/ml medium of [^{14}C]isoleucine, [^{14}C]leucine, [^{14}C]valine (ILV) (NEN, Germany). Labeling was done from 6 to 18 hr postinfection, and the infected cell proteins were extracted for immunoprecipitation in a 1% (v/v) Triton X-100-containing glycine–Tris buffer (pH 8.6) with 10^{-5} M TLCK and TPCK. The soluble proteins were prepared by centrifugation at 100,000g for 1 hr. Immunoprecipitation was done by incubation of 25 μl of protein extract with 50 μl of human serum. The immune complexes were isolated after binding to formalin-fixed *Staphylococcus* (Roizman *et al.*, 1982). The immunoprecipitate was solubilized and electrophoresed on SDS-polyacrylamide gels as described by Morse *et al.* (1978). Sera A and B are from two patients with *in situ* cervical carcinoma. ^{14}C-glucosamine-labeled HSV-1 proteins extracted from infected Vero cells are included as a marker in lane 3. The arrows indicate from the top the glycoproteins gC, gB^1, gB^2, and gD.

hyperimmune human sera tested in that study are listed in Table II, and it should be noted that only a few of the known nonglycosylated infected cell proteins are precipitated by the human sera.

In order to get a better measure of the number of HSV proteins that are immunogenic, we applied the immunoblotting technique for the analysis of the HSV antibodies present in human hyperimmune sera. The number of immunoreactive HSV-1 proteins demonstrated by this method was as high as 23, whereas 10–15 HSV-2 proteins at the most were reactive, as illustrated for five human sera in Fig. 3. It is clear that there are differences in the set of antibodies present in the different sera, and also that the demonstration of immunoreactivity is very dependent on the method used. The HSV-1 proteins that reacted most frequently with human sera from persons without recurrences are listed in Table III. Both the ICPs 32/33 and the two glycoproteins gB1 (ICP11) and gD (ICP29) reacted with 50% of the sera tested. Our data do not agree with the immunoprecipitation data of Gilman *et al.* (1981), who found that the HSV-1 proteins with molecular weights of 133, 99, 82, and 104K were detected most frequently after reaction with human antibodies.

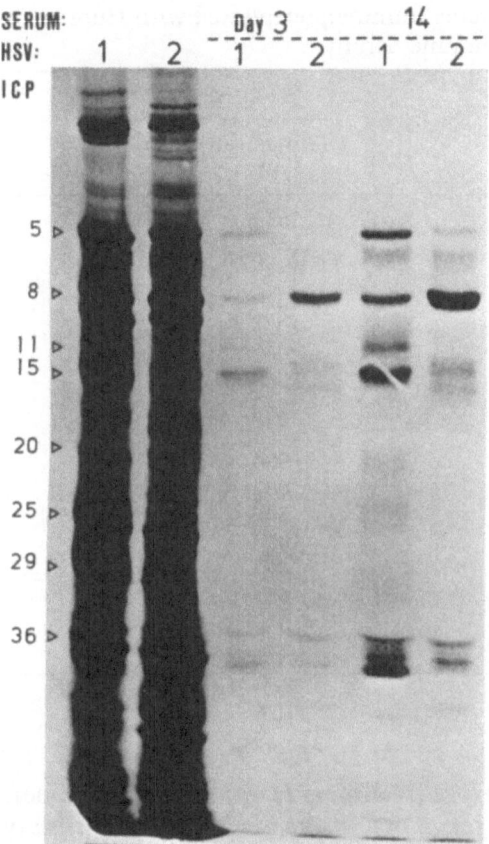

FIGURE 2. SDS-polyacrylamide gel electrophoresis of HSV-1 and HSV-2 proteins immunoprecipitated with human serum from a primary infection. The HSV-1 and HSV-2 protein extracts were labeled and prepared as described in Fig. 1. The HSV-1 or HSV-2 proteins were precipitated with serum collected on day 3 and on day 14 postlesion. Sera were kindly provided by Dr. J. Blomberg, Lund, Sweden.

In conclusion, the different studies done in order to demonstrate the immunogenicity of the individual HSV proteins have shown that most of the proteins are able to induce the production of specific antibodies during a productive infection.

X. MEASUREMENT OF HSV-SPECIFIC ANTIBODIES BELONGING TO THE IgM AND IgA CLASSES

HSV induces specific antibodies of the IgM, IgA, and IgG classes during a productive infection. The kinetics of appearance of the different immunoglobulins has been studied extensively for diagnostic purposes. As IgM was considered a specific marker for primary infections, human sera were screened for the appearance of specific IgM either by indirect immunofluorescence or by ELISA or RIA tests. Although it was reported in most studies that IgM is present in serum only in primary cases, some investigators found IgM to persist in serum (Kimmel *et al.*, 1982; Nahmias and Roizman, 1973c). In primary infections IgM was demonstrated in sera

TABLE II. HSV-1 and HSV-2 Proteins Immunoprecipitated with Human Hyperimmune Serum

HSV-1		HSV-2	
Molecular weight $\times 10^{-3}$	ICP[a]	Molecular weight $\times 10^{-3}$	ICP[a]
151	5	153	5
146	6	146	6
		138	7
128	8	127	8
119	(gB^2)	102	
114	11 (gB^1)	99	
110		96	
103		92.5	
77			
		90	
64	25	63	25
55	29 (gD)	54.5	29 (gD)
42.5	36	42.5	36
41.5		41.5	
40		35	
33.2		31.3	
22		21	

[a] According to Morse *et al.* (1978).

from five infants with neonatal herpes (Nahmias *et al.*, 1971a), and Doerr *et al.* (1976) found IgM in serum for many weeks after onset of primary CNS disease but not after localized recurrent diseases. Kalimo *et al.*, (1977) reported that IgM was induced in primary as well as in severe recurrent cases, and an elevated level of IgM was found in sera from patients with recent genital infections (El Falaky and Vestergaard, 1977). Antibodies of the IgA class were proposed to play a role in the reactivation of HSV (Tokumaru, 1966), but later studies showed that the IgA was induced during a primary infection where an increase in titer could be demonstrated but IgA persisted in human sera and therefore was neither of diagnostic value nor of significance in the development of recurrences (Friedman and Kimmel, 1982). HSV-specific IgA was also present in secretions and was found in vaginal secretions from women with genital herpes (Kalimo *et al.*, 1981; Nahmias *et al.*, 1971a) and from pregnant women, where an elevated IgA titer to HSV was related to risk of abortion (Grönroos *et al.*, 1983). The significance of IgA in secretions is not known, but it might be related to local immunity as suggested for secretory IgA in tears (Centifanto and Kaufman, 1970; Norrild *et al.*, 1982). Induction of locally produced antibodies in CSF in patients with encephalitis has also been followed, and HSV-specific antibodies of the IgG and IgA classes (although in small amounts) were demonstrated from 8 days after onset of the disease (Vaheri *et al.*, 1982; Vandvik *et al.*, 1982).

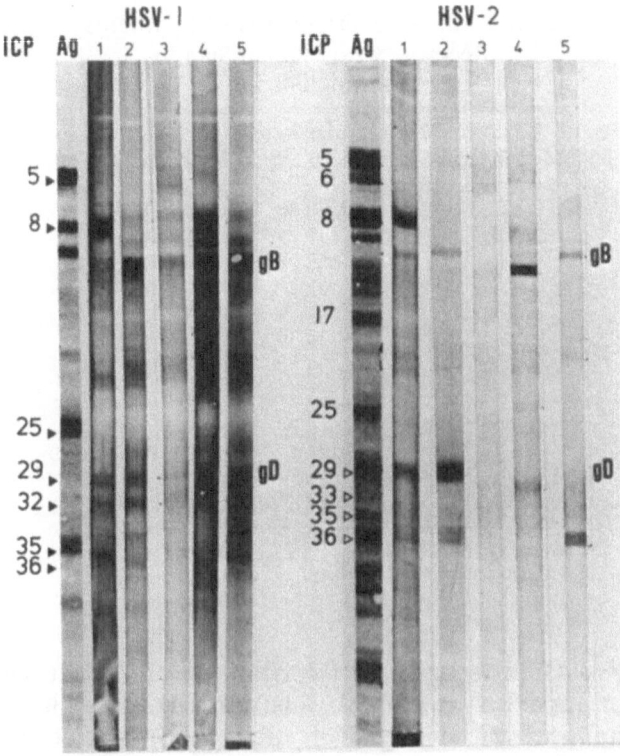

FIGURE 3. Analysis of human sera by immunoblotting. HSV-1 and HSV-2 proteins were ^{35}S-methionine-labeled from 6 to 8 hr postinfection, extracted, electrophoretically separated, and transferred to nitrocellulose paper as described by Braun *et al.* (1983). The strips were incubated with human sera diluted to an ELISA titer of 40, followed by binding of peroxidase-coupled rabbit anti-human IgG (DAKO, Copenhagen) used at a dilution of 1:300. The substrate used was *ortho*-dianisidine. The left panel shows the reactivity of five sera, 1–5, with HSV-1 proteins, and the right panel shows the reactivity of the same five sera with HSV-2 proteins. The strips labeled Ag show the autoradiographic picture of the HSV-1 and HSV-2 proteins, respectively, bound to the nitrocellulose paper. The ICPs are numbered to the left of each of the Ag strips and the leading edge of some of the proteins is marked with arrows for clarification. The position of the glycoproteins gB and gD is marked to the right of each panel. Serum No. 1 is from a patient with no clinical manifestation of herpes simplex; Nos. 2–5 are sera from patients with cervical cancer in stages I or II.

XI. FUNCTION OF HSV-SPECIFIC ANTIBODIES *IN VIVO*

It is well appreciated that recurrent HSV infections occur in the presence of circulating, specific antibodies, a finding that raises questions regarding the role of the humoral immune response *in vivo*. Inasmuch as circulating antibodies have both neutralizing and cytolytic potentials, and both complement and K cells are present in the peripheral blood, it could be expected that infectious virus and infected cells would be eliminated as soon as they were recognized, which was for infected cells pos-

TABLE III. Rank Order of HSV-1 Proteins
Reactive with Human Sera in
Immunoblotting

ICP[a]	Molecular weight $(\times 10^{-3})$	Sera positive[b]
32/33	51.5/50	16
11	114	15
29	55	15
20	77	14
18b	88	13
35a	45	12
36	42.5	12
18a	45	11
19	78	11
8	128	10
9	122	8
23	71	8

[a] Morse *et al.* (1978).
[b] Total number: 30.

sibly as early as 4 hr postinfection (Norrild *et al.*, 1980). It has been possible only in model systems to illustrate the possible function of antibodies in limiting the spread of virus and in protecting against a lethal dosage of HSV, experiments that were most often done in mice. Passive transfer of hyperimmune serum and leukocytes given intraperitoneally (i.p.) to newborn mice protected the animals against 10^6 PFU of virus also injected i.p. 1 day later (Kohl *et al.*, 1981). Monoclonal antibodies reactive with gC and gD of HSV-1 and injected i.p., protected mice against 10^6 PFU of HSV-1 and HSV-1/HSV-2, respectively, when challenged into the footpad (Dix *et al.*, 1981). Passive transfer of neutralizing polyclonal or monoclonal antibodies (made to gD) apparently interfered with the transport of HSV to the CNS, and when antibodies were injected intravenously before day 3 postinfection, virus was not transmitted to the ganglia or to the spinal cord of mice injected with virus in the ear pinna (Kapoor *et al.*, 1982b). In a similar study, monoclonal antibodies made to the HSV-2 glycoproteins also conferred protection when injected i.p. 3 hr before challenge with 10 times the lethal doses of HSV-2 given by the footpad route (Balachandran *et al.*, 1982). The authors conclude that an ADCC mechanism is responsible for the elimination of virus. Based on these studies there seems to be a specific effect of antibodies on the outcome of an HSV infection, but other reports do not support the findings obtained in the *in vivo* model system. In humans there is apparently no correlation between the presence of maternal HSV antibodies and the development of neonatal herpes (Whitley *et al.*, 1980). In an animal model, however, specifically designed to study the effect of maternal antibodies for the outcome of HSV infections in newborn mice, both immunization of the female mice and oral feeding of the newborn mice with human hyperimmune IgG given 24 and 4 hr before i.p. infection gave protection. Infection was done with 10^2 PFU on day 2 after birth (Hayashi *et al.*, 1983).

XII. CONCLUSION

One of the major trends in herpesvirus research is toward the development of a human vaccine that can prevent primary genital infections with HSV (Cappel *et al.*, 1982; Hilfenhaus *et al.*, 1982; Roizman *et al.*, 1982, 1984a; Skinner *et al.*, 1982). It was an early hypothesis that even though HSV-1 did not prevent later infections with HSV-2, it did prevent the development of cervical carcinoma etiologically related to HSV-2, and therefore a certain level of protection was obtained by the first HSV infection (Nahmias and Norrild, 1980). Although early studies found genital infections to be caused mainly by HSV-2, more recent studies have shown that a large number of primary genital infections are caused by HSV-1, especially in younger persons, infections that are less likely to give recurrence than HSV-2 infections (Kawana *et al.*, 1982; Ozaki *et al.*, 1980; Peutherer *et al.*, 1982). In one report recurrent HSV-2 infections were demonstrated more frequently in persons with high titers of neutralizing antibodies than in patients with low neutralizing titers, an observation that might reflect possible frequent, even silent recurrences giving rise to a stimulation of the antibody response (Reeves *et al.*, 1981).

The success of a vaccine will depend on its ability to confer immunity to HSV-1 and HSV-2 or to prevent recurrence of HSV already present in the CNS. Although a vaccine consisting of either a variant strain of HSV or purified protein may induce the production of specific antibodies, these are not alone likely to confer protection on the host organism. From animal studies the cell-mediated immune response is known also to be important for the elimination of infectious HSV (Kapoor *et al.*, 1982a,b; Sethi *et al.*, 1983), and cooperation between the humoral and the cellular immune systems seems necessary.

How, then, can we monitor the take of a vaccine? It will be optimal if both antibody and cellular parameters are followed, but preliminary studies in monkeys have shown that the level of precipitating antibodies induced by an HSV variant strain reflects the protection obtained against both HSV-1 and HSV-2 challenge (Roizman *et al.*, 1984).

Much has still to be learned about the status of HSV in an infected host before we will be able to understand how the antibodies and the stimulated lymphocytes can gain access to the infectious virus and to the infected cells in an organism.

ACKNOWLEDGMENTS. Financial support was obtained from the Danish Cancer Society and from P.O. Pedersens Foundation.

REFERENCES

Balachandran, N., Harmish, D., Killington, R. A., Bacchetti, S., and Rawls, W. E., 1981, Monoclonal antibodies to two glycoproteins of herpes simplex virus type 2, *J. Virol.* **39:**438.

Balachandran, N., Bacchetti, S., and Rawls, W. E., 1982, Protection against lethal challenge of BALB/c mice by passive transfer of monoclonal antibodies to five glycoproteins of herpes simplex virus type 2, *Infect. Immun.* **37:**1132.

Bauke, R. B., and Spear, P. G., 1979, Membrane proteins specified by herpes simplex viruses. V. Identification of an Fc-binding glycoprotein, *J. Virol.* **32:**779.

Braun, D. K., Pereira, L., Norrild, B., and Roizman, B., 1983, Application of denatured, electrophoretically separated, and immobilized lysates of herpes simplex virus-infected cells for detection of monoclonal antibodies and for studies of the properties of viral proteins, *J. Virol.* **46:**103.

Cappel, R., Sprecher, S., and de Cuyper, F., 1982, Immune responses to DNA free herpes simplex proteins in man, *Dev. Biol. Stand.* **52:**345.

Cassai, E. N., Sarmiento, M., and Spear, P. G., 1975, Comparison of the virion proteins specified by herpes simplex virus types 1 and 2, *J. Virol.* **16:**1327.

Centifanto, Y. M., and Kaufman, H. E., 1970, Secretory immunoglobulin A and herpes keratitis, *Infect. Immun.* **2:**778.

Cesario, T. C., Poland, J. D., Wulff, H., Chin, T. D. Y., and Wenner, H. A., 1969, Six years experience with herpes simplex virus in a children's home, *Am. J. Epidemiol.* **90:**416.

Christenson, B., 1978, Antibody-dependent cell-mediated cytotoxicity to herpes simplex virus type 2 infected target cells in the course of cervical carcinoma, *Am. J. Epidemiol.* **106:**126.

Cohen, G. H., Katze, M., Hydrean-Stern, C., and Eisenberg, R. J., 1978, Type-common CP-1 antigen of herpes simplex virus is associated with a 59,000-molecular-weight envelope glycoprotein, *J. Virol.* **27:**172.

Dix, R. D., Pereira, L., and Baringer, J. R., 1981, Use of monoclonal antibody directed against herpes simplex virus glycoproteins to protect mice against acute virus-induced neurological disease, *Infect. Immun.* **34:**192.

Doerr, H. W., Gross, H., Schmitz, H., and Enders, G., 1976, Neutralizing serum IgM antibodies in infections with *Herpes simplex* virus hominis, *Med. Microbiol. Immunol.* **162:**83.

Douglas, R. G., and Couch, R. B., 1970, A prospective study of chronic herpes simplex virus infection and recurrent herpes labialis in humans, *J. Immunol.* **104:**289.

Eberle, R., and Courtney, R. J., 1981, Assay of type-specific antibodies to herpes simplex virus types 1 and 2 in human sera, *Infect. Immun.* **31:**1062.

El Falaky, I. H., and Vestergaard, B. F., 1977, IgG-, IgA- and IgM-antibodies to herpes simplex virus type 2 in sera from patients with cancer of the uterine cervix, *Eur. J. Cancer* **13:**247.

Friedman, M. G., and Kimmel, N., 1982, Herpes simplex virus-specific serum immunoglobulin A: Detection in patients with primary or recurrent herpes infections and in healthy adults, *Infect. Immun.* **37:**374.

Fujinami, R. S., Oldstone, M. B. A., Wroblewska, Z., Frankel, M. E., and Koprowski, H., 1983, Molecular mimicry in virus infections: Crossreaction of measles virus phosphoprotein or of herpes simplex virus protein with human intermediate filaments, *Proc. Natl. Acad. Sci. USA* **80:**2346.

Geder, L., and Skinner, G. R. B., 1971, Differentiation between type 1 and type 2 strains of herpes simplex virus by an indirect immunofluorescent technique, *J. Gen. Virol.* **12:**179.

Gilman, S. C., Docherty, J. J., and Rawls, W. E., 1981, Antibody responses in humans to individual proteins of herpes simplex viruses, *Infect. Immun.* **34:**880.

Grönroos, M., Honkonen, E., Terho, P., and Punnonen, R., 1983, Cervical and serum IgA and serum IgG antibodies to *Chlamydia trachomatis* and herpes simplex virus in threatened abortion: A prospective study, *Br. J. Obstet. Gynaecol.* **90:**167.

Hampar, B., Notkins, A. L., Mage, M., and Keehn, M. A., 1968, Heterogeneity in properties of 7 S and 19 S rabbit-neutralizing antibodies to herpes simplex virus, *J. Immunol.* **100:**586.

Hayashi, Y., Wada, T., and Mori, R., 1983, Protection of newborn mice against herpes simplex virus infection by prenatal and postnatal transmission of antibody, *J. Gen. Virol.* **64:**1007.

Heine, J. W., Spear, P. G., and Roizman, B., 1972, Proteins specified by herpes simplex virus. VI. Viral proteins in the plasma membrane, *J. Virol.* **9:**431.

Heine, J. W., Honess, R. W., Cassai, E., and Roizman, B., 1974, Proteins specified by herpes simplex virus. XII. The virion polypeptides of type 1 strains, *J. Virol.* **14:**640.

Hilfenhaus, J., Moser, H., Herrmann, A., and Mauler, R., 1982, Herpes simplex virus subunit vaccine, characterization of the virus strain used and testing of the vaccine, *Dev. Biol. Stand.* **52:**321.

Hoggan, H. D., and Roizman, B., 1959, The isolation and properties of a variant of herpes simplex producing multinucleated giant cells in monolayer cultures in the presence of antibody, *Am. J. Hyg.* **70:**208.

Jordan, J., and Rytel, M. W., 1981, Detection of herpes simplex virus (HSV) type-1 IgG and IgM antibodies by enzyme-linked immunosorbent assay (ELISA), *Am. J. Clin. Pathol.* **76:**467.

Kalimo, K. O. K., Marttila, R. J., Granfors, K., and Viljanen, M. K., 1977, Solid-phase radioimmunoassay of human immunoglobulin M and immunoglobulin G antibodies against herpes simplex virus type 1 capsid, envelope, and excreted antigens, *Infect. Immun.* **15:**883.

Kalimo, K., Terho, P., Honkonen, E., Grönroos, M., and Halonen, P., 1981, Chlamidya trachomatis and herpes simplex virus IgA antibodies in cervical secretions of patients with cervical atypia, *Br. J. Obstet. Gynaecol.* **88:**130.

Kapoor, A. K., Ling, N. R., Nash, A. A., Buchan, A., and Wildy, P., 1982a, In vitro stimulation of rabbit T lymphocytes by cells expressing herpes simplex antigens, *J. Gen. Virol.* **59:**415.

Kapoor, A. K., Nash, A. A., Wildy, P., Phelan, J., Mclean, C. S., and Field, H. J., 1982b, Pathogenesis of herpes simplex virus in congenitally athymic mice: The relative roles of cell-mediated and humoral immunity, *J. Gen. Virol.* **60:**225.

Kawana, T., Kawagoe, K., Takizawa, K., Chen, J. T., Kawaguchi, T., and Sakamoto, S., 1982, Clinical and virologic studies on female genital herpes, *Obstet. Gynecol.* **60:**456.

Kieff, E. D., Bachenheimer, S. L., and Roizman, B., 1971, Size, composition, and structure of the deoxyribonucleic acid of herpes simplex virus subtypes 1 and 2, *J. Virol.* **8:**125.

Kieff, E., Hoyer, B., Bachenheimer, S., and Roizman, B., 1972, Genetic relatedness of type 1 and type 2 herpes simplex viruses, *J. Virol.* **9:**738.

Kimmel, N., Friedman, M. G., and Sarov, I., 1982, Enzyme-linked immunosorbent assay (ELISA) for detection of herpes simplex virus-specific IgM antibodies, *J. Virol. Methods* **4:**219.

Kohl, S., Loo, L. S., and Pickering, L. K., 1981, Protection of neonatal mice against herpes simplex viral infection by human antibody and leukocytes from adult, but not neonatal humans, *J. Immunol.* **127:**1273.

Kohl, S., Adam, E., Matson, D. O., Kaufman, R. H., and Dreesman, G. R., 1982, Kinetics of human antibody response to primary genital herpes simplex virus infection, *Intervirology* **18:**164.

Ludwig, H. O., Biswal, N., and Benyesh-Melnick, M., 1972, Studies on the relatedness of herpesviruses through DNA–DNA hybridization, *Virology* **49:**95.

Mann, D. R., and Hilty, M. D., 1982, Antibody response to herpes simplex virus type 1 polypeptides and glycoproteins in primary and recurrent infection, *Pediatr. Res.* **16:**176.

Morse, L. S., Pereira, L., Roizman, B., and Schaffer, P. A., 1978, Anatomy of herpes simplex virus (HSV) DNA. X. Mapping of viral genes by analysis of polypeptides and functions specified by HSV-1 × HSV-2 recombinants, *J. Virol.* **26:**389.

Nahmias, A. J., and Norrild, B., 1980, Oncogenic potential of herpes simplex viruses and their association with cervical neoplasia, in: *Oncogenic Herpesviruses*, Vol. II (F. Rapp, ed.), pp. 25–45, CRC Press, Boca Raton, Fla.

Nahmias, A. J., and Roizman, B., 1973a, Infection with herpes-simplex viruses 1 and 2 (First of Three Parts), *N. Engl. J. Med.* **289**:667.

Nahmias, A. J., and Roizman, B., 1973b, Infection with herpes-simplex viruses 1 and 2 (Second of Three Parts), *N. Engl. J. Med.* **289**:719.

Nahmias, A. J., and Roizman, B., 1973c, Infection with herpes-simplex viruses 1 and 2 (Third of Three Parts), *N. Engl. J. Med.* **289**:781.

Nahmias, A., DelBuono, I., Pipkin, J., Hutton, R., and Wickliffe, C., 1971a, Rapid identification and typing of herpes simplex virus types 1 and 2 by a direct immunofluorescence technique, *Appl. Microbiol.* **22**:455.

Nahmias, A. J., DelBuono, I., Schneweis, K. E., Gordon, D. S., and Thies, D., 1971b, Type-specific surface antigens of cells infected with herpes simplex virus (1 and 2), *Proc. Soc. Exp. Biol. Med.* **138**:21.

Nahmias, A. J., Naib, Z. M., and Josey, W. E., 1974, Epidemiological studies relating genital herpetic infection to cervical carcinoma, *Cancer Res.* **34**:1111.

Norrild, B., 1980, Immunochemistry of herpes simplex virus glycoproteins, *Curr. Top. Microbiol. Immunol.* **90**:67.

Norrild, B., Shore, S. L., Cromeans, T. L., and Nahmias, A. J., 1980, Participation of three major glycoprotein antigens of herpes simplex virus type 1 early in the infectious cycle as determined by antibody-dependent cell-mediated cytotoxicity, *Infect. Immun.* **28**:38.

Norrild, B., Pedersen, B., and Møller-Andersen, S., 1982, Herpes simplex virus specific secretory IgA in lacrimal fluid during herpes keratitis, *Scand. J. Clin. Lab. Invest.* **42**(Suppl. 161):29.

Norrild, B., Emmertsen, H., Krebs, H. J., and Pedersen, B., 1984, Antibody dependent immune mechanisms and herpes simplex virus infections, in: *Immune Mechanisms in Herpesvirus Infections* (B. Rouse, ed.), pp. 91–105, CRC Press, Boca Raton, Fla.

Notkins, A. B., 1971, Immunological injury of virus-infected cells by antiviral antibody and complement, in: *Progress in Immunology* (B. Amos, ed.), p. 779, North-Holland, Amsterdam.

Ohashi, M., and Ozaki, Y., 1981, Studies on the neutralizing antibody to herpes simplex virus. I. Effect of heterotypic antigenic stimulus on the type specificity of neutralizing antibody in rabbits, *Arch. Virol.* **67**:57.

Oldstone, M. B. A., 1977, Role of antibody in regulating virus persistence: Modulation of viral antigens expressed on the cell's plasma membrane and analyses of cell lysis, in: *Development of Host Defences* (M. D. Cooper and D. H. Dayton, eds.), pp. 223–235, Raven Press, New York.

Ozaki, Y., Ishiguro, T., Ohashi, M., and Kimura, E. M., 1980, Relationship between antigenic type of virus and antibody response in female patients with herpes genitalis, *J. Med. Virol.* **5**:249.

Pauls, F. P., and Dowdle, W. R., 1967, A serologic study of herpesvirus hominis strains by microneutralization tests, *J. Immunol.* **98**:941.

Pereira, L., Cassai, E., Honess, R. W., Roizman, B., Terni, M., and Nahmias, A., 1976, Variability in the structural polypeptides of herpes simplex virus 1 strains: Potential application in molecular epidemiology, *Infect. Immun.* **13**:211.

Peutherer, J. F., Smith, I. W., and Robertson, D. H. H., 1982, Genital infection with herpes simplex virus type 1, *J. Infect.* **4**:33.

Plummer, G., 1973, Review of the identification and titration of antibodies to herpes simplex viruses type 1 and type 2 in human sera, *Cancer Res.* **33**:1469.

Powell, K. L., Buchan, A., Sim, C., and Watson, D. H., 1974, Type-specific protein in herpes simplex virus envelope reacts with neutralizing antibody, *Nature (London)* **249**:360.

Powell, K. L., Mirkovic, R., and Courtney, R. J., 1977, Comparative analysis of polypeptides induced by type 1 and type 2 strains of herpes simplex virus, *Intervirology* **8**:18.

Rager-Zisman, B., and Bloom, B. R., 1974, Immunological destruction of herpes simplex virus infected cells, *Nature* **251**:542.

Rawls, W. E., 1973, Herpes simplex virus, in: *The Herpesviruses* (A. Kaplan, ed.), pp. 291–325, Academic Press, New York.

Rawls, W. E., and Tompkins, W. A. F., 1975, Destruction of virus-infected cells by antibody and complement, in: *Viral Immunology and Immunopathology* (A. L. Notkins, ed.), pp. 99–112, Academic Press, New York.

Rawls, W. E., Bacchetti, S., and Graham, F. L., 1977, Relation of herpes simplex viruses to human malignancies, *Curr. Top. Microbiol. Immunol.* **87**:71.

Rawls, W. E., Clarke, A., Smith K. O., Docherty, J. J., Gilman, S. C., and Graham, S., 1980, Specific antibodies to herpes simplex virus type 2 among women with cervical cancer, in: *Viruses in Naturally Occurring Cancers* (M. Essex, G. Todaro, and H. zur Hausen, eds.), pp. 117–133, Cold Spring Harbor Laboratory, New York.

Reeves, W. C., Corey, L., Adams, H. G., Vontver, L. A., and Holmes, K. K., 1981, Risk of recurrence after first episodes of genital herpes: Relation to HSV type and antibody response, *N. Engl. J. Med.* **305**:315.

Roizman, B., and Roane, P. R., Jr., 1961, Studies of the determinant antigens of viable cells. I. A method, and its application in tissue culture studies, for enumeration of killed cells, based on the failure of virus multiplication following injury by cytotoxic antibody and complement, *J. Immunol.* **87**:714.

Roizman, B., Spear, P. G., and Kieff, E. D., 1973, Herpes simplex viruses I and II: A biochemical definition, in: *Persistent Virus Infections* pp. 129–166, Academic Press, New York.

Roizman, B., Warren, J., Thuning, C. A., Fanshaw, M. S., Norrild, B., and Meignier, B., 1982, Application of molecular genetics to the design of live herpes simplex virus vaccines, *Dev. Biol. Stand.* **52**:287.

Roizman, B., Meignier, B., Norrild, B., and Wagner, J. L., 1984a, Bioengineering of herpes simplex virus variants for potential use as live vaccines, in: *Modern Approaches to Vaccines* (R. Chanock and R. Lerner, eds.), Cold Spring Harbor Laboratory, New York.

Roizman, B., Norrild, B., Chan, E., and Pereira, L., 1984b, Identification and preliminary mapping with monoclonal antibodies of a Herpes simplex virus 2 glycoprotein lacking a known type 1 counterpart, *Virol.*, **133**:242.

Rouse, B. T., and Lawman, M. J. P., 1980, Induction of cytotoxic T lymphocytes against herpes simplex virus type I: Role of accessory cells and amplifying factor, *J. Immunol.* **124**:2341.

Schneweis, K. E., 1962, Serologische Untersuchungen zur Typendifferenzierung des Herpesvirus hominis, *Z. Immunitaetsforsch.* **124**:24.

Schneweis, K. E., and Nahmias, A. J., 1971, Antigens of herpes simplex virus type 1 and 2—Immunodiffusion and inhibition passive hemagglutination studies, *Z. Immunitaetsforsch.* **141**:471.

Sethi, K. K., Omata, Y., and Schneweis, K. E., 1983, Protection of mice from fatal herpes simplex virus type 1 infection by adoptive transfer of cloned virus-specific and H-2-restricted cytotoxic T lymphocytes, *J. Gen. Virol.* **64**:443.

Shore, S. L., Black, C. M., Melewicz, F. M., Wood, P. A., and Nahmias, A. J., 1976, Antibody-dependent cell-mediated cytotoxicity to target cells infected with type 1 and type 2 herpes simplex virus, *J. Immunol.* **116**:194.

Shore, S. L., Milgrom, H., Wood, P., and Nahmias, A. J., 1977, Neonatal function of antibody-dependent cell-mediated cytotoxicity to target cells infected with herpes simplex virus, *Pediatrics* **59**:22.

Skinner, G. R. B., Woodman, C., Hartley, C., Buchan, A., Fuller, A., Wiblin, C., Wilkins, G., and Melling, J., 1982, Early experience with "antigenoid" vaccine Ac NFU$_1$ (S$^-$) MRC towards prevention or modification of herpes genitalis, *Dev. Biol. Stand.* **52**:333.

Smith, J. W., Adams, E., Melnick, J. L., and Rawls, W. E., 1972, Use of the ^{51}Cr release test to demonstrate patterns of antibody response in humans to herpesvirus types 1 and 2, *J. Immunol.* **109**:554.

Spear, P. G., 1976, Membrane proteins specified by herpes simplex viruses. I. Identification of four glycoprotein precursors and their products in type 1-infected cells, *J. Virol.* **17**:991.

Spear, P. G., and Roizman, B., 1972, Proteins specified by herpes simplex virus. V. Purification and structural proteins of the herpesvirion, *J. Virol.* **9**:143.

Subramanian, T., and Rawls, W. E., 1977, Comparison of antibody-dependent cellular cytotoxicity and complement-dependent antibody lysis of herpes simplex virus-infected cells as methods of detecting antiviral antibodies in human sera, *J. Clin. Microbiol.* **5**:551.

Tokumaru, T., 1965, Studies of herpes simplex virus by the gel diffusion technique. I. Distribution of precipitating antibodies among human sera, *J. Immunol.* **95**:181.

Tokumaru, T., 1966, A possible role of γA-immunoglobulin in herpes simplex virus infection in man, *J. Immunol.* **97**:248.

Vaheri, A., Keski-Oja, J., Salonen, E. M., and Kiskiniemi, M. L., 1982, Cerebrospinal fluid IgG bands and virus-specific IgG, IgM, and IgA antibodies in herpes simplex virus encephalitis, *J. Neuroimmunol.* **3**:247.

Vandvik, H., Vartdal, F., and Norrby, F., 1982, Herpes simplex virus encephalitis: Intrathecal synthesis of oligoclonal virus-specific IgG, IgA and IgM antibodies, *J. Neurol.* **228**:25.

Vass-Sørensen, M., Abeler, V., Berle, E., Pedersen, B., Davy, M., Thorsby, E., and Norrild, B., 1984, Prevalence of antibodies in herpes simplex virus and frequency of HLA-antigens in patients with preinvasive and invasive cervical cancer, *Gynecol. Oncol.* in press.

Vestergaard, B. F., 1979, Quantitative immunoelectrophoretic analysis of human antibodies against herpes simplex virus antigens, *Infect. Immun.* **23**:553.

Vestergaard, B. F., 1980, Herpes simplex virus antigens and antibodies: A survey of studies based on quantitative immunoelectrophoresis, *Rev. Infect. Dis.* **2**:899.

Vestergaard, B. F., and Grauballe, P. C., 1979, ELISA for herpes simplex virus (HSV) type-specific antibodies in human sera using HSV type 1 and type 2 polyspecific antigens blocked with type-heterologous rabbit antibodies, *Acta Pathol. Microbiol. Scand. Sect. B* **87**:261.

Vestergaard, B. F., and Norrild, B., 1978, Crossed immunoelectrophoresis of a herpes simplex virus type 1-specific antigen: Immunological and biochemical characterization, *J. Infect. Dis.* **138**:639.

Vestergaard, B. F., Hornsleth, A., and Pedersen, S. N., 1972, Occurrence of herpes- and adenovirus antibodies in patients with carcinoma of the cervix uteri: Measurement of antibodies to herpesvirus hominis (types 1 and 2), cytomegalovirus, EB-virus, and adenovirus, *Cancer* **30**:1201.

Wallis, C., and Melnick, J. L., 1971, Herpesvirus neutralization: The role of complement, *J. Immunol.* **107**:1235.

Wentworth, B. B., and Alexander, E. R., 1971, Seroepidemiology of infections due to members of the herpesvirus group, *Am. J. Epidemiol.* **94**:496.

Whitley, R. J., Nahmias, A. J., Visintine, A. M., Fleming, C. L., and Alford, C. A., 1980, The natural history of herpes simplex virus infection of mother and newborn, *Pediatrics* **66**:489.

Wildy, P., Field, H. J., and Nash, A. A., 1982, Classical herpes latency revisited, in: *Virus Persistence* (B. W. J. Mahy, A. C. Minson, and G. K. Darby, eds.), pp. 133–167, Cambridge University Press, London.

Yoshino, K., and Taniguchi, S., 1965, Studies on the neutralization of herpes simplex virus. I. Appearance of neutralizing antibodies having different grades of complement requirement, *Virology* **26**:44.

Yoshino, K., Hashimoto, M., and Shinkai, K., 1977, Studies on the neutralization of herpes simplex virus. VIII. Significance of viral sensitization for inactivation by complement, *Microbiol. Immunol.* **21**:231.

Yoshino, K., Isono, N., Tada, A., and Urayama, M., 1979, Studies on the neutralization of herpes simplex virus. X. Demonstration of complement-requiring neutralizing CRN and slow-reacting CRN (s-CRN) antibodies in late IgG, *Microbiol. Immunol.* **23**:975.

Zweerink, H. J., and Corey, L., 1982, Virus-specific antibodies in sera from patients with genital herpes simplex virus infection, *Infect. Immun.* **37**:413.

CHAPTER 4

The T-Cell-Mediated Immune Response of Mice to Herpes Simplex Virus

A. A. NASH, K.-N. LEUNG, AND P. WILDY

I. INTRODUCTION

The host response to infection with herpes simplex virus (HSV) in which both natural resistance and specific immunological responses become marshaled is clearly very complex. Natural resistance mechanisms, mediated for example by macrophages, NK cells, and interferon, represent a formidable early barrier to the virus (Lopez, Chapter 2, this volume), although the eventual resolution of the infectious process and the establishment of long-lasting antiviral defense is the property of the immune system. Whereas both humoral and cell-mediated responses are readily induced by the virus, it is clear that the action of T cells is central for recovery from the primary infection and in the control of recrudescent lesions. Much of this evidence comes from studies carried out in the mouse in which the precise nature of the T-cell response to herpes can readily be studied. In this respect the variety of inbred and congenic mouse strains has been invaluable in studying MHC restriction of T-cell responses; and the availability of monoclonal antibodies to specific T-cell subsets has enabled a detailed analysis of the antiviral T-cell response to be carried out.

The purpose of this review is to consider the available evidence on the role of T cells in experimental HSV infections. Included in the survey will be the nature of the T-cell subsets induced, their role in antiviral

A. A. NASH, K.-N. LEUNG, AND P. WILDY • Department of Pathology, University of Cambridge, Cambridge CB2 1QP, England.

immunity, and the regulation of immune responses by suppressor cells. (See the Appendix for notes on the immunological terms used herein.)

II. EVIDENCE FOR THYMUS-DEPENDENT RESISTANCE TO HERPES INFECTIONS

A. Use of Athymic Mice and Immunosuppression

In studying resistance mechanisms of animals to virus infections, the relevance of T-cell immunity has often been assessed by studying infections in athymic mice. Neonatally thymectomized mice were shown to be considerably more susceptible to HSV-1, whether previously immunized against the virus or not (Mori et al., 1967). On the other hand, adult thymectomy does not appear to predispose the animal to a more severe herpetic infection (A. Nash and P. Gell, unpublished observations). This is perhaps not too surprising as one assumes that a sufficient number of herpes-specific T cells have been seeded to the peripheral lymphoid tissues shortly after birth.

In mice with a congenital defect of the thymus (athymic nude mice), infection with HSV-1 or HSV-2 intradermally (i.d.), subcutaneously (s.c.), or intraocularly leads to a progressive infection of the peripheral and central nervous system, resulting in death some 2–3 weeks later (Kapoor et al., 1982b; Nagafuchi et al., 1979; Metcalf et al., 1979). Kapoor and colleagues made a detailed study of the progression of virus to the nervous system and showed that the passive transfer of neutralizing monoclonal antibodies to nude mice, 3 days after an i.d. injection of HSV-1, markedly reduced the infective virus titer in sensory ganglia and the spinal cord when assayed 11 days later. However, it was only following the transfer of immune T cells (taken from syngeneic donors 7 days after infection) that the infection was sufficiently controlled, resulting in the long-term survival of the mice.

Experiments akin to this have been described in mice immunosuppressed by either X-irradiation (Oakes, 1975) or cyclophosphamide (Rager-Zisman and Allison, 1976). Such treated animals are highly susceptible to herpes infection, but could be protected from death by the transfer of immune spleen cells. In the system described by Rager-Zisman and Allison, protection was abolished if the cell suspension was treated with anti-Thy-1 serum plus complement. Furthermore, treatment of mice with antilymphocyte or antithymocyte globulin predisposed the animals to severe herpes infections, resulting in a high mortality (Oakes, 1975). The experiments described above on athymic mice and the protection of immunocompromised mice by immune T cells clearly indicate the importance of thymus-derived lymphocytes in protection against HSV infections.

B. Use of B-Cell-Suppressed Mice

The B-cell response of mice can be inhibited by the regular administration of anti-IgM antibody, starting within the first 12 hr of birth (Gordon, 1979). Treated mice develop no detectable serum IgM, and have depleted B-cell-dependent areas in the spleen and lymph node. Splenic lymphocyte preparations fail to respond to B-cell mitogens, such as LPS; whereas the response to Con A (a T-cell mitogen) is unaffected. At 4 weeks of age, antibody-treated and untreated BALB/c mice were injected in the ear flap with 10^4 PFU HSV-1 (strain SC16). The virus titer in the ear was assayed at various times and found to be similar in both the anti-IgM-treated and untreated groups. The B-cell-suppressed mice failed to produce anti-HSV-1-specific antibodies, which implies that the control of the primary infection was not dependent upon antibody-mediated mechanisms (Kapoor *et al.*, 1982a). Observations similar to those described above were reported by Burns (1977).

Although B-cell-suppressed mice recovered from the primary infection, they were predisposed to a higher incidence of latent infection and appeared to have a more florid primary infection of the peripheral and central nervous system (Kapoor *et al.*, 1982a). Such evidence might accord with the view that a major role for T cells exists in controlling the cutaneous infection and a role for B cells and antibody in restricting access of virus to and from the nervous system.

III. T-CELL SUBSETS INDUCED IN HERPES INFECTIONS

A. Cytotoxic T Lymphocytes

Cytotoxic T lymphocytes (Tc) are observed in mice following injections of HSV-1 or HSV-2 via intravenous (i.v.), intraperitoneal (i.p.), or s.c. routes. Herpes-specific Tc cells are first detected in draining lymph nodes 4 days after s.c. injection of virus (Pfizenmaier *et al.*, 1977; Nash *et al.*, 1980b); peak activity is reached by day 6–7 beyond which the killer response declines, being undetectable after day 14 (Fig. 1). Following infection with HSV-1, the Tc precursor frequency increases from 1/250,000 up to 1/3500 to 1/15,000 (Rouse *et al.*, 1983). A feature of the Tc cells taken from herpes-infected mice is a requirement for 2–3 days' *in vitro* culture before becoming functional killer cells. The reason for this requirement is unclear. It has been suggested that suppressor cells inhibit the expression of Tc cells *in vivo*, for pretreatment of mice with cyclophosphamide abolishes the requirement for *in vitro* culture (Pfizenmaier *et al.*, 1977). This assumes that such suppressor activity is lost *in vitro* and Tc cells can develop following a stimulus from the culture medium or factors derived from other cells present. Herpes-specific Tc cells with direct killer activity (i.e., *in vitro* culture unnecessary) have been observed

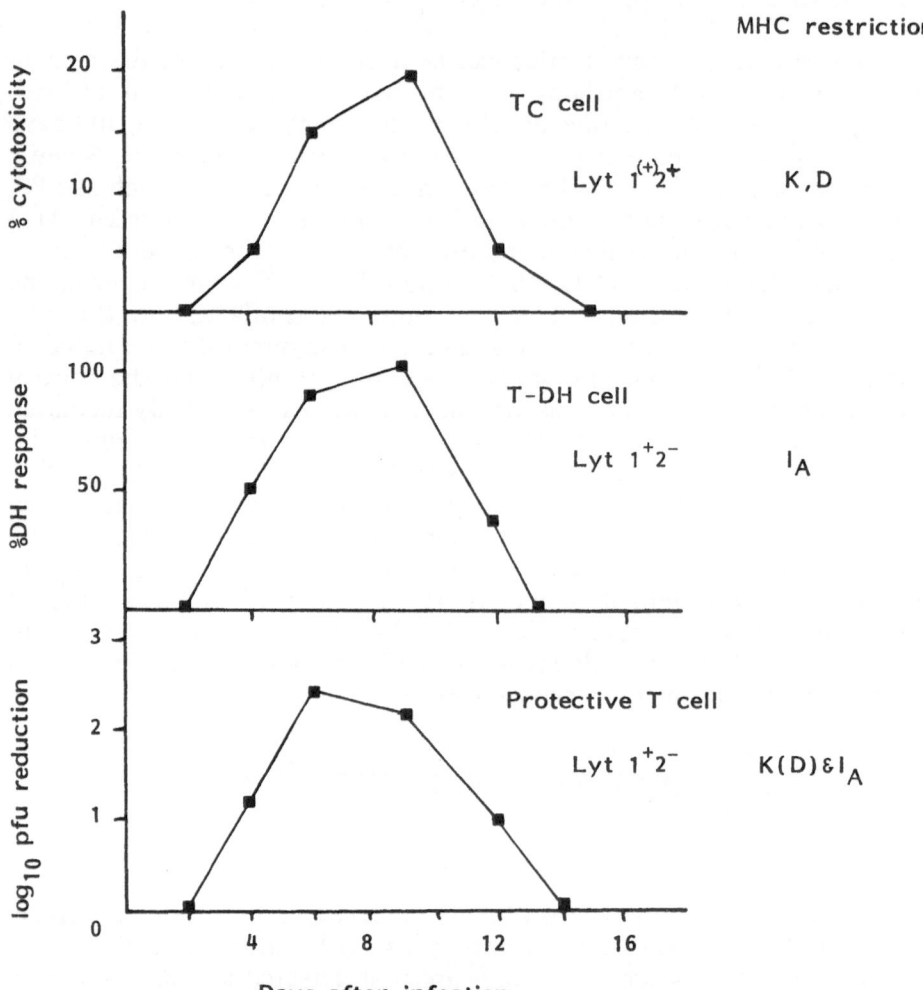

FIGURE 1. The properties of functional T-cell subsets obtained at various time intervals from the lymph node draining the ear pinna of mice injected with 10^5 PFU HSV-1. The Tc response shows the percentage specific killing of syngeneic targets by effector cells cultured for 3 days *in vitro*. T-DH cell response shows the percentage increase in ear thickness measured 2 days after transfer of 2×10^7 lymph node cells into syngeneic mice injected in the ear with 10^4 PFU HSV-1. The "protective" T-cell response shows the reduction in virus titer, compared to controls (no cell transfer), measured 4 days after transfer of 2×10^7 lymph node cells to mice injected in the ear with 10^4 PFU HSV-1. Lyt phenotyping of the DH and protective T-cell response is described in Nash and Gell (1983) and the MHC matching in Nash *et al.* (1981c).

in mice injected with monoclonal antibodies against gC or gD just after infection with HSV-1 (Sethi, 1983); the mechanism behind this effect is, however, difficult to explain.

Herpes-specific Tc cells are restricted to class I MHC gene products; for example, C57BL/10 (H-2^b) mice have a preference for recognizing K^b gene products in association with virus antigen (Jennings, et al., 1984). The precursor Tc cells are Lyt-1$^+$23$^+$ and the effector Tc cells can be destroyed with anti-Lyt-2 antibodies and complement (Rouse et al., 1983; A. K. Jayasuriya, personal communication). The induction of the Tc response is dependent upon injection of infective virus but not of inactivated (killed) virus (Rouse et al., 1983). Furthermore, the maintenance of Tc cell lines in vitro for long periods also requires the presence of infective virus and suitable presenting or stimulator cells and T-cell growth factor (Leung, personal communication).

Following induction in vivo, Tc cells will recognize both HSV-1- and HSV-2-infected syngeneic target cells (Nash and Ashford, 1982; Eberle et al., 1981; Carter et al., 1982). However, it has been shown that the total Tc cell response to virus is composed of lymphocyte clones that recognize either common or type-specific determinants. The type-specific Tc response is largely against gC (Eberle et al., 1981; J. Glorioso, personal communication). Indeed, recent work by Glorioso (personal communication) suggests that the bulk of the cytotoxic T-cell response is against gC on infected cells.

B. T Cells Mediating Delayed Hypersensitivity

The nature of the delayed hypersensitivity T-cell (T-DH) response to viruses has recently been reviewed (Nash and Gell, 1981). Herpes-specific T-DH cells are induced following injection of either infective or UV-inactivated virus s.c. or i.d. Other routes, by and large, fail to sensitize for these cells when measured by a subsequent skin test. In lymph nodes draining a site of infection, T-DH cells are first detected by day 4, as determined by adoptive transfer of cells to virus-infected recipients and measuring the increase in ear swelling 48 hr later. (This represents a classical tuberculin-type response, in which the infiltrate is composed almost entirely of mononuclear cells.) Maximum transferable activity is found after 6–7 days, followed by a decline until day 12 when the response is no longer transferable (Nash et al., 1980a) (Fig. 1) (but see Section VI on T-cell memory).

The adoptive transfer of DH is restricted by class II antigens (H-2IA), with no involvement of cells restricted by class I MHC genes (Nash et al., 1981c). The cell type transferring the response is Thy-1$^+$, Lyt-1$^+$23$^-$ and does not appear to involve a Lyt-1$^-$23$^+$ subpopulation. Mice injected with HSV-1 produce T-DH cells that are cross-reactive for type 2 virus (Nash et al., 1981a; Schrier et al., 1983a), although type-specific T-cell responses have been demonstrated using isolated gC from HSV-1 (Schrier *et al.*, 1983a). In these studies gC could both sensitize for and elicit a type-specific DH response. So, as with the Tc response, both type-specific and type-common T-DH clones exist in the infected animal.

C. T-Helper Lymphocyte Response

The induction of an antiherpes antibody response is thymus depen-dent (Burns *et al.*, 1975), as demonstrated by the failure of athymic nude mice to produce herpes-specific antibodies. Little information has been forthcoming on T–B-cell cooperation and in particular on the nature of the T-helper cell. Recently, however, Leung *et al.* (1984) have developed a number of T-cell clones specific for HSV some of which have helper activities. These clones are Lyt-1$^+$23$^-$ and I region restricted and are maintained in the presence of UV-inactivated-virus-infected stimulator cells and T-cell growth factor. The cloned line designated D7.1 produces a helper response for herpes-primed B cells when injected with virus into irradiated mice. The antibody response produced appears to be polyclonal, i.e., antibodies to a number of glycoproteins are produced, implying that the help in this case is nonspecific. The clone is, however, specific for herpes, although it does not distinguish between the two herpes types. Clone D7.1 also helps in the induction of DH *in vitro*. In this instance, normal spleen cells, clone D7.1, and appropriate X-irradiated feeders and growth factors are incubated for 5 days and viable cells are injected into the footpad of mice together with virus antigen. The DH produced is considerably increased when compared to the response produced in the absence of D7.1 (D7.1 by itself does not produce local footpad swelling). Clearly this particular clone is capable of helping in a variety of herpes-specific immunological responses.

T-cell help has also been reported for the induction of herpes-specific Tc responses. In this instance, helper cells are induced *in vivo* with in-fective virus, but not UV or heat-inactivated virus, and cultured with virus-infected stimulator cells and normal spleen cells *in vitro* (Schmid and Rouse, 1983). Tc cells are induced by this procedure after 4–5 days.

The T-helper cell therefore represents a key function in the rapid induction of specific antiviral defenses. Such a function is likely to be particularly relevant in secondary immune responses to recurrent infec-tions.

D. T-Suppressor Cells

The i.v. injection of infective or UV-inactivated HSV leads to the induction of T-suppressor (Ts) cells specific for herpes-induced DH re-sponses. At least two distinct Ts cell populations can be induced, one following injection of virus-infected cell sonicates, which produce Ts acting on the induction of DH responses (T-afferent suppression), and another Ts cell, which acts on an established DH response (T-efferent suppression) is induced following the injection of virus-infected, syngeneic spleen cell preparations (Nash et al., 1981b; Schrier et al., 1983b). T cells transferring suppressor activity to naive recipients appear in the spleens of donor animals 7 days after injection of virus. At this stage the cells suppressing the induction of DH to herpes are characterized as Thy-1.2$^+$, Lyt-1$^+$23$^-$, IJ$^+$, whereas at 28 days postinfection there appear to be two populations capable of transferring suppression: one, Lyt-1$^+$23$^-$, IJ$^+$ and another Lyt-1$^-$23$^+$, IJ$^+$ (Nash and Gell, 1983). The early Ts cell population once induced appears to be sensitive to cyclophosphamide treatment, whereas by day 28 the suppressor cells are resistant to the drug. Once induced, suppression appears to last for the life of the animal, an unusual feature as suppressor cells induced to other antigens are usually short-lived. However, this long-lived response might result from a constant or periodical stimulation by recurrent virus.

In contrast to other herpes-specific T-cell responses, Ts cells appear to be herpes type-specific. That is to say, following injection of type 1 herpes the Ts cells induced will suppress a type 1 DH response but not type 2 DH (Nash et al., 1981b). At present the epitope(s) involved is not known, although suppression can still be induced by gC-deficient virus mutants (A. Nash, unpublished observations).

Another intriguing feature of this model system is that suppression appears to be specific for DH responses; Tc cells and antiherpes antibodies are coinduced and the animals are resistant to infection. Consequently, a form of split-tolerance or suppression exists (Nash and Ashford, 1982). It would therefore seem that the route of infection influences the type of immunological response that is induced, and that this might reflect different modes of antigen presentation.

A summary of the properties of T lymphocyte subpopulations induced by HSV is presented in Table I.

IV. T CELLS INVOLVED IN ANTIHERPES RESPONSES *IN VIVO*

Before discussing the nature of the T-cell subsets involved in herpes infections of mice, it is important to define the model system used, for as already discussed the route of infection greatly influences the type of T-cell response induced. Where possible, animal models that have some relevance to the natural infection in man will be considered.

Of the various models illustrating the cutaneous HSV-1 infection, perhaps the best characterized is the mouse ear model (Hill et al., 1975). Briefly, the main features of this model are as follows: virus is inoculated s.c. into the ear flap, cells in the epidermis become infected, and the virus

TABLE I. Properties of Functional T-Lymphocyte Subpopulations Induced in HSV Infections

T-cell type	State of virus used in induction[a]	Route of induction	Membrane phenotype	MHC restriction	Specificity for virus type[b]
Tc	Infective	Various	$Lyt\text{-}1^-23^+$	Class I	Common
T-DH	Infective, noninfective	s.c. (predominantly)	$Lyt\text{-}1^+23^-$	Class II	Common
Th	Infective, noninfective	Various	$Lyt\text{-}1^+23^-$	Class II	Common
Ts	Infective, noninfective	i.v.	$Lyt\text{-}1^+23^-$, $Lyt\text{-}1^-23^+$	N.D.	Specific

[a] Noninfective here refers to UV-inactivated virus.
[b] Whether at a population level the T-cell response cross-reacts with heterologous virus type (common) or only "sees" the homologous virus type (specific).

spreads via the sensory nerves to the cervical dorsal root ganglia and into the CNS. Axonal trafficking of virus back to the epidermis occurs during the primary infection and probably represents the major route of virus spread. Infective virus is usually absent from the ear by day 8, although a latent infection can be established in the neurons of the sensory ganglia. Reactivation of virus can occur either spontaneously or upon traumatizing the ear, which in turn may result in a recurrent or recrudescent infection (reviewed by Wildy *et al.*, 1982). In all, this model displays the main pathogenic features associated with the natural infection in man.

In this mouse model, lymph nodes draining the site of infection contain T cells capable of adoptively transferring protection to naive recipients. The protective T cells are detected 4 days after infection, display maximum antiviral activity by day 6–8, and become undetectable beyond day 12 (Fig. 1). In the context of these studies, protection correlates with the rapid elimination of infective virus from the ear pinna. Protection can also be achieved by transferring immune cells taken from the draining lymph node 1 month after infection, but the kinetics of virus elimination is different, i.e., the clearance of virus is delayed (Nash *et al.*, 1980a). At 7 days postinfection, lymph node cell suspensions contain mixtures of T-DH and Tc cells, and their kinetics of induction and disappearance correlate with those of the protective T cell. The protection can be abrogated by treating the transferred cells with either anti-Thy-1.2 or anti-Lyt-1 plus complement (Nash and Gell, 1983). Similar findings have shown that Lyt-1^+23^- cells protect mice against "zosteriform" reactions (Nagafuchi *et al.*, 1982; A. Simmons, personal communication).

In order to investigate the significance of Lyt-1^-23^+ cells in immunity to herpes, adult-thymectomized CBA mice were injected with a rat anti-Lyt-2, Ig2b monoclonal antibody preparation (A. K. Jayasuriya, personal communication). This treatment markedly depletes Lyt-23^+ cells *in vivo* as determined by flow cytofluoroimetry, leaving only about 1% very weakly staining Lyt-2^+ cells. The depletion lasts for at least 7 months and is probably lifelong (Jayasuriya, personal communication). This treatment reduces the Tc cell response to background levels, but does not affect quantitatively the DH or antibody response to herpes. Furthermore, lymphocytes taken from untreated day 7 infected mice transfer antiviral activity to Lyt-2-depleted mice, despite the elimination of Lyt-23^+ cells from the transferred suspension. These findings favor a direct antiviral role for Lyt-1^+23^- cells which function in the absence of cells expressing Lyt-23^+ antigens. Indeed, anti-Lyt-2-treated mice survive an ear injection of 10^4 PFU HSV-1 (SC16), whereas anti-Lyt-1- or anti-Thy-1-treated mice show a 100% mortality by day 12 (A. Nash, unpublished observations). However, when Lyt-23^+-depleted mice were injected with 10^5 PFU HSV-1 (SC16) they were unable to control the infection and died by day 14 (A. K. Jayasuriya, personal communication). This observation implies that Lyt-23^+ cells become an important mechanism when the growth of virus reaches a high infectivity titer. More direct evidence supporting this latter observation is found in mouse models where protection from an otherwise lethal infection is used as a measure of antiviral activity. Larsen et al. (1983) demonstrated that herpes-primed Lyt-23$^+$ cells, upon restimulation with virus in vitro to produce specific cytotoxicity, could transfer protection to naive recipients coinjected with a lethal dose of HSV-1. Similarly, Sethi et al. (1983) protected cyclophosphamide-treated mice from a lethal herpes infection by transferring a herpes-specific T-cell clone that was Lyt-1$^-$23$^+$ and restricted to class I MHC antigens.

Further evidence suggesting that class I-restricted T-cell responses are involved in immunity to herpes comes from MHC matching experiments. In these experiments, herpes-immune T cells from one strain of mice are injected into another strain, which may share all, some, or none of the MHC antigens with the donor T cells. In this way, Howes et al. (1979) showed that either H-2K or D region-compatible T cells conferred protection to otherwise lethally infected recipients for up to 14 days; long-term protection was only achieved when cells and recipient were compatible at the H2-I region. However, in the mouse ear model the T cells involved in the rapid elimination of infective virus were restricted by both H-2IA and H-2K(D) antigens (Nash et al., 1981c). Compatibility at H-2K or D alone was nonprotective and H-2I alone only partially protective. Similar observations to these were made using an HSV-2 hepatitis model, in which compatibility at H-2K, I, and D was required for protection, whereas compatibility at H-2I alone allowed partial protection (S. Mogensen, personal communication). It would therefore seem that in the ear and hepatitis models, maximum antiherpes activity requires the action of more than one class of effector T cell; the precise nature of these cells remain to be determined.

The elimination of infective virus in primary cutaneous infections appears to correlate with cell-mediated inflammation (Kapoor et al., 1982b). Indeed, a strong correlation was found for the presence of T-DH cells and the rapid elimination of virus (Nash and Ashford, 1982). In man, a histological examination of a recrudescent lesion showed that the predominant T-cell subset infiltrating was Leu-3$^+$ (equivalent to Lyt-1$^+$23$^-$); in addition, the local production of interferon was detected (A. L. Cunningham and T. C. Merigan, personal communication). Interferon was also produced in large amounts following the injection of a Lyt-23$^+$ T-cell clone to herpes-infected mice (Sethi et al., 1983). As already discussed (see Section III.C) Lyt-1$^+$23$^-$, I region-restricted T-cell clones can mediate various helper activities and in addition produce factors that act on macrophages in vitro rendering them cytotoxic to a number of tumor cell lines (Leung et al., 1984). Clone D7.1 can also protect mice from a lethal infection, when given locally with the virus, and protection here does not correlate with the appearance of antiherpes antibodies. It would therefore seem likely that this clone functions in vivo by activating and arming macrophages.

In summary, we favor a central role for Lyt-1$^+$23$^-$ cells in recovery from cutaneous HSV infections. It is likely that they mediate this protective effect by releasing lymphokines, which could have a direct antiviral action or more likely are involved in recruiting and arming nonspecific effector cells. It would appear that for initial infectivity doses of virus less than 10^4 PFU (strain SC16), Lyt-23$^+$ cells play only a minimal protective role, whereas at doses greater than 10^4 PFU, the involvement of Lyt-23$^+$ cells in protection increases. In recurrent disease it is proposed that the same T-cell mechanisms are central in controlling infection of epidermal cells, although by this time antibody also contributes to the neutralization of virus.

V. REGULATION OF IMMUNE RESPONSES TO HERPES—A ROLE IN PREVENTION OF IMMUNOPATHOLOGY?

In Fig. 1 the effector phases of the Tc and T-DH response in the draining lymph node are shown to be transient, i.e., both are absent 2 weeks after infection. A transient DH response is at variance with the active elicitation of DH, which can occur throughout the lifetime of the mouse. This paradoxical situation is partially resolved when one studies the function of lymphoid cells in the draining lymph node beyond the second week. During this period a population of Ig + ve, Thy-1,2 − ve lymphocytes appears that can suppress an established DH response, i.e., when these cells are injected into herpes-sensitized recipients a depression in the intensity of the DH response occurs (Nash and Gell, 1980). These B-"Suppressor" cells appear to be localized to the lymph node draining the infection site; however, following adult thymectomy, an apparent redistribution occurs, for the cells are detected equally in the contralateral nodes and spleen. Indeed, if the suppressive cells from spleen are mixed with lymph node cells (day 7 immune cells) capable of adoptively transferring DH, the transfer of DH is markedly inhibited (A. Nash and P. Wildy, unpublished observations). Furthermore, the suppression is specific for herpes, i.e., the elicitation of vaccinia-specific DH is not inhibited. Interestingly, vaccinia-infected mice do not appear to induce a detectable B-suppressor cell response. As to how the suppressor cells mediate their inhibitory activity remains a puzzle. Hyperimmune anti-herpes globulin failed to inhibit the expression of DH and attempts to find suppressive antibodies or factors in the serum of mice containing active suppressor cells have proved unsuccessful (A. Nash, unpublished observation).

What is the functional significance of these cells in the immune response to the virus? It is possible that they represent an important homeostatic mechanism for regulating the intensity of DH reactions. Such reactions could be potentially damaging in organs such as the CNS. For example, demyelination appears to be a frequent event on the CNS side of the root entry zone of dorsal root ganglia following a primary infection of mice or rats (Hill, 1983; Kristensson *et al.*, 1979; Townsend, 1981). That T cells appear to be involved in this response is demonstrated by the work of Townsend (1981) who showed that demyelination in athymic nude mice was much less pronounced than that observed in immunocompetent littermates. Essentially, this kind of immunopathology is characteristic of a DH-type response. Although the damage is not prolonged and remyelination of nerves does occur, a potential environment is created for precipitating autoimmune reactions, e.g., T-cell responses to myelin basic protein. Other examples of cell-mediated immunopathology, particularly those involved in herpetic keratitis, have been reviewed by Rouse (this volume).

The role of Ts cells in herpes infections is less clear. The i.v. route of infection is not common in natural herpes infections. However, using the mouse ear model, recent evidence suggests that suppression of DH responses leads to a lower incidence of ear paralysis—a phenomenon associated with demyelination within the CNS (D. Altmann and W. A. Blyth, personal communication). This observation indicates a major role for Ts in regulating potentially harmful DH reactions. Analogous to this finding is the role of influenza-specific Ts cells in inhibiting lung consolidation induced by H-2I region-restricted T cells (Liew and Russell, 1983).

VI. T-CELL MEMORY

It is a feature of HSV infections in mice that both antibody- and cell-mediated immune responses to the virus are long-lived. Long-lived responses of this type may well reflect a constant or periodic exposure of the immune system to the virus (i.e., recurrent virus).

Memory T-cell responses as determined by proliferative assays are detectable in the lymph node draining the infection site by day 7–10; other lymphoid tissues such as spleen and contralateral lymph nodes become progressively populated between the second and third weeks after virus injection (A. Nash, unpublished observations). DH responses can be elicited throughout the lifetime of the animal and reinfection with the virus induces a brisk T-DH response that is once again transferable to naive recipients 2 days after infection. Similarly, following a secondary infection, Tc responses are detected within 2 days and are maximal by day 4–5. The secondary Tc responses also require a period of culture *in vitro*, before effectors are induced but in comparison with the primary response, maximal killing is achieved at lower effector to target cell ratios (Nash and Ashford, 1982). In keeping with the helper/cytotoxic T-cell responses, Ts cells once induced persist for long periods. However, interference with long-lived suppressor responses can be achieved following the injection of anti-Lyt-2 or anti-Thy-1.2 antibodies at the time of suppressor cell induction. Although early Ts responses are induced, in the absence of Lyt-23$^+$ cells, long-lived suppressor responses are not maintained. The significance of this observation for memory T cells in general, awaits future investigation.

Clearly, memory T-cell responses play an important role in surveillance against the reappearance of virus *in vivo*. The failure of the system, as seen in patients undergoing immunosuppression, leads to herpesvirus reactivations/recurrences, which may have serious consequences for the host.

VII. CONCLUSIONS

What can we learn from studies in animals that have relevance to herpes infection in man? It is true that at best animal models are second place to studying events in the natural host. However, it is likely that fundamental interactions between virus and host are similar across species. For example, neurotropism of herpes, herpes latency, and general immunological mechanisms are probably similar for mouse and man. Therefore, provided the animal model reflects the events in the natural infection, then it is reasonable to suppose the type of cell-mediated immune response will be comparable. In the mouse, recovery from a primary cutaneous herpes infection is mediated by Lyt-1$^+$23$^-$ cells, possibly via augmenting natural resistance mechanisms. We would argue for a similar role for OKT4$^+$ cells in controlling cutaneous herpes in man. What are the prospectives for vaccinating against herpesvirus infections? The first objective is to decide whether to prevent primary infection or in some way modify recurrent disease. Ideally, the former represents the logical goal although in reality we doubt very much that vaccination would completely inhibit a latent infection from subsequently becoming established. Consequently, the threat of recrudescent disease still remains. Better models of recurrent and of recrudescent herpes infections are urgently needed to investigate immunological control. Only then will it be possible to investigate whether defective immune responses are present, and consequently whether such deficiencies can be remedied prophylactically. The potential for corrective immunotherapy in patients with severe recrudescent disease remains the ultimate goal for herpes immunovirologists.

VIII. APPENDIX: IMMUNOLOGICAL TERMS USED

Lyt antigens: Lyt-1, 2, 3 are glycoproteins found on the membrane of mouse T cells. Lyt-1 appears on all T cells as determined by flow cytofluorimetry, but in high density on T-helper/inducer and T-DH cells (Lyt-1$^+$23$^-$) and low density on cytotoxic/suppressor cells (Lyt-1$^-$23$^+$).

Cell populations can be enriched for Lyt-1$^+$23$^-$ cells by treatment with anti-Lyt-2 plus complement and for Lyt-1$^-$23$^+$ cells with anti-Lyt-1 plus complement. this treatment also removes Lyt-1$^+$23$^+$ cells which function as precursors for other T-cell subsets.

OKT and Leu antigens: membrane glycoproteins found on human T cells (equivalent to Lyt antigens). OKT4/Leu-3 are found on helper/inducer T cells and OKT8/Leu-2 on cytotoxic/suppressor T cells.

Class I MHC products: membrane glycoproteins encoded by genes in the *K*, *D*, and *L* subregion of the *H-2* complex of the mouse (equivalent to the *HLA* complex).

Class II MHC products: membrane glycoproteins encoded by genes in the *I* (*I-A* and *I-E*) region of the *H-2* complex.

I-J determinants: antigens probably derived from the *I* region of the *H-2* complex and found in association with antigen-specific factors derived from suppressor T cells.

REFERENCES

Burns, W. H., 1977, Immune responses to herpes virus infections, in: *Progress In Immunology III* (T. E. Mandel, C. Cheers, C. S. Hosking, I. F. C. McKenzie, and G. J. V. Nossal, eds.) pp. 472–479. North-Holland, Amsterdam.

Burns, W. H., Billups, L. C., and Notkins, A. L. 1975, Thymus dependence of viral antigens, *Nature* **256**:654.

Carter, V. C., Rice, P. L., and Tevethia, S. S., 1982, Intratypic and intertypic specificity of lymphocytes involved in recognition of herpes simplex virus glycoproteins, *Infect. Immun.* **37**:116.

Eberle, R., Russell, R. G., and Rouse, B. T., 1981, Cell-mediated immunity to herpes simplex virus: Recognition of type-specific antigens by cytotoxic T cell populations, *Infect. Immun.* **34**:795.

Gordon, J., 1979, The B lymphocyte-deprived mouse as a tool in immunobiology, *J. Immunol. Methods* **25**:227.

Hill, T. J., 1983, Herpesviruses in the central nervous system, in: *Viruses and Demyelinating Diseases* (C. A. Mims, M. L. Cuznor, and R. E. Kelly, eds.), Academic Press, New York, pp. 29–45.

Hill, T. J., Field, H. J., and Blyth, W. A., 1975, Acute and recurrent infection with herpes simplex in the mouse: A model for studying latency and recurrent disease, *J. Gen. Virol.* **28**:341.

Howes, E. L., Taylor, W., Mitchison, N. A., and Simpson, E., 1979, MHC matching shows at least two T-cell subsets determine resistance to HSV, *Nature* **277**:67.

Jennings, S. R., Rice, P. L., Pan, S., Knowles, B. B., and Tevethia, S. S., 1984, Recognition of herpes simplex virus antigens on the surface of mouse cells of the H2^b haplotype by virus-specific cytotoxic T lymphocytes, *J. Immunol.* **132**:475.

Kapoor, A. K., Nash, A. A., and Wildy, P., 1982a, Pathogenesis of herpes simplex virus in B-cell suppressed mice: The relative roles of cell mediated and humoral immunity, *J. Gen. Virol.* **61**:127.

Kapoor, A. K., Nash, A. A., Wildy, P., Phelan, J., McLean, C. S., and Field, H. J., 1982b, Pathogenesis of herpes simplex virus in congenitally athymic mice: The relative roles of cell-mediated and humoral immunity, *J. Gen. Virol.* **60**:225.

Kristensson, K., Svennerholm, B., Persson, L., Vahlne, A., and Lycke, E., 1979, Latent herpes simplex virus trigeminal ganglionic infection in mice and demyelination in the central nervous system, *J. Neurol. Sci.* **43**:253.

Larsen, H. S., Russell, R. G., and Rouse, B. T., 1983, Recovery from lethal herpes simplex virus type 1 infection is mediated by cytotoxic T lymphocytes, *Infect. Immun.* **41:**197.

Leung, K. N., Nasti, A. A., Sia, D. Y., and Wildy, P., 1984, Clonal analysis of T cell responses to herpes simplex virus: isolation, characterization, and antiviral properites of an antigen-specific helper T cell clone, *Immunology*, in press.

Liew, F. Y., and Russell, S. M., 1983, Inhibition of pathogenic effect of effector T cells by specific suppressor T cells during influenza virus infection in mice, *Nature* **304:**541.

Metcalf, J. F., Hamilton, D. S., and Reichert, R. W., 1979, Herpetic keratitis in athymic (nude) mice, *Infect. Immun.* **26:**1164.

Mori, R., Tasaki, T., Kimura, G., and Takeya, K., 1967, Depression of acquired resistance against herpes simplex virus infection in neonatally thymectomized mice, *Arch. Gesamte Virusforsch.* **21:**459.

Nagafuchi, S., Oda, H., Mori, R., and Taniguchi, T., 1979, Mechanism of acquired resistance to herpes simplex virus infection as studied in nude mice, *J. Gen. Virol.* **44:**715.

Nagafuchi, S., Hayashida, I., Higa, K., Wada, T., and Mori, R., 1982, Role of Lyt-1 positive immune T cells in recovery from herpes simplex virus infection in mice, *Microbiol. Immunol.* **26:**359.

Nash, A. A., and Ashford, N. P. N., 1982, Split T-cell tolerance in herpes simplex virus-infected mice and its implication for anti-viral immunity, *Immunology* **48:**761.

Nash, A. A., and Gell, P. G. H., 1980, Cell-mediated immunity to herpes simplex virus-infected mice: Suppression of delayed hypersensitivity by an antigen-specific B lymphocyte, *J. Gen. Virol.* **48:**359.

Nash, A. A., and Gell, P. G. H., 1981, The delayed hypersensitivity T cell and its interaction with other T cells, *Immunol. Today* **2:**162.

Nash, A. A., and Gell, P. G. H., 1983, Membrane phenotype of murine effector and suppressor T cells involved in delayed hypersensitivity and protective immunity to herpes simplex virus, *Cell. Immunol.* **75:**348.

Nash, A. A., Field, H. J., and Quartey-Papafio, R., 1980a, Cell-mediated immunity to herpes simplex virus-infected mice: Induction, characterisation and antiviral effects of delayed type hypersensitivity, *J. Gen. Virol.* **48:**351.

Nash, A. A., Quartey-Papafio, R., and Wildy, P., 1980b, Cell-mediated immunity to herpes simplex virus-infected mice: Functional analysis of lymph node cells during periods of acute and latent infections, with reference to cytotoxic and memory cells, *J. Gen. Virol.* **49:**309.

Nash, A. A., Gell, P. G. H., and Wildy, P., 1981a, Tolerance and immunity in mice infected with herpes simplex virus: Simultaneous induction of protective immunity and tolerance to delayed-type hypersensitivity, *Immunology* **43:**153.

Nash, A. A., Phelan, J., Gell, P. G. H., and Wildy, P., 1981b, Tolerance and immunity in mice infected with herpes simplex virus: Studies on the mechanism of tolerance to delayed-type hypersensitivity, *Immunology* **43:**363.

Nash, A. A., Phelan, J., and Wildy, P., 1981c, Cell-mediated immunity in herpes simplex virus-infected mice: H-2 mapping of the delayed-type hypersensitivity response and the antiviral T cell response, *J. Immunol.* **126:**1260.

Oakes, J. E., 1975, Role for cell-mediated immunity in the resistance of mice to subcutaneous herpes simplex infection, *Infect. Immun.* **12:**166.

Pfizenmaier, K., Jung, H., Starzinski-Powitz, A., Rollinghoff, M., and Wagner, H., 1977, The role of T cells in anti-herpes simplex virus immunity. I. Induction of antigen-specific cytotoxic T lymphocytes, *J. Immunol.* **119:**939.

Rager-Zisman, B., and Allison, A. C., 1976, Mechanism of immunologic resistance to herpes simplex virus 1 (HSV-1) infection, *J. Immunol.* **116:**35.

Rouse, B. T., Larsen, H. S., and Wagner, H., 1983, Frequency of cytotoxic T lymphocyte precursors to herpes simplex virus type 1 as determined by limiting dilution analysis, *Infect. Immun.* **39:**785.

Schmid, D. S., and Rouse, B. T., 1983, Cellular interactions in the cytotoxic T lymphocyte response to herpes simplex virus antigens: Differential antigen activation requirements for the helper T lymphocyte and cytotoxic T lymphocyte precursors, *J. Immunol.* **131**:479.

Schrier, R. D., Pizer, L. I., and Moorhead, J. W., 1983a, Type-specific delayed hypersensitivity and protective immunity induced by isolated herpes simplex virus glycoprotein, *J. Immunol.* **130**:1413.

Schrier, R. D., Pizer, L. I., and Moorhead, J., 1983b, Tolerance and suppression of immunity to herpes simplex virus: Different presentations of antigens induce different types of suppressor cells, *Infect. Immun.* **40**:514.

Sethi, K. K., 1983, Effects of monoclonal antibodies directed against herpes simplex virus-specified glycoproteins on the generation of virus-specific and H-2 restricted cytotoxic T-lymphocytes, *J. Gen. Virol.* **64**:203.

Sethi, K. K., Omata, Y., and Schneweis, K. E., 1983, Protection of mice from fatal herpes simplex virus type 1 infection by adoptive transfer of cloned virus-specific and H-2 restricted cytotoxic T lymphocytes, *J. Gen. Virol.* **64**:443.

Townsend, J. J., 1981, The demyelinating effect of corneal HSV infections in normal and nude (athymic) mice, *J. Neurol. Sci.* **50**:435.

Wildy, P., Field, H. J., and Nash, A. A., 1982, Classical herpes latency revisited, in: *Virus Persistence* (B. W. J. Mahy, A. C. Minson, and G. K. Darby, eds.), pp. 133–167, Cambridge University Press, London.

CHAPTER 5

Immunopathology of Herpesvirus Infections

Barry T. Rouse

I. INTRODUCTION

In the absence of normal immune responsiveness, herpesvirus infections may be both more severe and more frequent (Shore and Nahmias, 1982). From numerous clinical observations in humans with primary or secondary immune deficiencies or from immunological manipulations of experimental infections in animal models, it has become evident that many aspects of immunity, both natural and adaptive, serve to protect the host at varous stages of the virus–host interaction (Rouse, 1984). The principal emphasis of studies on herpesvirus immunobiology has been to stress the protective aspects of various defense mechanisms. The logic behind many investigations was to set the stage for the design and rational use of antiherpetic vaccines. Nevertheless, with some exceptions in the veterinary field (Plowright, 1980), vaccines suitable for use against herpesvirus infections have not been widely acclaimed.

It is now abundantly clear that the adaptive immune response is finely regulated and that many cell types must interact to generate a response. This interaction proceeds by way of direct cell contact as well as by means of chemical mediators and involves both specific and nonspecific mechanisms. Regulation of responsiveness occurs at all levels from the antigen-presenting cells (APC) until the effector phase of the response (Fig. 1). Evidently, the APC are heterogeneous and their function is subject to regulation by factors released by other cell types such as interferon-γ (IFN-γ) from T cells. Interaction with this molecule, for example, causes APC

BARRY T. ROUSE • Department of Microbiology, College of Veterinary Medicine, University of Tennessee, Knoxville, Tennessee 37996.

103

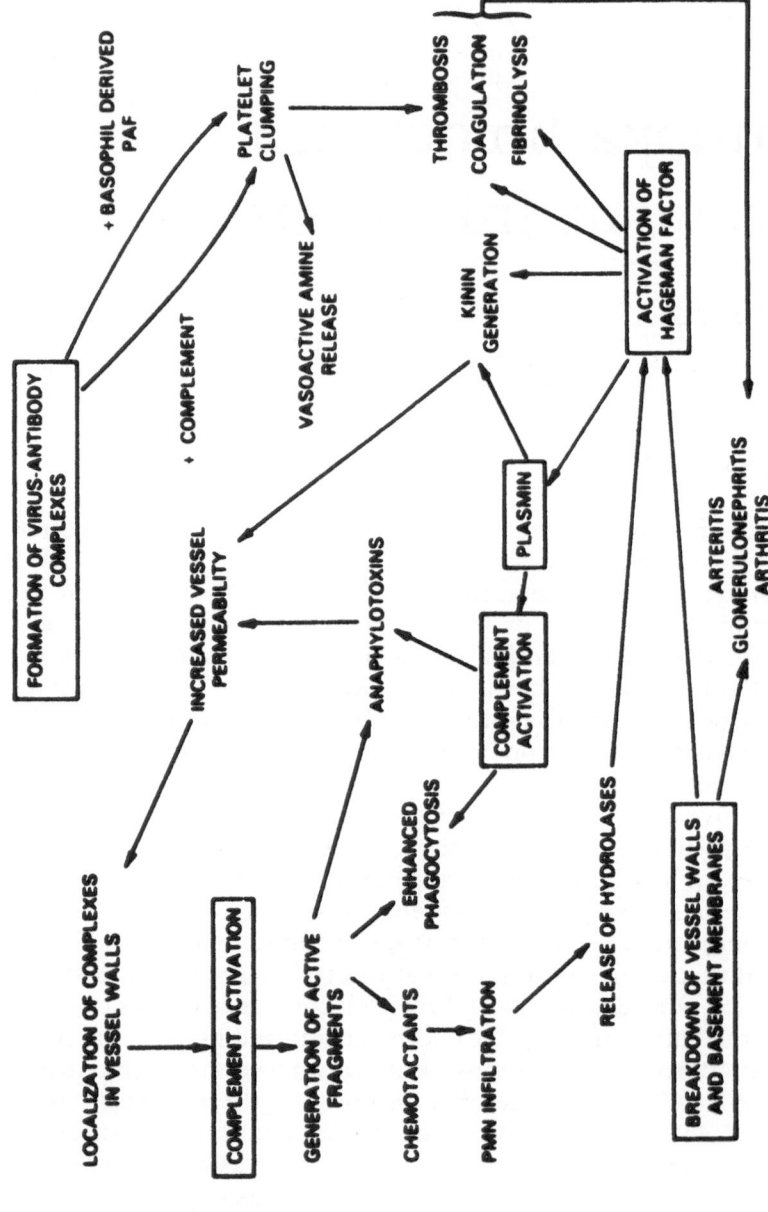

FIGURE 1. Immunoregulation and immunopathology. The antigen-presenting cell (APC) is known to play a central role and is depicted as a single cell with dividions. This is because the APC are heterogeneous and distinct cells may trigger the different responses. The bold arrows indicate the major regulatory pathways. The major mechanisms of immunopathology (IP) are indicated.

to preferentially express class II MHC antigens (Steeg *et al.*, 1982) whose presence is necessary to effectively trigger antigen-specific helper T-cell subsets (Schmid and Rouse, 1983). Other subsets of T cells, such as cytotoxic T-cell precursors, and B cells are probably triggered by different APC. Regulating the type of response that results as well as its magnitude depends upon interaction between several cell types, which include different subtypes of suppressor cells, with helper cells, B cells, and APC (Germain and Benacerraf, 1981). The antigen-specific effectors of the immune response, antibodies and certain subsets of T cells, usually mediate their protective function by the recruitment of nonspecific activities such as phagocytosis, cellular destruction by macrophages and their products as well as NK cells. Perhaps under most circumstances the outcome of the adaptive immune response is protection, but in some instances significant tissue damage occurs. Such immunopathological reactions may be the price to pay for ultimate protection in most instances although examples exist where in the absence of an immune response disease fails to occur (Oldstone and Dixon, 1976). The best example with viruses is lymphocytic choriomeningitis virus (LCMV) infection of mice. In this instance, only mice that generate immune responsiveness against the virus suffer untoward effects (Oldstone, 1979).

The mechanisms of immunopathology are manyfold and include a chronic response to a persistent antigen, inappropriate regulation of the immune response such as a failure of suppression or the triggering of the types of helper cells that permit the generation of antibodies such as IgE. Included among the effectors that damage tissues in immunopathological reactions are polymorphonuclear (PMN) leukocytes, activated macrophages, and NK cells.

II. IgE-MEDIATED IMMUNOPATHOLOGY

The tissue damage in this form of hypersensitivity results from the release of a range of mainly low-molecular-weight vasoactive molecules from mast cells, platelets, or basophils and is initiated by the reaction of antigen with cell-bound IgE antibodies. These reactions are common in parasitic infections but their role against viruses has been inadequately explored. Reports of IgE antibody production in animals infected with herpes simplex virus (HSV) have appeared but the biological role of such antibodies in immunity or pathology was not defined (Day *et al.*, 1976; Ida *et al.*, 1983). Efforts have been made to establish the role of anti-HSV IgE in human herpetic infections but so far without success. It is quite possible that IgE-mediated immunopathology could account for the erythema at the base of reactivated herpetic lesions particularly in those circumstances where recurrences are more aggressive than usual. However, attempts to demonstrate specific antiherpes IgE or to favorably influence reactions with antihistamines have not been undertaken. It will

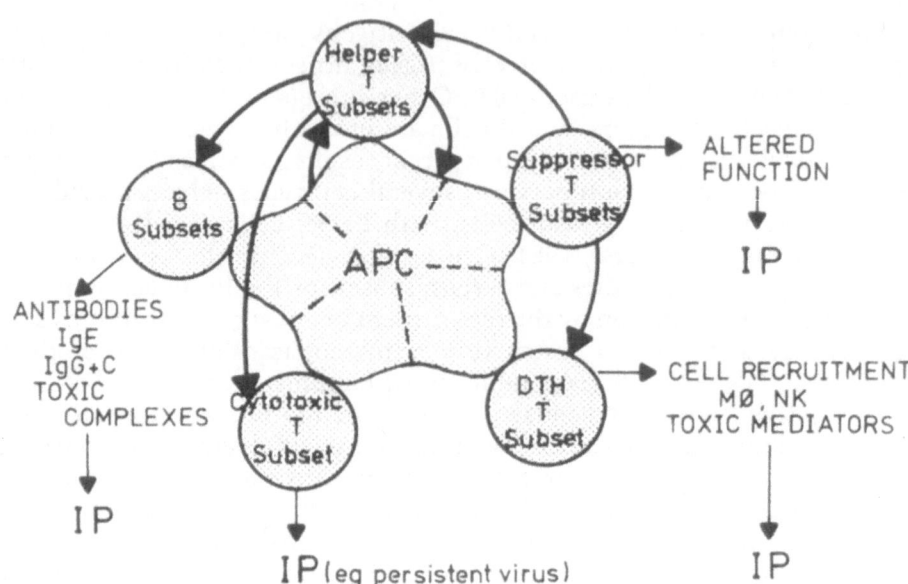

FIGURE 2. Diagram depicting some interrelated sequences of events in immune complex-mediated inflammatory responses. PAF, platelet-activating factor; PMN, polymorphonuclear leukocyte. From Rouse and Babiuk (1979) with permission.

be of interest to evaluate the IgE-mediated response in a suitable animal model of recrudescent disease.

The suggestion has been made that viruses could be associated with asthmatic attacks in children (Berkovich *et al.*, 1970; Minor *et al.*, 1974). Work with model systems has indicated that virus-induced interferon can increase the level of histamine released from IgE-sensitized cells (Ida *et al.*, 1977). It is also possible that IFN-γ induced from virus-specific T cells could react with mast cells bearing antiviral IgE to result in enhanced levels of histamine release and consequent immunopathology. Such a mechanism could be occurring in a local herpetic lesion where high levels of IFN may be present, and cells sensitized with anti-HSV IgE could occur.

III. IMMUNE COMPLEX IMMUNOPATHOLOGY

Immune complex-mediated immunopathology results from the deposition in tissues, usually in vessel walls or the glomerulus, of complexes of antigen and antibody. At certain size and ratios of reactants such as slight antigen or antibody excess, the complexes activate a range of effector enzyme systems of which the complement and blood-clotting cascades are the most prominent (Fig. 2). The result is an inflammatory tissue-damaging reaction, acute or chronic according to the duration of complex formation and persistence. In acute reactions it is the PMN neutrophils, attracted to the area by chemotactic factors such as those elaborated during the complement cascade, that appear to be directly responsible for the pathology. Thus, in the Arthus reaction, the role model for immune complex immunopathology, PMN neutrophil depletion is accompanied by marked suppression of the immunopathology (Cochrane et al., 1959). Immune complex lesions are especially likely to occur in situations of antigen excess. A prime example is afforded by LCMV infection, which if it occurs in the newborn is accompanied by a diminished immune response. However, low levels of antibody are formed and the antibody complexes with LCMV to give rise to immune complex-mediated lesions in the glomerulus (Oldstone and Dixon, 1976). Chronically infected mice ultimately succumb to kidney disease.

In persistent herpesvirus infections, the scenario is usually one of gross antibody excess so that the likelihood of immune complex immunopathology is low. However, situations have been reported where tissue destruction is seemingly immune complex mediated and the injection of antigen (vaccine) into an immune individual is expected on occasion to result in a local inflammatory lesion.

One good example of immune complex pathology involving herpesviruses is in recipients of renal transplants given large doses of immunosuppressive drugs and who suffer recurrent infections with cytomegalovirus (CMV) infection (Richardson et al., 1981). In such patients, immune complexes are frequently found in the glomerulus and although this probably fails to lead directly to transplant rejection, the complexes do render the transplant useless, as glomerulopathy is a frequent outcome. Thus far there is no definitive evidence that any CMV antigen forms part of the toxic complex although CMV can be found in many parts of the affected kidney. There is some evidence that CMV may cause glomerulonephritis in occasional cases of CMV infection in the adult (Ozawa and Stewart, 1979), and circulating immune complexes were also reported in congenital and natal CMV infection (Stagno et al., 1977). Finally, Epstein–Barr virus (EBV), which like CMV can cause infectious mononucleosis, may also be an infrequent cause of glomerulonephritis presumably associated with immune complex depostion (Wallace et al., 1974).

A second clinical situation involving herpesviruses where toxic immune complexes have been implicated is in erythema multiforme (EM). This dramatic self-limiting inflammatory disorder of the skin that tends to recur, is of unknown etiology but is likely a hypersensitivity reaction. In about 15% of cases, EM is proceeded by HSV recrudescence and in one instance the disease was triggered by the injection of inactivated HSV vaccine into the skin (Sheeley, 1967). Recently, circulating immune complexes were reported and deposits of immunoglobulin and complement components were observed by direct immunofluorescence in superficial dermal vessels or at the dermal–epidermal junction (Kazmierowski et al., 1982). In 8 of 12 patients with EM occurring after recurrent HSV infection,

HSV antigens were constituents of the circulating complexes. However, it is not known if HSV also forms part of complexes directly involved in the presumed immunopathological reaction of EM. It is also unclear why EM patients with circulating complexes failed to develop renal disease. Although evidence for the involvement of HSV–antibody complexes in the pathogenesis of EM is not solid, the possibility that such a reaction could occur should be considered in the design and use of vaccines.

Perhaps the best candidate syndrome involving herpesviruses in which the mechanism of pathogenesis is essentially immunopathological is that of disciform keratitis. This inflammatory reaction of the corneal stroma occurs following HSV recrudescence in the eye (Kaufman, 1978). Frequently the reaction can become more marked with each recrudescence and corneal blindness can result. In fact, HSV infection is a common cause of corneal blindness in the United States. Epithelial keratitis appears to be the direct result of virus replication in epithelial cells and responds very well to treatment with antivirals. Indeed, it was in the treatment of epithelial keratitis that antiviral drug treatment was first found to be clinically useful (Kaufman *et al.*, 1962). However, keratitis affecting the stroma responds less well to antiviral drugs and needs judicious treatment with corticosteroids for successful management. It is usually not possible to demonstrate infectious virus in the stroma, but from work done on clinical material and in animal models, viral antigens can often be detected in keratocytes (Metcalf and Kaufman, 1976). Perhaps the best model for herpetic keratitis is the rabbit model where after infection with certain strains both epithelial and stromal keratitis are induced. In herpetic stromal keratitis, some have observed, by means of electron microscopy, lymphocytes in association with HSV-containing keratocytes (Metcalf and Reichert, 1979). However, others have not observed this phenomenon (Meyers-Elliot *et al.*, 1980) but point out that stromal keratitis in the rabbit, particularly the acute form, is a reaction in which the principal inflammatory cell type is the PMN neutrophil (Meyers-Elliot and Chitjian, 1976). In fact, lesions were markedly diminished in rabbits treated either to deplete PMN neutrophils or complement (Meyers-Elliot and Chitjian, 1981) (Table I). The prominence of PMN neutrophils in the stromal lesion and evidence that their depletion ameliorates the condition suggest a resemblance to the Arthus reaction and that a principal mechanism of the immunopathology is immune complex mediated. Direct support for this was provided by Meyers and Pettit (1973) who could demonstrate by direct immunofluorescence HSV antigen, complement, and immunoglobulin in the corneal stromal tissues. Nevertheless, immune complex-mediated immunopathology seems not to be the only mechanism because immune lymphocytes, presumably T lymphocytes, can also be demonstrated in local lymph nodes and in inflammatory ocular exudates and such cells were assumed to play an additional role in immunopathogenesis (Meyers-Elliot and Chitjian, 1976). Furthermore, viral antigens persist after PMN neutrophils are no longer vis-

TABLE I. Clinical Course of Herpes Virus Keratitis[a]

Group (treatment)	Epithelial disease		Stromal disease	
	No. involved (eyes/total)	Clinical severity[b]	No. involved (eyes/total)	Clinical severity[c]
I (untreated control)	20/20	4+	12/20	3+
II (anti-PMN globulin)	12/12	3+	4/12	1+
III (anti-PMN globulin)	6/6	4+	6/6	3–4+
IV (normal globulin)	6/6	4+	6/6	3+
V (nitrogen mustard)	20/20	3+	4/20	1+

[a] Modified from Meyers-Elliot and Chitjian (1981).
[b] Score for clinical severity of epithelial disease: 0 = none; 1+ = lesion of any type (punctate, dendritic, or geographic) affecting one-quarter or less of total cornea surface; 2+ = lesions affecting more than one-quarter but less than one-half of cornea; 3+ = lesions affecting one-half or more but less than three-quarters of cornea; 4+ = lesions affecting three-quarters or more of cornea.
[c] Score for clinical severity of stromal disease (edema and opacity): 0 = none; 1+ = slight edema or surface irregularity with or without slight infiltrate; 2+ = hazy with edema; 3+ = translucent gross clouding with edema and infiltration; pupillary border seen indistinctly; 4+ = opaque cornea with edema and infiltration.

ible in the stroma. It has been suggested that antigen-expressing cells could be lysed by cytotoxic antibody (Sheppard and Smith, 1981), cytotoxic T lymphocytes (CTL), or perhaps nonimmune phagocytes by the antibody-dependent cell cytotoxicity phenomenon (Meyers-Elliot and Chitjian, 1976). All such reactions, although possibly protective to the host, could also be considered as immunopathological.

IV. IMMUNOPATHOLOGY MEDIATED BY T LYMPHOCYTES

It is firmly established that reactions of effector T cells with target cell antigens can be accompanied by tissue damage. These in vivo reactions are called delayed-type hypersensitivities (DTH). Such immunopathological reactions may be severe and extremely prolonged in situations where antigens persist, but probably under most circumstances antigens are removed rapidly and the tissue damage is a small investment for ultimate removal of the antigen. Although recognition by the effector T cells is antigen specific, a range of activities are triggered that are nonspecific in action. The cell types involved in DTH and the activities responsible for tissue damage vary according to tissue location and nature of the target. As shown by the elegant studies of MacKaness (1970), activated macrophages are prominent cells in DTH reactions and these cells release a range of tissue-damaging activities (Nathan et al., 1980). Other immunopathological effects include direct target destruction by CTL and inflammatory mediators released from damaged tissue cells. It is also likely that NK cells become activated by lymphokines produced such as interleukin-2 and IFN-γ (Yamamoto et al., 1982). Some have shown that

NK cells can act immunopathologically under some circumstances (C. Lopez, personal communication).

Both the generation and the expression of DTH reactions are regulated by the T-suppresor cell (Ts) system. Thus, suppressor cells that act on both the afferent and the efferent phase of DTH to HSV have been reported (Schrier *et al.*, 1983; Nash and Gell, 1981). More than likely, if Ts regulation is impaired, then the immunopathology may become more marked. In such circumstances, any immune intervention that restores Ts function becomes advantageous.

The mechanism of DTH reactions has been carefully examined, as well as its role in defense against viruses including HSV. Only certain subsets of T cells participate in the effector phase of DTH (Nash and Gell, 1981). In the case of inert antigens, these are often exclusively of the Lyt-1^+2^- subset in the mouse. Such cells are restricted during their activation and effector function by class II antigens of the MHC. Consequently, it is only possible to adoptively transfer the DTH response when donor and recipient share class II MHC genotypes. Against infectious agents, more than one subset of antigen-specific T cells may be involved in the DTH response. For example, after infection with infectious influenza virus, two subsets of T cells are induced that can mediate DTH (Ada *et al.*, 1981). These are Lyt-1^-2^+ and Lyt-1^+2^- subsets, which are restricted by class I and class II MHC antigens, respectively. By way of contrast, following immunization with noninfectious influenza virus, only the class II-restricted Lyt-1^+ cell type is induced. More of interest for present discussion were the effects observed following the transfer of different cell types to mice 24 hr after infection. Whereas mice receiving the Lyt-2^+ cell subset were protected and virus titers in lungs were markedly reduced, mice given the Lyt-1^+ subset died more rapidly than controls (Leung and Ada, 1982). Using the uptake of [^{125}I]iododeoxyuridine into rapidly dividing cells as a method of measuring monocyte infiltration into the infected lung, a greater cell influx was noted in mice receiving the Lyt-1^+ population, this response contributing substantially to the immunopathology. Very recently, Liew and Russell (1983) demonstrated the induction of a suppressor cell from animals given live virus that could reverse the pathological effects of Lyt-1^+ population. The implication of the work with influenza, and similar observations with Sendai virus (Ertl, 1981), is that certain forms of vaccines, such as those that are inactivated, could give rise to immunopathology rather than protection. For the latter effect, attenuated vaccines are preferable.

Nash *et al.* (1981a,b) have investigated the DTH response to HSV and its relevance in immunity. In contrast to influenza, only Lyt-1^+ MHC class II-restricted cells are generated against infectious HSV (Nash *et al.*, 1981a). Furthermore, Nash and colleagues have assembled considerable evidence that these cells play essentially a protective rather than an immunopathological role (see Nash *et al.*, this volume). This is because, as discussed above, HSV infection rapidly kills cells and so further destruction by some immune cell type may not add to the problem. However, there are situations where HSV antigens persist for long periods without replication and cell destruction. The best example is with herpesvirus latency but the role of DTH in sustaining or disrupting latency remains ill-defined. The second example is in stromal keratitis where viral antigens can persist for at least a month in keratocytes (Meyers-Elliot and Chitjian, 1976). In the murine model of herpetic stromal keratitis, accumlating evidence indicates an immunopathologic reaction mediated by T lymphocytes. For instance, Metcalf *et al.* (1979) showed that whereas infection of BALB/c mice with the RE strain of HSV produced an epithelial keratitis followed by a deep stromal reaction that often resulted in permanent scarring, when BALB/c athymic nude mice were infected the results were different. In this instance, mice developed a similar degree of epithelial keratitis, but this cleared up after a few days and was not followed by the stromal reaction. However, attesting to the importance of T-cell immunity, at least following certain routes of infection, athymic mice usually die from encephalitis whereas normal BALB/c mice do not.

We have also compared infections of the scarified corneas of athymic and normal mice. In confirmation of the findings of Metcalf *et al.* (1979), epithelial reactions were observed in both groups but whereas the corneas healed uneventfully in athymic mice, in immunocompetent mice a stromal necrotic reaction occurred that resulted in scarring (Table II).

To gain further evidence that the stromal reaction was immunopathological, the course of events was followed in HSV-infected athymic mice subsequently given adoptive cell transfers. One cell population used for transfer was spleen cells from HSV-immune mice that were stimulated *in vitro* with HSV antigen. This cell population, which we have extensively characterized in previous publications, consists predominantly of T cells, expresses high levels of HSV-specific H-2-restricted cytotoxicity (Lawman *et al.*, 1980), contains helper cell activity (Schmid and Rouse, 1983), and can mediate DTH reactions (Larsen *et al.*, 1983). If T cells were involved in the immunopathological reaction, the HSV-immune cell source was expected to elicit a prompt response in the athymic recipients providing some cells could find access to the site of infection. Indeed, athymic mice adoptively given such cells developed a necrotizing stromal reaction and scarring similar to those observed in euthymic BALB/c mice. Moreover, peak responses were evident from 6 days after transfer compared with a similar response in BALB/c mice at 12 days after infection (Table II). The adoptive transfer of the stromal response was abrogated upon removal of the T cells by negative selection with specific antisera and complement, further implicating the role of T cells in mediating the stromal response. Consequently, our results support and extend previous observations that herpetic stromal keratitis represents an immunopathological reaction. However, further experiments are required to delineate the actual mechanism of the immunopathology *in vivo*. For instance, the tissue damage could be mediated directly by

TABLE II. Histopathologic Course following Ocular Infection with HSV-1

Group	Days after infection of cornea with HSV-1					
	3	6	9	12	15	18
BALB/c						
Stromal necrosis	−[a]	+	++	+++	++++	++++
Inflammatory infiltrate	++	+++	++	+++	++++	++++
Scarring	−	−	−	−/[c]++	+++	++++
Athymic						
Stromal necrosis	−	−	−	−	−	−
Inflammatory infiltrate	++	+++	++	+	−	−
Scarring	−	−	−	−	−	−
Athymic given adoptive transfers of immune spleen cells[b]						
Stromal necrosis	−	+	++++	++++	++++	++++
Inflammatory infiltrate	++	+++	++++	++++	++++	++++
Scarring	−	−	−	+	++	++++

[a] Histopathology judged: −, inapparent; +, mild; ++, moderate; +++, marked; ++++, severe.
[b] 5 × 10⁷ spleen cells from animals infected in vivo 6 weeks previously with HSV followed by restimulation in vitro for 5 days with UV-inactivated virus. Cells were given 3 days after corneal infection. Methods described in Lawman et al. (1980).
[c] Values range from (−) to (+).

CTL as we demonstrated against HSV-infected cells *in vitro* (Lawman *et al.*, 1980). Secondly, HSV-specific T cells could be mediating an inflammatory DTH response. A third possibility is that T lymphocytes were acting as helper cells in reconstituting the ability of recipient mice to produce antibody, such antibody mediating pathology by an immune complex mechanism or antibody and complement lysis as was discussed above.

The adoptive cell transfers we employed contained helper cells and did reconstitute antibody-producing responsiveness (data not shown) and so it is not possible to fully discount antibody as playing a role in mediating the stromal reaction. One way of further resolving the putative mechanism of immunopathology operative *in vivo* would be to perform adoptive cell transfers with defined subsets of T lymphocytes such as Lyt-2^+ cytotoxic/suppressor cells and Lyt-1^+ helper/DTH-mediating T lymphocytes. In a preliminary experiment involving limited numbers of animals infected with HSV and adoptively given Lyt-1^+ or Lyt-2^+ HSV-immune lymphocytes, our results have indicated that both cell populations can confer the stromal reaction as judged histologically 15 days posttransfer. As the Lyt-2^+ population lacked helper activity for antibody production, this indicates that one mechanism of immunopathology is mediated directly by T cells, presumably the type also capable of specific cytotoxicity *in vitro*. However, the observation that the Lyt-1^+ population also conferred the stromal reaction means that multiple mechanisms are presumably operative. In fact, we have further evidence for this in preliminary studies using the DTH tolerance model described by Nash *et al.* (1981b). In this model, mice exposed to HSV via the intravenous route develop a state of immune deviation in that they become selectively incapable of generating a DTH response yet retain normal antibody responsiveness and other aspects of T-cell immunity. Such DTH-tolerized mice showed diminished stromal reactions in comparison with mice infected by the subcutaneous route or mice not previously infected. Such data implicate the Lyt-1^+ DTH-mediating cells as perhaps the principal participants in the immunopathological reaction. Metcalf (1983) has made similar observations.

Immunopathological reactions mediated by T lymphocytes could be involved in other aspects of the pathogenesis of herpetic infections. A good example is infectious mononucleosis (IM) resulting from EBV infection (Fig. 3). In this disease the splenomegaly and hepatitis that occur resemble graft-versus-host reactions (Purtilo, 1980). It has been hypothesized that these reactions represent T-cell proliferations and abnormal maturation. They occur in response to recognition of specific viral antigens and polyclonal activating antigens on EBV-transformed B cells (Evans and Niederman, 1982). Part of the T-cell response to such antigens results in development of CTL, which in turn are lytic to B cells and so accounting for the disappearance of transformed B cells after the early stage of the disease (Hutt *et al.*, 1975; Svedmyr and Jondal, 1975). Other

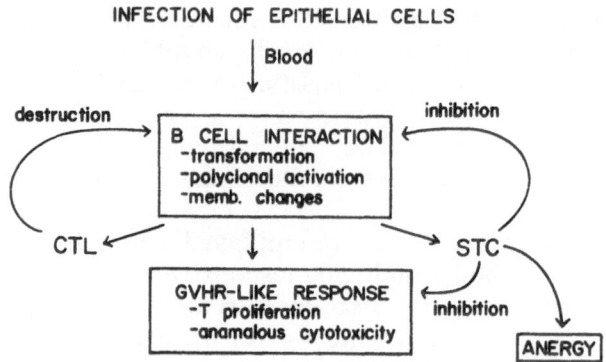

FIGURE 3. Possible immunopathological events occurring in infectious mononucleosis. CTL, cytotoxic T lymphocyte; STC, suppressor T cell. Modified from Evans and Niederman (1982).

activated T cells express anomalous nonspecific non-MHC-restricted cytotoxicity (Royston *et al.*, 1975). Included among the targets of such cells are EBV-infected cells and autologous infected cells such as hepatocytes (Sullivan *et al.*, 1983). It has been proposed that the anomalous cytotoxic response may proceed uncontrolled in some cases of fatal IM as occurs in the X-linked immunoproliferative disorder (XLP) (Sullivan *et al.*, 1983).

Another aspect of T-cell activity in IM is that of suppression (Tosato *et al.*, 1979; Haynes *et al.*, 1979). Such cells serve to ultimately restrict both B-cell and T-cell responses, and can perhaps be overactive in some circumstances, so accounting for anergy. Perhaps in other situations the anomalous and overreactive T-cell response resulting in immunopathology is the outcome of a failure of control by the Ts system. However, details of these processes are in need of elucidation.

The example of XLP illustrates a form of immunopathology where disease apparently results from uncontrolled regulation between lymphocytes. Such inappropriate lymphocyte interactions could be a feature in many disease syndromes and could account for instances of atypically severe patterns. Another example of topical interest is the acquired immune deficiency syndrome (AIDS), where changes in ratios between lymphocyte subsets with an increase in the suppressor cell population have been noted (Mildran *et al.*, 1982). There is also a suspicion that herpetic recrudescent disease results from some change in regulation in the Ts system and that Ts activity affects the severity of HSV lesions (Dannenberg *et al.*, 1980; Sheridan *et al.*, 1982; Rouse, 1984). However, these notions require further experimental verification.

V. MANAGEMENT OF IMMUNOPATHOLOGICAL REACTIONS

Study of the immune responses against herpesvirus infections has as its ultimate goal the development of new modalities of treatment that

TABLE III. Some Ways of Modulating Immune Responses

Immunization with selected oligopeptide
Substitution of certain groups into antigen
Coupling to selected carrier molecules
Use of adjuvants and liposomes
Selection of appropriate route of immunization
Construction of hybrid viruses containing genes for selected oligopeptide
Anti-idiotype immunization

will result in reduction of the severity of primary infection and/or the control of recrudescent disease. Determining the defense mechanisms responsible for the control of herpesvirus infection and then developing ways of augmenting the appropriate responses may yield the desired result. It is, however, important to remember that some of the immune responses that are triggered during virus infections can do almost as much damage to the host as does the virus itself. Consequently, attempts at immunotherapy or immunoprophylaxis must keep in mind the potential pathology that may be induced inappropriately by such treatments. The goal of vaccination is to mimic as far as possible those features that confer long-lasting protective immunity and induce preferentially, memory of those responses that are protective and cause least immunopathology. Such ideals beg the question of what is possible in terms of modulating immune responses in the desired direction. Alas, this topic is to date one that is very poorly understood. However, there are promising leads (summarized in Table III).

First, there is increasing evidence that certain oligopeptide sequences in antigens may be preferentially involved in certain activities. Thus, as shown by the elegant work of Sercatz's group with the simple protein egg white lysozyme, certain regions of the antigen are involved in the induction of suppression and different regions in the induction of helper T cells (Adorini et al., 1979). Manipulating the level of suppressors could hold the key to avoiding immunopathology. Along similar lines to the Sercatz work have been observations that modifying antigens with certain groups may influence the nature of the immune response induced. For instance, Parish (1972) showed that acetoacetylation converted flagellin from an antigen that induced antibody formation to a product that exclusively stimulated DTH. Similarly, Coon and Hunter (1973) showed that the addition of lipid groups to protein antigens both increased their immunogenicity and made them better able to stimulate DTH rather than antibody production.

Probably the types of results obtained by Coon and Hunter can be explained by changes made in the type of cells that handled the antigen and presented it to the appropriate T-cell type. Thus, it is now evident that there is marked diversity in APC and that T-cell subsets can only be triggered by the appropriate APC. It is well known that helper cells are only triggered by Ia-positive APC (Steinman and Nussenzweig, 1980;

Schmid *et al.*, 1982) and suppressor cell induction involves cells expressing IJ determinants (Nakamura *et al.*, 1982). Various APC types may show differential representation at different sites. This may explain why the route of antigen injection may influence the nature of the immune response induced. For example, as mentioned previously, exposure to HSV via the intravenous route fails to stimulate a DTH response and activates suppression (Nash and Gell, 1981). With reovirus, oral exposure to virus activates suppressor cells that modulate subsequent DTH responses to infectious virus (Rubin *et al.*, 1981).

The use of adjuvants has long been known to modulate the type and magnitude of the immune response but the mechanistic explanations for the action of adjuvant are still to be made. Choosing an appropriate adjuvant is still an inexact science. In the same vein, other formulations such as antigens presented in the form of liposomes or coupled to antibody, or some appropriate carrier protein may also change the nature of the response but the reasons for this are not clear.

Constructing hybrid viruses containing genes that code for the appropriate oligopeptide may be the method of choice in the future. Such viruses may lack pathogenicity but provide the adjuvanticity for the desired response. An example is the high antibody responses induced to hepatitis B virus surface antigen using vaccinia virus containing the hepatitis B gene (Smith *et al.*, 1983). Vaccinia viruses have been constructed containing the gene for thymidine kinase from HSV but not as yet with a glycoprotein gene (Nakano *et al.*, 1982).

The final topic arousing considerable excitement is the prospect of modulating immune responses by the use of anti-idiotype immunization (Fields and Greene, 1982). Thus, in the reovirus system, monoclonal antibodies have been developed that react with different determinants on the virus such as those responsible for neutralizing antibody production and those responsible for suppression. Furthermore, a second series of monoclonals could be developed that react with the binding region (idiotype) of the first series of monoclonals (for example with the suppressor-inducing determinant). The binding region of such anti-idiotype antibody mimics the determinant on the virus and, when injected into an animal, mimics this determinant. This powerful approach using anti-idiotype immunization could be used to select for the responses to different functional domains of the virus and to tailor the induction of desired responses. To date only reovirus has been studied but others including HSV need investigation.

REFERENCES

Ada, G. L., Leung, K. N., and Ertl, H., 1981, An analysis of T cell generation and function in mice exposed to influenza A or Sendai viruses, *Immunol. Rev.* **58**:5.

Adorini, L., Harvey, M. A., Miller, A., and Sercatz, E. E., 1979, Fine specificity of regulatory T cells. II. Suppressor and helper T cells are induced by different regions of hen white lysozyme in a genetically non-responder mouse strain, *J. Exp. Med.* **150**:293.

Berkovich, S., Millian, S. J., and Snyder, R. D., 1970, The association of viral and mycoplasma infections with recurrence of wheezing in the asthmatic child, *Ann. Allergy* **28**:43.

Cochrane, C. H., Weigle, W. O., and Dixon, F. J., 1959, The role of polymorphonuclear leukocytes in the inhibition and cessation of Arthus vasculitis, *J. Exp. Med.* **110**:481.

Coon, J., and Hunter, R., 1973, Selective induction of delayed hypersensitivity by a lipid conjugated protein which is localized in thymus dependent lymphoid tissue, *J. Immunol.* **110**:183.

Dannenberg, A. D., Chaikof, E., and Aurelian, L., 1980, Immunity to herpes simplex virus type 2: Cell mediated immunity in latently infected guinea pigs, *Infect. Immun.* **30**:99.

Day, R. P., Bienenstock, J., and Rawls, W. E., 1976, Basophil-sensitizing antibody response to herpes simplex viruses in rabbits, *J. Immunol.* **117**:73.

Ertl, H. C. J., 1981, Adoptive transfer of delayed type hypersensitivity to Sendai virus. II. Different modes of antigen presentation determine KD region or I region restriction of T cells mediating DTH to Sendai virus, *Cell. Immunol.* **63**:188.

Evans, A. S., and Niederman, J. C., 1982, Epstein–Barr virus, in: *Viral Infections of Humans: Epidemiology and Control* (A. S. Evans, ed.), pp. 253–282, Plenum Press, New York.

Fields, B. N., and Greene, M. I., 1982, Genetic and molecular mechanisms of viral pathogenesis: Implication for prevention and treatment, *Nature* **300**:19.

Germain, R. N., and Benacerraf, B., 1981, Hypothesis: A single major pathway of T lymphocyte interactions in antigen specific immune suppression, *Scand. J. Immunol.* **13**:1.

Haynes, B., Schooley, R. T., Payling-Wright, C. R., Grouse, J. E., Dolin, R., and Fauci, A. S., 1979, Emergence of suppressor cells of immunoglobulin synthesis during acute infectious mononucleosis, *J. Immunol.* **123**:2095.

Hutt, L. M., Huang, Y. T., Discomb, H. E., and Pagano, L. S., 1975, Enhanced destruction of lymphoid cell lines by peripheral blood leukocytes taken from patients with acute infectious mononucleosis, *J. Immunol.* **115**:243.

Ida, S., Hooks, J. J., Siragenian, R. P., and Notkins, A. L., 1977, Enhancement of IgE mediated histamine release from human basophils by viruses: Role of interferon, *J. Exp. Med.* **145**:892.

Ida, S., Siragenian, R. P., and Notkins, A. L., 1983, Cell-bound and circulating IgE antibody to herpes simplex virus, *J. Gen. Virol.* **64**:533.

Kaufman, H. E., 1978, Herpes keratitis, *Invest. Ophthalmol. Vis. Sci.* **17**:941.

Kaufman, H. E., Martola, E., and Dohlman, C., 1962, Use of 5-iodo-2'-deoxyuridine (IDU) in treatment of herpes simplex keratitis, *Arch. Ophthalmol.* **68**:235.

Kazmierowski, J. A., Peizner, D. S., and Wueppner, K. D., 1982, Herpes simplex antigen in immune complexes of patients with erythema multiforme: Presence following recurrent herpes simplex infection, *J. Am. Med. Assoc.* **247**:2547.

Larsen, H. S., Russell, R. G., and Rouse, B. T., 1983, Recovery from lethal herpes simplex virus type 1 infection is mediated by cytotoxic T cells, *Infect. Immun.* **41**:197.

Lawman, M. J., Rouse, B. T., Courtney, R. J., and Walker, R. D., 1980, Cell-mediated immunity against herpes simplex induction of cytotoxic T lymphocytes, *Infect. Immun.* **27**:133–139.

Leung, K. N., and Ada, G. L., 1982, Different functions of subsets of effector T cells in immune influenza virus infection, *Cell. Immunol.* **67**:312.

Liew, F. Y., and Russell, S. M., 1983, Inhibition of pathogenic effect of effector T cells by specific suppressor T cells during influenza virus infection in mice, *Nature (London)* **304**:541.

MacKaness, G. B., 1970, Cellular immunity, in: *Mononuclear Phagocytes* (R. van Furth, ed.), pp. 461–475, Blackwell, Oxford.

Metcalf, J. F., 1983, Role of cell mediated immunity in pathogenesis and prevention of herpetic stromal keratitis in mice, *Am. Vis. Res. Org. Proc.* Abstr. no. 200.

Metcalf, J. F., and Kaufman, H. E., 1976, Herpetic stromal keratitis: Evidence for cell mediated immunopathogenesis, *Am. J. Ophthalmol.* **82**:827.

Metcalf, J. F., and Reichert, R. W., 1979, Histological and electron microscopical studies of experimental herpes keratitis in the rabbit, *Invest. Ophthalmol. Vis. Sci.* **18**:1123.

Metcalf, J. F., Hamilton, D. S., and Reichert, R. W., 1979, Herpetic keratitis in athymic (nude) mice, *Infect. Immun.* **26**:1164.

Meyers, R. H., and Pettit, T. H., 1973, The pathogenesis of corneal inflammation due to herpes simplex virus. I. Corneal hypersensitivity in the rabbit, *J. Immunol.* **14**:1031.

Meyers-Elliot, R. L., and Chitjian, P. A., 1976, Immunology of herpesvirus infection: Immunity to herpes simplex virus in eye infections, *Surv. Ophthalmol.* **21**:194.

Meyers-Elliot, R. L., and Chitjian, P. A., 1981, Immunopathogenesis of corneal inflammation in herpes simplex virus stromal keratitis: Role of the polymorphonuclear leukocyte, *Invest. Ophthalmol. Vis. Sci.* **20**:784.

Meyers-Elliot, R. L., Pettit, T. H., and Maxwell, A., 1980, Viral antigens in the immune ring of herpes simplex stromal keratitis, *Arch. Ophthalmol.* **98**:897.

Mildran, D., Mathur, U., Eulow, R. W., Romain, P. L., Winchester, R. J., Colt, C., Fingman, H., Adelberg, B. R., and Spigland, I., 1982, Opportunistic infections and immune deficiency in homosexual man, *Ann. Intern. Med.* **97**:700.

Minor, T. E., Dick, E. C., DeMeo, N., Ouelette, J. J., Cohen, M., and Reed, C. E., 1974, Viruses as precipitants of asthmatic attacks in children, *J. Am. Med. Assoc.* **227**:292.

Nakamura, R. M., Tamaka, H., and Tokumega, T., 1982, In vitro induction of suppressor T cells in delayed type hypersensitivity to BCG and an essential role of IJ positive accessory cells, *Immunol. Lett.* **4**:295.

Nakano, E., Parriculi, D., and Paoletti, E., 1982, Molecular genetics of vaccinia virus: Demonstration of marker rescue, *Proc. Natl. Acad. Sci. USA* **79**:1593.

Nash, A. A., and Gell, P. G. H., 1981, The delayed hypersensitivity T cell and its interaction with other cells, *Immunol. Today* **2**:162.

Nash, A. A., Phelan, J., and Wildy, P., 1981a, Cell mediated immunity in herpes simplex virus infected mice: H-2 mapping of the delayed type hypersensitivity response and the antiviral T cell response, *J. Immunol.* **126**:1260.

Nash, A. A., Gell, P. G. H., and Wildy, P., 1981b, Tolerance and immunity in mice infected with herpes simplex virus: Simultaneous induction of protective immunity and tolerance to delayed type hypersensitivity, *Immunology* **43**:153.

Nathan, C. F., Murray, H. W., and Cohen, Z. A., 1980, The macrophage as an effector cell, *N. Engl. J. Med.* **303**:622.

Oldstone, M. B. A., 1979, Immune responses, immune tolerance and viruses, in: *Comprehensive Virology*, Vol. 15 (H. Fraenkel-Conrat and R. R. Wagner, eds.), pp. 1–36, Plenum Press, New York.

Oldstone, M. B. A., and Dixon, F. J., 1976, Immunopathology of viral infections, in: *Textbook of Immunopathology* (P. A. Miescher and H. J. Müller-Eberhard, eds.), p. 303, Grune & Stratton, New York.

Ozawa, T., and Stewart, J. A., 1979, Immune complex glomerulonephritis associated with cytomegalovirus infection, *Am. J. Clin. Pathol.* **72**:103.

Parish, C. R., 1972, The relationship between humoral and cell mediated immunity, *Transplant. Rev.* **13**:35.

Plowright, W., 1980, Vaccination against diseases associated with herpesvirus infections in animals: A review, in: *Oncogenesis and Herpesviruses III* (G. de Thé, W. Henle, and H. Rapp, eds.), p. 965, IARC, Lyon, France.

Purtilo, D. T., 1980, Immunopathogenesis and complications of infectious mononucleosis, in: *Pathology Annual 1980* (S. Somers and P. P. Rosen, eds.), pp. 253–299, Appleton–Century–Crofts, New York.

Richardson, W. P., Colvin, R. B., Cheeseman, S. H., Tolkoff-Rubin, N. E., Herrin, J. T., Cosimi, A. B., Collins, A. B., Hirsch, M. S., McCluskey, R. T., Russell, P. S., and Rubin, R. H., 1981, Glomerulopathy associated with cytomegalovirus viremia in renal allografts, *N. Engl. J. Med.* **305**:57.

Rouse, B. T., 1984, Role of adaptive immune defense mechanisms in herpes simplex resistance, in: *Immunobiology of Herpes Simplex Virus Infection* (B. T. Rouse and C. Lopez, eds.), CRC Press, Boca Raton, Fla., in press.

Rouse, B. T., and Babiuk, L. A., 1979, Mechanisms of viral immunopathology, *Adv. Vet. Sci. Comp. Med.* **23**:103.

Royston, I., Sullivan, J. L., Periman, P. O., and Perlin, E., 1975, Cell mediated immunity to Epstein–Barr virus transformed lymphoblastoid cells in acute infectious mononucleosis, *N. Engl. J. Med.* **293**:1159.

Rubin, D., Weiner, H. L., Fields, B. N., and Greene, M. I., 1981, Immunologic tolerance after oral administration of reovirus: Requirement for two viral gene products for tolerance induction, *J. Immunol.* **127**:1697.

Schmid, D. S., and Rouse, B. T., 1983, Celluar interactions in the cytotoxic T lymphocyte response to herpes simplex virus antigens: Differential antigen activation requirements for the helper T lymphocyte and cytotoxic T lymphocyte precursors, *J. Immunol.* **131**:479.

Schmid, D. S., Larsen, H. S., and Rouse, B. T., 1982, Role of Ia antigen expression and secretory function of accessory cells in the induction of cytotoxic T lymphocyte responses against herpes simplex virus, *Infect. Immun.* **37**:1138.

Schrier, R. D., Pizer, L. I., and Moorhead, J. W., 1983, Tolerance and suppression of immunity to herpes simplex virus: Different presentations of antigens induce different types of suppressor cells, *Infect. Immun.* **40**:514.

Sheeley, W. B., 1967, Herpes simplex virus as a cause of erythema multiforme, *J. Am. Med. Assoc.* **201**:71.

Sheppard, A. B., and Smith J. V., 1981, Antibody mediated destruction of keratocytes infected with herpes simplex virus, *Curr. Eye. Res.* **1**:397.

Sheridan, J. F., Dannenberg, A. D., Aurelian, L., and Elpern, D. J., 1982, Immunity to herpes simplex virus type 2. IV. Impaired lymphokine production during recrudescence correlates with an imbalance in T lymphocyte subsets, *J. Immunol.* **129**:326.

Shore, S. L., and Nahmias, A. J., 1982, Immunology of herpes simplex virus in: *Comprehensive Immunology*, Vol. 9 (R. A. Good and S. B. Day, eds.), Plenum Medical Book Co., New York, pp. 21–72.

Smith, G. L., Mackett, M., and Moss, B., 1983, Infectious vaccinia virus recombinants that express hepatitis B surface antigen, *Nature* **302**:490.

Stagno, S., Volanakis, J., Reynolds, D. W., Strous, R., and Alford, C. A., 1977, Immune complexes in congenital and natal cytomegalovirus infections of men, *J. Clin. Invest.* **60**:838.

Steeg, P. S., Moore, R. N., Johnson, H. M., and Oppenheim, J. J., 1982, Regulation of murine macrophage Ia antigen expression by a lymphokine with immune interferon activity, *J. Exp. Med.* **156**:1780.

Steinman, R. M., and Nussenzweig, M. C. 1980, Dendritic cells: Features and functions, *Immunol. Rev.* **53**:127.

Sullivan, J. L., Byron, K. S., Brewster, F. E., Baker, S. M., and Ochs, H. D., 1983, X linked lymphoproliferative syndrome: Natural history of the immunodeficiency, *J. Clin. Invest.* **71**:1765.

Svedmyr, E., and Jondal, M., 1975, Cytotoxic effector cells specific for B cell lines transformed by Epstein–Barr virus are present in patients with infectious mononucleosis, *Proc. Natl. Acad. Sci. USA* **72**:1622.

Tosato, G., Magrath, I., Koski, I., Dooley, N., and Blaese, M., 1979, Activation of suppressor T cells during Epstein–Barr virus induced infectious mononucleosis, *N. Engl. J. Med.* **301**:1133.

Wallace, M., Leet, G., and Rothwell, P., 1974, Immune complex mediated glomerulonephritis with infectious mononucleosis, *Aust. N.Z. Med. J.* **4**:192.

Yamamoto, J. K., Farrar, W. L., and Johnson, H. M., 1982, Interleukin 2 regulation of mitogen induction of immune interferon (IFN-gamma) in spleen cells and thymocytes, *Cell. Immunol.* **66**:333.

CHAPTER 6

Cell-Mediated Immunity in Cytomegalovirus Infections

GERALD V. QUINNAN, JR.

I. INTRODUCTION

The many interactions between cytomegalovirus (CMV) and host cellular immunity include some of the most clinically significant, complex, and fascinating problems in viral immunology. Most clinically important diseases caused by CMV occur in individuals with deficient cellular immunity (Meyers, this volume). Because of increasing numbers of individuals with deficiences of this type and the imminent potential for effective modulation of immune responses, there is a rather pressing need to understand the specific roles of the various components of the immune system in CMV infections. In addition, CMV itself can induce both profound immunosuppressive effects and significant immunopathology. As a result of recent progress in immunology, exciting opportunities have emerged to develop an understanding of these phenomena.

The traditional conceptualization of the immune system as consisting of two compartments, cellular and humoral, has become progressively difficult to apply, as increasing numbers of examples of combined and interdependent effects of cellular and humoral immunity have been demonstrated. Among the myriad actions of the cellular and soluble factors involved in immune responses, there are a number of functions that have direct effects on the extent of virus replication *in vivo*. The components of the immune response that mediate these effects include antibodies and certain types of leukocytes. The effector cells of cell-mediated immunity develop from precursor cells, a process that is dependent on precursor

GERALD V. QUINNAN, JR. • National Center for Drugs and Biologics, Division of Virology, Office of Biologics, Food and Drug Administration, Bethesda, Maryland 20205.

interactions with accessory cells and soluble factors derived from them, and modulated by immunoregulatory cells and their soluble factors. To understand how the immune system mediates recovery from CMV infection, we must understand what roles the various effector functions play. When they are deficient, it can be determined whether the deficiency relates to an abnormality of effector cell precursors, accessory cells, or immunoregulation. The immunosuppressive and immunopathological effects of CMV infection can potentially be understood using this same approach.

II. CHARACTERISTICS OF EFFECTOR CELL FUNCTIONS RELEVANT IN CMV INFECTIONS

The responses that have been most studied in CMV infections include antibodies, cytotoxic T cells, delayed-type hypersensitivity (DTH), natural killer (NK) cells, and antibody-dependent cell-mediated cytotoxicity (ADCC). Other possible effector responses have not been studied in detail. With the exception of antibodies functioning alone, each of these mechanisms is cell mediated in the sense that it involves the direct interaction between cells of the immune system and virus-infected cells.

The emphasis in this chapter on cellular, rather than antibody, effects is not meant to discount a significant role for antibodies in CMV infections. There is much evidence suggesting that antibodies are important. For example, premature infants with neonatal CMV infections are much more likely to die from their infections if they were born to nonimmune rather than to immune mothers (Yeager *et al.*, 1981). Nevertheless, even very high antibody levels are inadequate to mediate recovery from CMV infection in the absence of cellular immunity. For this reason, the cellular effector responses will be discussed in detail.

A. NK Cells and Other Large Granular Lymphocytes

The NK cell is a lymphocyte of the type referred to previously as null cells (Herberman and Holden, 1978; Herberman, 1980). It lacks surface immunoglobulin, differentiating it from B cells. NK cells do not adhere to nylon wool columns, differentiating them from both B cells and macrophages. They are distinguished from T cells in that they do not form rosettes with sheep erythrocytes at 29°C (they do form rosettes at 4°C), and they lack, or have very low density of, T-cell surface antigens (see also Lopez, this volume).

During the past few years, significant advances have been made in understanding this class of cells. Morphologically, NK cells are intermediate in size between typical T and B lymphocytes on the one hand, and monocytes on the other (Timonen and Saksela, 1980). They have indented or lobulated nuclei, a high cytoplasmic/nuclear ratio, and azurophilic cytoplasmic granules. As a consequence, they are referred to as large granular lymphocytes (LGL). NK cells have characteristic cell surface markers, including receptors for the Fc portion of IgG. There are at least two types of spontaneously cytotoxic cells in mice: NK cells and the cell type referred to as natural cytotoxic (NC) cells (Lattime et al., 1981). NC cells have not yet been identified in humans. In mice, NK cells are distinguished from NC cells by the presence of a high density of the sphingolipid, asialo-GM_1, on the surface of NK but not NC cells and by a difference in the target cells killed by each of the effectors. There are also monoclonal antibodies that appear to be specific for human NK cells, e.g., NK-8 (Nieminen et al., 1982) and HNK-1 (Abo and Baldi, 1981). In addition to their cytotoxic functions, LGL produce important lymphokines including interferon-α (IFN-α), interleukin-2 (IL-2), B-cell growth factor, and IL-1 (Trinchieri et al., 1978; Djeu et al., 1982; Kasahara et al., 1983). By virtue of these lymphokine-producing activities, LGL are required accessory cells for T-cell responses (Burlington et al., 1984) and they also self-regulate their own cytotoxic activity (Djeu et al., 1982). Thus, there are at least two effector functions and multiple accessory cell functions mediated by LGL.

The procedures used for measuring NK cell or NC cell activity involve testing the ability of these cells to kill different types of target cells. NK cells generally are highly active in killing of anchorage-independent tumor cell lines. The K562 erythroleukemia cell line is often used for measuring human NK cells because of its high sensitivity (West et al., 1977). NC cells are more active against anchorage-dependent cells, such as fibroblasts, and kill more slowly than NK cells. A variety of assay procedures are used, the most common of which is the chromium-release assay. Target cell proteins are radiolabeled with $Na_2 \, ^{51}Cr \, O_4$, and the amount of ^{51}Cr released from the target cells in the presence of effector cells is a measure of lytic activity. NK cell activity is enhanced by exposure to IFN (α, β, or γ) or IL-2 and is depressed by steroids, cyclophosphamide, or cyclosporin A (Gidlund et al., 1978; Djeu et al., 1979; Rook et al., 1983). NC cell activity in the mouse is enhanced by IL-3 (Djeu et al., 1983). The NK cells responding with increased cytotoxicity during CMV infection are not specific for CMV antigens, even though CMV-infected cells may be more susceptible to lysis by the NK cells than uninfected target cells (Quinnan and Manischewitz, 1979). NK cell-mediated killing ordinarily requires direct contact between effector cell and the target cell and is independent of antibodies and complement.

B. Antibody-Dependent Cell-Mediated Cytotoxicity

The effector cells that mediate ADCC probably include all leukocytes that possess receptors for the Fc portion of IgG. In human and murine CMV infections, the effector cells that have been found to mediate this effect are similar or identical to NK cells (Quinnan and Manischewitz, 1979; Kirmani *et al.*, 1981). Monocytes or neutrophils can also mediate ADCC in certain situations (Kohl *et al.*, 1977; Oleski *et al.*, 1977). Certain classes of antibodies are more active than others in ADCC, probably because they are more "cytophilic" and bind to Fc receptors more avidly (Okafor *et al.*, 1974). Like NK cell killing, ADCC requires direct contact between effector and target cell. ADCC is antigen specific by virtue of the fact that the effector cells are armed with specific antibodies. The addition of complement may sometimes enhance the killing that occurs in an ADCC assay (Rouse *et al.*, 1977).

The individual antigens recognized by CMV-specific ADCC are unknown. In the case of herpes simplex-specific ADCC, early antigens are recognized (Shore *et al.*, 1976). This specificity is probably important. It would allow for recognition of infected cells by the effector cells before infectious virus particles are formed, so that ADCC could be an efficient method of reducing the amount of virus replication and cell-to-cell spread of virus.

Assays for ADCC effector function may be designed to quantitate antibody-dependent killer cell activity, specific antibodies in sera that can participate in ADCC, or both (Kirmani *et al.*, 1981). Target cells sensitized with specific antibodies can be used as a measure of killer cell activity. Conversely, unknown sera can be tested for their ability to arm killer cells or sensitize target cells and participate in ADCC. In some cases, lymphocytes already armed with specific antibodies from the donor can be tested for cytotoxic activity. Thus, the results of a cytotoxicity assay using virus-infected target cells may reflect ADCC activity with or without the addition of virus-specific antisera.

C. Cytotoxic T Cells

Cytotoxic T cells for CMV-infected fibroblasts possess receptors for sheep erythrocytes (Quinnan *et al.*, 1981). They are also nonadherent cells that lack receptors for Fc of IgG and possess T-cell surface antigens (Quinnan *et al.*, 1982b; Rook and Quinnan, 1983). In mice they are Thy-1.2 and Lyt-23 positive (Quinnan *et al.*, 1978; Varho *et al.*, 1981). In humans they are usually OKT3 and OKT8 positive (Reinherz *et al.*, 1979). Cytotoxic T cells are typically antigen-specific: they kill target cells that possess the antigens that they were originally induced by. In most cases, cytotoxic T cells are also restricted in activity by antigens of the MHC (Zinkernagel and Welsh, 1976; McMichael, 1978). As a result, the target cells they kill must not only express the specific antigens against which the effector cell is directed, but must also have MHC antigens in common with the effector cells. The usual MHC antigens involved in this restric-

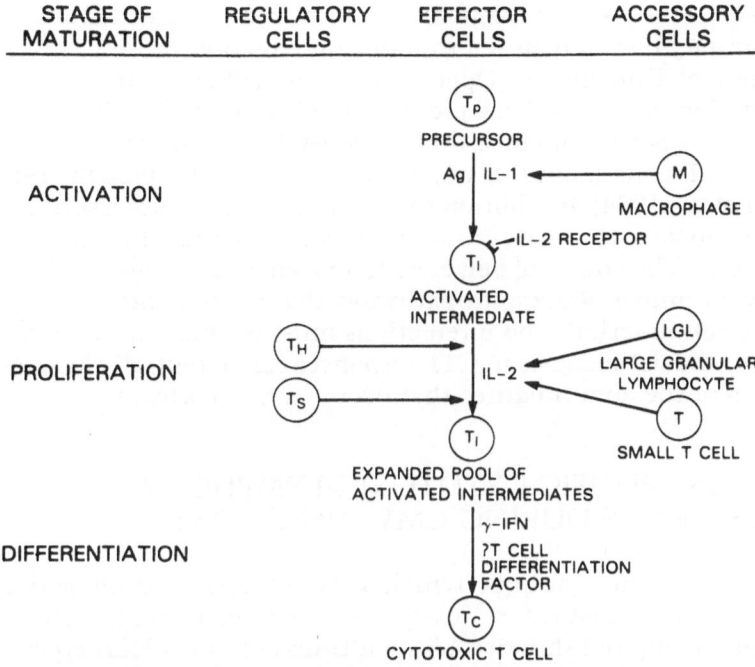

FIGURE 1. Proposed scheme for maturation of precursor T cells (T_p) into cytotoxic cells (T_c), through intermediate stages (T_I) of activation and proliferation. T_H and T_S indicate T-helper and T-suppressor cells, respectively. From Quinnan and Rook (1984).

tion are the class I antigens, HLA-A and B in humans and the analogous H-2K and D antigens in mice.

Lymphokines, such as IFN and IL-2, affect CTL activity, but in a somewhat different way than they enhance NK cell activity. NK cells are normally present in peripheral blood, and their activity is enhanced by these lymphokines. CTL, on the other hand, are not normally present in peripheral blood, but appear for a brief period of time several days after an appropriate antigenic exposure. During the course of such a response, CTL precursors, which circulate in peripheral blood, must undergo maturation into effector cells. The process by which this maturation probably occurs is summarized schematically in Fig. 1. Precursors appear to undergo a stepwise process of maturation into effector cells. The stages include precursor activation, proliferation, and differentiation. Each is dependent on lymphokines, including IL-1 in the activation stage (Oppenheim *et al.*, 1982), IL-2 in the proliferation stage (Farrar *et al.*, 1982), and IFN-γ [and possibly one or more additional T-cell differentiation factors (TCDF)] in the differentiation stage (Reddehase *et al.*, 1982). The lymphokines are produced by cells that are required as accessory cells for the process to occur. Each stage is dependent on the prior stage in the sense that the lymphokines involved at one stage induce production of or responsiveness to the lymphokines produced at the next stage, or both. For example, IL-1 is required to induce production of IL-2 as well as for activation of T precursors (Oppenheim et al., 1982) and IL-2 induces IFN-γ production as well as T-cell proliferation (Farrar et al., 1981). The known accessory cells that produce the necessary lymphokines are monocytes, LGL, and T cells (Farrar et al., 1981; Durum and Gershon, 1982; Burlington et al., 1984). In addition to the interactions of accessory cells with CTL precursors, the maturation process is modulated by immunoregulatory cells. The effects of helper cells and suppressor cells on this process involve a number of complex pathways that are not shown in Fig. 1. All of these cell functions and interactions must be intact for CTL responses to occur, and deficiencies in CTL responses can potentially be understood in terms of the abnormalities that occur in this pathway.

III. VIRUS-SPECIFIC CYTOTOXIC LYMPHOCYTE RESPONSES DURING CMV INFECTIONS

CMV-specific cytotoxic lymphocyte responses during active infection have been measured in ^{51}Cr-release assays using target cells prepared from human diploid skin fibroblast cultures of known HLA types. Several features of the cytotoxic response are shown in Fig. 2, which summarizes cytotoxic lymphocyte responses observed in a group of 88 bone marrow transplant recipients. The lower half of Fig. 2 indicates the times after transplantation at which viruses were isolated from specimens of blood, urine, throat washings, or biopsy specimens. Those marked with asterisks were the first isolates obtained from each patient. As is typical of CMV infections in bone marrow transplant recipients, the onset was usually between 4 and 12 weeks after transplantation. In the upper panel, the results of CMV-specific cytotoxicity assays on these same patients are shown. Lymphocytes tested early after transplantation, and before onset of CMV infection, have low levels of cytotoxicity against CMV-infected target cells, both HLA-matched and mismatched. The cells are generally more active in killing of infected than uninfected target cells. As will be discussed below, this nonrestricted killing is mediated by NK cells and/or ADCC. During the posttransplant period, when virus cultures are positive, there is an increase in cytotoxicity of lymphocytes from infected patients. In contrast to the cytotoxicity seen before onset of infection, however, this response to active infection is usually HLA-restricted (Quinnan et al., 1981, 1982b). There is increased killing of HLA-matched target cells, but little or no increase in CMV-specific lysis of HLA-mismatched target cells. This HLA-restricted response generally persists throughout the period of time when virus cultures are positive, the cytotoxicity then returns to the baseline level of nonrestricted activity. Thus, the HLA-restricted, CMV-specific cytotoxic lymphocyte response is a manifestation of acute infection.

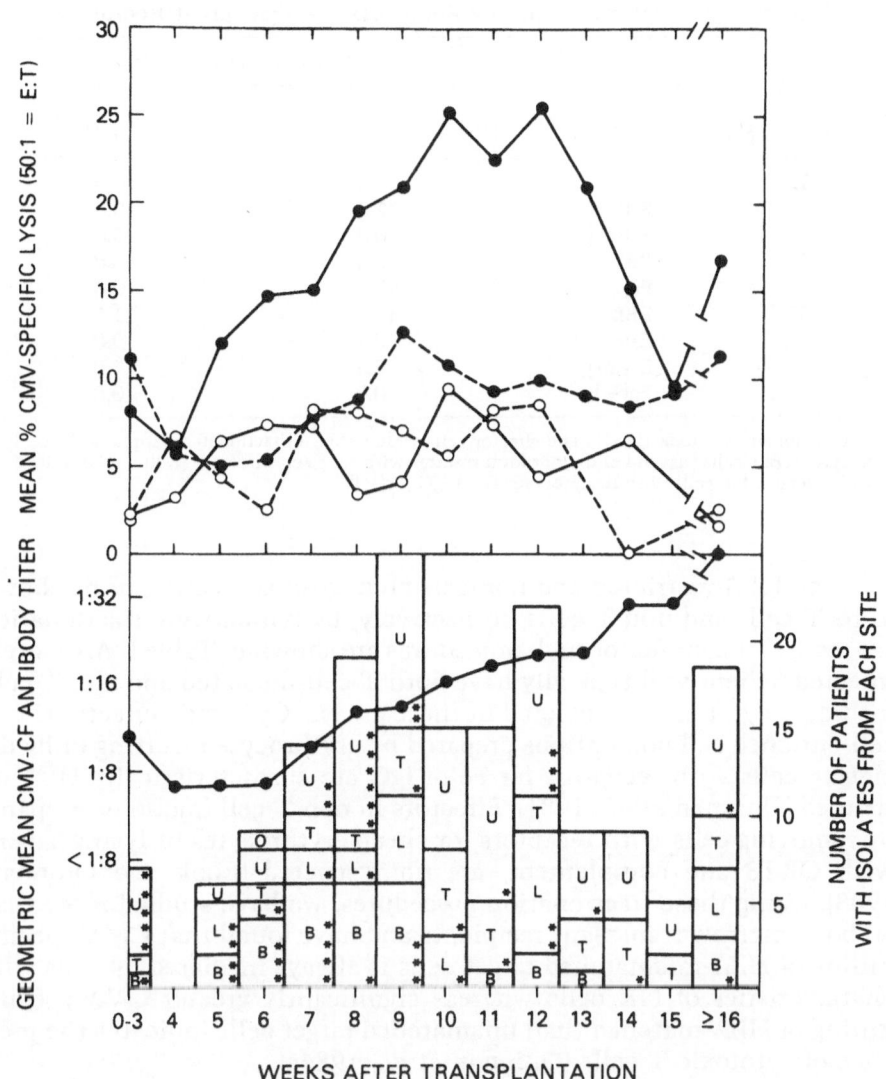

FIGURE 2. CMV-specific cytotoxic lymphocyte and antibody responses in relation to virus shedding in bone marrow transplant recipients. In the upper panel, results shown are CMV-specific cytotoxicity of lymphocytes from infected patients against HLA-matched (●———●) and mismatched (●– – – –●) target cells and CMV-specific cytotoxicity of lymphocytes from uninfected patients against HLA-matched (○———○) and mismatched (○– – – –○) target cells. In the lower panel are shown complement-fixing (CF) antibody responses (●) and, in the histogram, virus isolates. The total number of isolates each week is divided into the number from each site including (B) blood, (U) urine, (T) throat, (L) lung or liver, and (O) other sites. Asterisks indicate isolates that were the first ones obtained from individual patients. From Quinnan *et al.* (1984a).

TABLE I. CMV-Specific T- and Non-T-Cell-Mediated Cytotoxicity Early and Late during CMV Infection in a Bone Marrow Transplant Recipient[a]

| | | % CMV-specific lysis | |
| | | | HLA-mismatched target |
Day posttransplant	Effector	HLA-matched target cell	cell
39	PBL	7.3	0.0
	E-Ros(+)	22.3	0.0
	E-Ros(−)	0.0	0.0
46	PBL	25.3	0.0
	Fc(−)	27.1	0.0
69	PBL	15.7	11.5
	E-Ros(+)	0.2	0.0
	E-Ros(−)	5.3	7.0
	Fc(−)	0.0	0.0

[a] From Quinnan and Rook (1981). The effector cells were either unfractionated peripheral blood lymphocytes (PBL), cells that did or did not form rosettes with sheep erythrocytes [E-Ros(+) or E-Ros(−)], or PBL depleted of cells with receptors for Fc of IgG [Fc(−)].

The HLA-restricted and nonrestricted cytotoxic cells can be shown to be T cells and non-T cells, respectively, by lymphocyte fractionation procedures. Examples of such procedures are shown in Table I. An acutely infected patient will typically have both T-cell-mediated and non-T-cell-mediated cytotoxicity detected in these assays. Cytotoxic effectors in T-cell-enriched cell populations prepared by erythrocyte rosetting or by depleting cells with receptors for Fc of IgG are characteristically HLA-restricted (Quinnan et al., 1981). Effectors in non-T-cell fractions, prepared by removing cells with receptors for sheep erythrocytes or lysing T cells with OKT3 and complement, are nonrestricted (Rook and Quinnan, 1983). Using these fractionation procedures, we have studied more than 50 bone marrow transplant recipients and have found that CMV-specific killing of HLA-mismatched target cells is always mediated by cells with characteristics of NK cells whereas significantly greater CMV-specific killing of HLA-matched than mismatched target cells indicates the presence of cytotoxic T cells (Quinnan et al., 1984a).

The majority of bone marrow or renal transplant recipients develop CMV-specific cytotoxic lymphocyte responses during CMV infection (Quinnan et al., 1984; Rook et al., 1984). Based on the results of lymphocyte fractionation studies, it can be assumed that HLA-restricted cytotoxicity is an indication of cytotoxic T-cell activity. In most but not all cases, the responses that develop during infection are T cell mediated. The types of responses found in 46 infected patients are summarized in Table II. Among patients whose lymphocytes were adequately tested to define the types of effector cells involved, 89% had CTL responses, but a few had significant increases in cytotoxicity during infection that did not include a T-cell component.

TABLE II. Effector Cells Mediating CMV-Specific Cytotoxic Lymphocyte Activity in Peripheral Blood of Bone Marrow Transplant Patients[a]

| | | Type of CMV-specific cytotoxicity | |
Patient group	Number	HLA-restricted	Not HLA-restricted
CMV-infected	28	25	3
Uninfected	18	0	18

[a] From Quinnan et al. (1984a). Cytotoxicity was considered HLA-restricted if there was significantly greater CMV-specific lysis of all HLA-matched target cells tested than there was of the HLA-mismatched target cells.

The results of studies in healthy volunteers with and without CMV infections have confirmed that the responses observed in bone marrow transplant recipients are representative of the "normal" immune response. Peripheral blood lymphocytes from healthy, uninfected volunteers have CMV-specific cytotoxicity that is not HLA-restricted, and is mediated by cells with characteristics of NK cells, as shown in Table III. The activity measured is similar in magnitude in seropositive and seronegative volunteers (Kirmani et al., 1981). Repeated washing of lymphocytes from seropositive volunteers inconsistently reduces their cytotoxicity, indicating that it may sometimes be antibody independent. It was possible to demonstrate that addition of immune serum to lymphocytes from nonimmune volunteers enhanced their CMV-specific cytotoxicity, as shown in Table IV, demonstrating that ADCC might contribute to the CMV-specific cytotoxic response. During active CMV

TABLE III. Non-HLA-Restricted CMV-Specific Cytotoxicity of Peripheral Blood Lymphocytes from Normal, Uninfected Volunteers[a]

| | | CMV-specific lysis of target cells (% ± S.D.) | |
Volunteer number	Serum CMV antibody titer	HLA-matched target cells	HLA-mismatched target cells
1	1:160	11.7 ± 1.2	12.9 ± 2.0
2	1:160	10.6 ± 2.1	10.2 ± 1.4
3	1:1280	11.6 ± 1.7	16.9 ± 1.5
4	1:160	14.7 ± 2.0	15.1 ± 9.2
5	1:320	6.1 ± 2.8	14.5 ± 4.1
6	1:160	11.5 ± 3.7	16.4 ± 2.0
7	1:640	12.5 ± 1.1	13.9 ± 2.8
8	<1:40	11.6 ± 3.3	7.1 ± 1.6
9	<1:40	9.5 ± 5.2	14.0 ± 7.5
10	<1:40	19.9 ± 8.0	10.8 ± 1.3

[a] From Kirmani et al. (1981). CMV antibodies were measured by ELISA. Results shown for cytotoxicity tests are means ± S.D. for four replicates.

TABLE IV. Enhancement of CMV-Specific Cytotoxicity of Peripheral Blood
Lymphocytes from Normal Volunteers after Sensitization of Target Cells with
Immune Sera[a]

Volunteer number	CMV antibody status	Percent lysis of CMV-infected target cells[b]	
		Unsensitized	Antibody sensitized
1	1:160	6.2 ± 1.5	17.5 ± 4.9
2	1:160	−8.1 ± 2.2	31.4 ± 12.4
3	1:1280	1.0 ± 0.7	8.4 ± 2.9
4	<1:40	11.6 ± 3.3	52.8 ± 1.9
5	<1:40	6.9 ± 2.0	18.2 ± 4.7

[a] From Kirmani *et al.* (1981). Sera from volunteers were tested for antibodies by ELISA. The serum used
to sensitize target cells had an anti-CMV titer of 1:1280.
[b] Mean percent lysis ± S.D. of four replicates. There was no significant lysis of uninfected target cells
tested in parallel.

infection in experimentally infected healthy volunteers, an increase in
CMV-specific cytotoxicity occurs above the baseline NK cell/ADCC-me-
diated activity seen in uninfected volunteers. As shown in Fig. 3, this
response to acute infection is HLA-restricted (Quinnan *et al.*, 1984b). In
both transplant recipients and healthy volunteers, the CMV-specific cytotoxic response occurs early in infection, either prior to or coincident
with clinical manifestation of infection or virus shedding, and persists
until clinical manifestations have resolved (Quinnan *et al.*, 1984a,b).

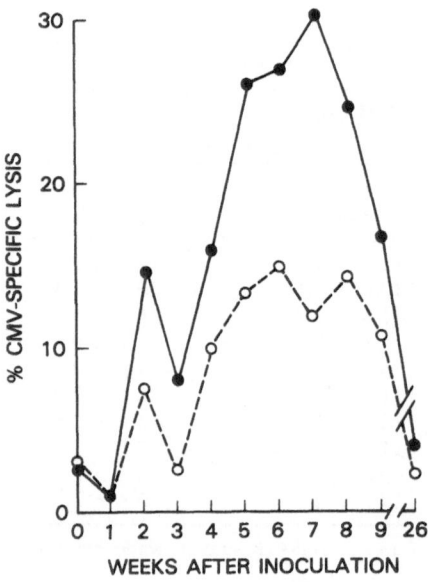

FIGURE 3. CMV-specific cytotoxic lymphocyte responses in healthy volunteers experimentally infected with a low-passage strain of
CMV (Toledo-1). From Quinnan *et al.* (1984b).

TABLE V. Correlation of Severity of CMV Infection in Bone Marrow Transplant Recipients with Failure to Develop a CMV-Specific Cytotoxic Lymphocyte Response[a]

CMV-specific cytotoxic responses	N	Interstitial pneumonitis	Cause of death	
			CMV infection	Other
Responders	34	9	0	7
		$p < 0.01$	$p < 0.001$	
Nonresponders	14	10	7	7

[a] From Quinnan *et al.* (1984a).

IV. EVIDENCE THAT VIRUS-SPECIFIC CYTOTOXIC LYMPHOCYTES DETERMINE THE OUTCOME OF CMV INFECTIONS

Studies in humans have consistently demonstrated a correlation between cytotoxic lymphocyte responses and a favorable outcome of CMV infection. In bone marrow transplant recipients with CMV infection, the occurrence of a CMV-specific cytotoxic response correlates with a decreased likelihood of interstitial pneumonitis and death, as shown in Table V (Quinnan *et al.*, 1982b). In these same individuals, pretransplant antibody status did not correlate with outcome. There was a tendency for patients who survived infection to develop higher antibody responses during infection, but the difference was not statistically significant. Similar results have been obtained in renal transplant recipients as summarized in Table VI. Both the severity of manifestations directly attributable to CMV infection, and the incidence of complications are greater in patients who fail to develop these responses (Rook *et al.*, 1984). In addition, renal transplant recipients with CMV infections who failed to develop virus-specific cytotoxic responses had clinical courses characterized by persistent viremia eventuating in death or loss of graft. Recovery from infection with conversion of cultures to negative did not occur in these nonresponders until after loss of the graft and discontinuation of immunosuppressive drug therapy.

Recently, we have found that patients with AIDS also fail to develop CMV-specific cytotoxic responses (Quinnan *et al.*, 1984c), and progress to fatal infection (Macher *et al.*, 1983) despite very high titers of antibodies to CMV in their serum. These results suggest that the cell-mediated immune responses are important in determining the outcome of infection.

Studies in mice also suggest that CMV-specific CTL are the most important effector cells in CMV infection. Mice develop CTL responses that are similar to those seen in humans, and are restricted by the MHC class I antigens, H-2D and K (Quinnan and Manischewitz, 1979). Nude mice, deficient in T-cell responses, are extremely susceptible to CMV

TABLE VI. Signs of Clinical Illness and Complications of CMV Infection in Relation to Responder and Nonresponder Status[a]

	Patient group	
	Responders ($N = 14$)	Nonresponders ($N = 6$)
Clinical findings		
Fever	4	5
Leukopenia	2	5
Thrombocytopenia	0	4
Elevated serum transaminases	4	5
Total with any 3 findings	0	5[b]
Complications		
Superinfection	0	3
Interstitial pneumonitis	0	1
Pancreatitis	0	1
Death	0	1
Total with any complication	0	5[b]
Viremia $\geq$ 3 weeks	0	5
Acute allograft dysfunction	1	4[c]

[a] From Rook et al. (1984).
[b] Findings and complications were more frequent in nonresponders than responders ($\chi^2 = 11.4$, $p < 0.001$).
[c] Allograft dysfunction was more common in nonresponders ($\chi^2 = 5.1$, $p = 0.02$).

infection (Starr and Allison, 1977). Ho (1980) has shown that adoptive transfer of CTL confers protection to mice of the same but not of a different H-2 type. Although the immune cells transferred undoubtedly included other types of effector cells, the finding that the protective effect was H-2-restricted suggests that the CTL determined recovery from infection.

V. EVIDENCE THAT NK CELLS AND OTHER LARGE GRANULAR LYMPHOCYTES ARE IMPORTANT IN CMV INFECTIONS

Increased NK cell activity during CMV infection was first demonstrated in mice (Quinnan and Manischewitz, 1979). This response is probably induced by IFN produced during the first few days of infection (Gidlund et al., 1978; Grundy et al., 1982; Quinnan et al., 1982a). Suppression of NK cell activity by hydrocortisone treatment of mice is accompanied by a decrease in the mononuclear cell infiltrate in lungs of mice with interstitial pneumonitis, and an increase in pulmonary virus replication (Quinnan et al., 1982a). These effects occur earlier in the course of infection than CTL or antibody responses, thus indicating that the suppression of NK cells may be the specific mechanism by which drug treatment is responsible for the increase in virus replication. Studies in mice ge-

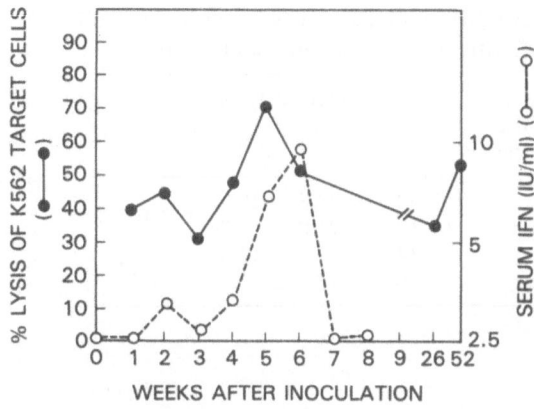

FIGURE 4. Natural killer cell and serum interferon response in healthy volunteers experimentally infected with a low-passage strain of CMV (Toledo-1). From Quinnan *et al.* (1984b).

netically susceptible or resistant to CMV (Grundy *et al.*, 1981) have provided further support for the hypothesis that NK cells are important determinants of the outcome of infection (Bancroft *et al.*, 1981). The mechanism by which NK cells might mediate such an effect is not certain. They lack viral antigen-specificity but are frequently more cytotoxic for infected than uninfected target cells (Diamond *et al.*, 1977; J. F. Manischewitz and G. V. Quinnan, unpublished observations). The most convincing evidence that NK cells mediate an antiviral effect in mice was demonstrated by Bukowski *et al.* (1983). They found that susceptibility of mice to infection was increased by administration of antibody to asialo-GM_1, a treatment that reduced NK cell activity without affecting CTL responses. The magnitude of the role of NK cells in murine CMV infection appears to be relatively small, however. When comparing mice that are susceptible or resistant based either on genetic background or on NK cell suppressive therapy, the differences in lethal dose of CMV were only 2- to 10-fold. By comparison, nude mice have normal NK cells but susceptibility is increased by 1000-fold or more (Starr and Allison, 1977). Evidence from studies in mice, therefore, indicates that NK cells probably have a specific but minor role in resistance to CMV infection.

NK cell activity also correlates with susceptibility to serious CMV infection in humans (Quinnan *et al.*, 1982b). NK cell cytotoxicity increased during human CMV infection of experimentally infected healthy volunteers (Fig. 4). This response was coincident with an increase in IFN production, similar to the response in mice, but it followed, rather than preceded, the CTL response (Quinnan *et al.*, 1984b).

In transplant recipients it has not been clearly demonstrated that this increase in NK cell activity occurs. However, the level of NK cell cytotoxicity measured during infection correlates with both the outcome of infection and cytotoxic T-cell responses. The latter is shown in Table VII, and similar comparisons are obtained in both bone marrow and renal transplant recipients (Quinnan *et al.*, 1982b; Rook *et al.*, 1984). NK cell

TABLE VII. Relationship of CMV-Specific and Natural Killer Cell
Cytotoxicity in CMV-Infected Transplant Recipients

Study group	CMV-specific cytotoxic responses	N	Percent lysis of K562 target cells (E:T = 50:1)	
BMTR[a]	Responders	32	42.3 ± 10.3	$p < 0.001$
	Nonresponders	13	21.0 ± 8.7	
RTR[b]	Responders	14	27.3 ± 12.6	$p < 0.01$
	Nonresponders	6	17.3 ± 7.9	

[a] From Quinnan et al. (1984c).
[b] From Rook et al. (1984).

cytotoxicity is also low in patients with AIDS (Siegal et al., 1981; Quinnan et al., 1984c). The consistency of the association of low NK cell activity with an unfavorable outcome suggests that these effectors may have a direct role in human CMV infections, but other possibilities should also be considered. For example, they may be depressed by the same factors (e.g., drug therapy) that depress CTL responses (Rook et al., 1984), or depressed NK activity may simply reflect a more generalized depression of the lymphokine-producing activity of LGL (Kasahara et al., 1983). The significance of the noncytotoxic functions of LGL has been emphasized recently in studies of patients with AIDS, as will be discussed below. The various possible explanations for the correlation of NK cell activity with outcome of CMV infection are not mutually exclusive. The extensive evidence from human and animal studies strongly suggests that one or more functions of LGL are vitally important for resistance to this infection.

VI. CAUSES OF DEPRESSED CYTOTOXIC T-CELL RESPONSES AND NK CELL ACTIVITY IN HIGH-RISK POPULATIONS

A principal objective of studies of immune responses to CMV infection is to develop concepts that can be turned to therapeutic advantage. With this objective in mind, we have attempted to define the reasons why some patients fail to develop CTL responses. In renal transplant recipients, steroid therapy has been correlated with nonresponsiveness in the CTL assay (Table VIII). When patients studied during CMV infection had received high-dose steroids prior to the time the testing was performed, CTL responses were frequently absent (Rook et al., 1984). This analysis was particularly revealing as the suppressive effect of steroids was not immediate. If the high-dose prednisone had been given within 4 days of the time that the blood was drawn, but not earlier, a normal cytotoxic response was observed. If the prednisone was given between 5

TABLE VIII. Methylprednisolone Effect on Cytotoxic T-Cell Precursors[a]

Interval between administration of methylprednisolone and assay date	CMV-specific cytotoxic responses (assays positive/total assays)
None given	10/12
0–4 days	4/4
5–10 days	2/10
10–14 days	3/10
>14 days	5/6

[a] From Rook et al. (1984). Intravenous methylprednisolone 1 g/day.

days and 2 weeks before the assay date, there was usually no detectable cytotoxicity. These results suggest that the steroids probably did not suppress mature CTL, but rather suppressed the precursors of these effector cells. These patients were all individuals who had antibodies to CMV before transplantation. The syndrome described by these studies was, therefore, one of graft rejection treated by high-dose prednisone followed by severe CMV infection with absent CTL responses and persistent viremia eventuating in death or graft loss. These findings indicate that judicious use of steroids is critical, and that a search for methods to reverse steroid effects may be rewarding.

The CTL responses observed in bone marrow transplant recipients are often greater in magnitude than in renal transplant recipients, as shown in Table IX. This difference was originally interpreted as evidence that the protective effect of CTL may be more significant in the former group. However, the recognition that steroids were important raised the possibility of an alternative explanation. The renal transplant patients were routinely treated with low to moderate doses of steroids continuously for weeks posttransplant, whereas bone marrow transplant recipients generally received steroids for specific treatment of graft-versus-host disease. When the magnitude of cytotoxic responses observed was compared after stratification for dose of steroids received, the responses found were remarkably similar (G. Quinnan et al., unpublished data). Likewise, it was found that nearly all bone marrow transplant recipients who had received no steroids developed CTL responses, whereas only about half of those on maintenance steroids for graft-versus-host disease developed these responses. The significance of the steroid therapy is further emphasized by the finding that there was no association between graft-versus-host disease and nonresponsiveness unless patients were still receiving steroids when CMV infection developed. Nevertheless, steroids are not the only factor of significance in this population, as some patients not on steroids fail to respond, and some individuals receiving steroids do respond. There must be at least one other determinant of the capacity of transplant patients to develop CTL. This determinant did not appear to be an immunoregulatory dysfunction, for the ability to develop cy-

TABLE IX. CMV-Specific Cytotoxicity in Infected and Uninfected Transplant Patients[a]

Study group	CMV infection	Percent with CMV-specific target cell lysis								
		<5%	≥5%	≥10%	≥15%	≥20%	≥25%	≥30%	≥35%	≥40%
BMTR	Infected	25	75	73	64	49	39	30	23	17
	Uninfected	35	65	27	15	7	3	3	0	0
RTR	Infected	30	70	40	20	10	5	0	0	0
	Uninfected	90	10	0	0	0	0	0	0	0

[a] Results abstracted from Quinnan et al. (1984a) and Rook et al. (1984).

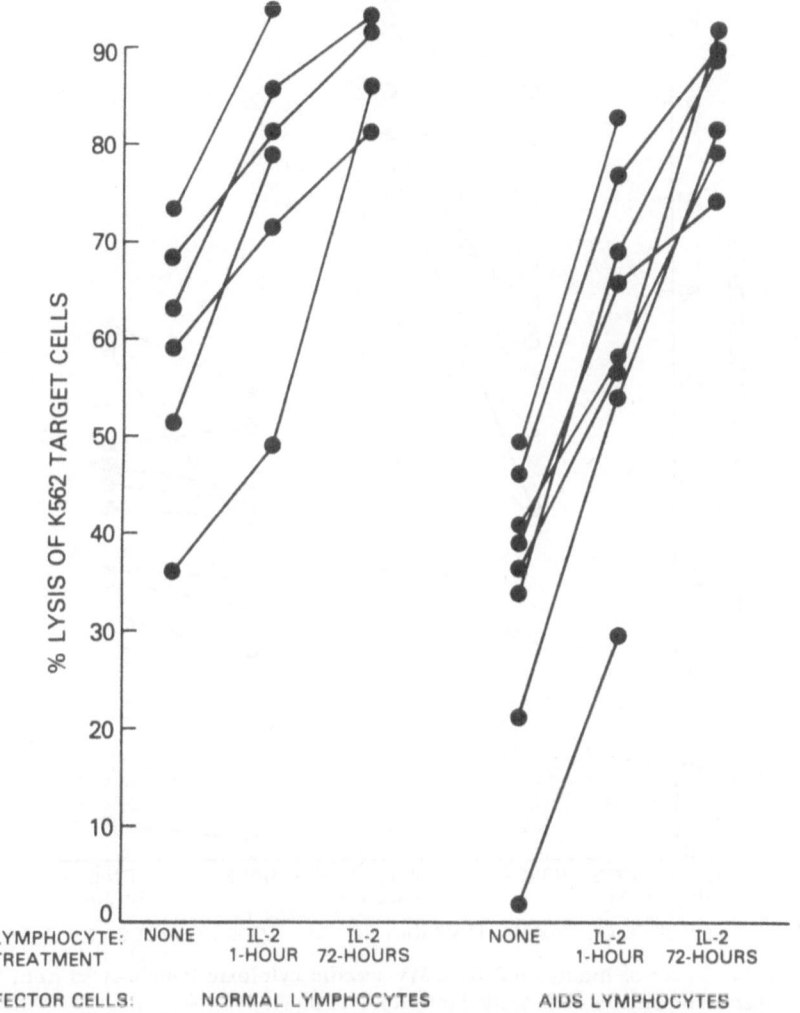

FIGURE 5. Effect of IL-2 on CMV-specific cytotoxicity of lymphocytes from healthy volunteers and patients with AIDS. From Rook et al. (1983).

totoxic T-cell responses did not correlate with abnormal numbers of helper or suppressor lymphocytes (G. Quinnan et al., unpublished observations).

The data pertaining to the cause of deficient CTL responses in AIDS patients are particularly revealing. These patients' lymphocytes have increased responsiveness to IL-2, as shown in Fig. 5(Rook et al., 1983). They do not respond to IFN-β, as shown in Fig. 6. Similar results were found in testing virus-specific cytotoxicity. Responsiveness of lymphocytes to IL-2 is dependent on their being activated by IL-1 (Larsson et al., 1980). In AIDS patients, therefore, there are apparently activated cytotoxic T-

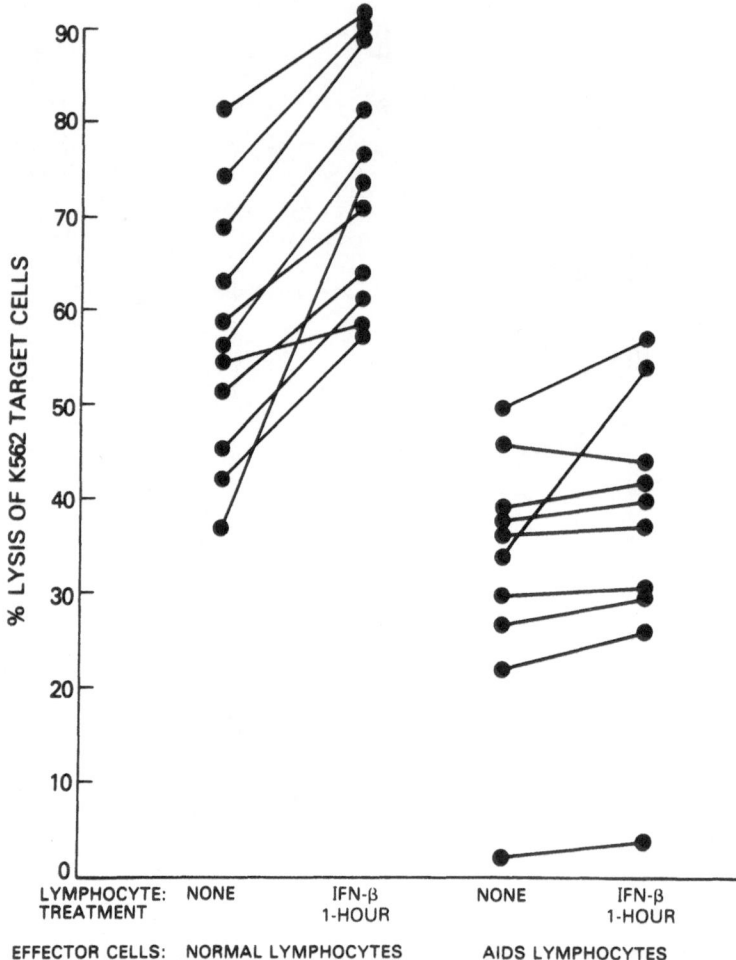

FIGURE 6. Effect of interferon-β on CMV-specific cytotoxic lymphocytes from healthy volunteers and patients with AIDS. From Rook *et al.* (1983).

cell precursors present, but the cells do not mature into effector cells *in vivo*. The fact that they mature readily under the influence of IL-2 localizes the maturation arrest to the proliferation stage and suggests that the activated precursors are not being exposed to IL-2 *in vivo*. A depression of production of at least one lymphokine, IFN-α, is also characteristic of AIDS (Lopez *et al.*, 1983) and a number of findings indicate that there is a generalized depression of LGL functions (Quinnan *et al.*, 1984c). Production of IL-2 can be induced *in vitro*, for the lymphocytes from these patients proliferate in response to Con A (Fig. 7). The lymphokine-producing cells must be present, therefore, but suppressed. Studies were performed to determine whether this suppression was mediated by suppressor cells or a soluble factor, and whether the inhibition was at the level of IL-2 production or effect. These studies revealed that the cause

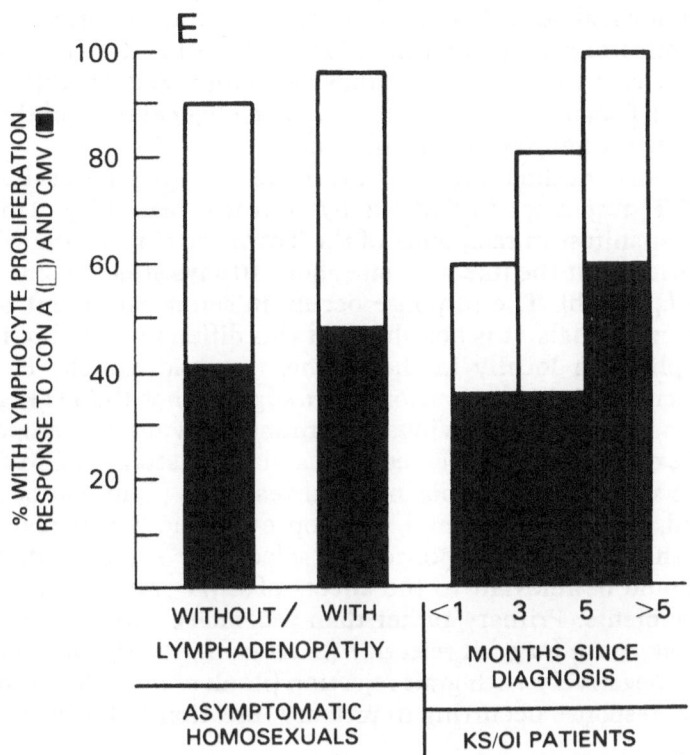

FIGURE 7. Proliferation responses to Con A and CMV antigens of lymphocytes from homosexual men who were healthy or had chronic lymphadenopathy or AIDS. From Quinnan *et al.* (1984c).

of the T-cell maturation arrest appeared to be a soluble factor in serum of AIDS patients that inhibited production of IL-2 (F. Siegal *et al.*, unpublished observations). The potential implications of this result for therapy are self-evident.

These studies have demonstrated that it is possible to define abnormalities of immune responses very specifically. Risk factors for severe infections can be determined more readily once the mechanism by which they produce their effects is understood, and appropriate modificatons in patient management can be made with greater confidence. The procedures involved in defining these abnormalities are simple, and the results are easily understood if the studies are performed in a systematic, stepwise fashion.

VII. DELAYED-TYPE HYPERSENSITIVITY RESPONSES TO CMV

In marked distinction to the current state of understanding of the roles of cytotoxic effector cells in defense against these infections, very little is known about DTH responses to CMV. It is important to recognize that the effector cells mediating DTH and T-cell cytotoxicity are distinct and are subject to differing immunoregulatory controls (Nash and Ashford, 1982). Because they are distinct, the biological roles of the two types of T-cell effectors may also be different.

The only method currently available that may be useful for measuring DTH responses to CMV is by skin test reactivity. This response has been manifest in recipients of the Towne strain vaccine who develop a local reaction at the injection site about 10 days after vaccination (Quinnan et al., 1984b). The response occurs in seronegative but not in seropositive individuals. It is possible that this difference results from greater virus replication locally in the seronegative individuals. However, we have recently made observations that suggest that the DTH response is actively suppressed in previously immune individuals. Seropositive volunteers experimentally infected with a virulent strain of CMV developed systemic virus infection, but no local reactions (Quinnan et al., 1984b). They did, on the other hand, develop cytotoxic T-cell responses, suggesting that the DTH response was selectively suppressed. This observation could be relevant to the effects of CMV infection in renal transplant recipients. Primary rather than secondary CMV infection appears to be a risk factor for graft rejection (Betts et al., 1977), and CTL responses correlate negatively with graft rejection (Rook et al., 1984). It may be that the DTH response occurring in primary infection induces or exacerbates rejection.

VIII. CMV-INDUCED IMMUNE SUPPRESSION

It is somewhat paradoxical that recovery from CMV infection is more dependent on cellular immunity than is the case with most other viruses yet CMV infection itself has profound suppressive effects on cellular immunity. The clinical relevance of these effects is manifest by increased susceptibility to other opportunistic infections (Rand et al., 1978) in patients infected with CMV. The specific mechanisms responsible for this increased susceptibility are unknown, but some potentially relevant effects have been described. CMV infection is frequently accompanied by a depressed lymphocyte response to T-cell mitogens (Howard et al., 1974; Booss and Wheelock, 1977; Rinaldo et al., 1980, Quinnan et al., 1981). This effect has been attributed, on the basis of in vitro studies, to direct effects of the virus on lymphocytes or to the effects of soluble factors in serum (Booss and Wheelock, 1977). There are also suppressor cell responses induced in patients with CMV infection. Bixler and Booss (1981) have reported evidence of an adherent cell suppressor response, possibly involving monocytes. Carney et al. (1981) have observed increased numbers of T cells of the suppressor phenotype in patients with CMV infection, observations confirmed by us (Fig. 8). An impressive aspect of this

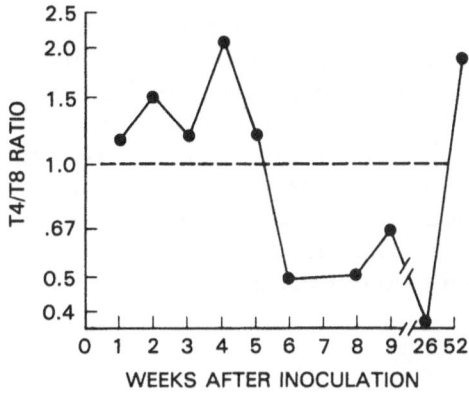

FIGURE 8. Ratio of helper cells (OKT4-positive) to suppressor cells (OKT8-positive) in healthy volunteers experimentally infected with a low-passage strain of CMV (Toledo-1). From Quinnan *et al.* (1984b).

suppressor response is its persistence. The results shown in Fig. 8 are from individuals with mild CMV infections (Quinnan *et al.*, 1984b.) They had persistent inversions of the normal ratio of helper cells to suppressor cells for 6 months after onset of infection. Which of these laboratory abnormalities is relevant to the clinically significant immunosuppression caused by CMV remains to be determined.

IX. CMV-INDUCED *IN VITRO* LYMPHOCYTE BLASTOGENESIS

The antigen-specific *in vitro* lymphocyte blastogenic response is a measure of immunological memory. Patients without *in vitro* blastogenic responses during infection, either seropositive or seronegative, can develop adequate effector cell responses, and the blastogenic response does not predict the outcome of infection (Quinnan *et al.*, 1982b). Nevertheless, this response does have potential relevance to understanding immunity to CMV if it is understood as a measure of the capacity for antigen-specific lymphocyte activation and proliferation.

X. CELLULAR IMMUNE RESPONSES TO CMV VACCINE

The complex nature of immunity to CMV has generated uncertainty with regard to whether a vaccine could be efficacious. It is not known which immune responses are responsible for protective immunity. As a result, there are no laboratory tests that predict reliably which individuals are resistant to infection. A comprehensive description of the immune responses to the vaccine is of value, so that the extent to which vaccine-induced immunity parallels that induced by natural infection can be approximated. Plotkin *et al.* (1976) have demonstrated that the Towne vaccine reliably induces both antibody and lymphocyte blastogenesis responses in healthy volunteers. We have recently demonstrated that the vaccine induces CTL responses. These responses were lower in magnitude and shorter in duration than those occurring during infection with a low-passage, virulent virus, but there were nevertheless distinct increases in HLA-restricted cytotoxicity (Quinnan et al., 1984b). The same individuals did not develop inversions of helper-to-suppressor cell ratios, indicating that the vaccine may induce the crucial cellular responses without causing immunosuppression. The effectiveness of these responses in preventing subsequent infection or disease in vaccinated individuals is still being studied.

XI. CONCLUSION

The importance of CMV as a cause of human disease in many different settings is more related to the overriding significance of cell-mediated immunity than it is to the inherent virulence of the agent itself. As a result of many recent advances in the general field of cellular immunology, it is now possible to define the effector cell functions that are important for defense of the host and to understand why they fail to develop in some individuals. Systematic study of cellular immunity in CMV infections should facilitate development of methods for modulating immune responses to the benefit of the patient while advancing our understanding of cellular immunity to virus infections in general.

REFERENCES

Abo, T., and Balch, C. M., 1981, A differentiation antigen of human NK and K cells identified by a monoclonal antibody (HNK-1), *J. Immunol.* **127**:1024.

Bancroft, G. J., Shellam, G. R., and Chalmer, J. E., 1981, Genetic influences on the augmentation of natural killer cells during murine cytomegalovirus infection: Correlation with patterns of resistance, *J. Immunol.* **126**:988.

Betts, R. F., Freeman, R. B., Douglas, R. G., and Talley, T. E., 1977, Clinical manifestations of renal allograft derived primary cytomegalovirus infection, *Am. J. Dis. Child.* **131**:759.

Bixler, G. S., and Booss, J., 1981, Adherent spleen cells from mice acutely infected with cytomegalovirus suppress the primary antibody response *in vitro*, *J. Immunol.* **127**:129.

Booss, J., and Wheelock, E. F., 1977, Progressive inhibition of T-cell function preceding clinical signs of cytomegalovirus infection in mice, *J. Infect. Dis.* **135**:478.

Bukowski, J. F., Woda, B. A., Habu, S., Okumura, K., and Welsh, R. M., 1983, Natural killer cell depletion enhances virus synthesis and virus-induced hepatitis *in vivo*, *J. Immunol.* **131**:1531.

Burlington, D. B., Djeu, J. Y., Wells, M., and Quinnan, G. V., 1984, Large granular lymphocytes provide an accessory function in the *in vitro* development of influenza A virus-specific cytotoxic T cells, *J. Immunol.* (in press).

Carney, W. P., Rubin, R. H., Hoffman, R. A., Hansen, W. P., Healey, K., and Hirsch, M. S., 1981, Analysis of T lymphocyte subsets in cytomegalovirus mononucleosis, *J. Immunol.* **126**:2114.

Diamond, R. D., Keller, R., Lee, G., and Finkel, D., 1977, Lysis of cytomegalovirus-infected human fibroblasts and transformed human cells by peripheral blood lymphoid cells from normal human donors, *Proc. Soc. Exp. Biol. Med.* **154**:259.

Djeu, J. Y., Heinbough, J. A., Vieira, W. D., Holden, H. T., and Herberman, R. B., 1979, The effect of immunopharmacological agents on mouse natural cell-mediated cytotoxicity and on its augmentation by poly I : C, *Immuno-pharmacology* **1**:231.

Djeu, J. Y., Heinbough, J. A., Vieira, W. D., Holden, H. T., and Herberman, R. B., 1982, Positive self regulation of cytotoxicity in human natural killer cells by production of interferon upon exposure to influenza and herpes-viruses, *J. Exp. Med.* **156**:1222.

Durum, S. K., and Gershon, R. K., 1982, Interleukin 1 can replace the requirement for I-A-positive cells in the proliferation of antigen-primed T cells, *Proc. Nat. Acad. Sci. USA* **79**:4747.

Farrar, J. J., Benjamin, W. R., Hilfiker, M. L., Howard, M., Farrar, W. L., and Fuller-Farrar, J., 1982, The biochemistry, biology and role of interleukin-2 in the induction of cytotoxic T cell and antibody-forming B cell responses, *Immunol. Rev.* **63**:129.

Farrar, W. L., Johnson, H. M., and Farrar, J. J., 1981, Regulation of the production of immune interferon and cytotoxic T lymphocytes by interleukin-2, *J. Immunol.* **126**:1120.

Gidlund, M., Orn, A., Wigzell, H., Senik, A., and Gresser, I., 1978, Enhanced NK cell activity in mice injected with interferon and interferon inducers, *Nature* **273**:759.

Grundy, J. E., Mackenzie, J. S., and Stanley, N. F., 1981, Influence of H-2 and non-H-2 genes on resistance to murine cytomegalovirus infection, *Infect. Immun.* **32**:277.

Grundy, J. E., Trapman, J., Allan, J. E., Shellam, G. R., and Melief, C. J. M., 1982, Evidence for a protective role of interferon in resistance to murine cytomegalovirus and its control by non-H-2 linked genes, *Infect. Immun.* **37**:143.

Herberman, R. B., and Holden, H. T., 1978, Natural cell-mediated immunity, *Adv. Cancer Res.* **17**:305.

Ho, M., 1980, Role of specific cytotoxic lymphocytes in cellular immunity against murine cytomegalovirus, *Infect. Immun.* **27**:767.

Howard, R. J., Miller, J., and Najarian, J. S., 1974, Cytomegalovirus induced immune suppression. II. Cell-mediated immunity, *Clin. Exp. Immunol.* **18**:119.

Kasahara, T., Djeu, J. Y., Dougherty, S. F., and Oppenheim, J. J., 1983, Capacity of human large granular lymphocytes to produce multiple lymphokines: Interleukin-2, interferon and colony stimulating factor, *J. Immunol.* **131**:2379.

Kirmani, N., Ginn, R. K., Mittal, K. K., Manischewitz, J. F., and Quinnan, G. V., 1981, Cytomegalovirus-specific cytotoxicity mediated by non-T lymphocytes from peripheral blood of normal volunteers, *Infect. Immun.* **34**:441.

Kohl, S., Starr, S. E., Oleske, J. M., Shore, S. L., Ashman, R. B., and Nahmias, A. J., 1977, Human monocyte-macrophage-mediated antibody-dependent cytotoxicity to herpes simplex virus infected cells, *J. Immunol.* **118**:729.

Lanza, E., Pastore, S., Hapel, A. J., and Djeu, J. Y., 1983, Growth of natural cytotoxic (NC) effector cells in interleukin-3, *Nature* **306**:788.

Larsson, E. L., Iscove, N. N., and Coutinho, A., 1980, Two distinct factors are required for the induction of T cell growth, *Nature* **283**:664.

Lattime, E. C., Pecoraco, G. A., and Stutman, O., 1981, Natural cytotoxic cells against solid tumors in mice. III. A comparison of effector cell antigenic phenotype and target cell recognition structures with those of NK cells, *J. Immunol.* **126**:2011.

Lopez, C., Fitzgerald, P. A., and Siegal, F. P., 1983, Severe acquired immune deficiency syndrome in male homosexuals: Diminished capacity to make interferon-alpha in vitro associated with severe opportunistic infections, *J. Infect. Dis.* **148**:962.

Macher, A., Reichert, C., Straus, S., Longo, D., Parillo, J., Lane, C., Fauci, A. S., Rook, A. H., Manischewitz, J., and Quinnan, G. V., 1983, Death in the AIDS patient: Role of cytomegalovirus, *N. Engl. J. Med.* **309**:1454.

McMichael, A., 1978, HLA-restriction of human cytotoxic T lymphocytes specific for influenze virus: Poor recognition of virus associated with HLA-A2, *J. Exp. Med.* **148**:1458.

Nash, A. A., and Ashford, N. P. N., 1982, Split T-cell tolerance in herpes simplex virus-infected mice and its implication for anti-viral immunity, *Immunology* **45**:761.

Nieminen, P., Paosivuo, R., and Saksela, E., 1982, Effect of a monoclonal anti-large granular lymphocyte antibody on the human NK activity, *J. Immunol.* **128**:1097.

Okafor, G. O., Turner, M. W., and Hay, F. C., 1974, Localization of monocyte binding site of human immunoglobulin G, *Nature* **248**:228.

Oleski, J. M., Ashman, R. B., Kohl, S., Shore, S. S., Starr, S. E., Wood, P., and Nahmias, A. J., 1977, Human polymorphonuclear leukocytes as mediators of antibody dependent cellular cytotoxicity to herpes simplex virus-infected cells, *Clin. Exp. Immunol.* **27**:446.

Oppenheim, J. J., Stadler, B. M., Siraganian, R. P., Mage, M., and Mathieson, B., 1982, Lymphokines: Their role in lymphocyte responses, *Fed. Proc.* **41**:257.

Ortaldo, J. R., Bonnard, G. D., Kind, P. D., and Herberman, R. B., 1979, Cytotoxicity by cultured human lymphocytes: Characteristics of effector cells and specificity of cytotoxicity, *J. Immunol.* **122**:1489.

Plotkin, S. A., Farquhar, J., and Hornberger, E., 1976, Clinical trials of immunization with the Towne 125 strain of human cytomegalovirus, *J. Infect. Dis.* **134**:470.

Quinnan, G. V., and Manischewitz, J. F., 1979, The role of natural killer cells and antibody-dependent cell-mediated cytotoxicity during murine cytomegalovirus infection, *J. Exp. Med.* **150**:1549.

Quinnan, G. V., and Rook, A. H., 1984, The importance of cytotoxic cellular immunity in the protection from cytomegalovirus infection, in: *CMV: Pathogenesis and Prevention of Human Infections* (S. A. Plotkin, ed.), p. 245, Liss, New York.

Quinnan, G. V., Manischewitz, J. F., and Ennis, F. A., 1978, Cytotoxic T lymphocyte response to murine cytomegalovirus infection, *Nature* **273**:514.

Quinnan, G. V., Kirmani, N., Esber, E., Saral, R., Manischewitz, J. F., Rogers, J. L., Rook, A. H., Santos, G. W., and Burns, W. J., 1981, HLA-restricted cytotoxic lymphocyte responses to cytomegalovirus infection of bone marrow transplant recipients, *J. Immunol.* **126**:2036.

Quinnan, G. V., Manischewitz, J. F., and Kirmani, N., 1982a, Involvement of natural killer cells in the pathogenesis of murine cytomegalovirus interstitial pneumonitis and the immune response to infection, *J. Gen. Virol.* **58**:173.

Quinnan, G. V., Kirmani, N., Rook, A. H., Manischewitz, J. F., Jackson, L., Moreschi, G., Santos, G. W., Saral, R., and Burns, W. J., 1982b, HLA-restricted T-lymphocyte and non-T-lymphocyte cytotoxic responses correlate with recovery from cytomegalovirus infection in bone-marrow-transplant recipients, *N. Engl. J. Med.* **307**:7.

Quinnan, G. V., Burns, W. H., Kirmani, N., Rook, A. H., Manischewitz, J., Jackson, L., Santos, G. W., and Saral, R., 1984a, HLA-restricted cytotoxic T lymphocytes are an important defense mechanism in cytomegalovirus infections, *Rev. Infect. Dis.* **6**:156.

Quinnan, G. V., Delery, M., Rook, A. H., Frederick, W. R., Epstein, J. S., Manischewitz, J., Jackson, L., Ramsey, K., Mittal, K., Plotkin, S., and Hilleman, M., 1984b, Avirulence and immunogenicity of the Towne strain in comparison to a non-attenuated strain of cytomegalovirus, *Ann. Intern. Med.* (in press).

Quinnan, G. V., Rook, A. H., Frederick, W. R., Manischewitz, J. F., Epstein, J. S., Siegel, J., Masur, H., Macher, A. M., and Deju, J. Y., 1984c, Prevalence, clinical manifestations, and immunology of herpesvirus infections in the acquired immunodeficiency syndrome, *Ann. NY Acad. Sci.* (in press).

Rand, K. H., Pollard, R. B., and Merigan, T. C., 1978, Increased pulmonary superinfection in cardiac transplant patients undergoing primary cytomegalovirus infection, *N. Engl. J. Med.* **298**:951.

Reddehase, M., Suessmuth, W., Moyers, C., Falk, W., and Droege, W., 1982, Interleukin 2 is not sufficient as helper component for the activation of cytotoxic T lymphocytes by synergizes with a late helper effect that is provided by irradiated I-region-incompatible stimulator cells, *J. Immunol.* **128**:61.

Reinherz, E. L., Kung, P. C., Goldstein, G., and Schlossman, S. F., 1979, Separation of functional subsets of human T cells by a monoclonal antibody, *Proc. Natl. Acad. Sci. USA* **76**:4061.

Rinaldo, C. R., Corney, W. P., Richter, B. S., Black, P. H., and Husch, M. S., 1980, Mechanisms of immunosuppression in cytomegaloviral mononucleosis, *J. Infect. Dis.* **141**:488.

Rook, A. H., and Quinnan, G. V., 1983, Cell mediated immunity to human cytomegalovirus, in: *Human Immunity to Viruses* (F. E. Ennis, ed.), pp. 241–246, Academic Press, New York.

Rook, A. H., Masur, H., Lane, H. C., Frederick, W., Kasahara, T., Macher, A. M., Djeu, J. Y., Manischewitz, J. F., Jackson, L., Fauci, A. S., and Quinnan, G. V., 1983, Interleukin-2 enhances the depressed natural killer and cytomegalovirus-specific cytotoxic activities of lymphocytes from patients with the acquired immune deficiency syndrome, *J. Clin. Invest.* **72**:398.

Rook, A. H., Quinnan, G. V., Frederick, W. F., Manischewitz, J. F., Jackson, L., Kirmani, N., Dantzler, T., Lee, B. B., and Courier, C. B., 1984, Importance of cytotoxic lymphocytes during cytomegalovirus infection of renal transplant recipients, *Am. J. Med.* **76**:385.

Rouse, B. T., Grewal, A. S., Babiuk, L. A., and Fujimiya, Y., 1977, Enhancement of antibody-dependent cell-mediated cytotoxicity of herpesvirus-infected cells by complement, *Infect. Immun.* **18**:660.

Shore, S. L., Cromeans, T. L., and Romano, T. J., 1976, Immune destruction of virus infected cells early in the infectious cycle, *Nature* **262**:695.

Siegal, F. P., Lopez, C., Hammer, G. S., Brown, A. E., Kornfeld, S. J., Gold, J., Hassett, J., Hirschman, S. Z., Cunningham-Rundles, C., Adelsberg, B. R., Parham, D. M., Siegal, M., Cunningham-Rundles, S., and Armstrong, D., 1981, Severe acquired immunodeficiency in male homosexuals, manifested by chronic perianal ulcerative herpes simplex lesions, *N. Engl. J. Med* **305**:1439.

Starr, W. E., and Allison, A. C., 1977, Role of T lymphocytes in recovery from murine cytomegalovirus infection, *Infect. Immun.* **17**:458.

Timonen, T., and Saksela, E., 1980, Isolation of human natural killer cells by density gradient centrifugation, *J. Immunol. Methods* **36**:285.

Trinchieri, G., Santoli, D., Dee, R. R., and Knowles, B. B., 1978, Antiviral activity induced by culturing lymphocytes with tumor-derived or virus-transformed cells: Identification of the anti-viral activity as interferon and characterization of the human effector lymphocyte subpopulation, *J. Exp. Med.* **147**:1299.

Varho, M., Lehman-Grube, F., and Simon, M. M., 1981, Effector T lymphocytes in lymphocytic chorimeningitis virus-infected mice. *J. Exp. Med.* **153**:992.

West, W. H., Cannon, G. B., Kaz, H. D., Bonnard, G. D., and Herberman, R. B., 1977, Natural cytotoxic reactivity of human lymphocytes against a myeloid cell line: Characterization of effector cells, *J. Immunol.* **118**:355.

Yeager, A. S., Grumet, F. C., Hafleigh, E. B., Arvin, A. M., Bradley, J. S., and Prober, C. G., 1981, Prevention of transfusion-acquired cytomegalovirus infections in newborn infants, *J. Pediatr.* **98**:281.

Zinkernagel, R. M., and Welsh, R. M., 1976, H-2 compatibility requirement for virus-specific T-cell-mediated effector functions *in vivo*, *J. Immunol.* **117**:1495.

CHAPTER 7

In Vivo Studies of Epstein–Barr Virus and Other Lymphotropic Herpesviruses of Primates

HARVEY RABIN

I. INTRODUCTION

The Epstein–Barr virus (EBV) of man is a member of a large group of lymphotropic herpesviruses of primates. These viruses have been isolated from prosimians, New World and Old World monkeys, apes, and man. The isolates fall into three groups. The first group includes EBV and related viruses of Old World monkeys and apes. The DNAs of EBV and these viruses of higher primates show a substantial degree of sequence homology (~ 40%). Moreover, their genomes share a common structure, and are colinear with each other. Their proteins show extensive antigenic cross-reactivity but differences among them can be observed especially in the case of their nuclear antigens. These viruses are B cell tropic and can immortalize B lymphocytes in tissue culture. The viruses of New World monkeys form a second distinct group. These viruses are not related to EBV and the EBV-like viruses but are related to each other antigenically and by DNA homology. They have a common genome structure, which is different from that of EBV, and are T cell tropic. These viruses are cytolytic for nonlymphoid cells *in vitro* and are also capable

HARVEY RABIN • DuPont Biomedical Products, North Billerica, Massachusetts 01862. By acceptance of this article, the publisher or recipient acknowledges the right of U.S. Government to retain a nonexclusive, royalty-free license in and to any copyright covering the article. Present address: DuPont Biomedical Products, 331 Treble Cove Rd, N. Billerica, MA 01862.

TABLE I. Lymphotropic Herpesviruses of Primates

Family	Virus (species of isolation)	Lympho-cyte tro-pism	*In vitro* transfor-mation	Disease association
Hominidae (human)	EBV (man)	B	+	Infectious mononucleosis, Burkitt's lymphoma, nasopharyngeal carcinoma, polyclonal lymphoproliferative disease/lymphoma, experimental lymphoproliferative disease/lymphoma
Pongidae (apes)	*H. pongo* (orangutan)	B	+	Isolation from B-cell culture of leukemic animal
	H. pan (chimpanzee)	B	+	NR[a]
	H. gorilla (gorilla)	B	+	NR
Cercopithecidae (Old World monkeys)	*H. papio* (baboon)	B	+	Lymphoid disease in Sukhumi baboon colony, experimental lymphoid disease in marmosets
	Undesignated EBV-like (macaque)	B	+	Isolation from cultured B cells of immunosuppressed organ transplant recipient with lymphoma
	AGM-EBV (African green monkey)	NR	NR	NR
Cebidae (New World monkeys)	*H. saimiri* (squirrel monkey)	T	+	Lymphoma in related species and in rabbits
	H. ateles (spider monkey)	T	+	Lymphoma in related species and in rabbits
Tupaiidae (prosimian)	THV-2 (tree shrew)	NR	NR	Lymphoma in tree shrews, thymic hyperplasia and thymoma in rabbits
	THV-3 (tree shrew)	NR	NR	Thymic hyperplasia and thymoma in rabbits

[a] NR, not reported.

of immortalizing T cells in tissue culture. Isolation of a third group of primate herpesviruses has been reported from tree shrews, *Tupaia glis belangeri*, a prosimian species. These viruses are differentiated by their characteristic genome structure.

As shown in Table I, EBV has been associated with several diseases of man. The difficulty of assigning a clear etiologic role to EBV for many of these diseases is well known. For many of the other lymphotropic herpesviruses, which, like EBV, appear to be also widely distributed in their host populations, the same difficulty pertains. Moreover, less is known about naturally occurring disease in most of these species than is known for man and, in general, relatively few isolates of each virus have been made. Therefore, at the present time, only limited significance can be given to the fact that several of these viruses have been isolated from cultures established from tumor-bearing animals. By the same token, the lack of apparent pathogenicity of certain viruses for their natural host species does not necessarily imply a total lack of the ability of these viruses to cause disease in these species. One circumstance that has afforded the opportunity for study of virus association with disease within a species was at the large baboon colony at the Institute for Experimental Pathology and Therapy at Sukhumi, USSR, where an outbreak of lymphoproliferative disease has been continuing for several years.

Studies of EBV in experimental hosts were undertaken to explore the oncogenic potential of this virus and to determine if a model for human disease could be developed in a nonhuman primate host. Similar studies have been performed with *Herpesvirus papio* and in particular with the T-cell-tropic herpesviruses of New World primates, *H. saimiri* and *H. ateles*. In this review, I will cover some aspects of experimental infection with EBV, describe the studies on the relationship between *H. papio* and lymphoproliferative disease in baboons, and discuss some of the newer findings on experimental T-cell disease induced by *H. saimiri*. I will also describe certain aspects of preliminary studies with other selected viruses, in particular an EBV-like virus of macaques and the tree shrew herpesviruses, THV-2 and THV-3.

II. EBV IN NEW WORLD PRIMATES

A. Host Range for Disease Induction

Earlier studies on tumor induction by EBV have been reviewed (Frank *et al.*, 1976; Miller, 1979) and I will attempt mainly to present some of the more recent findings. Table II shows those species of nonhuman primates in which some form of progressive, fatal lymphoproliferative disease has been induced. Gibbons have also been shown to be susceptible to EBV but progressive disease did not develop (Werner *et al.*, 1972). The species that seems most susceptible to specific EBV-induced disease is the cotton-topped marmoset (*Saguinus oedipus*). Several investigators have been able to induce disease in this species since the original observations by Shope *et al.* (1973). A variety of inocula have been used (cell-free transforming virus, various transformed cells) with positive results. EBV-positive cell lines can be established from these animals and EBV DNA or EBV nuclear antigen (EBNA) has been detected in tumor tissue. White-lipped marmosets (*S. fuscicollis, S. nigricollis*) have proved largely insusceptible to EBV. A tumor was produced in a single animal after repeated inoculation of the B95-8 strain of EBV by both intraperitoneal and intramedullary routes (Sundar *et al.*, 1981). EBV DNA was detected at a low level in an affected lymph node but cell lines could not be established. Falk *et al.* (1976b) reported the induction of fatal lymphoproliferative disease in seven out of eight common marmosets (*Callithrix jacchus*) inoculated with B95-8 virus. These monkeys developed antibody

TABLE II. Nonhuman Primates Susceptible to EBV-Induced
Lymphoproliferative Disease

Species	Antibody conversion[a]	Cell lines established	EBV DNA or EBNA in tumor or cell culture	Progressive disease	References
Cotton-topped marmoset	VCA, EA, EBNA, neutralizing	+	+	Lymphoproliferative disease/lymphoma (~ 30%)	Shope et al. (1973), Werner et al. (1975), Deinhardt et al. (1975), Rabin et al. (1976b, 1977b), Miller et al. (1977), Johnson et al. (1983)
White-lipped marmoset	VCA, EA	−	+	Lymphoma (1 animal)	Sundar et al. (1981)
Common marmoset	VCA, EA	+/−	+/−	Lymphoproliferative disease (~ 50%)	Falk et al. (1976b), Ablashi et al. (1978, 1983)
Owl monkey	VCA, EA, EBNA	+	+	Lymphoproliferative disease/lymphoma (1 animal)	Epstein et al. (1973a,b, 1975), Rabin et al. (1976b)

[a] VCA, viral capsid antigen; EA, early antigen; EBNA, Epstein–Barr virus nuclear antigen.

to viral capsid antigen (VCA) but not to early antigen (EA) or EBNA. The affected tissue appeared to be EBNA negative and continuous cell lines could not be established. Ablashi et al. (1983) reported a fatal lymphoproliferative disease in a common marmoset given two widely separated inocula of the Ag876 strain of EBV along with multiple inoculations of the tumor promoter 12-O-tetradecanoylphorbol-13-acetate (TPA). EBV DNA was apparently detected in affected tissue as was EBNA. Short-term cultures of affected lymph node were positive for VCA. A second marmoset inoculated with EBV but not TPA did not develop a tumor and a marmoset inoculated only with TPA similarly did not develop a tumor. Earlier attempts by this group to induce tumors with EBV (B95-8) in common marmosets in the absence of TPA were unsuccessful (Ablashi et al., 1978). Thus, it would seem that EBV can induce a lymphoproliferative disease in common marmosets, which, with the exception noted, may not involve stable virus transformation of B cells. A lymphoproliferative disease compatible histologically with certain forms of malignant lymphoma was induced in one out of three owl monkeys inoculated with a frozen and thawed suspension of EB3 cells (Epstein et al., 1973a). A cell line established from this animal was positive for both EBV DNA and EBNA (Epstein et al., 1973b,1975). Inoculation of cell-free owl monkey-passaged EB3 virus or B95-8 virus into six additional owl monkeys failed to produce progressive disease (Rabin et al., 1976b).

B. Clonality of Tumors in Cotton-Topped Marmosets

EBV, in man, has been associated with (1) monoclonal lymphoid disease, Burkitt's lymphoma, or polyclonal lymphoid disease and (2) infec-

TABLE III. Immunoglobulin Light Chain Expression on *in Vivo*-Derived EBV-Positive B Cells in Culture[a]

Marmoset/cell line	Immunoglobulin light chain (% positive)	
	κ	λ
79-E-1		
Peripheral lymph node	5	54
Mesenteric lymph node	19	42
Spleen	10	17
79-DP-2		
Peripheral lymph node	30	18
Mesenteric lymph node	13	23
Spleen	36	20
79-X-2		
Mesenteric lymph node	61	15

	Months in culture			
	4	16	4	16
1605				
Mesenteric lymph node	0	0	40	30
Liver	0	0	40	15
Spleen	0	0	40	2

[a] Data from Johnson *et al.* (1983) and Rabin *et al.* (1977b).

tious mononucleosis and various lymphomalike conditions (reviewed in Klein and Purtilo, 1981; Gilden and Rabin, 1982). It was of interest, therefore, to obtain information on the clonality of disease produced by EBV experimentally. Johnson *et al.* (1983) established several continuous cell lines from two cotton-topped marmosets that died of EBV-induced lymphoproliferative disease. Histologically, the lesions in the marmosets resembled fatal cases of human infectious mononucleosis. Staining of these lines and another line established from a marmoset with mild lesions for surface immunoglobulins revealed the presence of IgM-positive cells and a low percentage of IgD-positive cells. Both κ-positive and λ-positive cells were seen in each culture from all three marmosets (Table III). Moreover, both κ-positive and λ-positive chains were secreted into culture medium by these various lines. The lines failed to show evidence of a marker chromosome suggestive of clonal origin or of a marker similar to those characteristic of Burkitt's lymphoma. These cells, moreover, had a low cloning efficiency in soft agar medium and failed to induce tumors in nude mice by the subcutaneous route. Thus, these cells resembled lymphoblastoid cell lines rather than the clonal lines characteristic of Burkitt's lymphoma (Nilsson *et al.*, 1977). In contrast, cell lines were established from three widely separated tumors in another cotton-topped marmoset (Rabin *et al.*, 1977b). These cells exhibited both IgM- and IgG-positive cells but the only light chain detected was λ (Table III). Clonal sublines derived from all three parent lines also were positive only for λ. Moreover, each line showed the loss of one homolog of a pair of medium-sized metacentric chromosomes, which occurred in about 50% of the metaphase spreads in each line, suggesting that the lines shared a common origin. The liver and spleen lines were indistinguishable karyotypically but the lymph node line had a marker chromosome. Thus, these cells showed characteristics of being a uniclonal population. No evidence for a Burkitt's lymphoma-like translocation was seen. Therefore, it would appear from histopathologic evidence and from the limited observations on surface immunoglobulin and karyotype that both fatal polyclonal lymphoproliferative disease and disease more compatible with monoclonal lymphoma can be induced in cotton-topped marmosets by EBV.

C. Potential Use of Cotton-Topped Marmosets in EBV Vaccine Development

One other area of potential use for cotton-topped marmosets in EBV research should be mentioned and that is the use of these animals for vaccine development. It has been shown by several laboratories that monoclonal antibodies directed against EBV membrane antigens have virus-neutralizing activity (Hoffman *et al.*, 1980; Thorley-Lawson and Geilinger, 1980; Strnad *et al.*, 1982). These membrane antigens can be extracted and used as effective immunogens to induce EBV-neutralizing antibody (Thorley-Lawson, 1979). Recently, a highly purified membrane antigen, gp340 (340,000-dalton glycoprotein) was incorporated into liposomes and used as an immunogen in mice where strong neutralizing antibodies were induced (North *et al.*, 1982). Cotton-topped marmosets, which are susceptible to EBV-induced disease, would play an important role in further EBV vaccine development. Work of this type has recently been initiated in which several types of gp340 preparations are being tested in marmosets (Epstein and Morgan, 1983).

III. *Herpesvirus papio* AND LYMPHOPROLIFERATIVE DISEASE IN BABOONS

A. Background

Beginning in 1967, there has been an ongoing outbreak of lymphoid disease in baboons (*Papio hamadryas*) at the Institute of Experimental Pathology and Therapy in Sukhumi, USSR. The first cases were recorded after the introduction into the colony of animals that had been inoculated with various human leukemic material. The cases are confined to the main colony where the inoculated animals had been introduced. No cases have been observed in a geographically remote forest colony nor in separate areas at the main reservation where contact with diseased or inoculated animals has been avoided (Lapin, 1975). Mortality in the colony from lymphoid disease has varied between approximately 1.0 to greater than 2.0% per year. As the colony has several hundred baboons, this has amounted to up to 20 deaths per year (Neubauer et al., 1979). The majority of cases have been classified as non-Hodgkin's lymphoma of the lymphoid type but other types of lymphoid tumors have also been seen (Yakovleva et al., 1980). Most tumors have appeared to be comprised of B cells but T-cell and non-T-, non-B-cell tumors have also been reported (Yakovleva et al., 1980). Clinical signs of disease include marked splenomegaly and lymphadenopathy and other symptoms such as gingivitis, dermatitis, and increased skin pigmentation. There is also evidence of an immuno-depression associated with disease in terms of hyporesponsiveness of peripheral blood leukocytes to phytohemagglutinin and concanavalin A (Neubauer et al., 1980). Another feature of the disease is the high frequency of chromosome breaks and structural rearrangments, especially involving chromosomes 20 and 1, in cells from lymph nodes, spleen, and bone marrow of diseased baboons (Markaryon, 1980).

The epidemiologic features of the outbreak suggest an infectious etiology as all cases have occurred where there has been contact between either inoculated baboons or baboons with spontaneous disease and healthy baboons (Lapin, 1975). Cases have also arisen following inoculation of blood from one baboon to another (Falk et al., 1976a). Most cases have occurred in colony-born animals greater than 5 years old (Lapin, 1973).

B. Virus Isolates

Attempts to culture tumorous tissue, cocultivation of tumor tissue with various indicator cell lines, and the establishment of continuous B-cell lines have led to the isolation of several types of viruses from lymphomatous baboons. These isolates have included an endogenous type C retrovirus (Goldberg et al., 1974), a foamyvirus (Rabin et al., 1976a), a cytomegalovirus (H. Rabin and R. Hopkins, unpublished results), and an EBV-related herpesvirus, H. papio (Falk et al., 1976a; Rabin et al., 1977a). Moreover, electron microscopic examination of tumor tissue and plasma from affected monkeys has revealed both type C retrovirus particles and herpesvirus particles (Lapin, 1973, 1975, 1976).

C. Antibody Patterns to H. papio-Associated Antigens

Several B-cell lines were established from lymphomatous baboons (Agrba et al., 1975; Dyachenko et al., 1976; Falk et al., 1976a; Rabin et *al.*, 1977a). These lines were found to be good sources of *H. papio* antigens and transforming virus. Preliminary serologic screening of sera indicated that antibodies to *H. papio* VCA were present in sera of baboons from Sukhumi and elsewhere (Falk *et al.*, 1976a; Lapin, 1976; Rabin *et al.*, 1977a). Initial titrations suggested that antibody titers in the sera of diseased baboons and their contacts in the main colony were higher than in sera of the geographically separated and disease-free forest colony (Lapin, 1976). Based on these observations, in collaboration with Lapin and his associates, we undertook a serologic study of three groups of baboons from Sukhumi: (1) those with clinical disease, (2) baboons in contact with diseased animals but not, themselves, showing clinical disease, and (3) healthy baboons from the forest colony. We hoped to determine by this study if *H. papio* would show a serologic relationship to disease similar to that shown by EBV to Burkitt's lymphoma. The results showed that diseased baboons had, as a group, antibodies to more *H. papio* antigens, VCA, EA, soluble, and nuclear, and at higher titers than the contact or forest baboons (Neubauer *et al.*, 1979a). Differences among groups were evident in terms of the geometric mean titers (GMT) of anti-VCA (highest in the diseased group) and in the frequency and GMT of anti-EA (high in diseased and contact groups and low in the forest group). The frequency of antisoluble antigen and of antinuclear antigen were also highest in the diseased group. The frequency and titer of anti-EA also tended to show a correspondence with the anti-VCA titer in the diseased group whereas it did not in the contact group. Antibodies to two other virus isolates from Sukhumi baboons, a foamyvirus and a cytomegalovirus, failed to show disease-related patterns. These general serologic patterns with *H. papio* antigens were subsequently confirmed (Lapin *et al.*, 1982). The levels of significance in frequencies and titers between diseased and contact animals, where significant, are not as high ($p < 0.05$) as they are between diseased and forest animals ($p < 0.001$). This is not surprising as the disease may have a variable incubation period (Lapin, 1975) and thus, some contacts may be in a subclinical stage of the diseases and, as mentioned, there have been several types of lymphoid tumors described at Sukhumi (Yakovleva *et al.*, 1980) and *H. papio* may not necessarily play a role in all of them. Moreover, other factors, such as importation history and time after contact with diseased animals, have been reported to affect antibody titers to *H. papio* (Lapin *et al.*, 1980a). The high frequencies of anti-EA ($> 60\%$) in the main colony also suggested that much recent infection and possible reinfection may be taking place. This possibility is supported by the finding that it has proved relatively easy to recover infectious virus from oral swabs of animals in the main colony but not from those in the forest colony (Agrba *et al.*, 1980; Lapin *et al.*, 1982).

D. Presence of Viral DNA in Tumor Tissue and Spleen

Preliminary results with limited numbers of samples suggested that *H. papio* DNA could be detected in tumor tissue and spleens of lymphomatous baboons but not from uninvolved tissue or normal spleen (Neubauer et al., 1980; Lapin et al., 1980b). Subsequently, it was reported that H. papio DNA could be detected in spleen cells of diseased baboons and in spleens of animals in clinical remission (Dyachenko et al., 1981). Of seven healthy or nonlymphomatous monkeys, only one showed a low level of H. papio DNA in its spleen. These results again are suggestive that H. papio may play a role in the etiology of lymphoid disease in Sukhumi.

E. Experimental Disease in Cotton-Topped Marmosets

In addition to its ability to immortalize B cells in vitro, there is some evidence that H. papio can induce disease experimentally. Inoculation of large numbers ($1.3–5 \times 10^8$) of H. papio-producing baboon cells into adult cotton-topped marmosets induced an acute, fatal, widely disseminated lymphoproliferative disease. Affected tissue was negative for hybridization with an EBV DNA probe but several cell lines were established from inoculated animals. These lines all had marmoset karyotypes and some B-cell characteristics, but none was positive for viral antigens. One cell line was positive for EBV-related DNA in hybridization with EBV cRNA (Deinhardt et al., 1978). White-lipped marmosets did not develop disease with a similar inoculum nor did newborn cotton-topped marmosets. Autologous nonproducer cells failed to induce disease as did inoculation of large numbers of producer cells established from a healthy baboon. In other studies, inoculation of cell-free virus or H. papio-positive cell lines into several species of nonhuman primates (cotton-topped marmosets, yellow baboons, stump-tailed macaques, and rhesus monkeys) did not result in induction of progressive disease although, as judged by seroconversion and virus shedding, the recipients generally were infected (Gerber et al., 1977; Lapin et al., 1980b; H. Rabin, unpublished results). Inoculation of laboratory rodents and rabbits similarly did not induce disease (Lapin et al., 1980b).

F. Summary of Features of the Disease and Possible Similarity to Other Diseases of Captive Monkeys

To summarize the features of the outbreak at Sukhumi, it would appear that several types of lymphoid tumors have been occurring in the main colony for the past several years. Cases were first noted subsequent to the introduction into the colony of baboons that had been inoculated with human leukemic material. The disease has features of an infectious condition as all cases have been related to contact with either inoculated baboons or diseased baboons. Several viruses have been isolated from the colony and one of these, H. papio, shows a disease-related serologic pattern. This virus has also shown a certain capacity to induce lymphoproliferative disease in adult cotton-topped marmosets. Baboons with lymph-oid disease show signs of immunosuppression and a high degree of viral shedding. Certain features of the disease, its effect on the lymphoid system, its possible infectious nature, its sudden appearance, and its restriction to certain colonies resemble an acquired immune deficiency syndrome (simian AIDS) seen in other colonies of Old World monkeys (Henrickson et al., 1983; Letvin et al., 1983). A way to get further insight into the role of H. papio in the disease might be through a controlled vaccination study aimed at determining if prevention of infection would reduce the incidence of disease.

It is of course possible that viruses other than H. papio may also be involved in the etiology of the disease. As previously mentioned, a type C retrovirus was detected in lymphoid tissues and plasmas of diseased baboons and other types of viruses have been isolated from such baboons. Recently, antibody that cross-reacts with human T-cell leukemia virus (HTLV) has been detected in the sera of baboons from Sukhumi (C. Saxinger, personal communication). Thus, it would be important to evaluate the roles that these other viruses may play in the disease either individually or in some type of combination. It is also possible that viruses can be readily isolated from these baboons and certain antibody titers may be perturbed in them because of the immunosuppressive nature of the disease. In this case, the virologic findings would be secondary to a loss of normal immunologic function. Various cofactors, at present not recognized but perhaps related to crowding and stress, might also be important in the etiology of this disease. H. papio, however, remains a leading candidate for further study in relationship to the etiology of this outbreak especially in light of what is known of the pathogenic potential of EBV in immunosuppressed hosts.

IV. OTHER LYMPHOTROPIC HERPESVIRUSES OF OLD WORLD PRIMATES AND THEIR POSSIBLE RELATIONSHIP TO DISEASE

The other viruses of Old World primates have not shown definite disease-related associations. H. pongo was isolated from an aneuploid B-cell line established from an orangutan with monomyelocytic leukemia (Rasheed et al., 1977). This virus is B cell tropic, appears to be widely disseminated in orangutans, and antibody to it can be found in healthy animals. However, it is of interest that the B-cell line established from this leukemic orangutan grew in soft agar medium and transplanted to nude mice by the subcutaneous route (Rabin et al., 1978), indicating that it was not a typical lymphoblastoid cell line. Another EBV-like virus was recovered from a B-cell line established from a cynomolgus monkey (Macaca fascicularis) with an immunoblastic sarcoma (Heberling et al., 1982). This animal had been used in transplantation studies and had been treated with both irradiation and cyclosporin A. Attempts to establish cells in culture from this monkey were hindered by growth of foamyvirus and cytomegalovirus. A culture was established from an inguinal tumor. This culture was of B cell type and was positive for an EBV-related herpesvirus. The role that this virus may play in the etiology of lymphomas in immunosuppressed macaques is not clear but it is of considerable interest that monkeys receiving cyclosporin A particularly with other immunosuppressive treatment had a high incidence of lymphoma (Pennock *et al.*, 1980, 1981). The parallels between these macaque tumors and the presence of an EBV-like virus and of EBV and lymphomas arising in human transplant recipients (Hanto *et al.*, 1981) are of considerable interest. This system may have the potential to provide a model for the study of the relationships between immunosuppression, virus activation, and lymphoma. The other lymphotropic herpesviruses of Old World primates, *H. pan* (Gerber *et al.*, 1976, 1977), *H. gorilla* (Neubauer *et al.*, 1979), and AGM EBV (Böcker *et al.*, 1980), have shown no disease relationship to date.

V. T-CELL-TROPIC HERPESVIRUSES

A. Pathogenicity and Host Range

Isolates of T-cell-tropic herpesviruses have been obtained from two varieties of New World monkeys, *H. saimiri* from squirrel monkeys (*Saimiri sciureus*) and *H. ateles* from spider monkeys (*Ateles* spp.). The literature on these viruses has been reviewed extensively (Fleckenstein, 1979; Falk, 1980a,b; Deinhardt and Deinhardt, 1979; Rangan and Gallagher, 1979; zur Hausen, 1980; Neubauer and Rabin, 1979; Gilden and Rabin, 1982; Rabin, 1983). I will not attempt to duplicate this material here in any depth but will cover some of the more recent findings in terms of transformed cell phenotypes, tumor cell lines, and *in vitro* transformation.

H. saimiri and *H. ateles* have the capacity to induce T-cell tumors in several species of New World primates and in laboratory rabbits. Most studies have been conducted in cotton-topped or white-lipped marmosets and in owl monkeys and more work has been done with *H. saimiri* than with *H. ateles*. DNA from both *H. saimiri* and *H. ateles* is infectious and can induce tumors in cotton-topped marmosets (Fleckenstein *et al.*, 1978).

H. saimiri-induced disease in owl monkeys is an immunosuppressive condition characterized by a loss of mitogenic response of peripheral blood lymphocytes, appearance of lymphocyte infectious centers in the peripheral blood, rise in titer to viral EA and to virus-associated membrane antigen (as detected by antibody–dependent cell cytotoxicity), and the development of a suppressor cell population in the peripheral blood coincident with the rise in antibody titer to membrane antigen. These changes occur in about 75% of owl monkeys infected with the S295C strain of virus. The disease in these monkeys is fatal after a variable time course. Most, but not necessarily all, of these monkeys will develop peripheral blood leukemia as well as widely disseminated lymphoma. The other 25% of monkeys will develop an essentially disease-free chronic infection. For a review of *H. saimiri* pathology see Hunt (1978). The infection in marmosets is uniformly fatal and generally follows a shorter course than in owl monkeys.

H. saimiri has not been associated with spontaneous neoplastic disease in squirrel monkeys nor has experimental infection of squirrel monkeys with *H. saimiri* (strain S295C) led to disease even after immunosuppression (Falk *et al.*, 1973; Martin and Allen, 1975). However, there are a number of different strains of *H. saimiri*. These strains can be differentiated from each other by restriction enzyme fragmentation patterns of their DNAs (Desrosiers and Falk, 1982). It is possible that biologic activity, including pathogenic potential, may vary significantly among virus strains and information on this point would be of importance.

B. Lymphocyte Tropism and Antigen Modulation

The target cell for *H. saimiri* and *H. ateles* is the T lymphocyte. This was initially demonstrated by the formation of nonspecific sheep red blood cell (SRBC) rosettes by lymphocytes from *H. saimiri*-infected owl monkey tumor cell cultures (Wallen *et al.*, 1973). Later, it was shown that *H. saimiri* could be recovered preferentially from SRBC rosette-forming cells as opposed to non-rosette-forming cells of infected squirrel monkeys and marmosets (Wright *et al.*, 1976). More recently, it has been shown that monoclonal antibodies that define lymphocyte subpopulations in humans, define similar subpopulations in nonhuman primates (Neubauer *et al.*, 1981b, 1982, 1983a,b; Haynes *et al.*, 1982; Schooley *et al.*, 1983). Using such monoclonal antibodies, T cells, the helper/inducer subset of T cells, B cells, and HLA DR-positive cells could be readily identified in owl monkeys and certain species of marmosets. The association of *H. saimiri* with T cells was confirmed in chronically infected owl monkeys by infectious center analysis of sorted populations. Interestingly, analysis of virus association with T-cell subsets in three chronically infected owl monkeys indicated that the virus was restricted to the Leu-3a$^+$ (helper/inducer) population in two animals but was present in both the Leu-3a$^-$ and Leu-3a$^+$ populations in a third monkey. Continuous cell lines established from lymphomatous owl monkeys and marmosets were all T cells but were either all Leu-3a$^-$ or Leu-3a$^+$ depending on the cell line. A short term ($1\frac{1}{2}$ months) culture of an owl monkey *H. saimiri* tumor showed a mixture of Leu-3a phenotypes (65% Leu-3a$^+$ and 27% Leu-3a$^-$). Both types of cells were positive for infectious centers at nearly the same level (Neubauer et al., 1983a). Thus, while the virus seems confined to T cells, it does not appear to be restricted to a particular subset of T cells although it may be so in an individual animal. A monoclonal antibody specific for the cytotoxic/suppressor T-cell subpopulation in H. saimiri-susceptible monkeys would be a valuable reagent for further studies in this area. These observations confirm, in a general way, earlier findings that the virus could be found in T cells having receptors for the Fc portion of IgG and in cells lacking this receptor (Neubauer et al., 1981a) and that cultures initiated from H. saimiri tumors of chimeric marmosets contained both male and female cells (Chu and Rabson, 1972; Marczynska et al., 1973; Rabin et al., 1973).

Based on these findings on cell surface phenotypes in chronically infected monkeys and in cell culture, we recently initiated studies in white-lipped marmosets to determine if recognizable changes in cell phenotype would occur in individual monkeys as H. saimiri-induced disease progressed. A group of four marmosets was inoculated with H. saimiri (strain S295C) and at weekly intervals blood counts were performed and peripheral blood mononuclear leukocytes (PBL) were stained with monoclonal antibodies to lymphocyte cell surface markers. The stained cells were examined by flow cytometry. One of the marmosets developed leukemia as diagnosed at day 41 and died at day 72. At day 34 an increase in the staining intensity of the PBL with the monoclonal antibody T11, which blocks SRBC rosetting and thereby identifies T cells, was noted. This increased staining with T11 was detected repeatedly over the next 35 days with one exception at day 55. At day 69, just 3 days before death, the mean fluorescence intensity of PBL with T11 was about 40% greater than it was at the beginning of the study when the T11 staining profile appeared normal. Another leukemic marmoset from a second group of inoculated animals showed a similar pattern in terms of increased staining of PBL with T11. (R. Neubauer, E. Schechter, R. Massey, R. Schroff, and H. Rabin, unpublished results). Thus, it appears from these limited observations that at least one cell surface antigen modulates during the development of leukemia. The increased staining with T11 agrees with earlier work where it was found that lymphocytes from owl monkeys with H. saimiri-induced disease bound more SRBC per lymphocyte than did normal lymphocytes and that the rate constants for rosette formation were higher for lymphocytes from diseased monkeys than for lymphocytes from control monkeys (Wallen et al., 1973).

C. H. saimiri Tumor Cell Lines

Cell lines have been established from several H. saimiri-lymphomatous marmosets by several groups of investigators (Rabson et al., 1971; Falk et al., 1972; Fleckenstein et al., 1977) and also from an H. atelesinoculated marmoset (Falk et al., 1974b). Many of these cell lines have been in culture for several years and grow without the need for added IL-2. They have surface properties of T cells, exhibit NK-like activity (Johnson and Jondal, 1981), and secrete interferon (Wright *et al.*, 1974). Some of these lines, after continuous growth in culture, have become virus nonproducers. Examination of viral DNA in the *H. saimiri* cell lines by hybridization with an L-DNA probe indicated that *H. saimiri*-specific DNA was present in multiple copies. Hybridization of DNA from nonproducer cell lines showed a bimodal time course. This finding suggested that the viral genomes in the nonproducer cells were defective in that the entire viral genome was not present (Fleckenstein *et al.*, 1977). This observation was subsequently confirmed by denaturation mapping (Kaschka-Dierich *et al.*, 1982) and by Southern blot hybridization (Desrosiers, 1981). The viral DNA in the cell lines tested was present in large part as nonintegrated, circular DNA molecules (Werner *et al.*, 1977, Kaschka-Dierich *et al.*, 1982). In different cell lines, the deletions may range up to 40% of the unique L-DNA of the virus and in different lines may encompass very similar areas of the L-DNA. An *in vitro* transformed marmoset cell line was also found to have deleted regions in its L-DNA. Recent work using fragment-specific probes suggests that at least certain portions of viral DNA that are missing from episomal DNA in at least one cell line may be present in a nonepisomal form (Kaschka-Dierich *et al.*, 1981). There is also evidence that viral DNA in marmoset nonproducer cell lines may be methylated (Desrosiers *et al.*, 1979; Kaschka-Dierich *et al.*, 1981).

While several lines of cells were established from marmosets, no cell lines could be established in continuous culture from *H. saimiri*-infected owl monkeys, although a single cell line did persist for several months (Rabin *et al.*, 1973). Recently, three lines were established in culture from lymphomatous owl monkeys by culturing the cells in the presence of IL-2 (Brown *et al.*, 1982; Neubauer *et al.*, 1983a; G. Pearson, B. Fleckenstein, S. Schirm, and H. Rabin, unpublished results). These cells grow in the absence of IL-2 but do so at a relatively slow rate compared to their growth in IL-2. The addition of IL-2 to their culture medium induced cells to enter the S phase of the cell cycle with different kinetics than those shown by normal T cells after IL-2 addition (Brown *et al.*, 1982). More specifically, the tumor cells began to enter the S phase after a relatively short interval and continued to enter the S phase continually for several hours. These results suggest that the tumor cells, unlike normal cells, may be retarded at various points in G_1. It is of interest that certain marmoset tumor cells that grow in the absence of IL-2, unlike the owl monkey tumor cells, grow no better in the presence of IL-2 (H. Rabin and C. Briggs, unpublished results).

Recently, it has been found that one of the owl monkey lines, OMT-1, has a deletion in its L-DNA similar to that observed for nonproducer marmoset tumor cell lines (G. Pearson, S. Schirm, B. Fleckenstein, and H. Rabin, unpublished results). OMT-1 cells in early passage were virus producers but, in a fashion similar to the marmoset tumor lines, became nonproducers on continuous culture.

As has been mentioned, *H. saimiri*, like *H. ateles*, can also induce tumors in rabbits. One cell line that was established in culture with the aid of IL-2 can apparently be transplanted successfully in rabbits where fatal donor cell tumors are formed (Faggioni *et al.*, 1983). Successful transplantation provides evidence for the neoplastic nature of these cells. Earlier attempts to transplant tumor cells in marmosets had resulted in recipient cell tumors only (Marcyznska *et al.*, 1973).

D. *In Vitro* Transformation

H. ateles has been shown to transform cells from cotton-topped, white-lipped, and common marmosets (Falk *et al.*, 1974b, 1975, 1978). Both circulating lymphocytes and splenic lymphocytes could be transformed and transformation could be achieved both with cell-free virus and with cocultivation using X-irradiated producer lymphoid tumor cell lines. The transformed cells could be subcultured and grown continuously in culture, formed nonspecific rosettes with SRBC, and lacked surface immunoglobulin and receptors for activated complement. Transformation frequencies ranged from 49 to 81% in numerous attempts. Most cultures of transformed cells had cells that expressed viral antigens and were positive for infectious centers. Squirrel monkey lymphocytes, however, were refractory to transformation.

While initial attempts to transform lymphocytes with *H. ateles* were successful, similar attempts with *H. saimiri* were not (Falk *et al.*, 1975). An exception to this was a single marmoset *in vitro* transformed line (Desrosiers *et al.*, 1979, Kaschka-Dierich *et al.*, 1982; Schirm *et al.*, 1984). Recently, however, both owl monkey and marmoset lymphocytes have been transformed (Neubauer *et al.*, 1983b; Falk *et al.*, 1982, 1983; Rabin *et al.*, 1983, 1984). While it is not yet clear why success has been achieved recently while it was not previously, one aspect of the current transformation protocol has been the use of activated lymphocytes growing in IL-2 as target cells for virus transformation. Cocultivation of activated owl monkey lymphocytes with *H. saimiri*-infected monolayers of OMK or Vero cells has led to cells that can grow continuously without IL-2 within 4–6 weeks. In the case of marmoset cells cocultivated with irradiated producer lymphoid cells, IL-2-independent growth took longer to achieve. Both transformed owl monkey and marmoset cells have surface properties of T lymphocytes. The transformed owl monkey cells, while able to grow in the absence of IL-2, grow significantly better in its presence, suggesting that IL-2 receptors are still present on the cell surface. These cells apparently do not release IL-2 and efforts are currently under way to determine if they are positive for IL-2-specific mRNA. Thus, transformation with *H. saimiri* removes a major block to normal T-cell replication, the requirement for IL-2. The mechanism by which these transformed T cells can replicate continually in the absence of external signals is not clear but it is reasonable to speculate that the virus is supplying a gene product that is responsible for keeping the cells in continuous growth. Alternatively, the virus could be activating a cellular gene that induces cell growth. A search for activated oncogenes in *H. saimiri*-transformed T cells after various times in culture is of importance as is the search for growth-promoting regions of the viral genome. In regard to this latter point, the use of attenuated strains of virus such as *H. saimiri* att (Schaffer *et al.*, 1975) or of specific deletion mutants may be of value.

At least two other virus-transformation systems result in IL-2-independent growth. These are HTLV transformation of human T cells (Miyoshi *et al.*, 1981; Markham *et al.*, 1983) and radiation leukemia virus transformation of murine T cells (Haas *et al.*, 1983). Thus, both lymphotropic herpesviruses and lymphotropic retroviruses induce cells with a similar transformed phenotype. Whether the molecular mechanisms that underlie the transformed states in each case are similar or not remains to be determined. Whether or not this change in growth factor dependence, although a major one in terms of cell growth, represents immortalization (cells can grow continuously but cannot make tumors) or neoplastic transformation is also not clear for *H. saimiri*. In the case of murine lymphocytes, the neoplastic phenotype is accompanied by both IL-2 independence and aneuploidy (Haas *et al.*, 1983). As mentioned above, there is only one report of the successful transplantation of *H. saimiri* tumor cells (Faggioni *et al.*, 1983). More work on *H. saimiri*-transformed cells in terms of their growth properties, transplantability, and karyology is required before the nature of these cells becomes clear. Nonproducer cells, transformed *in vitro*, would be valuable for transplantation studies using autologous hosts. Prerequisites for such work, in addition to lack of virus production, would be specific cell markers so that any tumors produced could be related unequivocally to the transplanted cells.

In addition to independence from IL-2, the *H. saimiri in vitro* transformed cells resemble tumor-derived lines in several other respects (Rabin *et al.*, 1984). These characteristics include demonstration of NK-like activity, the release of interferon-γ, and deletions in viral L-DNA (Desrosiers, 1981; Kaschka-Dierich *et al.*, 1982). Thus, transformation *in vitro* with *H. saimiri* appears to establish a cell with a similar phenotype to *H. saimiri* tumor-derived cells.

VI. *Tupaia* HERPESVIRUSES

Four herpesviruses have been isolated from cultures established from tissues of tree shrews, *Tupaia* (Darai *et al.*, 1981). These viruses were designated as THV-1, 2, 3, and 4 and have been shown to be related to each other by DNA sequence homology, but can be distinguished from each other by restriction enzyme fragmentation patterns. The structure of their DNA is distinct from that of the other groups of primate herpesviruses and on this basis they comprise a separate group of viruses. All these viruses replicate well in tree shrew embryonic fibroblasts where they can be plaque assayed (Darai et al., 1979). THV-2 was isolated from cultures of a malignant lymphoma that occurred in a 7- to 8-year-old animal and THV-3 was recovered from cultures of a Hodgkin's disease-like tumor that also occurred in a 7- to 8-year-old female (Darai et al., 1982).

Both THV-2 and THV-3 induce thymic hyperplasia in rabbits, and newborn rabbits inoculated with these viruses develop malignant thymoma at low frequency (8%). Virus could be recovered from spleens of infected animals (Darai et al., 1983). In addition, THV-2 has produced malignant lymphoma after inoculation of tree shrews, and virus similar to THV-2 was recovered from the tumors. The oncogenic potential of THV-3 in tree shrews is currently under investigation (Darai et al., 1983). The pathogenic potential of THV-1 and THV-4 remains to be reported. It would be of interest if these viruses would induce tumors on a regular basis and thus serve as models for such tumors of man as Hodgkin's disease.

VII. SUMMARY AND CONCLUSIONS

EBV is related by DNA homology, genome structure, and antigenic cross-reactivity to several B-cell-tropic herpesviruses of apes and Old World monkeys. Lymphotropic herpesviruses have also been isolated from New World primates and from tree shrews, a prosimian species. The New World primate viruses are related to each other, but have a characteristic genome structure, and are T cell tropic. The tree shrew isolates, based on their distinctive genomes, form a third group.

EBV is infectious for several species of nonhuman primates and has induced lymphomas in at least three species of New World primates. The most susceptible species is the cotton-topped marmoset where about 30% of the inoculated animals have developed progressive EBV-positive lymphoproliferative disease. Histopathology, karyotypic examination, and staining for cell surface immunoglobulins suggest that some EBV-induced tumors are polyclonal while others may more nearly resemble monoclonal lymphomas. The susceptibility of cotton-topped marmosets to EBV-induced disease should enable this species to play a major role in the developmnt of a vaccine for EBV.

There is evidence that an EBV-related virus of baboons, H. papio, may be associated with a continuing outbreak of lymphoma among captive hamadryas baboons in the large colony at the Institute of Experimental Pathology and Therapy, Sukhumi, USSR. This outbreak began after baboons that had been inoculated with human leukemic material had been released into the colony. The spread of lymphoma within the colony suggested an infectious etiology as contact with diseased or inoculated animals seemed essential to developing a tumor. Diseased baboons generally showed antibodies to more *H. papio*-associated antigens and at higher titers than did control baboons. The presence of *H. papio* DNA has also been reported in spleens and tumors of lymphomatous baboons and in spleens of baboons in remission. *H. papio* was readily isolated from diseased baboons, which also showed signs of being immunosuppressed.

Several cynomolgus monkeys used as immunosuppressed organ transplant recipients developed lymphoma. An EBV-like virus was isolated from cultured cells of one such monkey. The relationship between this virus and lymphoma in this species is yet to be determined.

Both *H. saimiri* and *H. ateles* can induce T-cell lymphomas in several species of New World primates and in rabbits. The disease induced by *H. saimiri* is immunosuppressive and is accompanied by a rise in antibody titers to specific virus-associated antigens and by the appearance of a suppressor cell population. While confined to T lymphocytes in the peripheral blood, *H. saimiri* can, depending on the animal, infect one or more distinct T-cell subsets. During development of disease there is modulation of the SRBC receptor on leukemic T cells. Tumor cells cultivated *in vitro* differ from normal T cells in that they can grow in the absence of added IL-2. Tumor cells exhibit NK activity and secrete interferon. Several cell lines no longer produce virus and examination of their viral genomes has revealed large deletions and DNA methylation. T cells transformed *in vitro* with *H. saimiri* and *H. ateles* also grow without added IL-2, have NK activity, release interferon, and exhibit deletions in viral DNA similar to those shown in tumor cells.

Herpesviruses have also been isolated from lymphoid tumors of tree shrews. These viruses have exhibited oncogenic properties in rabbits and one of them in tree shrews, as well.

The lymphotropic herepsviruses of primates offer systems both for the study of experimental B-cell or T-cell neoplasms and for the study of *in vitro* transformation of these same cell types. In addition, *H. papio* may be associated with spontaneous lymphomas in captive baboons while another EBV-like virus may be associated with lymphomas arising in experimentally immunosuppressed hosts. The study of these viruses may be of importance in understanding the role of EBV in the etiology of human tumors and in providing significant models for basic research in cell transformation and tumorigenesis.

ACKNOWLEDGMENT. This work was supported by the National Cancer Institute DHHS under Contract NO-1-CO-23909 with Litton Bionetics, Inc.

The contents of this publication do not necessarily reflect the views or policies of the Department of Health and Human Services, nor does mention of trade names, commerical products, or organizations constitute endorsement by the United States government.

REFERENCES

Ablashi, D., Pearson, G., Rabin, H., Armstrong, G., Easton, J., Valerio, M., and Cicmanec, J., 1978, Experimental infection of *Callithrix jacchus* marmosets with *Herpesvirus ateles, Herpesvirus saimiri*, and Epstein–Barr virus, *Biomedicine* **29**:7.

Ablashi, D., Aulakh, G., Luetzeler, J., Sundar, K., Armstrong, G., and Faggioni, A., 1983, Fatal lymphoma proliferative disease in a common marmoset (*Callithrix jacchus*) following inoculation of Ag 876 strain of Epstein–Barr virus and a tumor-promoting agent: Preliminary report, *Comp. Immunol. Microbiol. Infect. Dis.* **6**:151.

Agrba, V., Yakovleva, L., Lapin, B., Sangulija, I., Timanovskaya, V., Markaryon, D., Chuvirov, G., and Salmonova, E., 1975, The establishment of continuous lymphoblastoid suspension cell cultures from hematopoietic organs of baboon (*Papio hamadryas*) with malignant lymphoma, *Exp. Pathol.* **10**:318.

Agrba, V., Lapin, B., Timanovskaya, V., Dzhachvliany, M., Kokosha, L., Chuvirov, G., and Dyachenko, A., 1980, Isolation of lymphotropic baboon herpesvirus (HVP) from oral swabs of hamadryas baboons of the Sukhumi monkey colony, *Exp. Pathol.* **18**:269.

Böcker, J., Tiedemann, K.-H., Bornkamm, G., and zur Hausen, H., 1980, Characterization of an EBV-like virus from African green monkey lymphoblasts, *Virology* **101**:291.

Brown, R., Griffith, R., Neubauer, R., and Rabin, H., 1982, The effects of T cell growth factor on the cell cycle of primate T cells, *J. Immunol.* **129**:1849.

Chu, E., and Rabson, A., 1972, Chimerism in lymphoid cell culture line derived from lymph node of marmoset infected with *Herpesvirus saimiri*, *J. Natl. Cancer Inst.* **48**:771.

Darai, G., Matz, B., Schroder, C., Flugel, R., Berger, U., Munk, K., and Gelderblom, H., 1979, Characterization of a tree shrew herpesvirus isolated from a lymphosarcoma, *J. Gen. Virol.* **43**:541.

Darai, G., Flugel, R., Matz, B., and Delius, H., 1981, DNA of *Tupaia* herpesviruses, in: *Herpesvirus DNA* (Y. Becker, ed.), pp. 345–361, Nijhoff, The Hague.

Darai, G., Zoller, L., Hofmann, W., Moller, P., Schwaier, A., and Flugel, R., 1982, Spontaneous malignomas in *Tupaia* (tree shrew), *Am. J. Primatol.* **2**:177.

Darai, G., Koch, H.-G., Moller, P., Hofmann, H., Gelderblom, H., and Flugel, R., 1983, Is the tree shrew a model system for the investigation of Hodgkin's disease?, in: *Viral and Immunological Diseases in Nonhuman Primates* (S. Kalter, ed.), pp. 235–238, Liss, New York.

Deinhardt, F., and Deinhardt, J., 1979, Comparative aspects: Oncogenic animal herpesviruses, in: *The Epstein–Barr Virus* (M. Epstein and B. Achong, eds.), pp. 373–415, Springer-Verlag, Berlin.

Deinhardt, F., Falk, L., Wolfe, L., Paciga, J., and Johnson, D., 1975, Response of marmosets to experimental infection with Epstein–Barr-virus, in: *Oncogenesis and Herpesviruses II* (G. de Thé, M. Epstein, and H. zur Hausen, eds.), Part 2, pp. 161–168, IARC, Lyon, France.

Deinhardt, F., Falk, L., Wolfe, L., Schudel, A., Nonoyama, M., Lai, P., Lapin, B., and Yakovleva, L., 1978, Susceptibility of marmosets to Epstein–Barr virus-like baboon herpesviruses, *Primates Med.* **10**:163.

Desrosiers, R., 1981, *Herpesvirus saimiri* DNA in tumor cells—Deleted sequences and sequence rearrangements, *J. Virol.* **39**:497.

Desrosiers, R., and Falk, L., 1982, *Herpesvirus saimiri* strain variability, *J. Virol.* **43**:352.

Desrosiers, R., Mulder, C., and Fleckenstein, B., 1979, Methylation of *Herpesvirus saimiri* DNA in lymphoid tumor cell lines, *Proc. Natl. Acad. Sci. USA* **76**:3839.

Dyachenko, A., Kakubava, V., Lapin, B., Agrba, V., Yakovleva, L., and Samilchuk, E., 1976, Continuous lymphoblastoid suspension cultures from cells of haematopoietic organs of baboons with malignant lymphoma—Biological characterization and biological properties of the herpesvirus associated with culture cells, *Exp. Pathol.* **12:**163.

Dyachenko, A., Kokosha, L., Lapin, B., Yakovleva, L., and Agrba, V., 1981, Detection of baboon herpesvirus DNA in the tissues of hemoblastosis diseased and healthy monkeys from Sukhumi nursery USSR, *J. Sov. Oncol.* **2:**48.

Epstein, M., and Morgan, A., 1983, Studies on cotton-top tamarins immunized with the 340 kd glycoprotein component of EB virus membrane antigen presented in various ways, in: *Leukemia Reviews International*, Vol. 1 (M. Rich, ed.), pp. 34–35, Dekker, New York.

Epstein, M., Hunt, R., and Rabin, H., 1973a, Pilot experiments with EB virus, in owl monkeys (*Aotus trivirgatus*). I. Reticuloproliferative disease in an inoculated animal, *Int. J. Cancer* **12:**309.

Epstein, M., Rabin, H., Ball, G., Rickinson, A., Jarvis, J., and Melendez, L., 1973b, Pilot experiments with EB virus in owl monkeys (*Aotus trivirgatus*). II. EB virus in a cell line from an animal with reticuloproliferative disease, *Int. J. Cancer* **12:**319.

Epstein, M., zur Hausen, H., Ball, G., and Rabin, H., 1975, Pilot experiments with EB virus in owl monkeys (*Aotus trivirgatus*). III. Serological and biochemical findings in a animal with reticuloproliferative disease, *Int. J. Cancer* **15:**17.

Faggioni, A., Torrisi, M., Ablashi, D., and Frati, L., 1983, Ultrastructural aspects of *Herpesvirus saimiri* producer and nonproducer rabbit lymphoid cells, in: *Leukemia Reviews International*, Vol. 1 (M. Rich, ed.) pp. 168–169, Dekker, New York.

Falk, L., Jr., 1980a, Biology of *Herpesvirus saimiri* and *Herpesvirus ateles*, in: *Viral Oncology* (G. Klein, ed.), pp. 813–832, Raven Press, New York.

Falk, L., Jr., 1980b, Simian herpesviruses and their oncogenic properties, in: *Oncogenic Herpesviruses* (F. Rapp, ed.), pp. 145–172, CRC Press, Boca Raton, Fla.

Falk, L., Wolfe, L., Marczynska, B., and Deinhardt, F., 1972, Characterization of lymphoid cell lines established from *Herpesvirus saimiri* (HVS)-infected marmosets, *Bacteriol. Proc.* **38:**191.

Falk, L., Wolfe, L., and Deinhardt, F., 1973, *Herpesvirus saimiri*: Experimental infection of squirrel monkeys (*Saimiri sciureis*), *J. Natl. Cancer Inst.* **51:**165.

Falk, L., Nigida, S., Deinhardt, F., Wolfe, L., Cooper, R., and Hernandez-Camacho, J., 1974a, *Herpesvirus ateles*: Properties of an oncogenic herpesvirus isolated from ciruclating lymphocytes of spider monkeys (*Ateles* sp.), *Int. J. Cancer* **14:**473.

Falk, L, Wright, J., Wolfe, L., and Deinhardt, F., 1974b, *Herpesvirus ateles*: Transformation *in vitro* of marmoset splenic lymphocytes, *Int. J. Cancer* **14:**244.

Falk, L., Wolfe, L., and Deinhardt, F., 1975, Transformation *in vitro* with *Herpesvirus ateles*, in: *Oncogenesis and Herpesviruses II* (G. de Thé, M. Epstein, and H. zur Hausen, eds.), Part 1, pp. 379–384, IARC, Lyon, France.

Falk, L., Deinhardt, F., Nonoyama, M., Wolfe, L., Bergholz, C., Lapin, B., Yakovleva, L., Agrba, V., Henle, G., and Henle, W., 1976a, Properties of a baboon lymphotropic herpesvirus related to Epstein–Barr virus, *Int. J. Cancer* **18:**798.

Falk, L., Deinhardt, F., Wolfe, L., Johnson, D., Hilgers, J., and de Thé, G., 1976b, Epstein–Barr virus: Experimental infection of (*Callithrix jacchus*) marmosets, *Int. J. Cancer* **17:**785.

Falk, L., Johnson, D., and Deinhardt, F., 1978, Transformation of marmoset lymphocytes *in vitro* with *Herpesvirus ateles*, *Int. J. Cancer* **21:**652.

Falk, L., Silva, D., Byington, R., and Schooly, R., 1982, Long-term propagation of marmoset T-lymphocytes in T-cell growth factor: Infection with T-lymphotropic herpesviruses and surface membrane markers, *Immunobiology* **163:**227.

Falk, L., Silva, D., Byington, R., and Schooly, R., 1983, *Herpesvirus saimiri* and *Herpesvirus ateles*: Transformation *in vitro* of cultured marmoset T-lymphocytes, *American Association of Cancer Research Abstracts*, p. 327.

Fleckenstein, B., 1979, Oncogenic herpesviruses of nonhuman primates, *Biochim. Biophys. Acta* **560**:301.

Fleckenstein, B., Muller I., and Werner, J., 1977, The presence of *Herpesvirus saimiri* genomes in virus-transformed cells, *Int. J. Cancer* **19**:546.

Fleckenstein, B. Daniel, M., Hunt, R., Werner, J., Falk, L., and Mulder, C., 1978, Tumor induction with DNA of oncogenic primate herpesviruses, *Nature (London)* **274**:57.

Frank, A., Andiman, W., and Miller, G., 1976, Epstein–Barr virus and nonhuman primates: Natural and experimental infection, *Adv. Cancer Res.* **23**:171.

Gerber, P., Pritchett, R., and Kieff, E., 1976, Antigens and DNA of a chimpanzee agent related to Epstein–Barr virus, *J. Virol.* **19**:1090.

Gerber, P., Kalter, S., Schidlovsky, G., Peterson, W., Jr., and Daniel, M., 1977, Biologic and antigenic characteristics of Epstein–Barr-virus-related herpesviruses of chimpanzees and baboons, *Int. J. Cancer* **20**:448.

Gilden, R., and Rabin, R., 1982, Mechanisms of viral tumorigenesis, *Adv. Virus Res.* **27**:281.

Goldberg, R., Scolnick, E., Parks, W., Yakovleva, L., and Lapin, B., 1974, Isolation of a primate type-C virus from a lymphomatous baboon, *Int. J. Cancer* **14**:722.

Haas, M., Altman, A., Rothenberg, L., and Jones, W., 1983, Transformation of factor-dependent T lymphoblastoma cells to autonomous T-lymphoma cells, in: *Leukemia Reviews International*, Vol. 1 (M. Rich, ed.), pp. 178–179, Dekker, New York.

Hanto, D., Frizzera, G., Purtilo, D., Sakamoto, K., Sullivan, J., Saemundsen, A., Klein, G., Summoric, R., and Najarian, J., 1981, Clinical spectrum of lymphoproliferative disorders in renal transplant recipients and evidence for the role of Epstein–Barr virus, *Cancer Res.* **41**:4253.

Haynes, B., Dowell, B., Hensley, L., Gore, I., and Metzgar, R., 1982, Human T cell antigen expression by primate T cells, *Science* **215**:298.

Heberling, R., Bieber, C., and Kalter, S., 1982, Establishment of a lymphoblastoid cell line from a lymphomatous cynomologous monkey, in: *Advances in Comparative Leukemia Research 1981* (D. Yohn and J. Blakeslee, eds.), Elsevier/North Holland, New York, pp. 385–386.

Henrickson, R., Osborn, K., Madden, D., Anderson, J., Maul, D., Sever, J., Ellingsworth, L., Lowenstine, L., and Gardner, M., 1983, Epidemic of acquired immunodeficiency in rhesus monkeys, *Lancet* **1**:388.

Hoffman, G., Lazarowitz, S., and Hayward, S., 1980, Monoclonal antibody against a 250,000-dalton glycoprotein of Epstein–Barr virus identifies a membrane antigen and a neutralizing antigen, *Proc. Natl. Acad. Sci. USA* **77**:2979.

Hunt, R., 1978, Herpesvirus oncogenesis in new world monkeys: History of herpesvirus oncogenesis and overview of *Herpesvirus saimiri*, *Recent Adv. Primatol.* **4**:117.

Johnson, D., and Jondal, M., 1981, *Herpesvirus ateles* and *Herpesvirus saimiri* transform marmoset T cells into continuously proliferating cell lines that can mediate natural killer cell-like cytotoxicity, *Proc. Natl. Acad. Sci. USA* **78**:6391.

Johnson, D., Wolfe, L., Levan, G., Klein, G.,Ernberg, I., and Aman, P., 1983, Epstein–Barr virus (EBV)-induced lymphoproliferative disease in cotton-topped marmosets, *Int. J.Cancer* **31**:91.

Kaschka-Dierich, C., Bauer, I., Fleckenstein, B., and Desrosiers, R., 1981, Episomal and nonepisomal herpesvirus DNA in lymphoid tumor cell lines, in: *Modern Trends in Human Leukemia* (R. Neth, R. Gallo, T. Graf, K. Mannweiler, and K. Winkler, eds.), pp. 197–203, Springer-Verlag, Berlin.

Kaschka-Dierich, C., Werner,F., Bauer, I., and Fleckenstein, B., 1982, Structure of nonintegrated, circular *Herpesvirus saimiri* and *Herpesvirus ateles* genomes in tumor cell lines and *in vitro*-transformed cells, *J. Virol.* **44**:295.

Klein, G., and Purtilo, D., 1981, Symposium on Epstein–Barr virus-induced lymphoproliferative diseases in immunosuppressed patients, *Cancer Res.* **41**:4209.

Lapin, B., 1973, The epidemiologic and genetic aspects of an outbreak of leukemia among hamadryas of the Sukhumi monkey colony, in: *Unifying Concepts of Leukemia* (R. Dutcher and L. Chieco-Bianchi, eds.), pp. 263–268, Karger, Basel.

Lapin, B., 1975, Possible ways viral leukemia spreads among the hamadryas baboons of the Sukhumi monkey colony, in: *Comp Leukemia Research, 1973* (Y. Ito and R. Dutcher, eds.), pp. 75–84, University of Tokyo Press, Tokyo/Karger, Basel.

Lapin, B., 1976, Epidemiology of leukemia among baboons of Sukhumi monkey colony, in: *Comp Leukemia Research 1975* (J. Clemmesen and D. Yohn, eds.), pp. 212–215, Karger, Basel.

Lapin, B., Agrba, V., Dyachenko, A., Kokosha, L., and Voevodin, A., 1980a, Epidemiology of hemoblastosis in the hamadryas baboons of the Sukhumi stock, in: *Advances in Comp Leukemia Research 1979* (D. Yohn, B. Lapin, and J. Blakeslee, eds.), pp. 417–418, Elsevier/North-Holland, Amsterdam.

Lapin, B., Agrba, V., Voevodin, A., Dyachenko, A., Yakovleva, L., Kokosha, L., Chuvirov, H., Rabin, H., Deinhardt, F., and Falk, L., 1980b, Characterization of the baboon herpesvirus (HVP) associated with malignant lymphoma in the Sukhumi hamadryas baboon colony, In: *Advances in Comp Leukemia Research 1979* (D. Yohn, B. Lapin, and J. Blakeslee, eds.), pp. 417–418, Elsevier/North-Holland, Amsterdam.

Lapin, B., Agrba, V., Voevodin, A., and Yakovleva, L., 1982, The biology of lymphotropic baboon herpesvirus (HVP) and its association with malignant lymphoma, in: *Advances in Comp Leukemia Research 1981* (D. Yohn and J. Blakeslee, eds.), pp. 395–397, Elsevier/North-Holland, Amsterdam.

Letvin, N., Eaton, K., Aldrich, W., Sehgal, P., Blake, B., Schlossman, S., King, N., and Hunt, R., 1983, Acquired immunodeficiency syndrome in a colony of macaque monkeys, *Proc. Natl. Acad. Sci. USA* **80**:2718.

Marczynska, B., Falk, L., Wolfe, L., and Deinhardt, F., 1973, Transplantation and cytogenetic studies of *Herpesvirus saimiri*-induced disease in marmoset monkeys, *J. Natl. Cancer Inst.* **50**:331.

Markaryon, D., 1980, Regular changes of karyotype with hemoblastosis in hamadryas baboons, in: *Advances in Comp Leukemia Research 1979* (D. Yohn, B. Lapin, and J. Blakeslee, eds.), pp. 415–416, Elsevier/North-Holland, Amsterdam.

Markham, P., Salahuddin, S., Kalyanaraman, V., Popovic, M., Sarin, P., and Gallo, R., 1983, Infection and transformation of fresh human umbilical cord blood cells by multiple sources of human T-cell leukemia-lymphoma virus (HTLV), *Int. J. Cancer* **31**:413.

Martin, L., and Allen, W., 1975, Response to primary infection with *Herpesvirus saimiri* in immunosuppressed juvenile and newborn squirrel monkeys, *Infect. Immun.* **12**:528.

Miller, G., 1979, Experimental carcinogenicity by the virus *in vivo*, in: *The Epstein–Barr Virus* (M. Epstein and B. Achong, eds.) pp. 351–372, Springer-Verlag, Berlin.

Miller, G., Shope, T., Coope, D., Waters, L., Pagano, J., Bornkamm, G., and Henle, W., 1977, Lymphoma in cotton-top marmosets after inoculation with Epstein–Barr virus: Tumor incidence, histologic spectrum, antibody responses, demonstration of viral DNA, and characterization of viruses, *J. Exp. Med.* **145**:948.

Miyoshi, I., Kubonishi, I., Yoshimoto, S., Akazi, T., Ohtsuki, Y., Shiraishi, Y., Nagata, K., and Hinuma, Y., 1981, Type C virus particles in a cord T cell line derived by co-cultivating normal human cord leukocytes and human leukemic T cells, *Nature* **294**:770.

Neubauer, R., and Rabin, H., 1979, Herpesvirus-induced lymphomas: Immunodepression and disease, in: *Naturally Occurring Biological Immunosuppressive Factors and Their Relationship to Disease* (R. Neubauer, ed.), pp. 203–231, CRC Press, Boca Raton, Fla.

Neubauer, R., Rabin, H., Strnad, B., Lapin, B., Yakovleva, L., and Indzie, E., 1979a, Antibody responses to *Herpesvirus papio* antigens in baboons with lymphoma, *Int. J. Cancer* **23**:186.

Neubauer, R., Rabin, H., Strnad, B., Nonoyama, M., and Nelson-Rees, W., 1979b, Establishment of a lymphoblastoid cell line and isolation of an Epstein–Barr-related virus of gorilla origin, *J. Virol.* **31**:845.

Neubauer, R., Rabin, H., Strnad, B., Nonoyama, M., Lapin, B., Dyachenko, A., Agrba, V., Indzhiia, L., and Yakovleva, L., 1980, Association between *Herpesvirus papio* and spontaneous lymphoma in baboons, in: *Advances in Comp Leukemia Research 1979* (D.

Yohn, B. Lapin, and J. Blakeslee, eds.), pp. 411–412, Elsevier/North-Holland, Amsterdam.

Neubauer, R., Dunn, F., and Rabin, H., 1981a, Infection of multiple T cell subsets and changes in lymphocyte functions associated with *Herpesvirus saimiri* infection of owl monkeys, *Infect. Immun.* **32:**698.

Neubauer, R., Levy, R., Strnad, B., and Rabin, H., 1981b, Reactivity of monoclonal antibodies against human leukocyte antigens with lymphocytes of nonhuman primate origin, *J. Immunogenet.* **8:**433.

Neubauer, R., Marchalonis, J., Strnad, B., and Rabin, H., 1982, Surface markers of primate B and T lymphoid cell lines, *J. Immunogenet.* **9:**209.

Neubauer, R., Briggs, C., Noer, K., and Rabin, H., 1983a, Identification of normal and transformed lymphocyte subsets of nonhuman primates with monoclonal antibodies to human lymphocytes, *J. Immunol.* **130:**1323.

Neubauer, R., Hopkins, R., III, and Rabin, H., 1983b, Monoclonal antibody identification of lymphoid populations transformed *in vitro* and *in vivo* with *Herpesvirus saimiri*, in: *Leukemia Reviews International*, Vol. 1 (M. Rich, ed.), pp. 109–110, Dekker, New York.

Nilsson, K. Giovanella, B., Stehling, J., and Klein, G., 1977, Tumorigenicity of human hematopoietic cell lines in athymic nude mice, *Int. J. Cancer* **19:**337.

North, J., Morgan, A., Thompson, J., and Epstein, M., 1982, Purified Epstein–Barr virus M_r 340,000 glycoprotein induces potent virus-neutralizing antibodies when incorporated in liposomes, *Proc. Natl. Acad. Sci. USA* **79:**7504.

Pennock, J., Reitz, B., Bieber, C., Jamieson, S., Burton, N., Ramey, A., Oyer, P., and Stinson, E., 1980, Lethal complications due to cyclosporin A immunosuppression with combination drug therapy in monkey heart and combined heart and lung transplantation, *Surg. Forum* **31:**375.

Pennock, J., Reitz, B., Bieber, C., Aziz, S., Oyer, P., Strober, S., Hoppe, R., Kaplan, H., Stinson, E., and Shumway, N., 1981, Survival of primates following orthotopic cardiac transplantation treated with total lymphoid irradiation and chemical immune suppression, *Transplantation* **32:**467.

Rabin, H., 1983, Lymphotropic herpesviruses of nonhuman primates, in: *Viral and Immunological Diseases in Nonhuman Primates* (S. Kalter, ed.), pp. 111–133, Liss, New York.

Rabin, H., Pearson, G., Chopra, H., Orr, T., Ablashi, D., and Armstrong, G., 1973, Characteristics of *Herpesvirus saimiri*-induced lymphoma cells in tissue culture, *in vitro* **9:**65.

Rabin, H., Neubauer, R., Woodside, N., Cicmanec, J., Wallen, W., Lapin, B., Agrba, V., Yakovleva, L., and Chuvirov, G., 1976a, Virological studies of baboon (*Papio hamadryas*) lymphoma: Isolation and characterization of foamyviruses, *J. Med. Primatol.* **5:**13.

Rabin, H., Pearson, G., Wallen, W., Neubauer, R., Cicmanec, J., and Levy, B., 1976b, Comparative studies with different strains of Epstein–Barr virus in owl monkeys and marmosets, in: *Comp Leukemia Research 1975* (J. Clemmesen and D. Yohn, eds.), pp. 326–330, Karger, Basel.

Rabin, H., Neubauer, R., Hopkins, R., III, Dzhikidze, E., Shevtsova, Z., and Lapin, B., 1977a, Transforming activity and antigenicity of an Epstein–Barr-like virus from lymphoblastoid cell lines of baboons with lymphoid disease, *Intervirology* **8:**240.

Rabin, H., Neubauer, R., Hopkins, R., III, and Levy, B., 1977b, Characterization of lymphoid cell lines established from multiple Epstein–Barr virus (EBV)-induced lymphomas in a cotton-topped marmoset, *Int. J. Cancer* **20:**44.

Rabin, H., Neubauer, R., Hopkins, R., III, and Nonoyama, M., 1978, Further characterization of a herpesvirus-positive orangutan cell line and comparative aspects of *in vitro* transformation with lymphotropic Old World primate herpesviruses, *Int. J. Cancer* **21:**762.

Rabin, H., Hopkins, R., III, Neubauer, R., Ortaldo, J., and Djeu, S., 1983, *In vitro* transformation of owl monkey T cells by *Herpesvirus saimiri* (HVS), *Eighth International Herpesvirus Workshop*, p. 79.

Rabin, H., Hopkins, R. F. III, Desrosiers, R. C., Ortaldo, J. R., Djeu, J. Y., and Neubauer, R. H., 1984, Transformation of owl monkey T cells *in vitro* with *Herpesvirus saimiri*, *Proc. Natl. Acad. Sci. USA*, **81**:4563.

Rabson, A., O'Conor, G., Lorenz, D., Kirschstein, R., Legallais, F., and Tralka, T., 1971, Lymphoid cell-culture line derived from lymph node of marmoset infected with *Herpesvirus saimiri*—Preliminary report, *J. Natl. Cancer Inst.* **46**:1099.

Rangan, S., and Gallagher, R., 1979, Tumors and viruses in nonhuman primates, *Adv. Virus Res.* **24**:1.

Rasheed, S., Rongey, R., Bruszweski, J., Nelson-Rees, W., Rabin, H., Neubauer, R., Esra, G., and Gardner, M., 1977, Establishment of a cell line with associated Epstein–Barr-like virus from a leukemic orangutan, *Science* **198**:407.

Schaffer, P., Falk, L., and Deinhardt, F., 1975, Attenuation of *Herpesvirus saimiri* for marmosets after successive passage in cell culture at 30°C, *J. Natl. Cancer Inst.* **55**:1243.

Schirm S, Müller, I., Desrosiers, R. C., and Fleckenstein, B., 1984, *Herpesvirus saimiri* DNA in a lymphoid cell line established by *in vitro* transformation, *J. Virol.* **49**:938.

Schooley, R., Byington, R., and Falk, L., 1983, Phenotypic analysis of New World primate mononuclear cell surface antigens, *J. Med. Primatol.* **12**:30.

Shope, T., Dechairo, D., and Miller, G., 1973, Malignant lymphoma in cotton-top marmosets after inoculation with Epstein–Barr virus, *Proc. Natl. Acad. Sci. USA* **70**:2487.

Strnad, B., Schuster, T., Klein, R., Hopkins, R., III, Witmer, T., Neubauer, R., and Rabin, H., 1982, Production and characterization of monoclonal antibodies against the Epstein–Barr virus membrane antigen, *J. Virol.* **41**:258.

Sundar, S., Levine, P., Ablashi, D., Leiseca, S., Armstrong, G., Cicmanec, J., Parker, G., and Nonoyama, M., 1981, Epstein–Barr virus-induced malignant lymphoma in a white-lipped marmoset, *Int. J. Cancer* **27**:107.

Thorley-Lawson, D., 1979, A virus-free immunogen effective against Epstein–Barr virus, *Nature (London)* **281**:486.

Thorley-Lawson, D., and Geilinger, K., 1980, Monoclonal antibodies against the major glycoprotein (gp 350/220) of Epstein–Barr virus neutralize infectivity, *Proc. Natl. Acad. Sci. USA* **77**:5307.

Wallen, W., Neubauer, R., Rabin, H., and Cicmanec, J., 1973, Nonimmune rosette formation by lymphoma and leukemia cells from *Herpesvirus saimiri*-infected owl monkeys, *J. Natl. Cancer Inst.* **51**:967.

Werner, F.-J., Bornkamm, G., and Fleckenstein, B., 1977, Episomal viral DNA in a *Herpesvirus saimiri*-transformed lymphoid cell line, *J. Virol.* **22**:794.

Werner, J., Pinto, C., Haff, R., Henle, G., and Henle, W., 1972, Responses of gibbons to inoculation of Epstein–Barr virus, *J. Infect. Dis.* **126**:678.

Werner, J., Wolf, H., Apodaca, J., and zur Hausen, H., 1975, Lymphoproliferative disease in a cotton-top marmoset after inoculation with infectious mononucleosis-derived Epstein–Barr virus, *Int. J. Cancer* **15**:1000.

Wright, J., Falk, L. A., and Deinhardt, F., 1974, Interferon production by simian lymphoblastoid cell lines, *J. Natl. Cancer Inst.* **53**:271.

Wright, J., Falk, L., Collins, D., and Deinhardt, F., 1976, Mononuclear cell fraction carrying *Herpesvirus saimiri* in persistently infected squirrel monkeys, *J. Natl. Cancer Inst.* **57**:959.

Yakovleva, L., Bukaeva, I., and Markova, T., 1980, Pathomorphological, ultrastructural, cytochemical, and immunological characterization of malignant lymphoma in baboons of the Sukhumi monkey colony, in: *Advances in Comp Leukemia Research 1979* (D. Yohn, B. Lapin, and J. Blakeslee, eds.), pp. 419–420, Elsevier/North-Holland, Amsterdam.

zur Hausen, H., 1980, Oncogenic herpesvirus, in: *DNA Tumor Viruses* (J. Tooze, ed.), pp. 747–795, Cold Spring Harbor Laboratory, New York.

CHAPTER 8

Cell-Mediated Immunity to EBV

Eva Klein

I. INTRODUCTION

Theoretically, potent oncogenic viruses may persist in their natural host species but must reach a symbiotic equilibrium that maintains infection without causing severe pathological conditions during or before the reproductive age. Tumor development may then be a biological accident, occurring when control mechanisms, which usually ensure a relatively harmless virus–host relationship, break down or fail to function properly.

Epstein–Barr virus (EBV)-carrying lymphoblastoid cell lines can be regularly established *in vitro* from the blood of individuals who have experienced EBV infection (Nilsson *et al.*, 1971). In the course of acute infectious mononucleosis (IM), the consequence of a primary EBV infection, such lines can be established with ease (Pope, 1967). This suggests that the patients carry EBV-infected B cells with the capacity to replicate indefinitely. B cells with proliferative potential appear soon after primary infection. For example, we have established a B_{EBV} cell line from an individual 12 days after assumed infection with EBV and 4 weeks before the symptoms of IM were manifested (Svedmyr *et al.*, 1984). The presence of virus-infected B lymphocytes can also be visually confirmed by detection of EBNA (EBV-determined nuclear antigen)-positive cells in the blood of IM patients (Klein *et al.*, 1976). Thus, the existence of EBV genome-carrying B cells in the blood is a well-proven fact. However, it is not known whether (1) they are immortalized cells that are kept from proliferation by the host control or (2) they are newly generated B_{EBV} cells,

EVA KLEIN • Department of Tumor Biology, Karolinska Institutet, S 104 01 Stockholm, Sweden.

produced by infection of B cells with virus from the oropharynx (Yao *et al.*, 1983). It is also possible that conditions allow proliferation of B_{EBV} *in vitro* whereas different conditions *in vivo* are not compatible with proliferation. For example, factors needed for the maturation of B cells may not be present *in vitro*. The mitogenic drive of EBV may be limited to a less mature stage and *in vivo*, due to the presence of the maturation signals, the B_{EBV} cells may readily pass that particular stage.

As the B lymphocyte, which is the target of transformation, is part of the immune system, its behavior is controlled by the regulatory circuits that govern the various immune compartments. As demonstrated in autologous mixed lymphocyte cultures *in vitro*, B cells can induce a T-cell response (Kuntz *et al.*, 1976). Responding T cells proliferate as well as acquire suppressor and cytotoxic potentials (Smith and Knowlton, 1979; Tomonari, 1980). Both mitogen- and EBV-induced B blasts are more potent stimulators of autologous T cells than are resting B cells (Klein *et al.*, 1981). The cytotoxicity of such stimulated T-cell cultures can be manifested against several targets that are sensitive to activated killer cells. The T-cell populations of IM patients exhibit a similar cytotoxic potential (Klein *et al.*, 1981).

Mononucleosis can present as an alarming lymphoproliferative disease and yet, in the vast majority of cases, it subsides without medical intervention. We assume that the pathology associated with this clinical syndrome is initiated by the T-cell response against B blasts and by the developing immunity against B_{EBV} cells and EBV (Klein and Masucci, 1982).

Although *in vitro* immortalization is often the consequence of the action of oncogenic agents, this phenomenon must be distinguished from malignant growth potential *in vivo*. The characteristics of EBV-positive B lines derived from normal blood lymphocytes, lymphoblastoid cell lines (LCL), and those derived from tumors differ considerably (Nilsson and Klein, 1982). The growth behavior of B cells in nude mice and as clones in agarose reflects their origin. Tumor-derived lines are monoclonal, are in a lower maturation stage, produce mainly surface rather than secreted immunoglobulin, do not carry surface IgD, and have a relatively low concentration of insulin receptors (Åman *et al.*, 1981). They also differ from LCL in expression of surface moieties detectable with monoclonal antibodies that define B-cell maturation markers (Fu *et al.*, 1980; Klein *et al.*, 1983a,b).

In addition to EBV infection, other cellular events contribute to malignant transformation. A regular cytogenetic change, i.e., translocation between the *myc* oncogene-carrying chromosome 8 and one of the immunoglobulin gene cluster-carrying chromosomes (14, 22, or 2), is found in all Burkitt's lymphomas (BL) (Klein, 1983). As similar changes also occur in EBV-negative BL and in some B-cell leukemias, EBV infection of the B cells is not the exclusive cause of this malignancy but one of the factors representing a step toward the neoplastic character. The regularity with which the specific chromosomal translocations have been found in BL cells suggests that they determine the malignant phenotype.

The mechanisms controlling the *in vivo* behavior of B_{EBV} cells belong to two essentially different immunological functions. One is the regulation of the various cell compartments of the immune system, the other is the immune response against EBV-induced antigens.

II. INHIBITION OF THE *IN VITRO* GROWTH OF B CELLS

As mentioned above, B_{EBV} cells can be grown directly from seropositive individuals and even more readily from IM patients. For studies on the immune response against the EBV-infected B cells, experimental infection is more suitable because it provides a predictable system. The final outcome of such experiments is influenced by the culture conditions. The density of the cells in culture is one particularly important factor. The continued growth of B_{EBV} cells and the establishment of permanent lines (LCL) is counteracted by T cells if the experiment is performed with seropositive donors. These results reflect the existence of an EBV-related cellular memory (reviewed by Rickinson and Moss, 1983). Regression of B-cell growth in the cultures containing T cells occurs more readily when the total cell density with which the cultures were initiated is increased. Analysis of experiments in which EBV-infected B cells were explanted alone or together with various subsets of autologous T cells suggested that the inhibition of B_{EBV} cell proliferation is a complex phenomenon. One contributing factor is the homeostatic regulation within the cellular compartments of the immune response. The experimental results on which this statement is based include phenomena that occur independently of the EBV serological status of the lymphocyte donor. Blastogenesis of T cells derived from seronegative and seropositive lymphocyte donors induced by autologous B cells, i.e., either mitogen- or EBV-induced B blasts, did not reveal any EBV-related pattern or memory (Weksler, 1976; Klein *et al.*, 1981).

Early after EBV infection of total lymphocyte populations, T cells can be seen attached to the B cells (Klein *et al.*, 1981). This sign for cell interaction did not differ in experiments performed with lymphocytes of seropositive and seronegative individuals.

The outgrowth of B_{EBV} cells was inhibited by Fcγ receptor-carrying T cells from both seropositive and seronegative individuals (Shope and Kaplan, 1979). These cells suppressed the proliferation of B cells even if they were allowed to be present only for 3 hr and thereafter removed. Their presence for a longer time (2 days) was not necessary and did not improve the growth inhibitory effect (Thorley-Lawson, 1980). Because addition of antibodies that neutralized interferon (IFN) blocked the inhibitory effect and IFN was shown to suppress B-cell growth, this prompt suppression appears to be mediated by IFN produced by this T cell subset (Thorley-Lawson, 1981).

The role of IFN in the control of B-cell growth has also been demonstrated in experiments with lymphocytes of patients suffering from rheumatoid arthritis (RA). T cells from these patients did not regulate the growth of B_{EBV} cells as efficiently as cells from healthy individuals (Tosato et al., 1981). In autologous MLR cultures of the RA patients, the IFN production was found to be suppressed (Hasler et al., 1983). The defect in the patients seemed to be confined to the autologous interactions because IFN was produced in regular amounts when mixed cultures were set up in allogeneic combinations. Adherent cells appeared to play a role in the defect of the RA patients because their removal lead to improved IFN-γ production.

We have studied the impact of various T-cell subsets, separated on the basis of density, on the transformation event. Experiments were performed with lymphocytes from seropositive individuals (Masucci et al., 1983). The growth of B cells was inhibited by the low-density subset after a few days of culture. This subset is active in NK assays and contains the IFN-secreting cells. However, if the B-cell cultures were observed for 3–4 weeks, the initial growth regressed even in the cultures containing high-density T cells.

An important difference was seen between the function of the low- and high-density cells with regard to radiosensitivity. The light cells with the early inhibitory effect were not inactivated by irradiation, while the delayed effect of the dense cells was abrogated, indicating that they proliferated in response to B_{EBV} cells. Each of these two effector cell systems contributed, to different extents, to the inhibitory effect of the total T population. Irradiated, unfractionated T cells did not cause regression of B_{EBV} cells. Consequently, the radiation-resistant effectors represented by the low-density cells, were not sufficient by themselves to inhibit completely the establishment of B_{EBV} cultures. This is probably due to the fact that there are only a few of these in unfractionated populations. Their effect was observable, however, because the growth kinetics of cultures containing separated high-density T cells were different from those containing the total T population.

The growth inhibition of the B cells by the T compartment was shown to occur also when the B cells were triggered for growth by pokeweed mitogen (PWM), a B-cell mitogen. Compared to the experiments with EBV-induced activation of B cells, the events in the mitogen activation system are more complex. In contrast to the EBV-induced proliferation, PWM-induced proliferation requires help of a T-cell subset. Thus, the conditions for the demonstration of T-cell growth inhibitory effect on the B cells require carefully designed experimental strategies. Such experiments were perforemd by Yen and Fairchild (1982) who exposed lymphocytes to low concentrations of PWM. These induced cell proliferation limited to two rounds of mitosis without inducing maturation of

B cells to plasmocytes. Analysis of the mitotic activity in the cultures established from the total population showed that in the presence of T cells, the proportion of B cells entering DNA synthesis declined earlier as compared to the cultures of purified B cells.

III. EBV-INDUCED CELL SURFACE ANTIGENS DETECTED BY LYMPHOCYTE CYTOTOXICITY

Because B-cell tumors are able to establish and grow in an immunocompetent host, one possibility is that these are less sensitive to an immune attack. It thus seemed a paradox that among the hemopoietic lines, the BL-derived ones have usually been more sensitive than the LCL in the natural cytotoxicity assay. In order to make a strict comparison, we have tested the NK sensitivity of EBV-carrying B-cell lines established from BL and normal B cells immortalized *in vitro* (LCL) from four patients (Torsteinsdottir *et al.*, 1984). Each pair of BL and LCL lines were maintained *in vitro* for similar periods (provided by G. Lenoir, IARC, Lyon) and tested simultaneously in the 4-hr ^{51}Cr release assay. Untreated and IFN-activated blood lymphocytes from healthy donors were the effectors. All lines were virtually resistant to the cytotoxic potential of unmanipulated lymphocytes. Lymphocytes exposed for 3 hr to 1000 IU IFN-α lysed all lines. There was some variation in the sensitivity of the four pairs but the corresponding LCL and BL-derived lines did not differ. Thus, the experiments did not reveal different NK sensitivity of the tumor-derived lines—a possibility that was proposed when the role of NK cells in immunosurveillance was formulated—nor did it provide any clue for the immunological basis of the control of B_{EBV} but not of the BL cells.

The first attempt to document cytotoxic T cells with specificity for EBV-related surface antigens was performed with blood lymphocytes of patients in the acute phase of IM (Svedmyr and Jondal, 1975). Cytotoxicity was demonstrated and the target panel suggested the role of an EBV-determined plasma membrane moiety in the effector–target interaction. As no antibodies were found against this target structure, the antigen was designated lymphocyte-defined membrane antigen (LYDMA). MHC restriction of kill, which has been found with other T-cell systems, was not found with cytotoxicity directed at LYDMA. We have proposed subsequently that dissection of EBV-specific cytotoxicity in IM blood cells is difficult because it may be "hidden" in the nonspecific cytotoxicity due to the activated state of the majority of T cells (Klein *et al.*, 1980, 1981; Klein and Masucci, 1982). In those *in vitro* systems that lead to the generation of cytolytic T cells such as (1) in EBV-infected lymphocyte cultures when the B_{EBV} cell growth regressed and (2) T cells cocultivated and restimulated with autologous LCL, the cytolytic T cells were MHC restricted, i.e., the effectors and targets had to share HLA-A or HLA-B antigens for cytotoxicity to be demonstrated. In addition, monoclonal

antibodies specific for common determinants on the HLA molecules inhibited the reactivity (reviewed by Rickinson and Moss, 1983). Also, EBV
specificity was shown by the lack of such response with cells of seronegatives and the lack of cytotoxicity against B blasts generated with
other stimulators. The molecular entity of LYDMA is still undefined.

Cytotoxic lymphocytes that recognize and lyse target cells expressing
EBV-envelope and EBV-induced "early antigen" on the membrane have
also been demonstrated (Sethi et al., 1981). The active T cells were generated in lymphocyte cultures infected with EBV and their effect was also
found to be MHC restricted. EBV-infected, mitogen-induced blasts were
used as targets. During the first few hours after infection, the reactivity
was probably due to recognition of the membrane-integrated viral envelope from the input virus. After this was removed by papain treatment,
the blasts lost sensitivity to the T cells but became sensitive if cultured
for a further 3–4 days. Inhibition of viral DNA synthesis did not abolish
the reappearance of sensitivity, indicating that an "early" membrane antigen was recognized by the T cells. This antigen is probably different
from LYDMA, which is defined operationally on EBNA-positive LCL or
B_{EBV} cells. The EBV-infected, mitogen-induced blasts used as targets were
not assayed for growth potential or expression of EBNA. However, on the
basis of results obtained in our laboratory, it is unlikely that they were
EBV-transformed cells. When human blood and tonsil B cells were separated into subsets according to their densities, only the small resting B
cells could be transformed by EBV. This was not due to a lack of receptors,
for the virus bound to and penetrated the low- as well as high-density B
cells (Åman et al., 1984). Furthermore, B cells activated by mitogens were
found to be refractory to the transforming property of EBV (Einhorn and
Klein, 1981; Åman et al., 1984). These data suggest that for transformation by EBV, the virus must initiate DNA synthesis.

Sensitivity to the effect of NK cells and reactivity with EBV-positive
sera detected by antibody-dependent cellular cytotoxicity was acquired
by EBV-positive LCL and some tumor-derived lines, i.e., Daudi and Raji
when induced to enter the viral cycle (Pearson et al., 1979; Blazar et al.,
1980).

The relationship between the various EBV-induced cell surface moieties detected by cell-mediated responses on the one hand and by antibodies on the other hand remains to be defined.

IV. IMMUNOLOGICAL MEMORY TO EBV-DETERMINED ANTIGENS DETECTED BY LYMPHOKINE PRODUCTION

The production of lymphokines is an early event of T-cell populations
stimulated with antigens to which they exhibit a memory. One of these
inhibits the movement of leukocytes, namely leukocyte migration inhibition factor (LIF), and can be used for assessing antigen recognition in

vitro. This assay has been used in studies of EBV-related cellular memory. Reactivity in this assay depended on the serological status of the donor in that only T lymphocytes from seropositive individuals responded with LIF production to extracts derived from EBV-carrying cell lines (Szigeti *et al.*, 1981, 1982, 1984a). In addition, the development of immunity to the various EBV-related antigens, to the extent it could be distinguished in the absence of purified antigens, during the acute phase of IM, also correlated with serology (Szigeti *et al.*, 1982). Cells expressing EBNA only, or EBNA and viral antigens (VCA and EA) were used as starting material for the antigen extracts. With these antigens, responsiveness against viral antigens was distinguished from that against EBNA or other EBV-determined cellular antigens expressed on transformed cells. Cellular recognition of EBNA was shown with extracts from EBNA-positive, VCA- and EA-negative cells and with partially purified EBNA preparations. This was substantiated by the parallelism of the reactivities with the presence or absence of anti-EBNA activity in the serum of the lymphocyte donors. However, in view of very recent experiments, this interpretation must be reevaluated. Solubilized plasma membrane preparations from Raji and EBV-converted sublines of Ramos elicited LIF production in lymphocytes of seropositive individuals (Szigeti *et al.*, 1984b). It seems, therefore, that an EBV-induced plasma membrane moiety, present in transformed cells, can also be detected by the lymphokine production assay. It remains to be seen whether this is the same antigen as that detected on EBV-positive cells in the T-cell cytotoxicity assay. Antibodies were found in BL and nasopharyngeal carcinoma patients that inhibited the antigen-induced effect and bound to the plasma membrane of EBV-positive lines.

The earlier results with extracts of EBNA-positive cells and the partially purified EBNA preparations may have also detected the same antigen. As the solubilized plasma membrane preparation was shown to be highly active, i.e., as little as 2 μg protein/ml was sufficient to elicit LIF production, small amounts present in the earlier preparations may have been recognized by the T cells. Furthermore, the parallel kinetics in the appearance of anti-EBNA serological response and LIF production by T cells of IM patients exposed to the extracts may be coincidental in that the plasma membrane antigen detected by lymphocytes could be expressed when the cells acquire EBNA antigen. Thus, the two antigens may become available for immune processing at the same time. A basis for this assumption is provided by the experiments of Moss *et al.* (1981) in which newly transformed cultures were followed for expression of EBNA and sensitivity for cytotoxic reactivity.

The lymphokine production assay detects T-cell memory to viral antigens, antigens present on the cell membrane of transformed cells and possibly EBNA. The plasma membrane antigen may be identical with the operationally defined LYDMA and thus the assay system may help to achieve its molecular definition.

V. EBV-RELATED IMMUNE PARAMETERS IN PATHOLOGICAL CONDITIONS

The results demonstrating EBV-related cell-mediated immunity *in vitro* correlate well with the serological status of healthy lymphocyte donors. Blastogenesis of T cells, lymphokine production, outgrowth inhibition, and generation of cytotoxic T cells have been shown with the lymphocytes of seropositive individuals. The question arises, however, whether disturbances in any of the *in vitro* immune parameters can be correlated with the *in vitro* situation. Also, it would be important to know whether B_{EBV} cell growth occurs in individuals who show impaired reactions.

Mononucleosis can be regarded as a model for the success of the host response in that its adequacy ensures that the B_{EBV} cells do not proliferate. In contrast, the X-linked lymphoproliferative (XLP) syndrome represents the consequences of an EBV infection that cannot be controlled. XLP is a maternally inherited, combined immunodeficiency (Purtilo, 1981; Sullivan, this volume). In these patients all the general immune parameters studied—unstimulated and IFN-activated NK activity, and EBV-specific antibodies and T-cell responses (including growth inhibition of B_{EBV} cells)—were found to be defective (Masucci *et al.*, 1981; Sullivan, this volume).

A relationship between experimental results and clinical experience is also provided by studies with cyclosporin A. Treatment of lymphocytes with this immunosuppressive drug eliminated the T-cell control of the growth of B_{EBV} cells *in vitro* (Bird *et al.* 1981). These results suggest that suppression of this mechanism may be responsible for the EBV-positive, progressive lymphoproliferative diseases occurring in renal and cardiac transplant patients treated with this drug (Saemundsen *et al.*, 1982).

Our understanding of the role of the various immune parameters controlling the EBV-infected cells is only fragmentary. There are some puzzling, presently unexplainable facts that may provide important clues in the future. One of these is the dissociation between the anti-EBNA and antiviral responses in immunodefective patients. During the development of anti-EBV-related immunity in acute mononucleosis, the response to VCA and EA viral antigens is measurable earlier than that to EBNA (Henle and Henle, 1979). The time difference is considerable, several weeks or months. Healthy seropositive individuals show both cellular and humoral immunity to these antigens. In T-cell immunodeficiencies the pattern or response is similar to that seen in early IM (Henle and Henle, 1981), i.e., the anti-EBNA reaction is low, often absent. In contrast, the antiviral immunity may be greater than in healthy individuals. Our present state of knowledge is insufficient to provide the reasons for the unusual EBV-related immune profile. It may be speculated that: (1) T cells or a T-cell-dependent mechanism is responsible for the destruction of EBNA-positive B_{EBV} cells and thus for the release of EBNA for immune processing, or (2) an increased proportion of B_{EBV} cells enter the viral cycle and EBNA disappears in the cells before VCA and EA is released.

We have selected for study patients suffering from three diseases exhibiting disturbances in T subsets and having an unusual EBV-related antibody profile: Hodgkin's lymphomas, non-Hodgkin's lymphomas, and ataxia telangiectasia (AT) patients (Masucci *et al.*, 1984a,b). Their blood T lymphocytes were tested for inhibitory effect in the B_{EBV} transformation system as well as for production of lymphokines (LIF) after exposure to EBV-determined antigen-containing extracts.

Among the lymphoma patients, some had exceptionally high anti-VCA titers (> 5120) and no rise in the anti-EBNA titers. Their OKT4/OKT8 ratio was abnormal (< 1).

In 6 of 9 Hodgkin's lymphoma patients with high antiviral titers, the lymphocytes performed with poor efficiency in the *in vitro* growth inhibition of B_{EBV} cells. In contrast, only 1 of 10 non-Hodgkin's lymphoma cases showed this impairment. Thirteen AT patients were selected for study who had low anti-EBNA titers relative to the antiviral titers. With 6 of these 13, growth inhibition by T cells was poor. It is important to note that none of these patients had EBV-positive malignancies. In the group of AT patients studied, one had an EBV-carrying lymphoma (Saemundsen *et al.*, 1981) but lymphocytes from this patient were not assayed in the B-cell growth inhibitory test.

Until recently, experimental findings in the various EBV-related disease states suggested that while the B_{EBV} cells are under immunological host control, once they acquire the translocation typical for BL, they escape and proliferate. In view of the occurrence of similar translocations in EBV-positive lymphomas in patients with AIDS, the validity of this assumption has to be questioned (Chaganti *et al.*, 1983; Ernberg *et al.*, 1984). AIDS is a syndrome characterized by severe immune dysregulation. The cytogenetics of the lymphomas that appear in transplant patients are largely unknown. Thus, the above-described distinction may not be sharp and it has to be assumed that B_{EBV} cells with translocations may appear in healthy individuals but are kept from proliferative potential by host immune mechanisms.

ACKNOWLEDGMENTS. The author's research was supported by PHS Grants RO1-CA-28380-03 and RO1-CA-30264-01 from the National Cancer Institute, and by the Swedish Cancer Society.

REFERENCES

Åman, P., Lundin, G., Hall, K., and Klein, G., 1981, Insulin receptors in human lymphoid cell lines of B-cell origin, *Cell. Immunol.* **65**:307.

Åman, P., Ehlin-Henriksson, B., and Klein, G., 1984, Epstein–Barr-virus susceptibility of normal human B-lymphocyte populations, *J. Exp. Med.* **159:**208.

Bird, A. G., McLachlan, S. M., and Britton, S., 1981, Cyclosporin-A promotes spontaneous outgrowth *in vitro* of Epstein–Barr virus induced B-cell lines, *Nature* **289:**300.

Blazar, B., Patarroyo, M., Klein, E., and Klein, G., 1980, Increased sensitivity of human lymphoid lines to natural killer cells after induction of the Epstein–Barr viral cycle by superinfection or sodium butyrate, *J. Exp. Med.* **151:**627.

Chaganti, R. S. K., Shanwar, S. C., Koziner, B., Arlin, Z., Mertelsmann, R., and Clarkson, B. D., 1983, Specific translocations characterize Burkitt's like lymphoma of homosexual men with the acquired immunodeficiency syndrome, *Blood* **61:**1269.

Einhorn, L., and Klein, E., 1981, EBV infection of mitogen-stimulated human B lymphocytes, *Int. J. Cancer* **27:**181.

Ernberg, I., Bjorkholm, M., Zech, L., Andersson, J., and Klein, G., 1984, Immunological characterization of a homosexual patient with an aleukemic EBV genome carrying acute lymphoblastic leukemia (in press).

Fu, S. M., Hurley, J. N. I., McCune, J. M., Kunkel, H. G., and Good, R. A., 1980, Pre-B cells and other possible precursor lymphoid cells derived from patients with X-linked agammaglobulinemia, *J. Exp. Med.* **152:**1519.

Hasler, F., Bluestein, H. G., Zwaifer, N. J., and Epstein, L. B., 1983, Analysis of the defects responsible for the impaired regulation of Epstein–Barr virus-induced B cell proliferation by rheumatoid arthritis lymphocytes. I. Diminished gamma interferon production in response to autologous stimulation, *J. Exp. Med.* **158:**173.

Henle, W., and Henle, G., 1979, Seroepidemiology of the virus, in: *The Epstein–Barr Virus* (M. Epstein and B. G. Achong, eds.), pp. 61–76, Springer-Verlag, Berlin.

Henle, W., and Henle, G., 1981, Epstein–Barr virus specific serology in immunologically compromised individuals, *Cancer Res.* **41:**4222.

Klein, E., and Masucci, M. G., 1982, Cell mediated immunity against Epstein–Barr virus infected B lymphocytes, *Springer Semin. Immunopathol.* **5:**63.

Klein, E., Masucci, M. G., Berthold, W., and Blazar, B. A., 1980, Lymphocyte mediated cytotoxicity toward virus-induced tumor cells: Natural and activated killer lymphocytes in man, in: *Viruses in Naturally Occurring Cancers* Cold Spring Harbor Conferences on Cell Proliferation, Vol. 7 (M. Essex, G. Todaro, and H. zur Hansen, eds.) *Cold Spring Harbor* Laboratory, Cold Spring Harbor, New York.

Klein, E., Ernberg, I., Masucci, M. G., Szigeti, R., Wu, Y. T., and Svedmyr, E., 1981, T cell response to B cells and Epstein Barr virus antigens in infectious mononucleosis, *Cancer Res.* **41:**4210.

Klein, G., 1983, Specific chromosomal translocation and the genesis of B cell derived tumors in man, *Cell* **32:**311.

Klein, G., Svedmyr, E., Jondal, M., and Persson, P. O., 1976, EBV-determined nuclear antigen (EBNA) positive cells in the peripheral blood of infectious mononucleosis patients, *Int. J. Cancer* **17:**21.

Klein, G., Ehlin-Henriksson, B., and Schlossman, S. F., 1983a, Induction of an activated B-lymphocyte associated surface moiety, defined by the B2 monoclonal antibody, by EBV-conversion of an EBV negative lymphoma line (Ramos): differential effect of transforming (B95-8) and nontransforming (P3HR-1) EBV-substrains, *J. Immunol.* **130:**1985.

Klein, G., Manneborg-Sandlund, A., Ehlin-Henriksson, B., Godal, T., Wiels, J., and Tursz, T., 1983b, Expression of the BLA antigen, defined by the monoclonal 38.13 antibody, on Burkitt lymphoma lines, lymphoblastoid cell lines, their hybrids and on other B-cell lymphomas and leukemias, *Int. J. Cancer* **31:**535.

Kuntz, M. M., Innes, J. B., and Weksler, M. E., 1976, Lymphocyte transformation induced by autologous cells, IV. Human T-lymphocyte proliferation induced by autologous or allogeneic non-T lymphocytes, *J. Exp. Med.* **143:**1043.

Masucci, G., Berkel, I., Masucci, M. G., Ernberg, I., Szigeti, R., Ersoy, F., Sanal, O., Yegin, O., Henle, G., Henle, W., Pearson, G., Aman, P., and Klein, G., 1984a, EBV-specific cell mediated.and humoral immune responses in ataxia telangiectasia patients (submitted for publication.)

Masucci, G., Mellstedt, H., Masucci, M. G., Szigeti, R., Ernberg, I., Bjorkholm, M., Tsukuda, U., Henle, G., Henle, W., Pearson, G., Holm, G., Biberfeld, P., Johansson, B., Svedmyr, E., and Klein, G., 1984b, Immunological characterization of Hodgkin's and non Hodgkin's lymphoma patients with high antibody titers against Epstein–Barr virus associated antigens, *Cancer Res.* **44**:1288.

Masucci, M. G., Szigeti, R., Ernberg, I., Masucci, G., Klein, G., Chessels, J., Sieff, C., Liel, S., Glomstein, A., Busino, L., Henle, W., Henle, G., Pearson, G., Sakamoto, K., and Purtilo, D. T., 1981, Cellular immune defects to Epstein–Barr virus-determined antigens in young males, *Cancer Res.* **41**:4284.

Masucci, M. G., Bejarano, M. T., Masucci, G., and Klein, E., 1983, Large granular lymphocytes inhibit the *in vitro* growth of autologous Epstein–Barr virus infected B cells, *Cell. Immunol.* **76**:311.

Moss, D. J., Rickinson, A. B., Wallace, L. E., and Epstein, M. A., 1981, Sequential appearance of Epstein–Barr virus nuclear and lymphocyte-detected membrane antigens in B cell transformation, *Nature (London)* **291**:664.

Nilsson, K., and Klein, G., 1982, Phenotypic and cytogenetic characteristics of human B lymphoid cell lines and their relevance for the etiology of Burkitt's lymphoma, *Adv. Cancer Res.* **37**:380.

Nilsson, K., Klein, G., Henle, W., and Henle, G., 1971, The establishment of lymphoblastoid lines from adult and fetal human lymphoid tissue and its dependence on EBV, *Int. J. Cancer* **8**:443.

Pearson, G. R., Qualtiere, L. F., Klein, G., Norin, T., and Bal, I. S., 1979, Epstein–Barr virus-specific antibody-dependent cellular cytotoxicity in patients with Burkitt's lymphoma, *Int. J. Cancer* **24**:402.

Pope, J. H., 1967, Establishment of cell lines from peripheral leukocytes in infectious mononucleosis, *Nature (London)* **216**:810.

Purtilo, D. T., 1981, Immune deficiency predisposing to Epstein–Barr virus induced lymphoproliferative diseases: The X linked lymphoproliferative syndrome as a model, *Adv. Cancer Res.* **34**:279.

Rickinson, A. B., and Moss, D. J., 1983, Epstein Barr virus induced transformation: Immunological aspects, in: *Advances in Viral Oncology* (G. Klein, ed.), Vol. 3, pp. 213–238, Raven Press, New York.

Saemundsen, A. K., Berkel, A. I., Henle, W., Henle, g., Anvret, M., Sanal, O., Ersoy, F., Caglar, M., and Klein, G., 1981, Epstein–Barr virus carrying lymphoma in a patient with ataxia teleangiectasia, *Br. Med. J.* **282**:425.

Saemundsen, A. K., Klein, G., Cleary, M., and Warnke, R., 1982, Epstein–Barr virus carrying lymphoma in cardiac transplant recipient, *Lancet* **2**:158.

Sethi, K. K., Stroehmann, I., and Brandis, H., 1981, Cytolytic human T-cell clones expressing Epstein–Barr virus specificity and HLA restriction, *Immunobiology* **160**:274.

Shope, T. C., and Kaplan, J., 1979, Inhibition of the *in vitro* outgrowth of Epstein–Barr virus infected lymphocytes by Tg lymphocytes, *J. Immunol.* **123**:2150.

Smith, J. B., and Knowlton, R. P., 1979, Activation of suppressor T cells in human autologous mixed lymphocyte culture, *J. Immunol.* **123**:419.

Svedmyr, E., and Jondal, M., 1975, cytotoxic effector cells specific for B cell lines transformed by Epstein–Barr virus (EBV) determined nuclear antigen (EBNA) in relation to the EBV-carrier status of the donor, *Cell. Immunol.* **58**:269.

Svedmyr, E., Ernberg, I., Seeley, J., Weiland, O., Masucci, M. G., Tsukuda, K., Szigeti, R., Masucci, G., Blomgren, H., Henle, W., and Klein, G., 1984, Virologic, immunologic and clinical observations on a patient during the incubation, acute and convalescent phases of infectious mononucleosis, *Clin. Immunol. Immunopathol.* **30**:437.

Szigeti, R., Luka, J., and Klein, G., 1981, Leukocyte migration inhibition studies with Epstein–Barr virus (EBV) determined nuclear antigen (EBNA) in relation to the EBV-carrier status of the donor, *Cell. Immunol.* **58**:269.

Szigeti, R., Masucci, M. G., Henle, W., Henle, G., Purtilo, D., and Klein, G., 1982, Effect of different Epstein–Barr virus determined antigen (EBNA), EA, and VCA on the leukocyte migration of healthy donors and patients with infectious mononucleosis and certain immunodeficiencies, *Clin. Immunol. Immunopathol.* **22**:128.

Szigeti, R., Masucci, G., Ehlin-Henriksson, B., Bendtzen, K., Henle, G., Henle, W., Klein, G., and Klein, E. 1984a, EBNA specific LIF production of human lymphocyte subsets, *Cell. Immunol.* **83**:136.

Szigeti, R., Sulitzenau, D., Henle, G., Henle, W., and Klein, G., 1984b, Detection of an EBV-associated membrane antigen in EBV-transformed virus nonproducer cells by leukocyte migration inhibition and blocking antibody, *Proc. Natl. Acad. Sci. USA* (in press).

Thorley-Lawson, D. A., 1980, The suppression of Epstein–Barr virus infection *in vitro* occurs after infection but before transformation of the cells, *J. Immunol.* **124**:745.

Thorley-Lawson, D. A., 1981, The transformation of adult but not newborn human lymphocytes by Epstein–Barr virus and phytohemagglutinin is inhibited by interferon: The early suppression by T cells of Epstein–Barr infection is mediated by interferon, *J. Immunol.* **126**:829.

Tomonari, K., 1980, Cytotoxic T cells generated in the autologous mixed lymphocyte reaction. I. Primary autologous mixed lymphocyte reaction, *J. Immunol.* **124**:1111.

Torsteinsdottir, S., Masucci, M. G., Lenoir, G., Klein, G., and Klein, E., 1984, Natural killer cell sensitivity of human lymphoid lines of B-cell origin does not correlate with tumorigenicity or with the expression of certain differentiation markers, *Cell Immunol.* **86**:278.

Tosato, G., Steinberg, A. D., and Blaese, R. M., 1981, Defective EBV-specific suppressor T-cell function in rheumatoid arthritis, *N. Engl. J. Med.* **305**:1238.

Weksler, M. E., 1976, Lymphocyte transformation induced by autologous cells. III. Lymphoblast-induced lymphocyte stimulation does not correlate with EB viral antigen expression or immunity, *J. Immunol.* **116**:310.

Yao, Q. Y., Rickinson, A. B., and Epstein, M. A., 1983, A reexamination of the virus carrier state, *in: Abstracts, VIIIth International Herpesvirus Workshop*, p. 204.

Yen, A., and Fairchild, D. G., 1982, T-cell control of B cell proliferation uncoupled from differentiation, *Cell. Immunol.* **74**:269.

CHAPTER 9

Immunobiology of EBV-Associated Cancers

GARY R. PEARSON

I. INTRODUCTION

The Epstein–Barr virus (EBV) has been studied as a candidate human cancer virus since its discovery in 1964. The herpesvirus was first isolated from a lymphoblastoid cell line established from an African Burkitt's lymphoma (BL) biopsy thereby stimulating interest in the role of this virus in the etiology of this disease (Epstein *et al.*, 1964). Subsequent studies by the Henle and Henle conclusively demonstrated that EBV was the causative agent of heterophil-positive infectious mononucleosis (Henle *et al.*, 1968; Henle and Henle, 1978). In addition, results were published by Old and associates suggesting that EBV might be a factor in the etiology of nasopharyngeal carcinoma (NPC), a high-incidence cancer in Africa and Southeast Asia (Old *et al.*, 1966). These initial disease associations, established largely from seroepidemiological findings, stimulated extensive research on the possible role of this virus in the etiology of these diseases as well as toward identifying other neoplastic disease states in which this virus might play an etiological role. This resulted in publications suggesting that EBV might also be an important cofactor in the etiology of such diseases such as Hodgkin's disease and CLL (Johannson *et al.*, 1970, 1971). In addition, this virus has been implicated as a factor in the etiology of B-cell lymphomas, which have been occurring at a relatively high frequency in patients with genetic-associated or drug-induced immunodeficiency states (Purtilo, 1976; Klein and Purtilo, 1981). More recently, results have been reported suggesting that EBV might also

GARY R. PEARSON • Section of Microbiology, Mayo Foundation, Rochester, Minnesota 55905.

be a causative factor of Kaposi's sarcoma in patients with AIDS (Quinnan *et al.*, 1984). Thus, from the etiological point of view, this virus continues to be of interest in relation to the etiology of certain types of human cancers.

Early studies on the association of EBV to human cancer concentrated on the use of seroepidemiology to identify disease states associated with infection by EBV. Results from these types of studies were instrumental in establishing the initial leads suggesting that EBV was a possible human cancer virus and have been reviewed in depth (Henle and Henle, 1978, 1979; Pearson, 1980). It soon became apparent, however, that the use of seroepidemiology by itself was not adequate for establishing an etiological relationship between a virus and a specific cancer. It was also considered necessary to directly demonstrate the presence of viral genetic information in candidate tumor cells. This was deemed necessary because it was subsequently established that reactivation of latent EBV infections also resulted in the presence of high antibody titers to EBV antigens (Henle and Henle, 1980). The search for virus information in tumor cells was approached by examining cells from candidate tumors for the presence of viral antigens and for viral DNA. By these approaches, it was observed that only biopsy cells from certain histopathological types of NPC and African BL were consistently positive for these viral markers (Adams, 1980; J. Pagano, personal communication). Cells from other types of cancers including biopsies from BL tumors obtained from low-incidence areas such as the United States were in general negative for these markers. Thus, from these types of studies, it was concluded that EBV was probably at minimum a necessary cofactor in the etiology of NPC and African BL.

More recently, immunological studies on EBV have focused on (1) applying the knowledge gained from the seroepidemiological studies to a clinical setting to aid the physician in the diagnosis and clinical management of patients with EBV-associated cancers; (2) identifying, purifying, and characterizing the individual proteins that compose the major EBV antigen complexes; (3) identifying the cellular components of the immune response active against EBV-transformed cells; and (4) developing immunlogical approaches for the prevention and treatment of EBV-associated cancers. This chapter will concentrate on advances in the first three areas. Advances in prevention and treatment will be covered by other participants in this symposium.

II. DIAGNOSTIC VALUE OF ANTIBODIES TO EBV ANTIGENS

In addition to the immunological investigations directed largely at establishing an etiological association between EBV and different neoplastic diseases, more recent studies have been directed at determining the clinical value of EBV serology in patients with EBV-associated cancers. The rationale for these investigations was based on findings from retrospective studies that indicated that (1) antibodies to certain EBV antigens were present more frequently and at higher titers in patients with NPC and African BL than in different control groups; (2) antibody titers increased with stage of disease and therefore reflected tumor load; and (3) antibodies to certain viral antigens exhibited disease-related patterns following treatment and therefore were of potential prognostic importance (Pearson, 1980).

To determine if antibodies to EBV antigens might be useful to the physician for the diagnosis of patients with EBV-associated cancers, a number of prospective studies were initiated over the past 4–5 years in different parts of the world to examine this question. These prospective studies have focused on patients with NPC. The results from such studies have provided fairly definitive evidence that some of the anti-EBV serological markers are of diagnostic value for certain histopathological types of NPC including the occult form (Ho *et al.*, 1976, 1978a; Coates *et al.*, 1978; Neel *et al.*, 1980, 1981; Pearson *et al.*, 1983b). Of particular importance to this question is the presence of serum IgA antibodies to EBV antigens. Henle and Henle originally established that sera from patients with NPC were frequently positive for IgA anti-EBV antibodies, in contrast to infected but nondiseased control populations (Henle and Henle, 1976). This observation was confirmed in a number of different studies (Desgranges *et al.*, 1977; Pearson *et al.*, 1978a). In general, it has been observed that 80–90% of the sera from patients with NPC were positive for IgA antibodies to EBV as opposed to 5–15% of the sera from various control populations. More recently, the serological profile of sera collected at diagnosis was related to tumor histopathology according to the WHO classification scheme (Shanmugaratnam and Sobin, 1978). By this approach, a striking difference was noted between the well-differentiated WHO 1 carcinomas versus the less-differentiated WHO 2 and WHO 3 histopathological types (Pearson *et al.*, 1983b). Sera from patients with WHO 2 and WHO 3 tumors were frequently positive for IgA anti-EBV antibodies. However, the serological profiles of patients with WHO 1 cancer were similar to those previously noted for different control populations. This is illustrated in Table 1 for IgA antibodies to viral capsid antigens (VCA) for North American NPC patients. In this study, approximately 83% of the sera from 249 patients with the less-differentiated types of NPC were positive for IgA anti-VCA antibodies at diagnosis as opposed to 10% of 51 patients with the WHO 1 histopathological types of NPC and 5–15% of the sera from various control populations. This specificity was even more apparent when the frequency of sera positive for both IgA anti-VCA and IgG anti-early antigen (EA) antibodies was compared among different disease categories as presented in Table II. By the method of analysis, 76% of the sera from patients with WHO 2 and WHO 3 tumors were positive for both antibodies as opposed to 8% of the

TABLE I. Frequency of IgA-Positive Sera in Different Disease Categories

Disease	No. sera positive/ No. tested	Percent positive
NPC		
WHO 1	5/51	10
WHO 2 and 3	206/249	83
Other head and neck cancers	51/324	15
Benign head and neck diseases	57/428	13
Other malignancies including lymphomas	7/73	10
Normals	18/405	5

patients with WHO 1 cancers and 2–11% of the various control sera. Thus, these results suggested that the IgA antibody assay was a potentially useful test for the diagnosis of WHO 2 and WHO 3 histopathological types of NPC but not WHO 1 when used either alone or in conjunction with the anti-EA assay. Because of this specificity, these assays have now proven useful for the diagnosis of the occult form of this disease (Ho *et al.*, 1976, 1978a; Coates *et al.*, 1978; Neel *et al.*, 1980, 1981) and also as screening tests for identifying individuals at risk for developing this disease (Zeng *et al.*, 1982). These serological findings, which are supported by results on the presence of virus DNA in biopsies from different histopathological types of this disease (Adams, 1980; J. Pagano, personal communication), also indicate that EBV might not be an etiological factor for the well-differentiated type of NPC. This relationship requires further examination.

With regard to the diagnostic value of these tests, it must be emphasized that these tests are not absolute and that there are "false-positive" and "false-negative" groups that have to be taken into account when interpreting the results. As pointed out above, approximately 15%

TABLE II. Frequency of Sera Positive for Both IgA Anti-VCA and IgG Anti-EA Antibodies in Different Disease Categories

Disease	No. sera positive/ No. tested	Percent positive
NPC		
WHO 1	4/51	8
WHO 2	189/249	76
Other head and neck cancers	35/324	11
Benign head and neck diseases	34/428	8
Other malignancies including lymphomas	5/73	7
Normals	8/405	2

of the sera from North American patients with NPC were negative for IgA antibodies to EBV at diagnosis. The reason for this is still unclear but this has to be taken into account when using these assays as aids to histopathology in the diagnosis of NPC. Similarly, sera from 5 to 15% of the individuals in the various control groups were also positive for IgA antibodies. Again the significance of this is unknown. Whether these individuals are at risk for the development of NPC or whether the presence of IgA antibodies to EBV reflects some cellular immune defect remains to be determined. With regard to this latter point, it is noteworthy that many individuals with immunodeficiencies also contain high levels of both IgA anti-VCA and IgG anti-EA antibodies in their sera (Klein and Purtilo, 1981). However, even with these exceptions, it is evident that these anti-EBV assays should be useful to clinicians for the diagnosis of certain histopathological types of NPC including the occult form of this disease.

Efforts to identify similar anti-EBV-specific antibody responses in patients with African BL have been unsuccessful. However, recent published results from a prospective study in Africa indicated that the presence, in predisease sera, of unusually high levels of antibodies to the different EBV antigens might identify those children at risk for developing this disease. Based on the results of a 5-year follow-up study on 42,000 children in Uganda bled sequentially from birth to 8 years of age, it was concluded that children with high levels of EBV activity, as reflected by high antibody titers, constituted the cohort at risk for developing BL (Geser et al., 1982). In fact, the increased risk of developing BL in this population was estimated to be approximately 30-fold higher in children having high titers as opposed to those with normal levels. These results were interpreted as supportive for the role of EBV in the etiology of BL.

III. PROGNOSTIC VALUE OF EBV SEROLOGY

EBV serology has also been evaluated prospectively as prognostic markers in patients with EBV-associated malignancies over the past few years. Interest in this question stemmed from results obtained largely through retrospective studies indicating that antibody titers to certain EBV antigens varied with the course of disease following treatment. This type of finding was interpreted as supportive of the role of EBV in the etiology of NPC and African BL. The results also suggested, however, that the monitoring of such antibodies was of potential clinical significance in these patient populations.

The results from the retrospective studies indicated that increases in antibody titers to certain EBV immunofluorescent antigens following treatment reflected the development of recurrent or metastatic disease (Henle and Henle, 1978; Pearson, 1980). These increases were frequently noted before there was clinical evidence of metastatic disease. In contrast, stable or decreasing titers tended to reflect a good prognosis. The antibodies that have been reported to show the greatest fluctuations with disease course include IgG antibodies to VCA and EA, IgA antibodies to VCA, and, more recently, antibodies to the EBV-induced DNase (Henle et al., 1973, 1977; Chan et al., 1977; Ho et al., 1978b; Cheng et al., 1980; Naegle et al., 1982). With regard to EA, the antibody of prognostic importance was identified to be directed against the restricted or R component in African BL and the diffuse or D component in NPC (Henle et al., 1971). The results from the various prospective studies now ongoing in different geographical locations in the world tend to support these conclusions.

A second assay that appears promising in relation to prognosis is the antibody-dependent cell-mediated cytotoxicity (ADCC) assay (Pearson and Orr, 1976). This test measures a cytotoxic antibody directed against a major EBV-induced membrane antigen determinant and has been of interest because published findings from retrospective studies indicated that ADCC titers determined at diagnosis were predictive of disease course (Pearson et al., 1978, 1979; Chan et al., 1979). In general, the results from these retrospective studies indicated that disease progression ultimately resulting in death occurred primarily in those patients with low ADCC titers at diagnosis whereas patients with high titers generally survived disease-free for significant periods of time.

To examine this further, a prospective study was initiated, using this assay, on North American patients with NPC. The current results from this study tend to support the findings from the retrospective studies. In general, disease progression ultimately resulting in death, in this patient population, has occurred mainly in individuals who presented with low ADCC titers at diagnosis. Patients with high titers in general have survived disease-free for significant periods of time following treatment. This is illustrated in Figs. 1 and 2. In Fig. 1, the data are presented as actuarial disease progression curves and only include North American patients with WHO 2 or WHO 3 carcinomas who were on study for at least 1 year. A similar pattern has not been noted with patients with the well-differentiated WHO 1 carcinomas. The patients were divided into high-titer (7680–15360) and low-titer (480–5760) groups based on ADCC titers determined at diagnosis. Disease progression was defined for this analysis as the presence of clinically evident metastatic disease. As shown in Fig. 1, approximately 75% of the patients in the high category have remained disease-free for 3 years or longer as opposed to approximately 30% of those individuals in the low-titer group. In the high-titer group, disease progression occurred primarily during the first $2\frac{1}{2}$ years following diagnosis and has since remained relatively stable. In contrast, the number of patients remaining disease-free in the low-titer group has continued to show a downward trend.

A second approach to the analysis of these data was through the calculation of "disease progression rates" at the different ADCC levels.

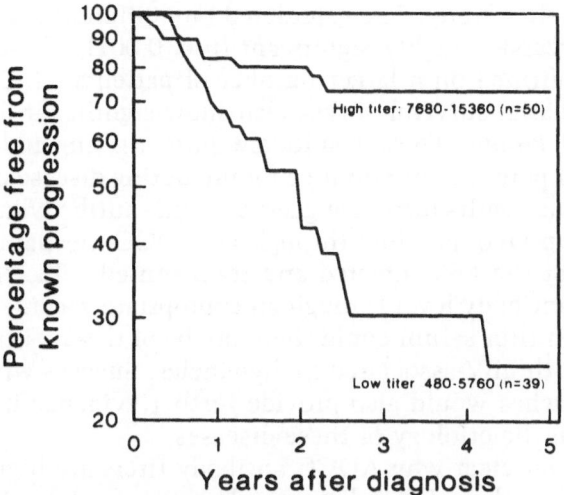

FIGURE 1. Actuarial disease progression curves for patients with low ADCC titers at diagnosis versus those with high ADCC titers. Only patients with WHO 2 or WHO 3 carcinomas were included in these calculations. The differences between these two groups at 3 years is significant at $p = 0.0008$.

The results are shown on Fig. 2. As an aid for this analysis, a "pooled" estimate of rates in the form of a "moving average" was also calculated. To accomplish this, the results for the first three titered groups were pooled and then plotted at midtiter (960), then the results for the second, third, and fourth titered group were pooled and plotted midtiter (1440), etc. The result is a smooth picture of trend. As shown in Fig. 2, the progression rates were very high at the low ADCC levels in comparison

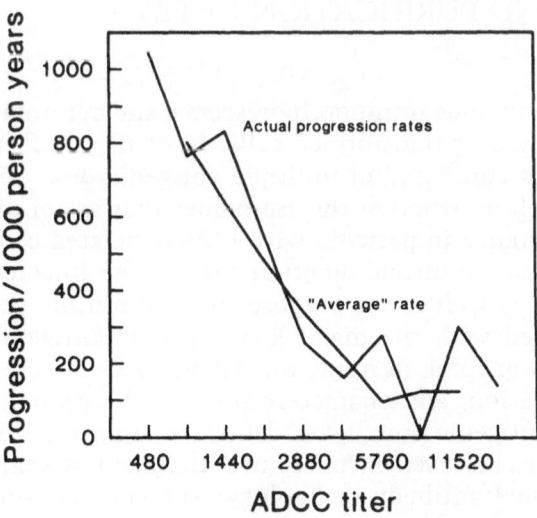

FIGURE 2. Actual and "average" disease progression rates at the different ADCC levels determined at diagnosis. These rates are based on disease progression during the first 3 years following the diagnosis of NPC.

with the high-titer group. The calculated rate difference (422 vs. 116 per 1000 person years) is highly significant ($p < 0.001$). These results have now been confirmed on a larger number of patients (G. Pearson *et al.*, 1984) and actuarial survival curves also show significant differences between the two groups. These results are quite striking and indicate that this assay is of potential importance for predicting disease course following therapy. The results further suggest that this antibody might be active in immunity *in vivo*, possibly through an ADCC mechanism including K cells, against the EBV-infected and transformed cells. Approaches to enhance this antibody level through an appropriate vaccine or by passive transfer of high-titer serum could therefore be of therapeutic importance for patients with EBV-associated malignancies. Success with these therapeutic approaches would also provide further evidence in support of a role for EBV in the etiology of these diseases.

It is still not clear why ADCC antibody titers are high in some patients and low in others. It has, however, been reported that IgA antibodies to EBV antigens block the ADCC reaction mediated by IgG antibodies (Mathew *et al.*, 1981). This could therefore be the explanation in NPC patients for, in general, patients with a poor prognosis also have high levels of IgA antibodies (Pearson, 1980). This would not be true, however, for patients with African BL. In these individuals, variation in ADCC levels could reflect differences in antibody avidity, the presence of antibody in immunoglobulin subclasses not capable of mediating ADCC, or lack of antibody to the appropriate ADCC determinant that is expressed on the major EBV membrane glycoprotein (Qualtiere *et al.*, 1982a). All of these factors have been shown to be important in the ADCC reaction (Pearson, 1978). Further studies are clearly indicated to resolve this question.

IV. IDENTIFICATION AND PURIFICATION OF EBV PROTEINS

The antigens used in the various immunofluorescence and cytotoxic assays are expressed in infected or transformed cells. Most if not all of these antigens are complexes composed of multiple antigenic determinants. It is not known, therefore, whether the antibodies that are of diagnostic or prognostic importance in patients with EBV-associated cancers are directed against one or multiple determinants in the infected cells. To refine these assays as well as for purposes of determining the biological activities associated with the major EBV-specified antigenic complexes, efforts have been ongoing in numerous laboratories directed at the identification, purification, and characterization of the proteins that compose these major antigenic complexes defined by immunofluorescence. Progress in this area has been excellent over the past few years with the advent of monoclonal antibody technology. Monoclonal anti-

TABLE III. EBV Polypeptides
Composing the Major EBV
Immunofluorescence Antigen
Complexes

Antigen	Polypeptides
VCA	160K[a]
	152K
	125K[a]
EA(D)	140K
	50/52K (47–60K)[a]
EA(R)	85K[a]
MA	gp300/350[a]
	gp200/250[a]
	gp90[a]
EBNA	81K
	73K
	70K
	65K
	48K

[a] Confirmed with monoclonal antibodies.

bodies have now been produced against a number of major antigenic determinants expressed in the VCA, EA, and MA complexes (Hoffman *et al.*, 1980; Thorley-Lawson and Geilinger, 1980; Mueller-Lantzsch *et al.*, 1981; Strnad *et al.*, 1982; Qualtiere *et al.*, 1982b; Pearson *et al.*, 1983a; Kishishita *et al.*, 1984). These monoclonal antibodies have also been useful for the purification and characterization of these proteins and for the development of new assays for monitoring antibodies to specific EBV-induced polypeptides. Such assays capable of monitoring antibodies directed against specific antigenic determinants might be more discriminating between individuals with and without EBV-associated malignancies than the current assays and therefore of greater value to the physician for the diagnosis and clinical management of individuals with these diseases.

A list of some of the polypeptides, identified by monoclonal antibodies or by biochemical purification, that compose the major immunofluorescence-defined EBV antigen complexes is shown in Table III. The major polypeptides composing the VCA complex as defined by immunofluorescence have molecular weights of 125, 152, and 160K. These are all intracellular proteins synthesized late in the virus replication cycle. Whether they are all associated with the viral capsid has yet to be determined. The major polypeptides associated with the D component of EA have molecular weights of 47–60 and 140K; the restricted component contains a major polypeptide with a molecular weight of 85K (Pearson *et al.*, 1983a). The major constituents of the 47–60K EA(D) complex have molecular weights of 50/52K (Pearson *et al.*, 1983a). However, these components become phosphorylated with time, shifting the molecular weights upward to 60K (Luka *et al.*, 1984). The 47K protein was originally identified by immunoprecipitation of an *in vitro* primary translation product using a monoclonal antibody reactive with the 50/52K polypeptides of the EA(D) complex (Pearson *et al.*, 1983a). By this approach, it was possible to map this EA(D) component to the EBV DNA *Bam*HI M fragment.

The MA complex has been characterized to a greater extent than the other EBV antigen complexes. This is largely due to the fact that these are the antigens of importance for the preparation of a subunit vaccine against this virus (Epstein, 1976). Three major glycoproteins of EBV have been identified as composing this complex (Qualtiere and Pearson, 1979, 1980; Thorley-Lawson and Edson, 1979; Strnad *et al.*, 1979; North *et al.*, 1980). These have recently been designated gp300/350, gp200/250, and gp85/90 to take into account the molecular weight variations for each of these glycoproteins reported by different laboratories. Neutralizing determinants are expressed on all three of these glycoproteins (Hoffman *et al.*, 1980; Thorley-Lawson and Geilinger, 1980; Strnad *et al.*, 1982; Qualtiere *et al.*, 1982a) while the determinants responsible for ADCC are expressed on gp300/350, possibly on gp200/250, but not on gp85/90 (Pearson *et al.*, 1979; Qualtiere *et al.*, 1982a). Interestingly, monoclonal antibodies to MA have been produced that recognize a membrane determinant expressed on cells producing transforming EBV but not lytic virus (Mueller-Lantzsch *et al.*, 1981; Qualtiere *et al.*, 1982b). Whether this determinant might be related to the transformation properties of EBV needs to be determined.

Efforts to produce monoclonal antibodies to the polypeptides composing EBNA have still not been successful for unknown reasons. However, proteins composing EBNA with molecular weights ranging from 48 to 81K have been identified by biochemical approaches (Luka *et al.*, 1978; Strnad *et al.*, 1981; Sculley *et al.*, 1983).

ELISA assays have now been established for measuring antibodies to these different polypeptides purified from immunoaffinity columns prepared with the appropriate monoclonal antibody or by biochemical techniques (Luka *et al.*, 1984). The sensitivity of these assays for detecting antibody has ranged from 2-fold to greater than 20-fold higher than immunofluorescence depending on the nature of the specific antigen. The assays are also applicable for detecting IgM and IgA antibodies to the different EBV-induced polypeptides (Pearson, 1984). It will now be of interest to determine whether the detection of antibodies to the individual polypeptides using the newly developed ELISA assays will show a greater specificity for individuals with EBV-associated cancers than the current immunofluorescence assays and therefore will be of greater clinical value. Such studies are now ongoing.

V. CELLULAR IMMUNITY

Progress in defining the role of cellular immunity against EBV in relation to disease progression has lagged behind the serological investigations. This is largely because of the lack of well-defined assays for monitoring cellular immune responses to viral antigens and the unavailability of well-defined antigens. Evidence has been reported, however, that clearly demonstrates that patients do mount a cellular immune response against their EBV-associated cancers (Pearson, 1980). In addition, HLA-restricted EBV-specific memory T cells have been identified in infected individuals as determined by the inhibition of outgrowth of EBV-transformed B cells (Rickinson et al., 1977; Thorley-Lawson et al., 1977; Moss et al., 1978; Misko et al., 1980). This T-cell population appears to be defective in individuals with severe immunodeficiencies and it has therefore been postulated that the relatively high frequency of EBV-associated lymphomas that occur in this population is the result of a defect in this memory T-cell population (Klein and Purtilo, 1981). The state of EBV-specific cellular immunity in patients with NPC and African BL, however, is still relatively unresolved. Evidence has been reported indicating that NPC and African BL patients do mount a T-cell-specific response against their tumor and that the target antigen is induced by EBV (Pearson, 1980). It has also been reported that virus-specific cellular immunity was decreased in patients with NPC and African BL as measured by skin testing and in vitro cytotoxicity assays (Fass et al., 1970; Nkrumah et al., 1972; Levine et al., 1977). The phenotype of the suppressed T-cell population in these patients as well as the nature of the factors responsible for the immunosuppression, however, have still not been defined. Interestingly, with regard to this latter point, a factor was recently identified in the sera of patients with NPC that inhibited the in vitro response of sensitized lymphocytes to EBV antigens (Sundar et al., 1982). This factor possibly responsible for this immunosuppression was reported to be associated with the presence of IgA antibodies possibly in the form of immune complexes.

The results discussed above therefore do support the concept that specifically immune T lymphocytes play a major role in the control of EBV-transformed cells and that a defect in this cell population will allow transformed cells to replicate and develop into a full-blown malignancy. In addition, data published recently suggest that NK cells might also function actively in resistance to EBV infection. Blazar and co-workers have reported that activation of the virus replication cycle increases the susceptibility of these cells to NK-mediated cytotoxicity (Blazar et al., 1980; Patarroyo et al., 1980). Furthermore, it was more recently observed in our laboratory that EBV genome-positive cells were more susceptible to NK cytotoxicity than their EBV genome-negative counterparts and that this susceptibility also increased in parallel with virus expression (G.

TABLE IV. Increased Susceptibility of TPA-Activated Cells to the Cytotoxicity Mediated by NK Cells

Target cell	No. lymphocytes	cpm $\pm$ S.D.[a]	Cytotoxicity (%)
A. Ramos	5.0×10^5	343 ± 26	3
	2.5×10^5	322 ± 10	0
B. B-95 Ramos	5.0×10^5	283 ± 39	16
	2.5×10^5	208 ± 4	9
	1.25×10^5	183 ± 7	7
C. TPA-Ramos	5.0×10^5	601 ± 18	12
	2.5×10^5	581 ± 28	9
D. TPA-B-95 Ramos	5.0×10^5	421 ± 40	26
	2.5×10^5	332 ± 26	19
	1.25×10^5	253 ± 40	13

[a] Total ^{51}Cr incorporation per 5×10^3 cells was: Ramos, 1111 cpm; TPA-Ramos, 1232 cpm; B-95 Ramos, 1228 cpm; TPA-B-95 Ramos, 1345 cpm.

Pearson, unpublished results). This is illustrated in Table IV. As shown in there, the EBV genome-negative Ramos cell line was less susceptible to cytotoxicity mediated by NK cells than its EBV genome-positive (B-95 Ramos) counterpart. In addition, treatment of the cells with tumor-promoting agent (TPA), which activates the virus replication cycle, increased the susceptibility of both cell lines but the effect was more pronounced with the EBV genome-positive B-95 Ramos cell line. These types of findings raise the possibility that NK cytotoxicity might be active *in vivo* in the control of EBV-infected or -transformed cells. The nature of the target antigen expressed in the genome-positive cells is still unknown but the results of inhibition assays suggest that it is a cellular antigen induced by the EBV infection and not a virus-specified antigen (G. Pearson, unpublished results).

VI. CONCLUSIONS

It is apparent that immunology has played a major role in establishing an etiological relationship between EBV and NPC, African BL, and B-cell lymphomas in immunodeficient populations. Of equal importance, however, was the identification of disease-related immune response patterns indicating that EBV serology might be useful for the diagnosis and clinical management of patients with EBV-associated cancers. This has been an area of emphasis over the past few years and the results from a number of prospective studies tend to confirm this speculation. A summary of the more important observations on EBV serology in relation to disease course is presented in Table V. These include the presence of high pre-disease anti-EBV titers in children at risk for the development of African BL; the specificity of the IgA anti-EBV antibodies for the less-differentiated forms of NPC; the expression of EBNA in cancer cells characteristic

TABLE V. Conclusions Concerning EBV Immunobiology

High anti-EBV titers identify individuals at risk for the development of African BL

The most specific antibody response for the diagnosis of the less-differentiated histopathological types of NPC is the presence of serum IgA anti-EBV antibodies. Detection of this antibody has proven useful to the clinician for the diagnosis of NPC including the occult form

The presence of EBNA or EBV DNA in biopsy material is also diagnostic for EBV-associated diseases

Antibody titers to EBV antigens detected by immunofluorescence generally increase in parallel with the development of metastatic disease. Decreasing or stable titers tend to reflect a good prognosis

Low antibody titers at diagnosis as determined with the ADCC assay in patients with EBV-associated malignancies in general reflect a poor prognosis

Control of the growth of EBV-transformed cells *in vivo* is mediated mainly by cytotoxic T cells and also possibly by NK and K cells

of these two diseases; the variation of antibody titers in relation to disease course; and the association between ADCC titers determined at diagnosis and disease course. These disease-related immune parameters not only provide additional support for an etiological association between EBV and these malignant diseases but also provide new and important laboratory markers to the clinician for the diagnosis and clinical management of patients with these cancers. In addition, the results of these immunological investigations have identified potentially important new approaches for treatment of these EBV-associated cancers. Success with such antiviral treatment approaches would provide direct proof that EBV is a major factor in the etiology of these diseases.

There has also been good progress in the past few years directed toward the identification, purification, and characterization of the major components composing the EBV-induced antigen complexes defined by immunofluorescence. This should lead to, among other things, the development of new and more specific assays for the detection of antibodies to EBV. These could be of more clinical relevance than the standard immunofluorescence tests and therefore efforts to develop such assays with purified virus-specified proteins should be encouraged.

Less is known about the role of cell-mediated immunity in relation to disease course although existing data do implicate cytotoxic T lymphocytes, NK and K cells as possible important effector mechanisms of immunity (Table V). It is clear that more work is needed in this area not only in relation to effector cell populations but also in regard to the nature of the antigens recognized in the cell-mediated immunity assays and the disease-associated factors that alter this cellular immunity. Little is currently known about the effect of the tumor on these cell-mediated immune responses. This information is needed before it will be realistically possible to develop and evaluate new forms of treatment for these diseases involving the immune system. This area of research should be of the highest priority in future studies involving the immunology of EBV-associated cancers. Hopefully, advances in this and related areas over the next few years will provide new leads for the prevention, treatment, and eventual eradication of these EBV-associated cancers.

REFERENCES

Adams, A., 1980, Molecular biology of the Epstein–Barr virus, in: *Viral Oncology* (G. Klein, ed.), pp. 683–711, Raven Press, New York.

Blazar, B., Patarroyo, M., Klein, E., and Klein, G., 1980, Increased sensitivity of human lymphoid lines to natural killer cells after induction of the Epstein–Barr viral cycle by superinfection or sodium butyrate, *J. Exp. Med.* **151**:614.

Chan, S. H., de Thé, G., and Goh, E. H., 1977, Prospective study of Epstein–Barr virus antibody titers and survival in patients with nasopharyngeal carcinoma, *Lancet* **1**:948.

Chan, S. H., Levine, P. H., Mulroney, S. E., de Thé, G. B., Lavoue, M. F., and Goh, E. H., 1979, A comparison of the prognostic value of antibody dependent lymphocyte cytotoxicity and other EBV antibody assays in Chinese patients with nasopharyngeal carcinoma, *Int. J. Cancer* **23**:181.

Cheng, Y. C., Chen, J. Y., Glaser, R., and Henle, W., 1980, Frequency and levels of antibodies to Epstein–Barr virus-specific DNase are elevated in patients with nasopharyngeal carcinoma, *Prox. Natl. Acad. Sci. USA* **77**:6162.

Coates, H. L., Pearson, G. R., Neel, H. B., III, Weiland, L. H., and Devine, K., 1978, An immunologic basis for the detection of occult primary malignancies of the head and neck, *Cancer* **41**:912.

Desgranges, C., de Thé, G., and Ho, J. H. C., 1977, Neutralizing EBV-specific IgA in throat washings of nasopharyngeal carcinoma (NPC) patients, *Int. J. Cancer* **19**:627.

Epstein, M. A., 1976, Epstein–Barr virus—Is it time to develop a vaccine program?, *J. Natl. Cancer Inst.* **56**:697.

Epstein, M. A., Achong, B. G., and Barr, Y. M., 1964, Virus particles in cultured lymphocytes from Burkitt's lymphoma, *Lancet* **1**:702.

Fass, L., Herberman, R., and Ziegler, J. L., 1970, Delayed cutaneous reactions to autologous extracts of Burkitt's lymphoma cells, *N. Engl. J. Med.* **292**:776.

Geser, A., de Thé, G., Lenoir, G., Day, N. E., and Williams, E. H., 1982, Final case reporting from the Ugandan prospective study of the relationship between EBV and Burkitt's lymphoma, *Int. J. Cancer* **29**:397.

Henle, G., and Henle, W., 1976, Epstein–Barr virus-specific IgA serum antibodies as an outstanding feature of nasopharyngeal carcinoma, *Int. J. Cancer* **17**:1.

Henle, G., Henle, W., and Diehl, V., 1968, Relationship of Burkitt's tumor associated herpestype virus to infectious mononucleosis, *Proc. Natl. Acad. Sci. USA* **59**:94.

Henle, G., Henle, W., and Klein, G., 1971, Demonstration of two distinct components in the early antigen complex of Epstein–Barr virus-infected cells, *Int. J. Cancer* **8**:272.

Henle, W., and Henle, G., 1978, The immunological approach to study virus-induced human malignancies using the Epstein–Barr virus as example, *Prog. Exp. Tumor Res.* **21**:19.

Henle, W., and Henle, G., 1979, Seroepidemiology of the virus, in: *The Epstein–Barr Virus* (M. A. Epstein and B. G. Achong, eds.), pp. 61–78, Springer-Verlag, Berlin.

Henle, W., and Henle, G., 1980, Consequences of presisting Epstein–Barr infections, in: *Viruses in Naturally Occurring Cancers*, Cold Spring Harbor Conferences on Cell Proliferation, Vol. 27 (M. Essex, G. Todaro, and H. zur Hausen, eds.). Cold Spring Harbor Laboratory, New York

Henle, W., Henle, G., Gunven, P., Klein, G., Clifford, P., and Singh, S., 1973, Patterns of antibodies to Epstein–Barr virus-induced early antigens in Burkitt's lymphoma: Comparison of dying patients with long term survivors, *J. Natl. Cancer Inst.* **50**:1163.

Henle, W., Ho., J. H. C., Henle, G., Chan, J. C. W., and Kwan, H. C., 1977, Nasopharyngeal carcinoma: Significance of changes in Epstein–Barr virus-related antibody patterns following therapy, *Int. J. Cancer* **20**:663.

Ho, J. H. C., Ng, M. H., Kwan, H. C., and Chan, J. C. W., 1976, Epstein–Barr virus-specific IgA and IgG serum antibodies in nasopharyngeal carcinoma, *Br. J. Cancer* **34**:655.

Ho, J. H. C., Kwan, H. C., Ng, M. H., and de Thé, G., 1978a, Serum IgA antibodies to Epstein–Barr virus capsid antigen preceding symptoms of nasopharyngeal carcinoma, *Lancet* **1**:436.

Ho, J. H. C., Ng, M. H., and Kwan, H. C., 1978b, Factors affecting serum IgA antibody to Epstein–Barr viral capsid antigens in nasopharyngeal carcinoma, *Br. J. Cancer* **37**:356.

Hoffman, G. L., Lazarowitz, S.-B., and Hayward, S. D., 1980, Monoclonal antibodies against a 250,000-dalton glycoprotein of Epstein–Barr identifies a membrane antigen and a neutralizing antigen, *Proc. Natl. Acad. Sci. USA* **77**:2979.

Johannson, B., Klein, G., Henle, W., and Henle, G., 1970, Epstein–Barr virus (EBV)-associated antibody patterns in malignant lymphoma and leukemia, *Int. J. Cancer* **6**:450.

Johannson, B., Klein, G., Henle, W., and Henle, G., 1971, Epstein–Barr virus (EBV)-associated antibody patterns in malignant lymphoma and leukemia. II. Chronic lymphocytic leukemia and lymphocytic lymphoma, *Int. J. Cancer* **8**:475.

Kishishita, M., Luka, J. Vroman, B., Poduslo, J. F., and Pearson, G. R., 1984, Production of monoclonal antibody to a late intracellular Epstein–Barr virus-induced antigen, *Virology* **133**:363.

Klein, G., and Purtilo, D., 1981, Symposium on Epstein–Barr virus-induced lymphoproliferative diseases in immunodeficient patients, *Cancer Res.* **41**:4209.

Levine, P. H., Wallen, W. C., Ablashi, D. V., Granlund, D. J., and Connelly, R., 1977, Comparative studies on immunity to EBV-associated antigens in NPC patients in North America, Tunisia, France and Hong Kong, *Int. J. Cancer* **20**:332.

Luka, J., Lindahl, T., and Klein, G., 1978, Purification of the Epstein–Barr virus-determined nuclear antigen (EBNA) from EB virus transformed human lymphoid cell lines, *J. Virol.* **27**:604.

Luka, J., Chase, R. C., and Pearson, G. R., 1984, A sensitive enzyme linked immunosorbent (ELISA) against the major EBV-associated antigens. I. Correlation between ELISA and immunofluorescence titers using purified antigens, *J. Immunol.* **67**:145.

Mathew, G. D., Qualtiere, L. F., Neel, H. B., III, and Pearson, G. R., 1981, IgA antibody, antibody dependent cellular cytotoxicity and prognosis in patients with nasopharyngeal carcinoma, *Int. J. Cancer* **27**:175.

Misko, I. S., Moss, D. J., and Pope, J. H., 1980, HLA-antigen-related restriction of T-lymphocyte cytotoxicity to Epstein–Barr virus, *Proc. Natl. Acad. Sci. USA* **77**:1247.

Moss, D. J., Rickinson, A. B., and Pope, J. H., 1978, Long-term T-cell cell mediated immunity to Epstein Barr virus in man. 1. Complete regression of virus-induced transformation in cultures of seropositive donor leucocytes, *Int. J. Cancer* **22**:662.

Mueller-Lantzsch, N., Georg-Fries, B., Herbst, H., zur Hausen, H., and Braun, D. G., 1981, Epstein–Barr virus strain- and group-specific antigenic determinants detected by monoclonal antibodies, *Int. J. Cancer* **28**:321.

Naegle, R. F., Champion, J., Murphy, S., Henle, G., and Henle, W., 1982, Nasopharyngeal carcioma in American children: Epstein–Barr virus-specific antibody titers and prognosis, *Int. J. Cancer* **29**:209.

Neel, H. B., III, Pearson, G. R., Weiland, L. H., Taylor, W. F., Goepfert, H. H., Pilch, B. Z., Lanier, A. P., Huang, A. T., Hyams, V. J., Levine, P. H., Henle, G., and Henle, W., 1980, Anti-EBV serologic tests for nasopharyngeal carcinoma, *Laryngoscope* **12**:1981.

Neel, H. B., III, Pearson, G. R., Weiland, L. H., Taylor, W. F., and Goepfert, H. H., 1981, Immunologic detection of occult primary cancer of the head and neck, *Otolaryngol. Head Neck Surg.* **89**:230.

Nkrumah, F. K., Herberman, R., Bigger, R., and Perkins, I. V., 1972, Sequential evaluation of cutaneous delayed hypersensitivity to recall and lymphoid cell line antigens in Burkitt's lymphoma, *Int. J. Cancer* **10**:142.

North, J. R., Morgan, A. J., and Epstein, M. A., 1980, Observations on the EB virus envelope and virus-induced membrane antigen (MA) polypeptides, *Int. J. Cancer* **26**:231.

Old, L. J., Boyse, E. A., Oettgen, H. F., de Harven, E., Geering, G., Williamson, B., and Clifford, P., 1966, Precipitating antibody in human sera to an antigen present in cultured Burkitt's lymphoma cells, *Proc. Natl. Acad. Sci. USA* **56**:1699.

Patarroyo, M., Blazar, B., Pearson, G., Klein, E., and Klein, G., 1980, Induction of the EBV cycle in B-lymphocyte derived lines is accompanied by increased natural killer (NK) sensitivity and the expression of EBV-related antigen(s) detected by the ADCC reaction, *Int. J. Cancer* **2**:365.

Pearson, G. R., 1978, *In vitro* and *in vivo* investigations on antibody-dependent cellular cytotoxicity, *Curr. Top. Microbiol. Immunol.* **80**:65.

Pearson, G. R., 1980, Epstein–Barr virus: Immunology, in: *Viral Oncology* (G. Klein, ed.), pp. 739–767, Raven Press, New York.

Pearson, G. R., and Orr, T. W., 1976, Antibody-dependent lymphocyte cytotoxicity against cells expressing Epstein–Barr virus antigens, *J. Natl. Cancer Inst.* **56**:485.

Pearson, G. R., Coates, H. L., Neel H. B., III, Levine, P., Ablashi, D., and Easton, J., 1978a, Clinical evaluation of EBV serology in American patients with nasopharyngeal carcinoma, *in: Nasopharyngeal Carcinoma: Etiology and Control* (G. de Thé and Y. Ito, eds.), pp. 439–448, IARC, Lyon, France.

Pearson, G. R., Johannson, B., and Klein, G., 1978b, Antibody-dependent cellular cytotoxicity against Epstein–Barr virus-associated antigens in African patients with nasopharyngeal carcinoma, *Int. J. Cancer* **22**:120.

Pearson, G. R., Qualtiere, L. F., Klein, G., Norin, T., and Bal, I. S., 1979, Epstein–Barr virus-specific antibody-dependent cellular cytotoxicity in patients with Burkitt's lymphoma, *Int. J. Cancer* **24**:402.

Pearson, G. R., Vroman, B., Chase, B., Sculley, T., Hummel, M., and Kieff, E., 1983a, Identification of polypeptide components of the Epstein–Barr virus early antigen complex with monoclonal antibodies, *J. Virol.* **47**:193.

Pearson, G. R., Weiland, L. H., Neel, H. B., III, Taylor, W., Earle, J., Mulroney, S. E., Goepfert, H., Lanier, A., Talvot, M. L., Pilch, B., Goodman, M., Huang, A., Levine, P. H., Hyams, V., Moran, E., Henle, G., and Henle, W., 1983b, Application of Epstein–Barr virus (EBV) serology to the diagnosis of North American nasopharyngeal carcinoma, *Cancer* **51**:260.

Pearson, G. R., Neel, H. B., III, Weiland, L. H., Mulroney, S. E., Taylor, W., Goepfert, H., Huang, A., Levine, P., Lanier, A., Pilch, B., and Goodman, M., 1984 Antibody-dependent cellular cytotoxicity and disease course in North American patients with nasopharyngeal carcinoma: a prospective study, *Int. J. Cancer* **33**:777.

Purtilo, D., 1976, Pathogenesis and phenotypes of an X-linked lymphoproliferative syndrome, *Lancet* **2**:882.

Qualtiere, L. F., and Pearson, G. R., 1979, Epstein–Barr virus-induced membrane antigens: Immunochemical characterization of Triton X-100 solubilized viral membrane antigens from EBV superinfected Raji cells, *Int. J. Cancer* **23**:808.

Qualtiere, L. F., and Pearson, G. R., 1980, Radioimmune precipitation study comparing the Epstein–Barr virus membrane antigens expressed on P_3HR-1 virus-superinfected Raji cells to those expressed on cells in a B-95 virus-transformed producer culture activated with tumor-promoting agent (TPA), *Virology* **102**:360.

Qualtiere, L. F., Chase, R., and Pearson, G. R., 1982a, Purification and biologic characterization of a major Epstein–Barr virus-induced membrane glycoprotein, *J. Immunol.* **129**:814.

Qualtiere, L. F., Chase, R., Vroman, B., and Pearson, G. R., 1982b, Identification of Epstein–Barr virus strain differences with monoclonal antibody to a membrane glycoprotein, *Proc. Natl. Acad. Sci. USA* **79**:616.

Quinnan, G. V., Jr., Masur, H., Rook, A. H., Armstrong, G., Frederick, W., R., Fauci, A. S., Ames, J., Epstein, J., Lane, C. H., Macher, A. M., Manischewitz, J. F., Jackson, L., Smith, H. A., Parker, M., Pearson, G. R., Parrillo, J., and Straus, S., 1984, Prevalence and possible relationship of herpesvirus infections to the etiology of the acquired immunodeficiency syndrome, *JAMA* (in press).

Rickinson, A. B., Crawford, D., and Epstein, M. A., 1977, Inhibition of the in vitro outgrowth of Epstein–Barr virus-transformed lymphocytes by thymus-dependent lymphocytes from infectious mononucleosis patients, *Clin. Exp. Immunol.* **28:**72.

Sculley, T. B., Kreofsky, T., Pearson, G. R., and Spelsberg, T. C., 1983, Partial purification of the Epstein–Barr virus nuclear antigen(s), *J. Biol. Chem.* **258:**3974.

Shanmugaratnam, K., and Sobin, L., 1978, Histological typing of upper respiratory tract tumors, in: *International Typing of Tumors*, No. 19, pp. 32–33, WHO, Geneva.

Strnad, B. C., Neubauer, R. H., Rabin, H., and Mazur, R. A., 1979, Correlation between Epstein–Barr virus membrane antigen and three large cell surface glycoproteins, *J. Virol.* **32:**885.

Strnad, B. C., Schuster, T. C., Hopkins, R. F., Neubauer, R. H., and Rabin, H., 1981, Identification of an Epstein–Barr virus nuclear antigen by fluoroimmunoelectrophoresis and radioimmunoelectrophoresis, *J. Virol.* **38:**996.

Strnad, B. C., Schuster, T., Klein, R., Hopkins, R. F., III, Witmer, T., Neubauer, R. H., and Rabin, H., 1982, Production and characterization of monoclonal antibodies against the Epstein–Barr virus membrane antigen, *J. Virol.* **41:**258.

Sundar, K. S., Ablashi, D. V., Kamaraju, L. S., Levine, P. H., Faggioni, A., Armstrong, G. R., Pearson, G. R., Krueger, G. R. F., Hewetson, J. F., Bertram, G., and Menezes, J., 1982, Sera from patients with undifferentiated nasopharyngeal carcinoma contain a factor which abrogates specific Epstein–Barr virus antigen-induced lymphocyte response, *Int. J. Cancer* **29:**407.

Thorley-Lawson, D. A., and Edson, C. M., 1979, Polypeptides of the Epstein–Barr virus membrane antigen complex, *J. Virol.* **32:**458.

Thorley-Lawson, D. A., and Geilinger, K., 1980, Monoclonal antibodies against the major glycoprotein (gp 350/220) of Epstein–Barr virus neutralize infectivity, *Proc. Natl. Acad. Sci. USA* **77:**5307.

Thorley-Lawson, D. A., Chess, L., and Strominger, J. L., 1977, Suppression of in vitro Epstein–Barr virus infection, *J. Exp. Med.* **146:**495.

Zeng, Y., Zhang, L. G., Li, H. Y., Jan, M. G., Zhang, Q., Wu, Y. C., Wang, Y. S., and Su, G. R., 1982, Serological mass survey for early detection of nasopharyngeal carcinoma in Wuzhou City, China, *Int. J. Cancer* **29:**139.

CHAPTER 10

Cytomegalovirus Infection after Organ Allografting
Prospects for Immunoprophylaxis

JOEL D. MEYERS

I. INTRODUCTION

Cytomegalovirus (CMV) infection is an important determinant of success following organ allografting. Exposure to an infection with CMV is common in normal persons, with the age-specific prevalence of antibody to CMV exceeding 50% by adulthood in most parts of the world (Wentworth and Alexander, 1971). In the normal host and especially in children, infection with CMV is often asymptomatic, although it may be the cause of atypical mononucleosis syndrome during adolescence and early adulthood (Klemola and Kaariainen, 1965). Uncommonly, it is the cause of more serious illnesses such as pneumonia, encephalitis, or Guillain–Barre syndrome. With the exception of congenital CMV infection, however, CMV is rarely the cause of severe disease in immunologically normal persons.

   The situation is quite different in the immunocompromised host. Severe disseminated CMV disease, including pneumonia, was recognized soon after the advent of kidney allografting (Rifkind *et al.*, 1964; Hedley-Whyte and Craighead, 1965), although the full spectrum of disease attributable to CMV infection did not become apparent until more recently and is undoubtedly still expanding. CMV infection has been associated with syndromes of fever and leukopenia, hepatitis, arthralgias and arthritis, retinitis, ulcerative disease of the esophagus, stomach, and bowel,

JOEL D. MEYERS • Fred Hutchinson Cancer Research Center and the University of Washington School of Medicine, Seattle, Washington 98104.

pneumonia, and encephalitis in allograft recipients as well as in other compromised hosts including neonates. Pneumonia has been an especially frequent manifestation of CMV infection following marrow allografting (Neiman *et al.*, 1977; Meyers *et al.*, 1983a). Apparently because CMV infection is itself immunosuppressive, bacterial, fungal, and protozoan superinfections have also been associated with CMV infection (Chatterjee *et al.*, 1978; Rand *et al.*, 1978). Both the primary manifestations of CMV infection and the secondary complications (i.e., superinfection) increase the morbidity and mortality and decrease the success of organ allografting.

II. PATHOGENESIS OF CMV DISEASE

Infection with CMV is a necessary, but not sufficient, element in the development of severe CMV disease. An incidence of CMV infection of 50–100% has been described in prospective studies of renal (Betts *et al.*, 1977; Peterson *et al.*, 1980), cardiac (Pollard *et al.*, 1982), and marrow transplant patients (Neiman *et al.*, 1977; Meyers *et al.*, 1980b). In many of these patients CMV infection appears to be truly asymptomatic even though viral excretion may be detectable for years after transplant (Cheeseman *et al.*, 1979a; Meyers, 1984).

Primary infection (i.e., a first infection in a previously seronegative patient) appears to be more severe than either infection due to reactivation of latent endogenous virus or reinfection with a new CMV strain in a previously seropositive individual (nonprimary infection) in some studies. Among kidney transplant patients, viremia is more common and more prolonged following primary infection (Betts *et al.*, 1975), and viremic infections are more commonly symptomatic (Betts *et al.*, 1977; Cheeseman *et al.*, 1979b). Thus, most studies of renal transplant patients have found a higher incidence of symptomatic CMV infection among patients with primary infections (Spencer, 1974; Ho *et al.*, 1975; Betts *et al.*, 1977; Pass *et al.*, 1978), although this has not been a universal finding (Fryd *et al.*, 1980). Following cardiac transplant, clinically apparent CMV disease is also more common among seronegative patients experiencing primary infection (Rand *et al.*, 1978; Pollard *et al.*, 1982). In marrow transplant recipients the concept of primary infection is more complex since the responding immune system is that of the marrow donor and, in contrast to renal and cardiac transplant patients, the occurrence of serious CMV disease (i.e., pneumonia) following marrow transplant is higher among patients who are seropositive to CMV before transplant (Meyers *et al.*, 1982c).

In addition to active CMV infection, the necessary feature in the development of CMV disease is profound immunosuppression. As mentioned, it is rare for immunologically normal individuals to develop serious manifestations of CMV infection such as pneumonia. Among allograft recipients, the severity of CMV disease appears to be related to the intensity of immunosuppression. For example, use of antithymocyte globulin in addition to corticosteroids and azathioprine increases the severity of CMV infection following renal transplantation (Pass *et al.*, 1981; Cheeseman *et al.*, 1979b). After cardiac transplant the lower dosages of antithymocyte globulin and corticosteroids made possible by the use of cyclosporin have been associated with less severe CMV disease (Preiksaitis *et al.*, 1983). After allogeneic marrow transplant, the use of antithymocyte globulin or total body irradiation for pretransplant conditioning among patients with aplastic anemia, or the use of cytotoxic agents in addition to cyclophosphamide for leukemics, significantly increases the risk of CMV pneumonia (Meyers *et al.*, 1983a).

The immune response that determines the severity of CMV disease as well as ultimate recovery is not defined. Although patients with fatal disease often do not make antibody against CMV (Simmons *et al.*, 1977; Neiman *et al.*, 1977), this observation is complicated by the rapid demise of some of these patients and thus only a short period during which antibody production is possible. It is clear that production of antibody does not itself ensure survival (Betts, 1984; Meyers *et al.*, 1982c). Suppression of lymphocyte proliferation responses and interferon (IFN) production *in vitro* have been found among allograft recipients during the period of maximum risk of CMV infection (Pollard *et al.*, 1978, 1982; Pass *et al.*, 1981; Levin *et al.*, 1978; Meyers *et al.*, 1980b), although there is a recent report of detectable IFN-γ production *in vivo* in marrow graft recipients with CMV infection (Rhodes-Feuillette *et al.*, 1983). The use of agents such as antithymocyte globulin specifically abrogates some of these responses (Meyers *et al.*, 1980b; Pass *et al.*, 1981). However, despite observations in individual patients associating restoration of the lymphocyte proliferation response to CMV antigen *in vitro* with recovery from CMV disease (Meyers *et al.*, 1982a; Wade *et al.*, 1982a), it has been difficult to prove a clear relationship between proliferative responses and the severity of disease (Meyers *et al.*, 1980b; Quinnan *et al.*, 1982). T-cell subsets, defined phenotypically by the use of monoclonal antibodies, have been examined serially in patients undergoing renal transplant (Schooley *et al.*, 1983). Among patients with cadaveric transplants, those with virologically proven herpesvirus infection [CMV, herpes simplex, or Epstein–Barr virus (EBV)] had inversion of the ratio of T-helper and T-suppressor cells, due to both a relative increase in T-suppressor cells and a decrease in T-helper cells. Similar, though less pronounced, changes were seen among patients receiving kidneys from living related donors. Patients who received antithymocyte globulin were more likely to have both herpesivrus infections and inverted T-subset ratios. The ratio remained inverted 100 days after transplantation in 11 of 12 patients with prolonged monitoring. These T-cell subset alterations also appeared to be associated with opportunistic superinfections.

TABLE I. Occurrence of CMV-Specific Cytotoxic Responses in Relation to Outcome of CMV Infection[a]

Outcome of infection	Frequency of occurrence of effector-cell response		
	Cytotoxic T cells[b]	Other[c]	Fraction responding[d]
	(No. of patients)		
Nonfatal	9	9	18/18
Fatal	0	2	2/10
Death from other causes	0	3	3/6

[a] From Quinnan et al. (1982).
[b] Includes all patients who were proven to have cytotoxic T-cell responses by testing of fractionated lymphocytes, with or without concurrent non-T-cell responses.
[c] Includes patients with responses mediated by non-T lymphocytes only and those whose effector cells were not fully characterized by fractionated studies.
[d] Five patients who survived CMV infection, two who died from CMV infection, and two who died from other causes were not tested until 3 or more weeks after the onset of infection and are not included.

Recent attention has also been directed to the role of cytotoxic effector cells in the acquisition of and recovery from CMV infection. Both nonspecific cytotoxic cells [i.e., NK cells (Starr and Garrabrant, 1981) and cells active in antibody-dependent cellular cytotoxicity (Kirmani et al., 1981)] and cytotoxic T lymphocytes can lyse CMV-infected target cells in vitro. Studies in marrow transplant patients have associated suppression of NK activity with risk of CMV infection (Dokhelar et al., 1981; Quinnan et al., 1982), although not all studies agree (Livnat et al., 1980). These studies have used tumor cell targets (e.g., K562) rather than CMV-infected targets and need to be repeated with CMV-infected targets for confirmation. More recently, development of either specific cytotoxic T-lymphocyte or nonspecific cytotoxic activity against CMV-infected target cells has been associated with recovery from CMV infection after marrow transplant (Quinnan et al., 1982) (Table I). Such studies should be extended to patients receiving renal or cardiac allografts. It has been suggested that cytotoxic T lymphocytes (and perhaps other transient immune responses) are important in recovery from active CMV infection while nonspecific effector cells (such as antibody-dependent cytotoxic cells) are important in maintaining the latency of CMV in seropositive patients (Kirmani et al., 1981). It is likely that both maintenance of latency and recovery from CMV infection are due to a complex and interrelated series of immune responses, and definition of a single response as being most important may not be possible. Many of these responses are intentionally suppressed after organ allografting to prevent rejection of the transplanted tissue, and it is not surprising that serious CMV disease occurs in association with increasing degree of immunosuppression.

Superinfection and Graft Rejection

The pathogenesis of superinfection or graft rejection temporally associated with CMV infection is also not clear. In normal patients with CMV mononucleosis, suppression of lymphocyte proliferation responses and IFN production (Levin *et al.*, 1979), increase in suppressor cell activity (Rinaldo *et al.*, 1980), and imbalance of T-cell subsets (Carney *et al.*, 1981) have been shown. Suppression of antibody production during acute CMV infection has also been shown in mice (Howard and Najarian, 1974). In humans, transient leukopenia, possibly due to the effect of suppressor cells on marrow progenitor cells, has often been associated with CMV infection. Thus, as with other viral infections, primary CMV infection itself appears to be immunosuppressive and superinfection with bacterial, fungal, or protozoan agents is not surprising although the specific mechanisms are not yet clearly defined.

Kidney allograft rejection temporally associated with active CMV infection has also become an accepted phenomenon (Lopez *et al.*, 1974; Simmons *et al.*, 1977; Betts *et al.*, 1977; Rubin *et al.*, 1977; Fryd *et al.*, 1980), although it is not yet clear that rejection or its treatment does not itself reactivate CMV in some cases. Some patients with acute rejection have glomerulopathy with antibody and complement found in the kidney (Richardson *et al.*, 1981; Baldwin *et al.*, 1982), and earlier studies demonstrated CMV antigen by immunofluorescence in specimens from some rejected kidneys (Payne *et al.*, 1974). Study of T-cell subset alterations after renal transplant has associated both CMV infection and T-cell subset alteration with glomerulopathy, whereas patients with acute rejection without glomerulopathy tended to have neither CMV infection nor T-cell alterations (Schooley *et al.*, 1983). Although the pathogenesis of the glomerulopathy is not yet fully explained, these studies do provide additional evidence that CMV infection is related to at least one form of kidney graft rejection.

III. EPIDEMIOLOGIC RISK FACTORS

CMV infection acquired during childhood in normal persons is presumably due to exposure to infected salivary secretions or urine, while during adulthood an additional source of infection appears to be sexual activity. CMV infection may be acquired at any age through exposure to blood products that contain latent CMV. In the early period after organ allografting, sexual exposure seems unlikely. Exposure to saliva or urine containing CMV remains possible but also seems less likely due to the relative isolation under which immunosuppressed patients are usually treated. In addition, although spread of CMV has been shown in the neonatal nursery (Spector, 1983), nosocomial transmission has not yet been

TABLE II. Relationship of CMV Infection to Patient and Granulocyte Donor CMV Antibody Status[a,b]

CMV serology before transplant	No granulocytes	Prophylactic granulocytes		Therapeutic granulocytes	
		GD-negative	GD-positive	GD-negative	GD-positive
Seronegative ($< 1:8$)	37/113 (33%)	16/46 (35%)	21/28 (75%)	2/11 (18%)	3/3 (100%)
Seropositive ($> 1:8$)	57/92 (62%)	7/13 (54%)	25/44 (57%)	3/4 (75%)	7/9 (78%)

[a] From Hersman et al. (1982).
[b] GD-negative = granulocyte donor seronegative for CMV; GD-positive = granulocyte donor seropositive for CMV. Numbers in parentheses represent percentage of patients affected.

demonstrated in transplant situations (Huang et al., 1981; Meyers, 1984). The two likely sources of infection after organ allografting are therefore blood products and transmission of virus in the transplanted tissue.

A. Infection Due to Blood Products in Seronegative Patients

Transmission of CMV in blood products was first described following open-heart surgery (Kreel et al., 1960). This continues to be a major source of infection in patients receiving fresh blood products such as patients requiring trauma surgery or neonates (Yeager et al., 1981). The risk of transmission increases with both the number of units (Prince et al., 1971) and the volume of blood (Yeager et al., 1981) transfused. However, the "dose" of blood (or leukocytes) required to transmit CMV is unknown. Studies in neonates showed that quantities as small as 50–99 ml were sufficient to transmit CMV in 3 of 18 (17%) babies (Yeager et al., 1981). However, even these 3 neonates received blood from 8 or more different donors, and it is not known if the number of donors or the volume transfused is more important in transmitting infection.

CMV infection in seronegative marrow, cardiac, and renal transplant patients has been associated with use of blood products. The situation is best defined following marrow transplant, after which one-third of seronegative patients who receive no granulocyte transfusions acquire CMV infection attributed to leukocyte-containing platelet transfusions (Hersman et al., 1982) (Table II). Seronegative patients who develop CMV infection receive slightly larger quantities of transfusions than those who remain uninfected (Meyers et al., 1983b). Considering that marrow transplant patients may receive platelet transfusions from 50 to 100 or more random donors, the incidence of 33% may be surprisingly low. Granulocyte tranfusions, as one type of blood product, are very effective in transmitting CMV infection. In contrast to the 33% incidence of infection

TABLE III. Risk of CMV Infection Defined as Seroconversion or Virus Shedding Posttransplant[a]

Pretransplant serology		Frozen blood only	No. infected/total (%)
Donor	Recipient		
−	−	Yes	1/46 (2)
−	−	No	25/40 (63)[b]
+	−	Yes	21/30 (70)
?	+	?	62/70 (89)

[a] Adapted from Betts (1984).
[b] Data from Dr. Robert Rubin, Massachusetts General Hospital.

among seronegative patients not receiving granulocytes, seronegative patients who received granulocytes from single seropositive donors developed CMV infection at incidences of 75% (Hersman et al., 1982; Table II) and 87% (Meyers et al., 1983b). These analyses are complicated by the fact that these patients also received platelet transfusions from random donors. The observation of a higher incidence of infection following granulocyte transfusions in which a larger number of leukocytes are transfused from single donors, compared with the lower incidence following platelet transfusions in which fewer leukocytes are given from a larger number of donors, suggests that the volume of blood transfused is the more important determinant.

Analogous data are available from studies of renal and cardiac transplant patients. Seronegative patients who receive kidney grafts from seronegative donors have been reported to have an incidence of CMV infection as high as 63% (25 of 40) if they receive fresh blood from random donors, compared with a 2% (1 of 46) incidence if only frozen blood is used (Betts, 1984) (Table III); frozen deglycerolyzed blood has previously been shown not to transmit CMV (Tolkoff-Rubin et al., 1978). Following cardiac transplant, a 20% incidence of infection (1 of 5) has been reported among seronegative patients receiving hearts from seronegative donors when unscreened blood was used compared with no infections among 8 patients when CMV antibody-negative blood was used (Preiksaitis et al., 1983). Thus, blood products appear to be an important source of infection for all three major groups of allograft recipients. Although an increase in risk has been observed as both volume of blood and number of different donors increase, the magnitude of risk associated with single donors or single units is unknown. Data derived from the use of granulocytes in marrow transplant patients suggest that blood products from virtually every seropositive donor can transmit CMV if sufficient blood is given to a sufficiently "susceptible" host. At this time, screening of blood to reduce the risk of transmitting CMV must rely on detection of antibody as no techniques presently exist for detection of antigen in latently infected individuals.

B. Transmission of Virus in Transplanted Tissue

Acquisition of CMV from the allograft itself was first apparent following kidney allografting (Ho et al., 1975; Betts et al., 1975). The incidence of CMV infection following transplantation of a kidney from a seropositive donor into a seronegative recipient has varied from 56% to 83% in a number of reports (Table III); this variability may be due in part to differences in intensity of surveillance. The site or state of the virus in the transplanted kidneys is unknown. Attempts to culture the virus from biopsy specimens have generally been unsuccessful (Orsi et al., 1978), and it appears that the virus is in a "latent," nonreplicating state not detectable by standard tissue culture techniques. One hypothesis is that CMV is actually in blood cells within the kidney vasculature (Betts, 1984). As with CMV transmitted in leukocytes, the virus may not be activated until an allogeneic reaction (i.e., graft rejection) occurs, thus explaining the association between rejection and CMV infection. The occurrence of primary CMV infection due to transmission of CMV in the kidney has led to proposals to select donor–recipient pairings based on CMV serologies whenever possible.

Following cardiac transplantation, the infection rate among seronegative recipients attributable to CMV carried within the transplanted organ has been reported to be as high as 92% (11 of 12) Preiksaitis et al., 1983). In contrast, risk of infection associated with marrow from seropositive donors following marrow transplantation is still unproven. In several studies the incidence of CMV infection has been slightly higher among seronegative patients receiving marrow from seropositive compared to seronegative donors (Meyers et al., 1980b; Hersman et al., 1982), but the difference has not been statistically significant. In other studies no difference was found (Meyers et al., 1983b), and the risk of CMV pneumonia is not increased among recipients of seropositive marrow (Meyers et al., 1982c). Of interest, however, CMV infection was apparently not preventable by the use of an immune globulin against CMV in seronegative patients receiving marrow from seropositive donors whereas infection in seronegative patients attributable to blood products was prevented (Meyers et al., 1983b; see below). The high risk of CMV infection due to leukocyte-containing blood products may make a small increase in risk due to donor marrow difficult to demonstrate in some of these studies. Despite the interest in donor selection based on CMV serologies, in many instances—especially cardiac or liver transplantation—such a choice is not available.

C. Reactivation of Latent Virus

The incidence of infection following allografting among seropositive patients has been greater than 50%, and sometimes as high as 100%, in

TABLE IV. Frequency of Development of Cytolytic Anti-CMV Antibody[a] to CMV-Infected Cells in Patients Grouped by Donor Status and Posttransplant Virus Shedding[b]

Pretransplant serology		Virus posttransplant	No. positive/No. tested
Donor	Recipient		
−	−	−	0/41
−	+	+	1/14
+	−	−	0/9
+	−	+	20/22
+	+	+	12/23[c]

[a] IgM and not IgG mediates this reaction.
[b] Adapted from Betts (1984).
[c] Data from Rochester and Massachusetts General Hospital.

virtually all studies (see Tables II and III). As noted, diagnosis of CMV infection is dependent upon intensity of surveillance, especially among patients with asymptomatic infection. Thus, it may be that all seropositive allograft recipients could be shown to reactivate CMV at some time during their posttransplant course if the intensity of virologic surveillance was sufficient. In some studies the severity of CMV infection appears to be lower among previously seropositive patients, though in more immunosuppressed kidney graft and marrow allograft recipients these nonprimary infections do not appear to be of lesser severity. The ability of preexistent antibody or cellular immune responses to modify the severity of subsequent nonprimary infection is of importance in considering the use of vaccines or immunoglobulins and will be discussed below.

Of additional interest, especially when considering the use of vaccines, is the question of reinfection of seropositive patients by new CMV strains. Epidemiologic studies are generally not able to distinguish reactivation of latent CMV from reinfection with new strains. For example, seropositive marrow graft recipients who also receive granulocytes from seropositive donors do not have a higher incidence of infection compared with seropositive patients receiving no granulocytes (Hersman et al., 1982) (Table II). There are at least three studies that specifically address this issue, two suggesting that reinfection may occur and one that it does not. Betts (1984) has reported that seropositive patients who receive kidneys from seropositive donors more often develop IgM cytolytic antibody than do seropositive patients receiving kidneys from seronegative donors (Table IV). It was previously shown that IgM cytolytic activity is formed in primary infection (Betts and Schmidt, 1981), suggesting that the seropositive patients who form IgM antibody may be doing so in response to infection with a new virus strain. In vaccine trials using Towne 125 CMV vaccine, previously seronegative patients who received Towne vaccine and later shed CMV after transplant were shown to be infected with non-Towne, wild-type CMV (Glazer *et al.*, 1979). Although reinfection after immunization with an attenuated virus strain is not entirely analogous to reinfection after primary infection with virulent wild-type virus, this observation does suggest that reinfection can occur. In contrast, the pre- and posttransplant CMV isolates from three marrow transplant patients who were shedding virus before transplant were found to be the same by restriction endonuclease analysis (Huang *et al.*, 1981), indicating in a very small number of patients that reinfection did not occur. Studies of the molecular epidemiology of CMV infection after organ allografting must be expanded before any firm conclusions can be made. Considerations about the control of CMV infection after organ allografting may be much different in seronegative and seropositive patients.

IV. PREVENTION OF PRIMARY INFECTION

Based on the known epidemiology of CMV infection, prevention of primary infection in the seronegative patient could be achieved either by reduction in exposure to exogenous sources of virus or by immunoprophylaxis; these will be discussed in turn. The use of antiviral agents for prophylaxis of primary infection will be discussed in later sections.

A. Donor Selection

Consideration of donor selection for seronegative patients has both practical and scientific aspects. For heart (and presumably liver) transplantation, the donor pool is sufficiently small and the logistics of organ preservation and transportation sufficiently complex that selection based on the donor's serologic characteristics would seem most difficult. In marrow transplantation, the opportunity to select between seropositive and seronegative donors who are otherwise equally appropriate (i.e., similar histocompatibility matches as well as other medical considerations) is uncommon, in part because there is a concordance of CMV serologies among family members of similar ages. Furthermore, the risk of CMV infection associated with use of marrow from seropositive donors is undefined and therefore the possible benefits accruing from "serologic matches" uncertain. Nevertheless, when the situation arises, the seronegative rather than seropositive donors should be chosen for seronegative marrow transplant patients. No data have been developed showing this choice has had clinical importance.

In kidney transplantation the opportunity to select donors based on CMV serologies is more frequent and usually involves patients requiring cadaveric transplants; this situation has been discussed most recently by Betts (1984). In the United States the major influences on the serologic status to CMV are age and socioeconomic group. The occurrence of primary infection is therefore said to be more common among parent-to-child (living related) transplants, for children are more likely to be seronegative and their parent-donors seropositive (Betts, 1984). In contrast, donors of cadaveric kidneys are often seronegative precisely because of their younger age. The serologic status of cadaveric donors can be ascertained by the rapid screening tests now available (Yeager, 1979) and the organs from seronegative donors reserved for seronegative recipients whenever possible.

B. Control of Blood Products

Despite the epidemiologic data associating CMV infection with the use of leukocyte-containing blood products, with one exception (Diosi *et al.*, 1969) it has not been possible to recover CMV from the blood of normal, asymptomatic blood donors by standard tissue culture techniques. Moreover, although molecular probes appear to be more sensitive than culture in the detection of viremia in actively infected patients (Spector *et al.*, 1983), these techniques have not yet detected CMV in blood from normal persons. The apparent explanation is that CMV is not in an active, replicating state in these blood cells but rather in a latent state, producing neither infectious virions nor antigens nor nucleic acid detectable by presently available techniques. These is evidence in mice that B cells carry murine CMV that can be reactivated *in vitro* by an allogeneic reaction (Olding *et al.*, 1975). The *in vivo* analog is the production of acute CMV infection after transfusion of allogeneic blood in mice (Cheung and Lang, 1977). Similar data are lacking in humans at present. In the absence of methods that can identify potentially infectious blood products, the only methods available to reduce or eliminate this risk is either to eliminate all blood products from seropositive donors or to remove the leukocytes, which are the putative reservoir for CMV.

The use of leukocyte-poor or frozen, deglycerolyzed blood products has been shown to reduce infection rates after open-heart surgery and dialysis, respectively (Lang *et al.*, 1977; Tolkoff-Rubin *et al.*, 1978). Use of seronegative blood products for transfusion of seronegative neonates eliminated CMV infection in that patient group (Yeager *et al.*, 1981), and that practice has been adopted in many neonatal nurseries.

In a small number of seronegative cardiac transplant patients, use of solely seronegative blood products also appeared to eliminate CMV infection (Preiksaitis *et al.*, 1983) (Table V). Study of additional patients is needed to confirm this observation. In renal transplant patients, Betts (1984) reported that only 1 of 46 seronegative patients who received only frozen blood products developed CMV infection (Table III). Such studies have not yet been reported among marrow transplant patients, who have much larger transfusion requirements, especially of platelets. Providing seronegative blood products for marrow transplant patients will increase

TABLE V. The Role of the Donor Heart in the Transmission of Infection Due to CMV[a,b]

CMV serologic status (recipient–donor)	No. patients infected/No. tested (%)	Transfusion requirements during first 3 postoperative months		
		Blood	Platelets	Fresh frozen plasma
Patients receiving unscreened blood				
Negative–negative	1/5 (20)[c]	10.4 ± 5.9	3.6 ± 2.7	7.4 ± 6.0
Negative–positive	6/6 (100)[d]	26.5 ± 17.8	17.8 ± 16.6	16.5 ± 14.4
Positive–negative	10/10 (100)	16.2 ± 5.2	6.4 ± 7.9	13.0 ± 14.9
Positive–positive	7/7 (100)	19.7 ± 19.1	30.3 ± 61.1	15.2 ± 9.5
Patients receiving screened blood				
Negative–negative	0/8[e]	30.3 ± 38.8	24.9 ± 45.4	13.9 ± 15.2
Negative–positive	5/6 (83)[f]	32.3 ± 29.7	47.0 ± 60.0	12.3 ± 15.2
Positive–negative	8/8 (100)	24.8 ± 16.4	17.5 ± 12.1	15.1 ± 12.8
Positive–positive	4/4 (100)	15.3 ± 6.2	21.0 ± 36.8	5.3 ± 3.6

[a] From Preiksaitis *et al.* (1983).
[b] Transfusion requirements are expressed as the mean ± S.D. units. Blood includes whole blood and/or packed red blood cells.
[c] One patient died at 6 weeks.
[d] $p < 0.01$ by χ^2 analysis.
[e] One patient died at 4 weeks.
[f] $p < 0.005$ by χ^2 analysis.

the already substantial burden on blood banking facilities as 40–50% of available donors may be eliminated because of seropositivity to CMV. These efforts are warranted, nevertheless, for the occurrence of serious or fatal CMV disease is one of the limiting factors in allogeneic marrow transplantation.

C. Passive Immunoprophylaxis with Plasma or Globulin

Passive immunization with specific antibody has been used for treatment and prevention of a number of infectious diseases. The closest parallel to prevention of CMV infection may be the use of zoster immune globulin or varicella–zoster immune globulin for the prevention of varicella after exposure. Important differences between varicella and CMV infection include the limited period of exposure to varicella virus and the likelihood that CMV is transmitted in a latent, nonreplicating, cell-associated form, whether in leukocytes or in other tissues in the transplanted organ.

Winston *et al.*, (1982b) first reported a study of passive immunization for the prevention of CMV infection in the compromised host. They gave high-titer human plasma to marrow transplant patients beginning before

TABLE VI. Results of Prophylactic Trial of CMV Immune Plasma after Marrow Transplantation[a]

	Plasma recipients	Control patients	p
All patients			
All CMV infection	12/24	15/24	0.56
Symptomatic CMV infection	5/24	12/24	0.07
All interstitial pneumonia	5/24	11/24	0.17
CMV pneumonia	3/24	8/24	0.17
Seronegative patients[b]			
All CMV infection	NR[c]	6/15	—
Symptomatic CMV infection	0/13	5/15	0.04
All interstitial pneumonia	1/13	7/15	0.04
CMV pneumonia	0/13	NR	—
Seropositive patients			
All CMV infection	NR	5/5	—
Symptomatic CMV infection	1/4	5/5	0.05
All interstitial pneumonia	0/4	3/5	0.16
CMV pneumonia	0/4	NR	—

[a] Data derived from Winston et al. (1982a).
[b] These patients received no granulocyte transfusions.
[c] NR, not reported.

transplant and continuing for 120 days after transplant, with most transfusions concentrated during the later posttransplant period when CMV pneumonia is most common rather than during the early period when the majority of transfusions are given. Evaluation of that study is complicated by the inclusion of patients with aplastic anemia, who have a significantly lower incidence of CMV pneumonia (Gale et al., 1981; Meyers et al., 1982c), and of patients seropositive for antibody to CMV before transplant. In the entire study group there was a slightly, but not significantly, lower incidence of CMV infection and CMV pneumonia among plasma recipients (Table VI). Among seronegative patients who did not receive granulocyte transfusions, the incidence of symptomatic CMV infection and of all interstitial pneumonia was significantly lower among plasma recipients; however, the incidence of all CMV infection, and of CMV pneumonia specifically, was not reported. A reduction in the incidence of symptomatic CMV infection and of all pneumonia was also reported among patients seropositive for antibody before transplant. Despite the difficulties in study design, the authors conclude that high-titer plasma given prophylactically modified but did not prevent CMV infection after marrow transplant.

The use of an intravenous globulin prepared from a high-titer plasma was described by O'Reilly et al. (1983). In this study globulin with an ELISA antibody titer of 1:3200 was given at a dose of 200 mg/kg body wt on days 25, 50, and 75 after marrow transplant, which includes the major risk period of CMV pneumonia. A randomized comparison group

TABLE VII. Incidence of CMV Infection and Interstitial Pneumonia after Marrow Transplant[a]

Globulin	Total patients	Interstitial pneumonia	Culture or serologic evidence of CMV infection
CMV hyperimmune	17	0	0
		$p = 0.23$	$p = 0.019^b$
CMV-antibody-deficient	18	3	6
		$p = 0.022^b$	$p = 0.01^b$
None	20	6	10

[a] From O'Reilly et al. (1983).
[b] Fisher's exact test. Comparisons are with patients receiving the CMV immune globulin.

received globulin with no antibody activity against CMV, and a third, concurrent but nonrandomized group received neither globulin. Evaluation of this study is also hindered by the inclusion of both seropositive and seronegative patients in the study groups and by the fact that at least two patients who received high-titer globulin shed CMV before the beginning of prophylaxis. Six of eighteen recipients of "CMV-antibody-deficient" plasma had evidence of CMV infection during the period of prophylaxis compared with none of 17 high-titer globulin recipients; this difference is significant at $p = 0.02$ (Table VII). One of three recipients of CMV-antibody-deficient globulin who developed pneumonia had CMV pneumonia. In the concurrent comparison group who received no globulin, both CMV infection and CMV pneumonia occurred at the expected high rates. These data also suggest that antibody can modify the manifestations of CMV infection even when given after the period of presumed exposure to infected blood products.

Finally, an immune globulin produced from a plasma pool with an even higher titer of antibody to CMV (globulin titer $= 1:16,000$ by ELISA) was given intramuscularly to marrow allograft recipients beginning before transplant and continuing through day 77 after transplant (Meyers et al., 1983b). Only seronegative patients were included in this study, with stratification both for use of prophylactic granulocyte transfusions and for marrow donor serology. Among patients who received no granulocyte transfusions, globulin recipients developed significantly fewer CMV infections than control patients (Table VIII). No effect was seen among patients who received granulocytes from seropositive donors; the number of patients receiving granulocytes from seronegative donors was insufficient for firm conclusions. Interestingly, among those who received no granulocytes, a protective effect was seen only among those with seronegative marrow donors (Table IX), possibly suggesting a dose–response effect. Among the eight control patients who received no granulocytes and who developed CMV infection, one each had CMV pneu-

TABLE VIII. Incidence of CMV Infection among Seronegative Globulin Recipients and Control Patients by Use of Prophylactic Granulocyte Transfusions[a]

Granulocyte use	Globulin recipients	Control patients
Seropositive granulocytes[b]	7/8[c]	6/7
Seronegative granulocytes[b]	1/5	0/6
No granulocytes	2/17 (12%)[d]	8/19 (42%)[d]

[a] From Meyers *et al.* (1983b).
[b] Seropositive and seronegative granulocytes refer to the serologic status of the granulocyte donor.
[c] Number infected/total number in group.
[d] Difference significant at $p = 0.05$ by one-sided Fisher's exact test and $p = 0.03$ by Mantel–Cox test.

monia, CMV esophagitis, and CMV viremia compared to no such occurrences among the two infected globulin recipients.

Although study design and therefore comparability of results are different in these three studies, the combined experience suggests that CMV infection can indeed be prevented or modified in seronegative patients who are given antibody either before exposure to CMV or before the onset of clinically manifest illness. The mechanism of this protection is unknown but presumably involves an ability of antibody (with or without immunocompetent cells) to interfere with the infectiousness or virulence of CMV. These studies are continuing with both licensed "nonimmune" globulins (Winston *et al.*, 1982a) and high-titer immune globulins given intravenously (Snydman *et al.*, 1982). Additional, careful study will be required to determine the effect of immune globulin prophylaxis in seropositive patients and in patients receiving renal, cardiac, or liver transplants as well as to determine the effect of combining passive immunization with the use of screened blood products.

D. CMV Vaccine

An alternate approach is the use of CMV vaccine in seronegative allograft recipients. While the use of passive immunization may temporarily restore the humoral arm of the immune system, active immunization with a live vaccine may provide both humoral and cellular mech-

TABLE IX. CMV Infection Rates in Nongranulocyte Recipients by Serology of Marrow Donor

	Globulin recipients	Controls
Seropositive marrow donor	2/6	3/7
Seronegative marrow donor	0/11[a]	5/12 (42%)[a]

[a] $p = 0.02$ by one-sided Fisher's exact test.

anisms of protection. The usefulness of vaccine will depend not only on its ability to stimulate an immune response but also on the durability of these responses during periods of intensive immunosuppression and on the ultimate safety of the vaccine, as latency and reactivation of the vaccine virus remains a theoretical possibility. A beneficial effect would seem least likely among marrow transplant recipients if vaccine is given before transplant because marrow allografting effectively ablates most active immune responses and because patients seropositive before transplant are not protected against subsequent CMV disease. Use of the vaccine to restore or initiate an immune response to CMV *after* marrow transplant is intriguing, but use of a live virus vaccine early after transplant may be hazardous. Immunization of the marrow donor is also intriguing because the responding immune system after transplant is derived from the marrow donor; however, there is no present evidence that the donor's antiviral immune responses are transferred to the recipient.

Among recipients of cardiac, liver, or renal allografts, in whom the intensity of immunosuppression is generally less, an immune response to CMV sufficient to protect against serious manifestations of CMV infection may persist throughout the peritransplant period. The target group for initial vaccine trials has been seronegative renal transplant patients in whom the risk of CMV-associated kidney graft rejection following primary infection is highest. The Towne 125 strain of live CMV vaccine has been shown to be safe and immunogenic in these patients. Both humoral and cellular (lymphocyte proliferation) responses have been produced with preservation of antibody responses for up to 46 weeks after transplant (Glazer *et al.*, 1979). Some of these patients have shed CMV after transplant, which has been shown to be due to infection with wild-type strains rather than vaccine virus. Thus, immunity against infection following exogenous exposure is incomplete. However, the more important issue is whether the serious manifestations of CMV disease can be prevented by immunization of seronegative patients, and this will only be apparent through randomized treatment trials already under way. As in marrow transplant patients, a future consideration might be the restoration or boosting of both cellular and humoral responses through the use of a vaccine given after transplant. In this case a killed or subunit vaccine may also be useful and may circumvent any potential problems with vaccine-associated illness.

V. PREVENTION OF VIRUS REACTIVATION

The problem of preventing active virus infection among seropositive patients may be more difficult than preventing primary infection among seronegative patients. Seropositive patients presumably harbor and become infected with their own latent CMV, although molecular evidence proving that reinfection with exogenous strains does not also occur is lacking at this time. The incidence of CMV infaction among seropositive patients (nonprimary infection) is generally higher than among seronegative patients. The period of time over which prophylaxis must be effective may be longer in seropositive patients than among seronegative patients in whom exposure to exogenous strains may be relatively limited in time (e.g., during the limited period of blood transfusions). The mechanisms important in preventing virus reactivation may also be different. For example, it is already clear that antibody (or a cellular mechanism using antibody) is not effective in preventing virus reactivation, nor does the presence or production of antibody prevent the development of severe CMV disease in some patients. Prevention of virus reactivation may therefore be more difficult than preventing primary infection. Conversely, modalities effective in preventing either virus reactivation or its serious manifestations should also be effective in the prevention of primary infection. The use of an effective antiviral agent to prevent virus replication and disease after initial reactivation may be the most straightforward approach. Ultimately, a combination of modalities, such as use of screened blood products to prevent exogenous exposure, an effective antiviral agent, and passive or active immunization with globulin or vaccine, may be most effective for both seronegative and seropositive patients.

A. Interferon

IFN was first recognized by its ability to prevent virus replication, even though its immunologic and antitumor activities are presently of equal or greater interest. There are few demonstrations of efficacy as an antiviral agent in humans. Among these is the ability of human leukocyte (α) IFN to delay or prevent the manifestations of CMV infection among renal allograft recipients. Cheeseman et al., (1979b) first reported that partially purified Cantell IFN-α, given at the time of transplant and twice weekly for 6 weeks after transplant, could delay the onset of CMV excretion and decrease CMV viremia (Table X) among patients receiving cadaveric or living related transplants. An effect on viremia was not seen among patients receiving antithymocyte globulin, nor did IFN-α reduce the expression of CMV disease among patients whether or not they received antithymocyte globulin. Subsequent publications have reported an effect on EBV though not on papovavirus infection (Cheeseman et al., 1980a,b). Although some patients had reversible thrombocytopenia and leukopenia, these possible toxicities did not interfere with IFN use in most cases.

In an extension of this initial study, Hirsch et al. (1983) gave Cantell IFN to seropositive recipients of cadaveric or living related transplants for a total of 14 weeks. IFN was given three times a week for the first 6 weeks and then twice weekly for the remaining 8 weeks, with 20 IFN

TABLE X. Incidence of Viremia in CMV-Infected Patients after Renal Transplant[a]

Treatment	Antithymocyte globulin	No antithymocyte globulin	Combined[b]
Interferon	5/7	0/4[c]	5/11[d]
Placebo	6/6	3/4[c]	9/10[d]
Combined[b]	11/13[e]	3/8[e]	14/21

[a] From Cheeseman et al. (1979b).
[b] 95% confidence intervals for combined groups were: interferon, 0.20–0.75; placebo, 0.65–0.99; antithymocyte globulin, 0.57–0.97; and no antithymocyte globulin, 0.11–0.71.
[c] $p = 0.07$, interferon versus placebo by Fisher's exact test.
[d] $p = 0.04$, interferon versus placebo.
[e] $p = 0.04$, antithymocyte globulin versus no antithymocyte globulin.

recipients compared to 22 placebo recipients in a randomized, blind trial. In contrast to the previous study, timing of CMV infection was not affected by IFN use except among patients receiving antithymocyte globulin. However, clinical syndromes attributed to CMV were significantly reduced by IFN use, occurring among 7 of 22 placebo recipients but only 1 of 20 IFN recipients ($p = 0.03$) (Table XI). Most syndromes occurred in recipients of cadaveric kidneys who were also receiving antithymocyte globulin. Once again, IFN was well tolerated and apparent marrow suppression was equivalent in the two groups. Similar trials in seronegative patients are still under way.

In contrast to these findings in renal transplant patients, Meyers et al. (1982b) reported preliminary results of a prophylactic IFN trial among marrow allograft recipients. Although a slight delay in the onset of CMV excretion or seroconversion was found, there was no apparent effect on the overall incidence of CMV infection or on CMV pneumonia. An important difference between the trials in renal and in marrow allograft

TABLE XI. Incidence of CMV Syndromes after Renal Transplant According to Randomization Subgroup[a]

Subgroup	No. with syndrome/No. in subgroup		Placebo	Total[b]
	Interferon			
Antithymocyte globulin	0/9	$p = 0.02^c$	5/10	5/19
No antithymocyte globulin	1/11		2/12	3/23
Total[b]	1/20	$p = 0.03^c$	7/22	8/42
Cadaver donor	0/12	$p = 0.01^c$	7/15	7/27
Living related donor	1/8		0/7	1/15
Total[b]	1/20	$p = 0.03^c$	7/22	8/42

[a] From Hirsch et al. (1983).
[b] 95% confidence intervals: interferon, 0.20–0.25; placebo, 0.17–0.54; antithymocyte globulin, 0.13–0.50; no antithymocyte globulin, 0.05–0.34; cadaver donor, 0.14–0.45; living related donor, 0.02–0.32.
[c] By Fisher's exact test.

recipients is that IFN was started at the time of renal transplantation but not until approximately $2\frac{1}{2}$ weeks after marrow infusion, possibly too late to interfere with the initial stages of reactivation or primary infection. Alternatively, marrow transplant recipients may be too immunosuppressed to benefit from IFN use.

With the wider availability of IFN produced by recombinant DNA techniques, these initial trials should be confirmed in renal transplant patients and extended to other allograft recipients. Despite the greater purity of these cloned IFNs, it appears that hematologic (and possibly liver) toxicity will continue to be a limiting factor in the use of IFN. An important question will be whether IFN is effective only in preventing virus reactivation or whether it can prevent primary infection as well. The mechanism of protection—antiviral or immunologic—also remains to be established.

B. Antiviral Agents

Few studies of the prophylactic use of antiviral agents other than IFN have been conducted, in part because of the poor activity of available antiviral agents against CMV. Vidarabine, given intermittently at a dose of 5 mg/kg per day for periods of up to 10 days after marrow transplant, did not substantially change the occurrence of CMV infection (Kraemer et al., 1978). Acyclovir has not been used in a trial specifically designed for the prevention of CMV infection. With one exception (Gluckman et al., 1983), acyclovir has not been shown to have an effect on CMV infection when used at doses and durations designed for the prevention of herpes simplex infection after marrow transplant (Saral et al., 1981; Hann et al., 1983; Wade et al., 1982b). Other agents with increased activity against CMV in vitro, such as FIAC (fluoro-iodo-arabinosyl-cytosine) or 2'-nor-2'-deoxyguanosine (Tocci et al., 1983), an acyclovir derivative, will be of future interest. Additionally, the additive or synergistic effects of combinations of antiviral agents including IFN against CMV may be of relevance for the design of future prophylactic trials (Levin and Leary, 1981; Spector and Kelley, 1983). Although it would seem logical to use antiviral agents for the prevention of virus reactivation, prolonged use should also be effective in preventing primary infection. Indeed, prophylaxis, like treatment, may be more effective when the viral load is still small.

VI. IMMUNOSUPPRESSIVE REGIMEN

As reviewed above, more severe CMV disease has been associated with the intensity of immunosuppression in renal, cardiac, and marrow transplant patients. In the case of IFN, it appeared that otherwise successful prophylaxis may be inhibited by the use of antithymocyte globulin after renal transplant (Cheeseman *et al.*, 1979b). The development of immunosuppressive agents or regimens that prevent graft rejection or graft-versus-host disease, but which are otherwise less broadly immunosuppressive, may reduce the significance of virus infection. Optimally, such immunosuppression would allow specific tolerance of the allograft without affecting other immune functions, a goal that may be difficult to achieve.

There are initial data that current regimens that include the agent cyclosporin (thus allowing the use of lower doses of other agents such as corticosteroids, azathioprine, or antithymocyte globulin) may indeed be associated with a lower incidence of posttransplant infections, at least after cardiac transplantation (Preiksaitis *et al.*, 1983). Such changes were not found in preliminary studies at another center, however (Ho *et al.*, 1982), and additional study of the impact of newer immunosuppressive regimens on the occurrence of severe CMV disease is needed.

VII. TREATMENT OF CMV INFECTION

As suggested above, presently available antiviral agents have little activity *in vivo* against CMV. Although some trials (Balfour *et al.*, 1982) suggest clinical improvement among patients treated for CMV infection with acyclovir, most trials with IFN (Meyers *et al.*, 1980a), vidarabine (Ch'ien *et al.*, 1974; Marker *et al.*, 1980), acyclovir (Wade *et al.*, 1982a), or combinations of these agents (Meyers *et al.*, 1982a; Wade *et al.*, 1983) indicate little efficacy against clinically manifest disease.

VIII. CONCLUSIONS

Based on the known epidemiology of CMV infection, prevention of primary CMV infection after organ allografting is possible at this time. Reduction in the most likely sources of exogenous exposure may be accomplished through selection of seronegative organ donors for seronegative patients and elimination of blood products potentially capable of transmitting CMV. Although donor selection may present practical difficulties or may not be possible at all in some situations such as cardiac or liver transplantation, the potential consequences of CMV infection are sufficiently severe after kidney transplantation that some investigators have suggested that it may be preferable to transplant seronegative patients with kidneys from seronegative cadaveric donors rather than from seropositive, living related donors (Betts, 1984).

Elimination of blood products that can transmit CMV may be accomplished either through elimination of seropositive donors or through the sole use of frozen blood; depletion of leukocytes is technically more difficult and appears to be less effective (Lang *et al.*, 1977). Among patients who require large quantities of blood products, such as marrow transplant patients, elimination of seropositive donors from the donor pool may have great practical consequences for blood banks, which supply a large number of platelet transfusions for these patients. As an alternative, passive immunoprophylaxis with high-titer plasma or globulin appears to be effective either in preventing CMV infection altogether or in reducing the severity of CMV disease. The use of a globulin rather than pooled plasma has recognized advantages. It remains to be determined whether globulins produced from unscreened donor pools with relatively lower antibody titers to CMV will be as effective as globulins produced from high-titer pools. Intravenous rather than intramuscular globulins would also be preferable, at least for usage at high doses and during the earlier phases of hospitalization, whereas intramuscular globulins may have a role during later outpatient phases of treatment.

Active immunization with CMV vaccines may be superior to passive immunization in that both humoral and cellular mechanisms would be stimulated and immunity would be longer-lasting. Although the CMV vaccine presently being tested in the U.S. (Towne 125) appears to be immunogenic in renal transplant patients, issues of safety and efficacy remain to be determined in ongoing randomized trials. Vaccines may be more effective in patients with lower degrees of immunosuppression postgrafting (e.g., renal transplant patients) compared to profoundly immunosuppressed marrow transplant patients. In the specific case of marrow transplant patients, immunization of the marrow donor or restoration of immunity *after* transplant with either live or subunit vaccines are intriguing possibilities that will require careful study.

Among seropositive patients, the prospects of preventing CMV reactivation appear less certain with presently available approaches. Preexistent antibody clearly does not prevent virus reactivation, and thus passive immunization would seem unlikely to reduce CMV infection in seropositive patients. However, there have been suggestions that passive immunization may reduce the severity of CMV *disease*, and additional study of passive immunization in seropositive patients is warranted. Similarly, in the absence of data clearly implicating reinfection as a source of clinically important CMV infection in seropositive patients, screening of blood products may not affect the significance of CMV disease. These issues will require additional study.

Restoration or boosting of immunity among seropositive patients with either live or subunit vaccines after transplantation is not presently being studied. The possibility that such maneuvers would reduce the severity of CMV *disease* without affecting the incidence of CMV *infection* is worthy of consideration. Concern that the immune response may play a role in the pathogenesis of CMV disease cannot be ignored, however. Furthermore, the safety of live vaccines given to immunosuppressed patients is also of concern, and thus subunit or killed vaccines may be more appropriate in this context. Restoration of immunity through other manipulations such as specific transfer factors may also be possible.

IFN, as an immunologically active agent as well as an antiviral agent, appears to be effective in modifying CMV infection in seropositive patients after renal transplant. Its efficacy in seronegative patients remains to be demonstrated, as does the efficacy of both natural and cloned IFNs in recipients of other organs.

Control of CMV infection among seropositive patients therefore may await the development of new antiviral agents effective against human CMV strains. Although control of primary CMV infection appears to be achievable at this time, control of nonprimary infection will require greater understanding of the immunology and pathogenesis of CMV disease in seropositive patients as well as the development of new approaches to treatment and prophylaxis.

ACKNOWLEDGMENT. This investigation was supported by NIH Grants CA-18029, AI-15689, CA-30924, and CA-26966.

REFERENCES

Baldwin, W. M., Van Es, A., Valentijn, R. M., Van Gemert, G. W., Daha, M. R., and Vanes, L. A., 1982, Increased IgM and IgM immune complex-like material in the circulation of renal transplant recipients with primary cytomegalovirus infections, *Clin. Exp. Immunol.* **50**:515–524.

Balfour, H. H., Jr., Bean, B., Mitchell, C. D., Sachs, G. W., Boen, J. R., and Edelman, C. K., 1982, Acyclovir in immunocompromised patients with cytomegalovirus disease, *Am. J. Med.* **73**:241–248.

Betts, R. F., 1984, The relationship of epidemiology and treatment factors to infection and allograft survival in renal transplantation, in: *Pathogenesis and Prevention of Human Cytomegalovirus Infection* (S. A. Plotkin, ed.), Alan R. Liss, New York, pp. 87–100.

Betts, R. F., and Schmidt, S. G., 1981, Cytolytic IgM antibody to cytomegalovirus in primary cytomegalovirus infection in humans, *J. Infect. Dis.* **143**:821–826.

Betts, R. F., Freeman, R. B., Douglas, R. G., Jr., Talley, T. E., and Rundell, B., 1975, Transmission of cytomegalovirus infection with renal allograft, *Kidney Int.* **8**:387–394.

Betts, R. F., Freeman, R. B., Douglas, R. G., Jr., and Talley, T. E., 1977, Clinical manifestations of renal allograft derived primary cytomegalovirus infection, *Am. J. Dis. Child.* **131**:759–763.

Carney, W. P., Rubin, R. H., Hoffman, R. A., Hansen, W. P., Healey, K., and Hirsch, M. S., 1981, Analysis of T lymphocyte subsets in cytomegalovirus mononucleosis, *J. Immunol.* **126**:2114–2116.

Chatterjee, S. N., Fiala, M., Weiner, J., Stewart, J. A., Stacey, B., and Warner, N., 1978, Primary cytomegalovirus and opportunistic infections: Incidence in renal transplant recipients, *J. Am. Med. Assoc.* **240**:2446–2449.

Cheeseman, S. H., Stewart, J. A., Winkle, S., Cosimi, A. B., Tolkoff-Rubin, N. E., Russell, P. S., Baker, G. P., Herrin, J., and Rubin, R. H., 1979a, Cytomegalovirus excretion 2–14 years after renal transplantation, *Transplant. Proc.* **11**:71–74.

Cheeseman, S. H., Rubin, R. H., Stewart, J. A., Tolkoff-Rubin, N. E., Cosimi, A. B., Cantell, K., Gilbert, J., Winkle, S., Herrin, J. T., Black, P. H., Russell, P. S., and Hirsch, M. S., 1979b, Controlled clinical trial of prophylactic human-leukocyte interferon in renal transplantation: Effects on cytomegalovirus and herpes simplex virus infections, *N. Engl. J. Med.* **300:**1345–1349.

Cheeseman, S. H., Black, P. H., Rubin, R. H., Cantell, K., and Hirsch, M. S., 1980a, Interferon and BK papovavirus—Clinical and laboratory studies *J. Infect. Dis.* **141:**157–161.

Cheeseman, S. H., Henle, W., Rubin, R. H., Tolkoff-Rubin, N. E., Cosimi, A. B., Cantell, K., Winkle, S., Herrin, J. T., Black, P. H., Russell, P. S., and Hirsch, M. S., 1980b, Epstein–Barr virus infection in renal transplant recipients: Effects of antithymocyte globulin and interferon, *Ann. Intern. Med.* **93:**39–42.

Cheung, K.-S., and Lang, D. J., 1977, Transmission and activation of cytomegalovirus with blood transfusion: A mouse model, *J. Infect. Dis.* **135:**841–845.

Ch'ien, L. T., Cannon, N. J., Whitley, R. J., Diethelm, A. G., Dismukes, W. E., Scott, C. W., Buchanan, R. A., and Alford, C. A., Jr., 1974, Effect of adenine arabinoside on cytomegalovirus infections, *J. Infect. Dis.* **130:**32–39.

Diosi, P., Moldovan, E., and Tomescu, N., 1969, Latent cytomegalovirus infection in blood donors, *Br. Med. J.* **4:**660–662.

Dokhelar, M.-C., Wiels, J., Lipinski, M., Tetaud, C., Devergie, A., Gluckman, E., and Tursz, T., 1981, Natural killer cell activity in human bone marrow recipients, *Transplantation* **31:**61–65.

Fryd, D. S., Peterson, P. K., Ferguson, R. M., Simmons, R. L., Balfour, H. H., Jr., and Najarian, J. S., 1980, Cytomegalovirus as a risk factor in renal transplantation, *Transplantation* **30:**436–439.

Gale, R. P., Ho, W., Feig, S., Champlin, R., Tesler, A., Arenson, E., Ladish, S., Young, L., Winston, D., Sparkes, R., Fitchen, J., Territo, M., Sarna, G., Wong, L., Paik, Y., Bryson, Y., Golde, D., Fahey, J., and Cline, M., 1981, Prevention of graft rejection following bone marrow transplantation, *Blood* **57:**9–12.

Glazer, J. P., Friedman, H. M., Grossman, R. A., Starr, S. E., Barker, C. F., Perloff, J., Huang, E.-S., and Plotkin, S. A., 1979, Live cytomegalovirus vaccination of renal transplant candidates, *Ann. Intern. Med.* **91:**676–683.

Gluckman, E., Lotsberg, J., Devergie, A., Zhao, X. M., Melo, R., Gomez-Morales, M., Mazeron, M. C., and Perol, Y., 1983, Oral acyclovir prophylactic treatment of herpes simplex infection after bone marrow transplantation, *J. Antimicrob. Chemother.* **12 (Suppl. B):**161–168.

Hann, I. M., Prentice, H. G., Blacklock, H. A., Ross, M. G. R., Brumfitt, W., Hoffbrand, A. V., Brigden, D., Rosling, A. E., Burke, C., and Crawford, D. H., 1983, Acyclovir prophylaxis against herpes virus infections in severely immunocompromised patients: Randomized double blind trial, *Br. Med. J.* **287:**384–388.

Hedley-Whyte, E. T., and Craighead, J. E., 1965, Generalized cytomegalovirus inclusion disease after renal homotransplantation, *N. Engl. J. Med.* **272:**473–475.

Hersman, J., Meyers, J. D., Thomas, E. D., Buckner, C. D., and Clift, R., 1982, The effect of granulocyte transfusions upon the incidence of cytomegalovirus infection after allogeneic marrow transplantation, *Ann. Intern. Med.* **96:**149–152.

Hirsch, M. S., Schooley, R. T., Cosimi, A. B., Russell, P. S., Delmonico, F. L., Tolkoff-Rubin, N. E., Herrin, J. T., Cantell, K., Farrell, M.-L., Rota, T. R., and Rubin, R. H., 1983, Effects of interferon-alpha on cytomegalovirus reactivation syndromes in renal-transplant patients, *N. Engl. J. Med.* **308:**1489–1493.

Ho, M., Suwansirikul, S., Dowling, J., Youngblood, L. A., and Armstrong, J. A., 1975, The transplanted kidney as a source of cytomegalovirus infection, *N. Engl. J. Med.* **293:**1109–1112.

Ho, M., Hardy, A. M., Gui, X. E., and Dummer, J. S., 1982, Effect of cyclosporin A on cytomegalovirus infection and natural killer cells in transplant patients, in: *Program and Abstracts of the Twenty-Second Interscience Conference on Antimicrobial Agents and Chemotherapy*, 4–6 October 1982, Miami Beach, p. 124 (Abstract No. 333).

Howard, R. J., and Najarian, J. S., 1974, Cytomegalovirus-induced immune suppression. I. Humoral immunity, *Clin. Exp. Immunol.* **18:**109–118.

Huang, E.-S., Winston, D. J., Miller, M. J., Ho, W . G., and Gale, R. P., 1981, Molecular analysis of cytomegalovirus isolates from bone marrow transplant recipients, in: *Program and Abstracts of the Twenty-First Interscience Conference on Antimicrobial Agents and Chemotherapy*, 4–6 November 1981, Miami Beach (Abstract No. 817).

Kirmani, N., Ginn, R. K., Mittal, K. K., Manischewitz, J. F., and Quinnan, G. V., Jr., 1981, Cytomegalovirus-specific cytotoxicity mediated by non-T lymphocytes from peripheral blood of normal volunteers, *Infect. Immun.* **34:**441–447.

Klemola, E., and Kaariainen, L., 1965, Cytomegalovirus as a possible cause of a disease resembling infectious mononucleosis, *Br. Med. J.* **2:**1099–1102.

Kraemer, K. G., Neiman, P. E., Reeves, W. C., and Thomas, E. D., 1978, Prophylactic adenine arabinoside following marrow transplantation, *Transplant. Proc.* **10:**237–240.

Kreel, I., Zaroff, L. I., Canter, J. W., Krasna, I., and Baronofsky, I. D., 1960, A syndrome following total body perfusion, *Surg. Gynecol. Obstet.* **111:**317–321.

Lang, D. J., Ebert, P. A., Rodgers, B. M., Boggess, H. P., and Rixsl, R. S., 1977, Reduction of postperfusion cytomegalovirus-infections following the use of leukocyte depleted blood, *Transfusion* **17:**391–395.

Levin, M. J., and Leary, P. L., 1981, Inhibition of human herpesviruses by combinations of acyclovir and human leukocyte interferon, *Infect. Immun.* **32:**995–999.

Levin, M. J., Parkman, R., Oxman, M. N., Rappeport, J. M., Simpson, M., and Leary, P. L., 1978, Proliferative and interferon responses by peripheral blood mononuclear cells after bone marrow transplantation in humans, *Infect. Immun.* **20:**678–684.

Levin, M. J., Rinaldo, C. R., Jr., Leary, P. L., Zaia, J. A., and Hirsch, M. S., 1979, Immune response to herpesvirus antigens in adults with acute cytomegalovirus mononucleosis, *J. Infect. Dis.* **140:**851–857.

Livnat, S., Seigneuret, M., Storb, R., and Prentice, R. L., 1980, Analysis of cytotoxic effector cell function in patients with leukemia or aplastic anemia before and after marrow transplantation, *J. Immunol.* **124:**481–490.

Lopez, C., Simmons, R. L., Mauer, S. M., Najarian, J. S., Good, R. A., and Gentry, S., 1974, Association of renal allograft rejection with virus infections, *Am. J. Med.* **56:**280–289.

Marker, S. C., Howard, R. J., Groth, K. E., Mastri, A. R., Simmons, R. L., and Balfour, H. H., Jr., 1980, A trial of vidarabine for cytomegalovirus infection in renal transplant patients, *Arch. Intern. Med.* **140:**1441–1444.

Meyers, J. D., 1984, Cytomegalovirus infection following marrow transplantation: Risk, treatment and prevention, in: *Pathogenesis and Prevention of Human Cytomegalovirus Infection* (S. A. Plotkin, ed.), Alan R. Liss, New York, pp. 101–120.

Meyers, J. D., McGuffin, R. W., Neiman, P. E., Singer, J. W., and Thomas, E. D., 1980a, Toxicity and efficacy of human leukocyte interferon for treatment of cytomegalovirus pneumonia after marrow transplantation, *J. Infect. Dis.* **141:**555–562.

Meyers, J. D., Flournoy, N., and Thomas, E. D., 1980b, Cytomegalovirus infection and specific cell-mediated immunity after marrow transplant, *J. Infect. Dis.* **142:**816–824.

Meyers, J. D. McGuffin, R. W., Bryson, Y. J., Cantell, K., and Thomas, E. D., 1982a, Treatment of cytomegalovirus pneumonia after marrow transplant with combined vidarabine and human leukocyte interferon, *J. Infect. Dis.* **146:**80–84.

Meyers, J. D., McGuffin, R. W., and Thomas, E. D., 1982b, Prophylactic human leukocyte interferon (HLI) after allogeneic marrow transplant: Immunologic and virologic effects, in: *Program and Abstracts of the Twenty-Second Interscience Conference on Antimicrobial Agents and Chemotherapy*, 4–6 October 1982, Miami Beach, p. 99 (Abstract No. 190).

Meyers. J. D., Flournoy, N., and Thomas, E. D., 1982c, Nonbacterial pneumonia after allogeneic marrow transplantation: A review of ten years' experience, *Rev. Infect. Dis.* **4:**1119–1132.

Meyers, J. D., Flournoy, N., Wade, J. C., Hackman, R. C., McDougall, J. K., Neiman, P. E., and Thomas, E. D., 1983a, Biology of interstitial pneumonia after marrow transplantation, in: *Recent Advances in Bone Marrow Transplantation* (R. P. Gale, ed.), pp. 405–423, Liss, New York.

Meyers, J. D., Leszczynski, J., Zaia, J. A., Flournoy, N., Newton, B., Snydman, D. R., Wright, G. G., Levin, M. J., and Thomas, E. D., 1983b, Prevention of cytomegalovirus infection by cytomegalovirus immune globulin after marrow transplantation, *Ann. Intern. Med.* **98:**442–446.

Neiman, P. E., Reeves, W., Ray, G., Flournoy, N., Lerner, K. G., Sale, G. E., and Thomas, E. D., 1977, A prospective analysis of interstitial pneumonia and opportunistic viral infection among recipients of allogeneic bone marrow grafts, *J. Infect. Dis.* **136:**754–767.

Olding, L. B., Jensen, F. C., and Oldstone, M. B. A., 1975, Pathogenesis of cytomegalovirus infection. I. Activation of virus from bone marrow-derived lymphocytes by in vitro allogeneic reaction, *J. Exp. Med.* **141:**561–572.

O'Reilly, R. J., Reich, L., Gold, J., Kirkpatrick, D., Dinsmore, R., Kapoor, N., and Condie, R., 1983, A randomized trial of intravenous hyperimmune globulin for the prevention of cytomegalovirus (CMV) infections following marrow transplantation: Preliminary results, *Transplant. Proc.* **15:**1405–1411.

Orsi, E. V., Howard, J. L., Baturay, N., Ende, N., Ribot, S., and Eslami, H., 1978, High incidence of virus isolation from donor and recipient tissues associated with renal transplantation, *Nature* **272:**372–373.

Pass, R. F., Long, W. K., Whitley, R. J., Soong, S.-J., Diethelm, A. G., Reynolds, D. W., and Alford, C. A., Jr., 1978, Productive infection with cytomegalovirus and herpes simplex virus in renal transplant recipients: Role of source of kidney, *J. Infect. Dis.* **137:**556–563.

Pass, R. F., Reynolds, D. W., Whelchel, J. D., Diethelm, A. G., and Alford, C. A., 1981, Impaired lymphocyte transformation response to cytomegalovirus and phytohemagglutinin in recipients of renal transplants: Association with antithymocyte globulin, *J. Infect. Dis.* **143:**259–265.

Payne, J. E., Fiala, M., Spencer, M., Chatterjee, S. N., and Berne, T. V., 1974, Cytomegalovirus antigen–antibody complexes in biopsy specimens in renal allograft rejection, *Surg. Forum* **25:**273–275.

Peterson, P. K., Balfour, H. H., Jr., Marker, S. C., Fryd, D. S., Howard, R. J., and Simmons, R. L., 1980, Cytomegalovirus disease in renal allograft recipients: A prospective study of the clinical features, risk factors and impact on renal transplantation, *Medicine (Baltimore)* **59:**283–300.

Pollard, R. B., Rand, K. H., Arvin, A. M., and Merigan, T. C., 1978, Cell-mediated immunity to cytomegalovirus infection in normal subjects and cardiac transplant patients, *J. Infect. Dis.* **137:**541–549.

Pollard, R. B., Arvin, A. M., Gamberg, P., Rand, K. H., Gallagher, J. G., and Merigan, T. C., 1982, Specific cell-mediated immunity and infections with herpes viruses in cardiac transplant recipients, *Am. J. Med.* **73:**679–687.

Preiksaitis, J. K., Rosno, S., Grumet, C., and Merigan, T. C., 1983, Infections due to herpesviruses in cardiac transplant recipients: Role of the donor heart and immunosuppressive therapy, *J. Infect. Dis.* **147:**974–981.

Prince, A. M., Szmuness, W., Millian, S. J., and David, D. S., 1971, A serologic study of cytomegalovirus infections associated with blood transfusions, *N. Engl. J. Med.* **284:**1125–1131.

Quinnan, G. V., Jr., Kirmani, N., Rook, A. H., Manischewitz, J. F., Jackson, L., Moreschi, G., Santos, G. W., Saral, R., and Burns, W. H., 1982, Cytotoxic T cells in cytomegalovirus infection: HLA-restricted T-lymphocyte and non-T-lymphocyte cytotoxic responses correlate with recovery from cytomegalovirus infection in bone-marrow transplant recipients, *N. Eng. J. Med.* **307:**6–13.

Rand, K. H., Pollard, R. B., and Merigan, T. C., 1978, Increased pulmonary superinfections in cardiac-transplant patients undergoing primary cytomegalovirus infection, *N. Engl. J. Med.* **298:**951–953.

Rhodes-Feuillette, A., Canivet, M., Champsaur, H., Gluckman, E., Mazeron, M. C., and Peries, J., 1983, Circulating interferon in cytomegalovirus infected bone-marrow-transplant recipients and in infants with congenital cytomegalovirus disease, *J. Interferon Res.* **3**:45–52.

Richardson, W. P., Colvin, R. B., Cheeseman, S. J., Tolkoff-Rubin, N. E., Herrin, J. T., Cosimi, A. B., Collins, A. B., Hirsch, M. S., McCluskey, R. T., Russell, P. S., and Rubin, R. H., 1981, Glomerulopathy associated with cytomegalovirus viremia in renal allografts, *N. Engl. J. Med.* **305**:57–63.

Rifkind, D., Starzl, T. E., Marchioro, T. L., Waddell, W. R., Rowlands, D. T., Jr., and Hill, R. B., Jr., 1964, Transplantation pneumonia, *J. Am. Med. Assoc.* **189**:808–812.

Rinaldo, C. R., Jr., Carney, W. P., Richter, B. S., Black, P. H., and Hirsch, M. S., 1980, Mechanisms of immunosuppression in cytomegalovirus mononucleosis, *J. Infect. Dis.* **141**:488–495.

Rubin, R. H., Cosimi, A. B., Tolkoff-Rubin, N. E., Russell, P. S., and Hirsch, M. S., 1977, Infectious disease syndromes attributable to cytomegalovirus and their significance among renal transplant recipients, *Transplantation* **24**:458–464.

Saral, R., Burns, W. H., Laskin, O. L., Santos, G. W., and Lietman, P. S., 1981, Acyclovir prophylaxis of herpes-simplex-virus infections: A randomized, double-blind, controlled trial in bone-marrow-transplant recipients, *N. Engl. J. Med.* **305**:63–67.

Schooley, R. T., Hirsch, M. S., Colvin, R. B., Cosimi, A. B., Tolkoff-Rubin, N. E., McClusky, R. T., Burton, R. C., Russell, P. S., Herrin, J. T., Delmonico, F. L., Giorgi, J. V., Henle, W., and Rubin, R. H., 1983, Association of herpesvirus infections with T-lymphocyte-subset alterations, glomerulopathy, and opportunistic infections after renal transplantation, *N. Engl. J. Med.* **308**:307–313.

Simmons, R. L., Matas, A. J., Rattazzi, L. C., Balfour, H. H., Jr., Howard, R. J., and Najarian, J. S., 1977, Clinical characteristics of the lethal cytomegalovirus infection following renal transplantation, *Surgery* **82**:537–546.

Snydman, D. R., McIver, J., Lesczczynski, J., Grady, G. F., Berardi, V. P. Wright, G. G., Cho, S., and Logerfo, F., 1982, Pharmacokinetics and safety of an intravenous cytomegalovirus immune globulin (CMVIG-IV) in renal transplant recipients, in: *Program and Abstracts of the Twenty-Second Interscience Conference on Antimicrobial Agents and Chemotherapy*, 4–6 October 1982, Miami Beach, p. 124 (Abstract No. 332).

Spector, S. A., 1983, Transmission of cytomegalovirus among hospitalized infants documented by restriction endonuclease digestion analyses, *Lancet* **1**:378–381.

Spector, S. A., and Kelley, E. A., 1983, Inhibition of human cytomegalovirus by combined acyclovir and vidarabine, in: *Program and Abstracts of the Twenty-Third Interscience Conference on Antimicrobial Agents and Chemotherapy*, 24–26 October 1983, Las Vegas, p. 253 (Abstract No. 920).

Spector, S. A., Rua, J. A., Spector, D. H., and McMillan, R., 1983, Rapid diagnosis of cytomegalovirus viremia in bone marrow transplant patients by DNA–DNA hybridization, in: *Program and Abstracts of the Twenty-Third Interscience Conference on Antimicrobial Agents and Chemotherapy*, 24–26 October 1983, Las Vegas, p. 252 (Abstract No. 914).

Spencer, E. S., 1974, Clinical aspects of cytomegalovirus infection in kidney-graft recipients, *Scand. J. Infect. Dis.* **6**:315–323.

Starr, S. E., and Garrabrant, T., 1981, Natural killing of cytomegalovirus infected fibroblasts by human mononuclear leucocytes, *Clin. Exp. Immunol.* **46**:484–492.

Tocci, M. J., Livelli, T. J., Perry, H. C., Crumpacker, C. S., and Field, K., 1983, Inhibition of human cytomegalovirus replication by 2′-nor-2′-deoxyguanosine (2′-NDG), in: *Conference on Pathogenesis and Prevention of Human Cytomegalovirus Infection*, 20–22 April 1983, Philadelphia (abstract).

Tolkoff-Rubin, N. E., Rubin, R. H., Keller, E. E., Baker, G. P., Stewart, J. A., and Hirsch, M. S., 1978, Cytomegalovirus infection in dialysis patients and personnel, *Ann. Intern. Med.* **89**:625–628.

Wade, J. C., Hintz, M., McGuffin, R. W., Springmeyer, S. C., Connor, J. D., and Meyers, J. D., 1982a, Treatment of cytomegalovirus pneumonia with high-dose acyclovir, *Am. J. Med.* **73**:249–256.

Wade, J. C., Newton, B., Flournoy, N., and Meyers, J. D., 1982b, Oral acyclovir prophylaxis of herpes simplex virus infections after marrow transplant, in: *Program and Abstracts of the Twenty-Second Interscience Conference on Antimicrobial Agents and Chemotherapy*, 4–6 October 1982, Miami Beach, p. 98 (Abstract No. 184).

Wade, J. C., McGuffin, R. W., Springmeyer, S. C., Newton, B., Singer, J. W., and Meyers, J. D., 1983, Treatment of cytomegalovirus pneumonia with high-dose acyclovir and human leukocyte interferon, *J. Infect. Dis.* **148:**557–562.

Wentworth, B. B., and Alexander, E. R., 1971, Seroepidemiology of infections due to members of the herpesvirus group, *Am. J. Epidemiol.* **94:**496–507.

Winston, D. J., Ho, W. G., Rasmussen, L. E., Lin, C.-H., Chu, C. L., Merigan, T. C., and Gale, R. P., 1982a, Use of intravenous immune globulin in patients receiving bone marrow transplants, *J. Clin. Immunol.* **2:**42S–47S.

Winston, D. J., Pollard, R. B., Ho, W. G., Gallagher, J. G., Rasmussen, L. E., Huang, S. N.-Y., Lin, C.-H., Gossett, T. G., Merigan, T. C., and Gale, R. P., 1982b, Cytomegalovirus immune plasma in bone marrow transplant recipients, *Ann. Intern. Med.* **97:**11–18.

Yeager, A. S., 1979, Improved indirect hemagglutination test for cytomegalovirus using human O erythrocytes in lysine, *J. Clin. Microbiol.* **10:**64–68.

Yeager, A. S., Grumet, F. C., Hafleigh, E. B., Arvin, A. M., Bradley, J. S., and Prober, C. G., 1981, Prevention of transfusion-acquired cytomegalovirus infections in newborn infants, *J. Pediatr.* **98:**281–287.

CHAPTER 11

Epstein–Barr Virus and the X-Linked Lymphoproliferative Syndrome

JOHN L. SULLIVAN

I. INTRODUCTION

The importance of understanding the mechanisms responsible for control of Epstein–Barr virus (EBV) infection in man was highlighted with the description in 1974 and 1975 of three families in which an X-linked immunodeficiency to EBV resulted in fatal infectious mononucleosis (IM) in young male children (Bar *et al.*, 1974; Provisor *et al.*, 1975; Purtilo *et al.*, 1975). These descriptions occurred 10 years after the discovery of EBV by Epstein (1964) and represent the initial description of a new X-linked primary immunodeficiency syndrome. (The other X-linked immunodeficiency disorders include: Bruton's X-linked agammaglobulinemia, X-linked severe-combined immunodeficiency, Wiskott–Aldrich syndrome, chronic granulomatous disease, and immunodeficiency with increased IgM.) Since these original descriptions, several new kindreds have been described throughout the world and intensive immunological and virological studies of these families have added to our understanding of the complex immune responses provoked by EBV. Furthermore, it is now well documented that EBV infections in the appropriate immunocompromised host can induce a lymphoproliferative disorder that is indistinguishable clinically and pathologically from fulminant malignancy.

JOHN L. SULLIVAN • Department of Pediatrics, University of Massachusetts Medical School, Worcester, Massachusetts 01605.

II. IMMUNOPATHOGENESIS OF EBV INFECTION IN THE NORMAL HOST

The immunological events occurring during acute EBV infection (IM) have been thoroughly covered in an earlier work in this series (Henle and Henle, 1982). To summarize:

EBV infection of the host oropharyngeal tissue appears to involve the initial infection of a nasopharyngeal epithelial cell (Sixbey et al., 1983a). Through the epithelial cell intermediate, the virus gains access to lymphocyte-rich tonsillopharyngeal tissue. Infection and virus replication in B lymphocytes is initiated and EBV-infected lymphocytes are disseminated throughout the lymphoid system with large numbers of EBV-infected B lymphocytes appearing in the peripheral blood. Nearly 20% of all B cells in the peripheral blood may be infected with the virus (Robinson et al., 1980). The majority of these B cells are nonpermissively infected and mature virus is not produced. The onset of clinical symptoms (fever, sore throat, lymphadenopathy, splenomegaly, and malaise) is associated with the presence of large numbers of atypical lymphocytes in the peripheral blood. The majority of atypical lymphocytes are T lymphocytes (Jondal and Klein, 1973). The atypical T lymphocytes belong to the T cytotoxic/suppressor cell population and express HLA-DR antigens (Reinherz et al., 1980; De Waele et al., 1981). During this phase of the immune response, there is a marked depression in cell-mediated immunity that can be clinically demonstrated by anergy to delayed cutaneous hypersensitivity skin tests (Mangi et al., 1974). In vitro lymphocyte responses to plant mitogens, alloantigens, and soluble antigens are markedly depressed. These observations are thought to reflect already committed immunologically active T lymphocytes in the circulation. Functional studies of atypical lymphocytes during acute stages of IM have demonstrated that cytotoxic/suppressor T lymphocytes can kill EBV-infected B lymphocytes (Svedmyr and Jondal, 1975; Royston et al., 1975; Hutt et al., 1975) and suppress the immunoglobulin secretion of polyclonally activated B lymphocytes (Tosato et al., 1979). The fact that T lymphocytes exhibiting cytotoxicity against EBV-infected cell lines are not restricted by class I histocompatibility antigens would suggest that they are not classical cytotoxic T cells but a type of NK cell. NK cells can lyse EBV-superinfected B-lymphoblastoid cell lines; however, EBV-infected B cells appearing in the peripheral blood and growing out in tissue culture appear to be resistant to NK cell activity (Blazar et al., 1980). NK cells are present during acute IM (Svedmyr and Jondal, 1975; Sullivan et al., 1980) and interferons (IFN), which are potent inducers of NK cell activity, have been shown to inhibit the outgrowth of EBV-infected lymphocytes in vitro (Thorley-Lawson, 1981). IFN is not usually detectable in the serum during acute EBV infection, although analyses of mononuclear cell lysates obtained during acute IM have demonstrated marked elevation of 2',5'-

TABLE I. Phenotypes Expressed in X-Linked Lymphoproliferative Syndrome[a]

Phenotype	Cases (%)	Mortality (%)
Infectious mononucleosis	75	90
With bone marrow aplasia	25	100
With lymphoma	15	—
With hypogammaglobulinemia	10	—
Lymphoma alone	15	65
Hypogammaglobulinemia alone	10	—

[a] Adapted from Purtilo et al. (1982).

oligoadenylate synthetase activity (Brewster and Sullivan, 1983). It is obvious from the events described that the cellular immune response to EBV is extremely complex and involves several different immunoregulatory events, each of which represents an important step in host recovery. It is hypothesized that these intense immunological reactions, which are occurring throughout the lymphoid system, are responsible for the clinical symptoms of acute IM.

III. X-LINKED LYMPHOPROLIFERATIVE SYNDROME

A. General Characteristics

Since the first description of families with the X-linked lymphoproliferative syndrome (XLP) in 1974 and 1975, over 25 kindreds have been reported, with more than 100 affected males. The most recent report of the XLP registry established by Purtilo and colleagues (Purtilo et al., 1982) gives details on approximately 100 cases of XLP. Table I summarizes the phenotypes and approximate mortalities seen in families with XLP. The most common presentation of XLP is severe fatal IM occurring in 75% of patients. The mean age of presentation is 6½ years, which in itself suggests an aberrant response to EBV because the IM syndrome as a manifestation of primary EBV infection in childhood is unusual. The majority of patients with XLP who present with IM die as a result of hepatic failure and its consequences, which include bleeding and infection. Massive hepatic necrosis is frequently observed at autopsy. Many patients experiencing fatal IM will also demonstrate an additional phenotypic manifestation such as bone marrow aplasia (25%), lymphoma (15%), or hypogammaglobulinemia (10%). Retrospective studies have suggested that 15% of XLP patients may develop lymphoma alone or become hypogammaglobulinemic following EBV infection. Prospective studies are needed to assess the role of EBV in these cases and to rule out other causes. The complexities involved in the pathogenesis of XLP where a herpesvirus infection, immunodeficiency, and lymphoproliferative disorder are all intertwined may be further clarified through prospective studies.

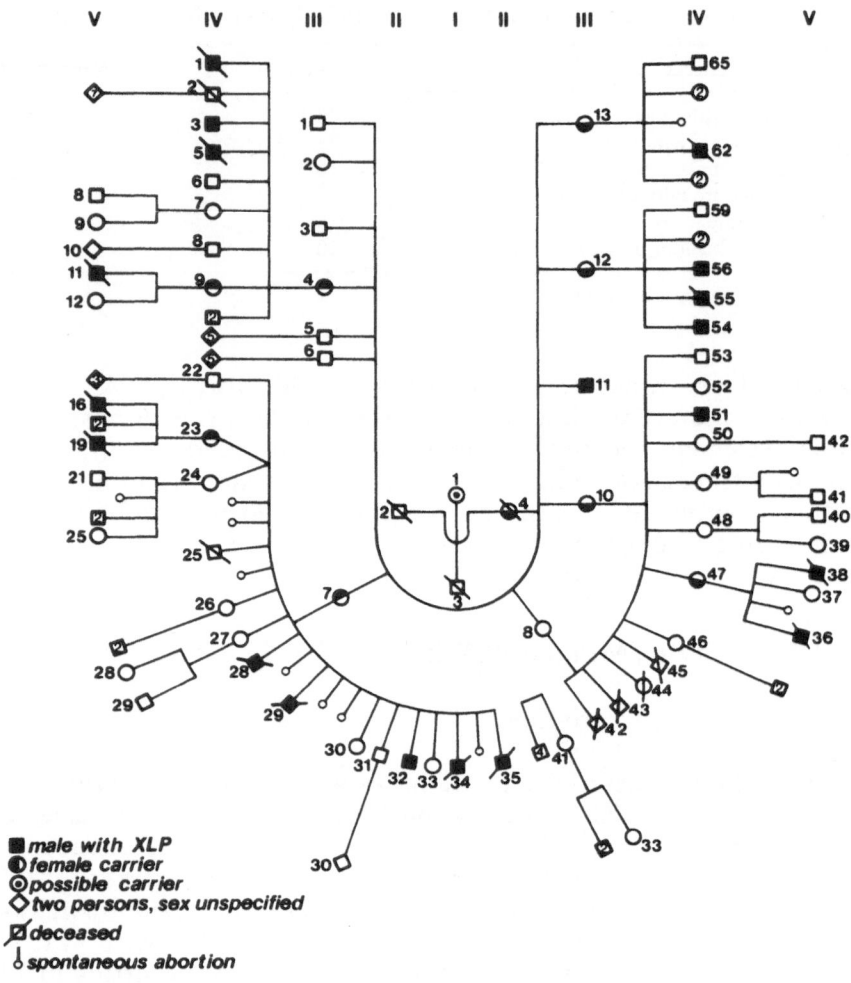

FIGURE 1. Pedigree of a family with the X-linked lymphoproliferative syndrome. The family history is detailed by Purtilo *et al.* (1977).

B. Immunologic and Virologic Studies in Males before and during Fatal EBV Infection

We have studied six members of a large family where 20 males have been affected with XLP (Purtilo *et al.*, 1977). An updated pedigree is shown in Fig. 1. Two members of this family were studied before their encounter with EBV and during the acute phase of fatal EBV-induced IM. Although both of these cases have been reported in detail (Sullivan *et al.*, 1982, 1983), Case 2 (V-19 in Fig. 1) will be presented here to illustrate the clinical picture.

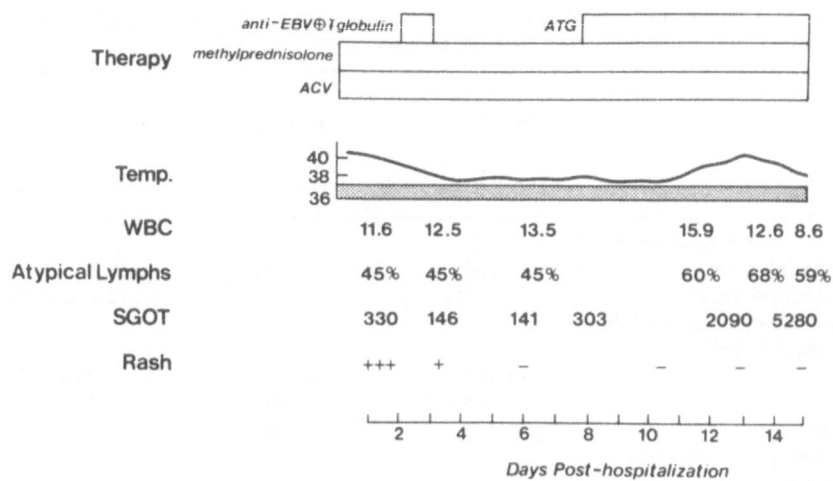

FIGURE 2. Clinical course of Case 2. ATG, antithymocyte γ-globulin; ACV, acyclovir.

Case 2 (V-19)

The clinical course is summarized in Fig. 2. A 10-year-old white male was admitted to the hospital with a diagnosis of IM. The patient was admitted with a 7-day history of fever, cough, headache, and abdominal pain. Lymphadenopathy, atypical lymphocytosis, elevated liver enzymes suggested acute IM, which was confirmed with a positive heterophil antibody test. EBV infection was proved unequivocally by EBV-specific serology demonstrating IgM antibody against EBV capsid antigen (VCA). Immunological studies demonstrated vigorous K cell activity against a panel of EBV-infected and noninfected target cells. It was decided to combine immunosuppressive therapy (methylprednisolone) and antiviral therapy (acyclovir). Acyclovir therapy was given at a dose of 500 mg/m^2 every 8 hr combined with methylprednisolone 5 mg/kg per day in three divided doses. During the first several days of therapy, the patient's rash faded, temperature dropped toward normal, and the SGOT (141) began to fall. A leukocytosis with marked atypical lymphocytes (45%) persisted. The patient received 1 unit of plasma obtained from early antigen (EA), and EBNA. On day 7 of hospitalization, he was still irritable and the SGOT (303) began to rise. Methylprednisolone was increased to 10 mg/kg per day in four divided doses. On day 8, the patient remained irritable and the SGOT increased further. Antithymocyte globulin (ATG) (ATGAM, Upjohn Pharmaceutical Co., Kalamazoo, Mich.) was administered because immunological studies performed on days 5 and 7 showed moderate levels of cytotoxic T lymphocytes to be present in the circulation. ATG was begun at 15 mg/kg per day in a single daily intravenous dose. Over the next week, his condition deteriorated, fever developed, and liver function tests demonstrated severe hepatitis. On day 13, pneu-

TABLE II. Immunological Assessment before and during Acute EBV Infection in a Male with XLP[a]

	Before infection	During infection	
	Case 2 Age 8 years	Case 2 Age 11 years	Normal values
Lymphocyte surface markers[b]			
T lymphocytes	63%	64%	65.0 ± 2.9%
B lymphocytes	8%	16%	10.5 ± 0.7%
Quantitative immunoglobulins (mg/100 ml)			
IgG	1,430	1,825	423–1657
IgA	348	693	27–472
IgM	270	706	28–113
Lymphocyte proliferation studies[c]			
PHA	110,352	42,929	71,450–82,414
Con A	206,950	1,242	81,847–106,414
PWM	64,135	1,706	27,164–38,194
MLR	169,384	11,148	153,460–181,134
Streptolysin O	70,757	3,525	41,495–75,858
Anti-immunoglobulin	4,758	15,073	5,248–11,803
Cytotoxicity studies[d]			
Target K562	62%	85%	>50%
8392	0%	35%	<10%

[a] Adapted from Sullivan et al. (1983).
[b] T lymphocytes determined by E-rosettes or fluorescence with monoclonal pan-T-cell antibody, T101; B lymphocytes determined by immunofluorescence with antiheterologous antisera. Normal values represent mean ± S.E. for 17 normal young adults.
[c] Results expressed as mean net counts per minute of triplicate cultures that showed maximal stimulation to graded doses of mitogens and antigens. MLR, using a B-lymphoblastoid cell line as a stimulator. Normal values represent 95% confidence interval from 17 normal young adults.
[d] Cytotoxicity studies were performed at an effector:target cell ratio of 50:1 in a 4-hr ^{51}Cr-release assay. Results are expressed as a cytotoxic index. K562 is a myeloid target sensitive to NK cell activity and 8392 is an EBV-infected B-lymphoblastoid cell sensitive to EBV-specific T-cell-mediated cytotoxicity.

mococcal bacteremia was diagnosed and treated with antibiotics. His clinical condition continued to deteriorate over the next 2 days; he became jaundiced, and expired on day 15 with fulminant liver failure.

An autopsy showed widespread evidence of lymphoproliferation. The liver was enlarged and massively infiltrated with immunoblasts consistent with the diagnosis of immunoblastic sarcoma. Liver necrosis was also present.

a. Immunological Studies

Table II summarizes cellular and humoral immune studies in Case 2 before and during acute EBV infection. Normal numbers of T and B lymphocytes were present before EBV infection, and functional studies demonstrated normal lymphocyte proliferative responses to phytohemagglutinin, concanavalin A, and pokeweed mitogen. Responses to soluble antigen and alloantigens were also normal. Humoral immune function was normal as demonstrated by normal quantitative immunoglobulins, B-lymphocyte numbers, and proliferative responses to anti-immunoglobulin. Further evidence of intact cellular and humoral immune responses included the clinical history of uncomplicated varicella infection at age 6½ years. Case 1, not reported here, also demonstrated intact cellular humoral immune competence prior to EBV infection (Sullivan et al., 1982).

NK cell activity was also studied prior to EBV infection at age 8 years and at an effector to target cell ratio of 50:1, the patient's mononuclear cells demonstrated 62% specific immune release against the K562 target cell.

During acute EBV infection, the patient demonstrated suppressed responses to mitogens and antigens. These responses were typical of those observed in normal individuals with acute IM (Mangi et al., 1974). Table III summarizes cytotoxic studies performed during acute EBV infection. Vigorous cytotoxic responses against a panel of cell lines susceptible to lysis by NK cells and EBV-induced cytotoxic T cells were observed. Cell lines K562 and Daudi are susceptible NK targets. Cell line 839 is an EBV-infected B-cell line that is lysed by cytotoxic T cells present in the acute phase of EBV-induced IM. These killer T cells are not restricted to EBV-infected target cells sharing class I HLA antigens. The T-cell line 8402, which is HLA identical to 8392, was not lysed to the same degree as the EBV-infected B-cell line 8392. These data parallel those observed in individuals with acute IM (Royston et al., 1975). These cytotoxic T cells probably represent anomalous killer (AK) cells, which kill several types of target cells including fresh tumor cells and can be generated in vitro by alloactivation (Seeley and Golub, 1978), coculture with tumor cells (Vankey et al., 1982), or inoculation with interleukin-2 (Grimm et al., 1982). Table IV shows the results of cell surface marker studies and cytotoxicity experiments during the course of Case 2 illness. Initial studies before immunosuppressive therapy showed marked increase in the OKT8 cytotoxic/suppressor cells, most of which expressed HLA-DR antigens as determined by staining with the p23, 30 antisera. These cells were capable of killing an autologous EBV-infected B-cell line as well as a primary hepatocyte culture obtained from autopsy tissue. During therapy with methylprednisolone, little change was observed in NK or AK activity as well as lymphocyte surface markers. Following treatment with ATG, there was a marked decrease in cytotoxic T-cell numbers and NK cell activity fell from 85% to 10% specific immune release at an effector to target ratio of 50:1. Concomitantly, there was a marked increase in κ and λ light-chain-bearing B lymphocytes.

b. Virological Studies

Table V summarizes the virological studies in Case 2. Acute serum demonstrated the presence of VCA-IgM antibodies as well as antibodies

TABLE III. Cytotoxicity Studies during Acute EBV Infection in a Male with XLP[a]

Effector cells	Day of study	Target cells[b]						
		K562	Daudi	8392	8402	Allo LCL	Fib	HSV Fib
Case 2	1	85	30	36	2	73	42	50
	5	72	14	17	0	27	15	18
	7	53	ND	3	ND	5	ND	ND
	12	10	ND	6	ND	17	ND	ND
							ND	
Infectious mononucleosis [N = 5][c]		84.2 ± 4.4	ND	38.4 ± 11.4	21.8 ± 9.4	43.8 ± 22.9	25.2 ± 6.5	ND
Normal controls [N = 12][d]		85.2 ± 12.8	24.8 ± 7.7	6.0 ± 7.0	4.1 ± 5.1	59[e]	22[e]	26[e]

[a] Adapted from Sullivan et al. (1983).

[b] Target cells studies include: K562, myeloid cell line, a very sensitive NK target; Daudi, Burkitt's lymphoma line, mildly sensitive NK target; 8392, EBV-infected B LCL, relatively resistant to NK and sensitive to EBV-specific killer cells; 8402, a T LCL relatively resistant to NK and syngeneic to 8392; allo LCL, an EBV-infected B LCL sensitive to NK; Fib, allogenic fibroblasts mildly sensitive to NK; HSV Fib, herpes simplex virus-infected fibroblasts mildly sensitive to NK. All results expressed as a cytotoxic index fololowing a 4-hr ^{51}Cr release assay at an effector : target cell ratio of 50 : 1. ND, not done.

[c] Five normal individuals with EBV-induced infectious mononucleosis, results expressed as mean ± S.D. of cytotoxic indices.

[d] Twelve normal adult controls, results expressed as mean ± S.D. of cytotoxic indices.

[e] Results of one normal control, mean cytotoxic index.

TABLE IV. Lymphocyte Surface Markers and T-Cell-Mediated Cytotoxicity in Fatal EBV Infection[a]

Day of Study	Therapy	Cell surface markers[b]				Autologous target cells[c]		
		OKT4	OKT8	p23,30		EBV + LCL	Fib	Hep
1	None	3	61	74	17/9	18	ND	18
5	Methylprednisolone	4	68	98	11/10	28[d]	20[d]	ND
15	Antithymocyte globulin	7	38	ND	50/46	ND	ND	ND

[a] Adapted from Sullivan et al. (1983).
[b] Results are expressed as percentage of stained cells: OKT4 helper/inducer T-cell subset; OKT8 cytotoxic/suppressor subset; p23, 30 HLA-DR; light chains; ND, not done.
[c] Autologous target cells include EBV + LCL, spontaneous EBV-infected LCL; Fib, fibroblasts; Hep, hepatocytes. Results are expressed as cytotoxic index at an effector:target ratio of 25:1.
[d] Results of a T-cell culture expanded from day 5 PBL with purified interleukin-2.

TABLE V. Virological Studies in a Male with XLP and Fatal EBV Infection

Time of study	EBV antibody titers			Spontaneous EBV + LCL[a]	EBNA[+] cells
	VCA (IgM)	EA	EBNA		
Case 2					
Preinfection	<1:10	<1:10	<1:10	ND[b]	ND
Day 1 hospital	1:160 (1:160)	1:160	<1:5	+	+ (blood)
Day 4 (after γ-globulin)	ND	ND	ND	+	ND
Day 15	ND	ND	ND	+	+ (blood, spleen, liver, lymph node)

[a] EBV = positive cell line established.
[b] ND, not done.

against early antigens. EBNA-positive cells were demonstrable in the peripheral blood at hospitalization and an EBNA-positive lymphoblastoid cell line was established from the peripheral blood on the day of hospitalization. A peripheral blood smear stained for EBNA obtained on the last day of life is shown in Fig. 3. In this smear 40–50% of the mono-

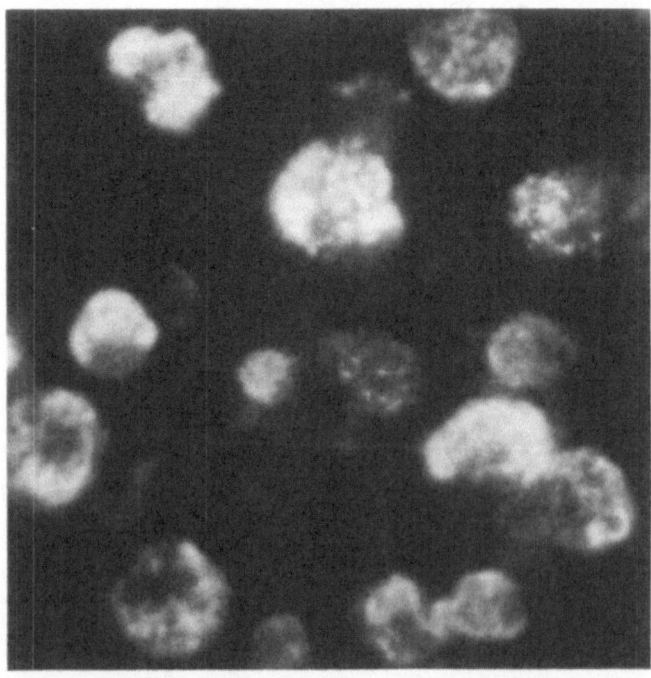

FIGURE 3. Case 2, peripheral blood mononuclear cells stained for EBV nuclear antigen. × 400. From Sullivan *et al.* (1983).

nuclear cells were EBNA positive. In addition, touch preparations of liver, spleen, thymus, and lymph nodes obtained at autopsy all revealed large numbers of EBNA-positive cells. Peripheral blood lymphocytes obtained on the last day of life when 30–50% of the cells were EBV infected were examined for the presence of VCA or EA and found to be negative, indicating a nonpermissive EBV infection. Hepatocyte cultures were obtained from liver tissue stained for EBNA and found to be negative. Lymphoblastoid cell lines were placed in culture and supernatants were checked for transforming virus and found to be negative. EBV-infected lymphoblastoid cell lines from XLP patients have been used as target cells with cytotoxic effector cells derived from normal individuals in the acute phase of IM. In each experiment, spontaneous lymphoblastoid cell lines from XLP patients were lysed to a similar degree as the standard target cells. This finding mitigates against defective expression of LYDMA (Rickinson *et al.*, 1980) being the defect in XLP. Finally, we have induced both IFN-α and IFN-γ production from lymphocytes of acutely infected XLP patients, and the assay of 2′,5′-oligoadenylate synthetase activity in fresh peripheral blood mononuclear cells was similar to that observed in patients with acute IM (Brewster and Sullivan, 1983).

C. Immunological Studies in Surviving Males

We have performed comprehensive immunological studies on 14 affected males with XLP who survived their initial encounter with EBV and/or demonstrate another phenotypic expression of XLP (hypogammaglobulinemia or malignancy) without evidence of EBV infection. The affected males were members of seven well-characterized kindreds (Hamilton *et al.*, 1980). Cellular and humoral studies are summarized in Figs. 4–8. Lymphocyte surface marker analysis revealed normal percentages of T and B lymphocytes. Enumeration of lymphocyte subsets by Seeley *et al.* (1981) demonstrated increased numbers of the OKT8 (cytotoxic/suppressor) T-cell subset, and OKT4/T8 ratios were less than 1.0 in 4 of 10 XLP patients studied. We have found similar results (Sullivan *et al.*, 1983). Assessments of quantitative immunoglobulins and polyclonal B-cell activation demonstrated abnormalities of one or more isotypes and many patients with XLP meet the criteria for common variable immunodeficiency with normal numbers of B lymphocytes and decreased IgG levels (Lindsten *et al.*, 1982; Sullivan *et al.*, 1983). Lymphocyte proliferative responses to graded amounts of mitogens and antigens were significantly deficient in males affected with XLP (Fig. 5). In addition to these T-cell defects, Harada *et al.* (1982) have shown 5 out of 8 surviving XLP males to have deficient EBV-specific T-memory cells (three XLP patients had normal numbers of EBV-specific T-memory cells). NK cell activity was decreased in 10 of 12 patients studied. These results are shown in Fig. 6 (Sullivan *et al.*, 1980). IFN treatment of the effector cells

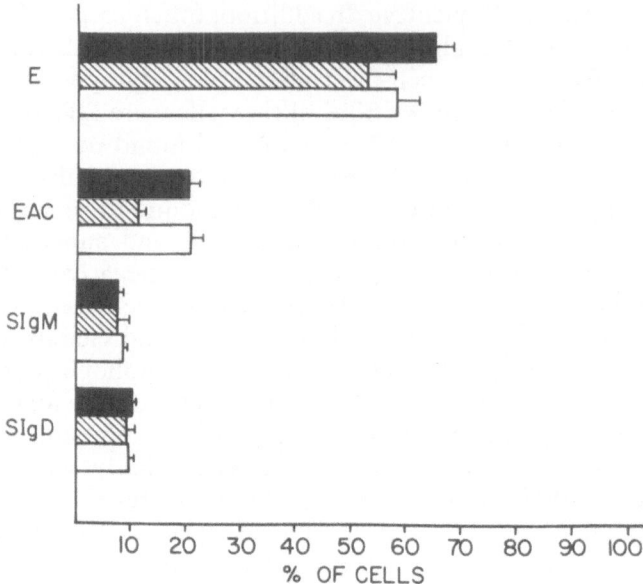

FIGURE 4. Surface marker analysis in XLP patients. (E) T-lymphocytes bearing sheep erythrocyte receptor, (EAC) complement-bearing lymphocytes; (SIgM) IgM membrane-bearing B lymphocytes, (SIgD) IgD membrane-bearing B lymphocytes. (■) Normal subjects (N = 17); (▨) XLP carriers (N = 9); (□) XLP patients (N = 14). Bars indicate S.E.

resulted in a small increase in cytotoxicity but normal activity was never achieved. Deficient NK cell activity against both K562 and herpes simplex virus (HSV)-infected target cells has been demonstrated. Lymphocyte populations identified with the monoclonal antibodies HNK1 and OKM1, both of which react with human NK cell populations, were normal in three XLP individuals with deficient NK activity (Sullivan et al., 1983).

Evaluation of a primary humoral immune response was performed on five affected males from three kindreds immunized with bacteriophage ϕX174 (Sullivan et al., 1983; Ochs et al., 1983). These results are shown in Fig. 7. Four of five XLP patients responded with deficient primary immune responses; however, all five showed normal anamnestic responses after secondary immunization. Qualitative assessment of the secondary immune response is shown in Fig. 8. All five XLP patients were found to be severely deficient in IgG production to bacteriophage ϕX174. These results are consistent with a defect in the T-inducer lymphocyte subset. In summary, extensive immunological evaluation of surviving males with XLP have demonstrated global immune defects encompassing T-cell, NK, and B-cell responses.

D. Immunological and Virological Studies in Carrier Females

Virological studies carried out on surviving males with XLP have been reported by Sakamoto et al. (1980). The majority of males with XLP

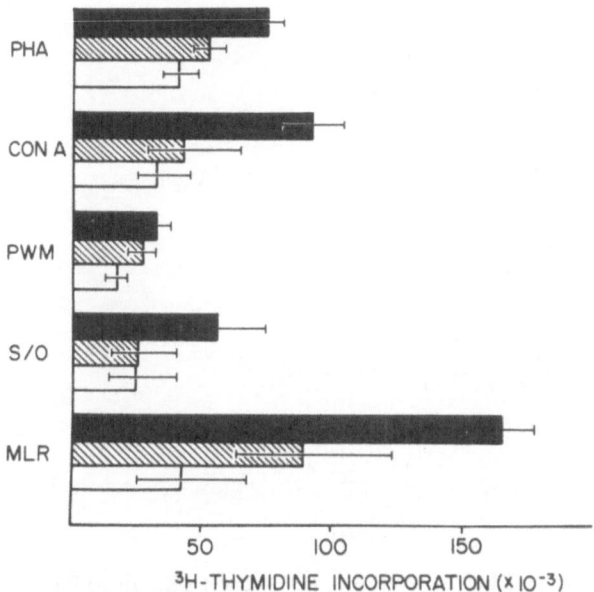

FIGURE 5. Mitogen, antigen, and mixed leukocyte reaction responses in XLP patients PHA, phytohemagglutinin; CON A, concanavalin A; PWM, pokeweed mitogen; S/O, streptolysin O; MLR, mixed leukocyte reaction. Responses indicate maximal net responses to varied doses of mitogens and antigens. (■) Normal subjects (N = 17); (▨) XLP carriers (N = 9); (□) XLP patients (N = 14).

will mount some antibody responses against EBV-specific antigens. It is likely that if prospective studies were possible, most if not all individuals with XLP who experience EBV infection would demonstrate detectable EBV-specific antibodies (Sullivan *et al.*, 1983). Following primary EBV infection, if the patient with XLP survives, it is possible that EBV-specific antibodies may disappear and anti-EBNA responses, which are frequently absent, may reflect defective T-cell immunoregulation.

Results of cellular immune studies carried out in nine carrier females are also shown in Figs. 4–6. T- and B-lymphocyte numbers and lymphocyte proliferative responses to mitogens and antigens were not found to be significantly different than normal controls (Figs. 4 and 5). NK cell activity was also not found to be significantly different than normal controls (Fig. 6). In addition, Seeley *et al.* (1981) have shown that obligate carrier females have normal numbers of cytotoxic/suppressor cells and helper/inducer T-cell subsets. In summary, immunological surveys of known carrier females with XLP have not revealed a constant significant immunological abnormality.

The studies of EBV-specific antibodies in carrier females with XLP carried out by Sakamoto *et al.* (1982) are of importance. Half (6 of 12) of the carrier females studied demonstrated persistent titers to EA and two had persistent IgM antibodies to VCA. These results are suggestive of an

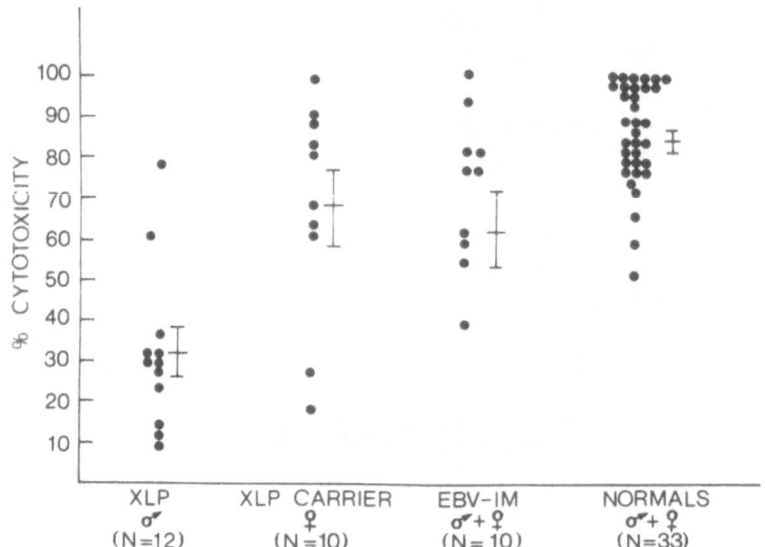

FIGURE 6. Activity of NK cells against K562 target cells in XLP-affected males, carrier females, individuals with acute EBV-induced infectious mononucleosis, and normal controls. Peripheral blood leukocytes (5×10^5) were cultured in round-bottom microtissue dishes for 4 hr at 37°C with K562 target cells (1×10^4) prelabeled with ^{51}Cr in a standardized chromium-release assay at lymphocyte:target cell ratios of 10:1, 25:1, and 50:1. The reaction was terminated by centrifugation for 10 min at 250g, and 100 μl of supernatant was removed from all wells for γ counting. The percentage of cytotoxicity was calculated as $[(E - SR)/max] \times 100$, where E is counts per minute per well, SR is spontaneous release from targets incubated in medium alone, and max is maximal incorporation of ^{51}Cr by target cells. Cytotoxicity indices expressed are those obtained at a lymphocyte:target cell ratio of 50:1. Each experiment was performed three times; each data point represents the mean index. Seven XLP-affected males were studied on more than one occasion; the average response is given for each of these individuals. Means ± S.E. are shown for each group. From Sullivan et al. (1980).

increased frequency of reactivation of EBV infection. All carrier females were asymptomatic, suggesting that if immunoregulatory disturbances are present, they are subtle and not significant enough to cause symptomatic EBV reactivation syndromes (see Tobi et al., 1982).

IV. PATHOGENESIS OF THE X-LINKED LYMPHOPROLIFERATIVE SYNDROME

Initial thoughts concerning the pathogenesis of XLP focused on a generalized X-linked immune defect resulting in the predisposition to B-cell lymphoproliferative disorders triggered by EBV infections or B-cell aplasia when T-cell immunoregulation was disturbed by measles virus or other infectious agents (Purtilo, 1976). However, the rather benign clinical histories of several affected males with XLP prior to experience

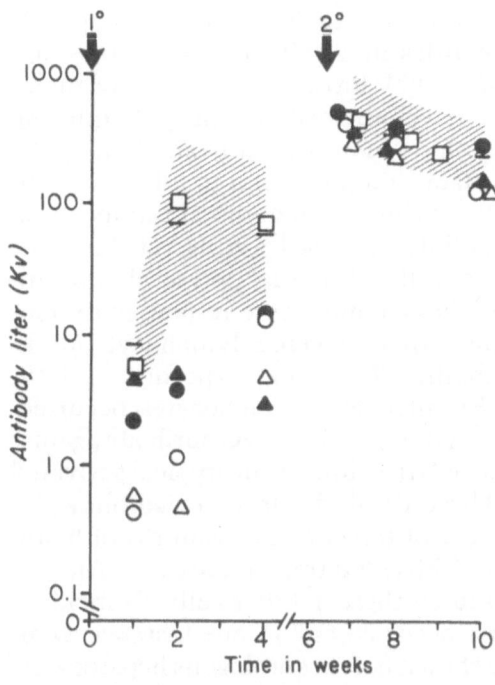

FIGURE 7. Primary (1°) and second-ary (2°) antibody responses to bacter-iophage $\phi \times 174$. The K_v of each in-dividual is indicated by a symbol. The ranges of K_v from 10 normal male con-trols are indicated by the hatched area, the geometric mean by the hor-izontal bar.

with EBV led one editorialist to refer to XLP as a "limited immunode-ficiency" (Editorial, 1978). Much has been learned in the past few years concerning the immunology and virology of XLP and we can now criti-cally reflect on these ideas.

Initial immunological studies of XLP focused almost entirely on af-fected males who survive their initial encounter with EBV. These studies have demonstrated abnormal humoral and cellular responses to EBV (Sak-

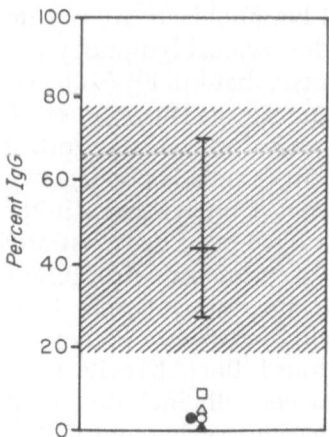

FIGURE 8. Percent IgG of the antibody to bacteriophage $\phi \times 174$ at 2 weeks after secondary immunization. The geometric mean of 10 normal male controls is indicated by the horizontal bar, the 66% confidence limit by the vertical bar, and the range by the hatched area.

amoto *et al.*, 1980; Harada *et al.*, 1982); severe deficiencies of NK cells (Sullivan *et al.*, 1980); and abnormalities in T-cell subsets and function (Seeley *et al.*, 1981; Lindsten *et al.*, 1982). Prospective studies of males at risk for the development of XLP have provided new insights into the specific immune defects present in this syndrome (Sullivan *et al.*, 1982, 1983). Patients studied in early and late childhood prior to infection with EBV have clearly demonstrated intact cellular and humoral immune systems. Normal *in vivo* responses to poliovirus, measles virus, rubella virus, mumps virus, and varicella virus, as well as bacterial polysaccharide antigens have been documented. *In vitro* T-lymphocyte responses to nonspecific mitogens and antigens were normal as were B-lymphocyte markers and quantitative immunoglobulins. Following exposure to EBV, infection and dissemination of EBV-infected B lymphocytes occurred, which initiated a *normal* humoral response (IgM and IgG antibody against EBV VCA) as well as the appearance of large numbers of atypical activated T lymphocytes in the circulation. These T cells displayed a vigorous killer cell activity directed against a variety of target cells regardless of histocompatibility type and expression of EBV-determined antigens. Autologous EBV-infected B cells were lysed by these T-killer cells. Coincident with this response, severe hepatitis occurred and in one instance, lymphocytes from an XLP patient were shown to lyse autologous hepatocytes. These observations provide strong evidence that the immunological defect in XLP is limited to or triggered by infection with EBV. There is little hard evidence to suggest that this immunodeficiency predisposes affected males to severe infections with other infectious agents prior to EBV infection. What then is the explanation for such "limited immunodeficiency"?

EBV infection replicates strictly in B lymphocytes. Infection of B cells results in the transformation to lymphoblastoid cells and EBV-infected lymphoblasts express virus-determined antigens (LYDMA), as well as poorly characterized blast cell neoantigens, which are potent polyclonal stimulators of autologous or allogeneic T lymphocytes (Bird *et al.*, 1981). Polyclonal T-cell responses to EBV-infected B lymphoblasts are not dependent on sensitization with EBV antigens. The atypical lymphocytosis characteristic of IM includes some cytotoxic T cells that kill EBV-infected lymphoid targets in the absence of shared HLA antigens (Royston *et al.*, 1975). This cytotoxic T-cell response is nonclassical in that recognition of shared HLA antigens between effector and target cells is not required as prescribed in the Doherty–Zinkernagel model (Doherty and Zinkernagel, 1974). HLA-restricted cytotoxic T cells have been demonstrated *in vivo* during influenza (Ennis *et al.*, 1981), acute cytomegalovirus (Quinnan *et al.*, 1982), and mumps (Kreth *et al.*, 1982) infections in man. Nonclassical T-cell-mediated cytotoxicity has recently been described in several *in vitro* systems. Originally termed anomalous killer (AK) cells, these cytotoxic T cells, which kill several types of target cells including fresh tumor cells, can be generated during alloactivation (Seeley and Golub,

1982), coculture with tumor cells (Vankey *et al.*, 1982), or inoculation with interleukin-2 (Grimm *et al.*, 1982). Recent cell surface marker analysis of AK cells has shown them to bear the pan-T-cell marker and the OKT8 antigen (Grimm *et al.*, 1982). Abo and Balch (1982) and Abo *et al.* (1982) have recently shown that AK cells can be distinguished from NK cells. AK cells do not express the HNK1 antigen found on NK cells. The observed cytotoxic responses in two patients with XLP experiencing acute EBV infection could be the result of two populations of cytotoxic T cells: AK cells (T8$^+$ and HNK1$^-$) and NK cells (T8$^+$ and HNK1$^+$). Furthermore, these cytotoxic cells were found to have anti-"self" activity. Fulminant liver failure with hepatic necrosis is an important cause of death in XLP. We have demonstrated that cytotoxic T cells obtained from patients with XLP can lyse autologous primary hepatocytes.

Studies in surviving males with XLP who have experienced their EBV infections years before have demonstrated multiple defects in the immune system. All have demonstrated some abnormality in quantitative immunoglobulin levels and many are receiving monthly γ-globulin injections. Each of these individuals has demonstrated one or more of the following abnormalities: normal numbers of T lymphocytes but increased numbers of OKT8-bearing cells, depressed responses to T-cell mitogens and antigens, and deficient NK cell activity to K562 or HSV-infected target cells. Immunization with bacteriophage φX174 has demonstrated near-normal primary and secondary immune responses, but only small amounts of IgG antibody are produced after secondary immunizations. These findings suggest global defects in NK and T-cell function.

Normal immune function before infection with EBV in affected males with XLP and demonstration of major defects in the immune system of surviving affected males is strong evidence for an aberrant immune response specifically triggered by EBV. Immunological studies reported to date have not demonstrated a constant deficiency in any cellular or humoral immune response normally observed in acute EBV-induced IM in normal individuals. In addition, we have not observed any abnormality in the virus–B-lymphocyte relationship, i.e., increased production of infectious virus, or resistance to killing by cytotoxic T cells. We have observed what appears to be an uncontrolled activation of anomalous and NK cells that can mediate lysis of autologous lymphoid and nonlymphoid targets. Our findings are consistent with the hypothesis that uncontrolled AK and NK activity results in fatal IM or immune deficiency in those affected males surviving EBV infections. This mechanism would explain the variation in the severity of the immune deficiency observed within affected kindreds (Purtilo *et al.*, 1975; Hamilton *et al.*, 1980). These secondary immune defects may predispose surviving males with XLP to opportunistic infections and perhaps to lymphoreticular malignancies. Further studies are in progress to determine normal regulatory mechanisms for AK and NK cells activated during acute EBV-induced IM in normal individuals.

V. TREATMENT OF FULMINANT EBV INFECTIONS

The use of corticosteroids in severe cases of EBV infection and IM is recommended in many standard textbooks and review articles (Chang, 1980). Controlled studies have demonstrated a shortened duration of fever in corticosteroid-treated IM (Bolden, 1972). Controlled studies of the efficacy of corticosteroids and the treatment of severe life-threatening EBV infections, including those associated with severe pharyngeal involvement, neurological involvement, thrombocytopenia, and myocarditis, have not been performed. Immunological interactions appear to be important for a normal host recovery and the use of corticosteroids in non-life-threatening EBV infections should be condemned. It is possible that steroid-related effects on the host–virus relationship may take months or years to develop and long-term epidemiological studies of steroid-treated patients have not been reported. Recently, extreme interest has been generated in clinical trials of agents that will be efficacious in acute EBV infections. A series of nucleoside analogs have been synthesized that have marked antiviral activity in animal and certain human herpesvirus infections. Acycloguanosine (acyclovir) nucleoside analog, is activated in a herpesvirus-infected cell by phosphorylation by a herpes-specific thymidine kinase (Elion et al., 1977). The phosphorylated acycloguanosine has greater inhibitory activity for viral DNA than for host DNA. Acyclovir has been shown to inhibit EBV replication in vitro, even though EBV lacks the thymidine kinase enzyme necessary for phosphorylation (Colby et al., 1980). Acyclovir treatment of two patients with life-threatening EBV infections (including one patient with XLP) has been reported (Sullivan et al., 1982). In each case, objective evidence of clinical improvement was not apparent. More recently, we have treated a third patient with XLP suffering from fatal IM (Sullivan et al., 1983; Case 2 reported herein). This patient died with disseminated EBV-infected B lymphocytes throughout the lymphoreticular organs in spite of a 2-week course of acyclovir (1500 mg/m^2 per day). Virological studies in this patient revealed that the virus in infected B lymphocytes was in a nonproductive state and mature virus particles were not being produced. EBV-infected B lymphocytes in this patient failed to express VCA and EA, which is consistent with the nonproductive state. In vitro studies have only shown efficacy of acyclovir during productive EBV infection when viral replication is taking place. In view of these results, it is likely that acyclovir will prove to be efficacious only in those patients suffering unusual productive EBV infections. A double-blind randomized study is currently under way in individuals with complicated IM to determine the efficacy of acyclovir in acute EBV infection (Sixbey et al., 1983b).

IFN is another candidate for the treatment of EBV infection. In addition to its general effects on the prevention of cellular infection, recent studies have shown that IFNs can inhibit EBV-induced transformation of B lymphocytes (Thorley-Lawson, 1981) and are strong stimulators of NK cell activity. In a double-blind placebo-controlled prophylactic trial of human leukocyte IFN in renal transplant recipients, EBV excretion diminished during IFN treatment (Cheeseman et al., 1980). Further studies of OFN in the treatment of EBV infections are in progress.

Epidemiological studies have demonstrated that maternal IgG antibodies against EBV appear to protect the neonate and young infant from primary infection (Bigger et al., 1978a,b). The development of an EBV-specific immunoglobulin preparation is currently in progress and may have value in the prevention and treatment of EBV infection in selected individuals. Recent developments in hybridoma and molecular biology technology may hasten the development of a subunit DNA-free vaccine. Thorley-Lawson (1979) has already succeeded in preparing a DNA-free protein that is capable of inducing a strong neutralizing antibody response against EBV in animals. Further progress in this area is eagerly awaited.

REFERENCES

Abo, T., and Balch, C. M., 1981, A differentiation antigen of human NK and K cells identified by a monoclonal antibody (HNK-1), *J. Immunol.* **127**:1024.

Abo, T., and Balch, C. M., 1982, Characterization of HNK-1 (Leu-7) human lymphocytes. II. Distinguishing phenotypic and functional properties of natural killer cells from activated NK-like cells, *J. Immunol.* **129**:1758–1761.

Abo, T., Cooper, M. D., and Balch, C. M., 1982, Characterization of HNK-1 (Leu-7) human lymphocytes. I. Two distinct phenotypes of human NK cells with different cytotoxic capability, *J. Immunol.* **129**:1752–1757.

Bar, R. S., Delor, C. J., Clausen, K. P., Hurtubise, W., Henle, W., and Heweston, J. F., 1974, Fatal infectious mononucleosis in a family, *N. Engl. J. Med.* **290**:363.

Biggar, R. J., Henle, G., Bocker, J., Lennett, E., Fleisher, G., and Henle, W., 1978a, Primary Epstein–Barr virus infections in African infants. II. Clinical and serological observations during seroconversion, *Int. J. Cancer* **22**:244–250.

Biggar, R. J., Henle, W., Fleisher, G., Bocker, J., Lennett, E., and Henle, G., 1978b, Primary Epstein–Barr virus infections in African infants. I. Decline of maternal antibodies and time of infection, *Int. J. Cancer* **22**:239–243.

Bird, A. G., Britton, S., Ernberg, I., and Nilsson, K., 1981, Characteristics of Epstein–Barr virus activation of human B lymphocytes, *J. Exp. Med.* **154**:832–839.

Blazar, B., Patarroyo, M., Klein, E., and Klein, G., 1980, Increased sensitivity of human lymphoid lines to natural killer cells after induction of the Epstein–Barr viral cycle by superinfection of sodium butyrate, *J. Exp. Med.* **151**:614.

Bolden, K. J., 1972, Corticosteroids in the treatment of infectious mononucleosis, *J. R. Coll. Gen. Practit.* **22**:87–97.

Brewster, F. E., and Sullivan, J. L., 1983, Interferon and 2'-5'-oligoadenylate during acute Epstein–Barr virus infection, *Fed. Proc.* **42**:966.

Chang, R. S., 1980, *Infectious Mononucleosis*, G. K. Hall Medical Publishers, Boston.

Cheeseman, S. H., Henle, W., Rubin, R. H., Tolkoff-Rubin, N. E., Cosimi, A. B., Cantell, K., Winkle, S., Herrin, J. T., Black, P. H., Russell, P. S., and Hirsch, M. S., 1980, Epstein–Barr virus in renal transplant recipients: Effects of interferon and antithymocyte globulin, *Ann. Intern. Med.* **93**:39.

Colby, B. M., Shaw, J. E., Elion, G. B., and Pagano, J. S., 1980, Effect of acyclovir (9-(2-hydroxyethoxymethyl) guanine) on Epstein–Barr virus DNA replication, *J. Virol.* **34**:560–568.

De Waele, M., Theilemans, C., and Van Camp, B. K. G., 1981, Characterization of immunoregulator T cells in Epstein–Barr virus-induced infectious mononucleosis by monoclonal antibodies, *N. Engl. J. Med.* **304**:460.

Doherty, P. C., and Zinkernagel, R. M., 1974, T-cell-mediated immunopathology in viral infections, *Transplant Rev.* **19**:89–120.

Editorial, 1978, Limited immunodeficiency, *Lancet* **1**:132.

Elion, G. B., Furman, P. A., Fyfe, J. A., DeMiranda, P., Beauchamp, L., and Schaeffer, H., 1977, Selectivity of action of antiherpetic agent 9-(2-hydroxyethoxymethyl) guanine, *Proc. Natl. Acad. Sci. USA* **74**:5716–5720.

Ennis, F. A., Hua, W. Y., Riley, D., Rook, A. H., Schild, G. C., Pratt, R., and Potter, C. W., 1981, HLA-restricted virus-specific cytotoxic T-lymphocyte responses to live and inactivated influenza vaccines, *Lancet* **2**:887–891.

Epstein, M. A., and Barr, Y. M., 1964, Cultivation *in vitro* of human lymphoblasts from Burkitt's malignant lymphoma, *Lancet* **1**:252.

Grimm, E. A., Mazumder, A., Zhang, H., and Rosenberg, S., 1982, Lymphokine-activated killer cell phenomenon: Lysis of natural killer-resistant fresh solid tumor cells by interleukin 2-activated autologous human peripheral blood lymphocytes, *J. Exp. Med.* **155**:1823.

Hamilton, J. K., Paquin, L., Sullivan, J., Maurer, H., Cruzi, F., Provisor, A., Steuber, P., Hawkins, E., Yawn, D., Cornet, J., Clausen, K., Finkelstein, G., Landing, B., Grunnet, M., and Purtilo, D., 1980, X-linked lymphoproliferative syndrome registry report, *J. Pediatr.* **96**:669.

Harada, S., Sakamoto, K., Seeley, J. K., Lindsten, T., Bechtold, T., Yetz, J., Rogers, G., Pearson, G., and Purtilo, D. T., 1982, Immune deficiency in the X-linked lymphoproliferative syndrome. 1. Epstein–Barr virus-specific defects, *J. Immunol.* **129**:2532.

Henle, W., and Henle, G., 1982, Immunology of Epstein–Barr virus, in: *The Herpesviruses* Vol. 1 (B. Roizman, ed.), pp. 209–252, Plenum Press, New York.

Hutt, L. M., Huang, Y. T., Discomb, H. E., and Pagano, J. S., 1975, Enhanced destruction of lymphoid cell lines of peripheral blood leukocytes taken from patients with acute infectious mononucleosis, *J. Immunol.* **115**:243.

Jondal, M., and Klein, G., 1973, Surface markers on human B and T lymphocytes. I. Presence of Epstein–Barr virus receptors on B lymphocytes, II. *J. Exp. Med.* **138**:1365.

Kreth, H. W., Kress, L., Kress, H. G., Ott, H. F., and Eckert, G., 1982, Demonstration of primary cytotoxic T cells in venous blood and cerebrospinal fluid of children with mumps meningitis, *J. Immunol.* **128**:2411–2415.

Lindsten, T., Seeley, J. K., Ballow, M., Sakamoto, K., St. Onge, S., Yetz, J., Pierre, A., and Purtilo, D. T., 1982, Immune deficiency in the X-linked lymphoproliferative syndrome, *J. Immunol.* **129**:2536.

Mangi, R. J., Niederman, J. C., Kelleher, J. E., Dwyer, J. M., Evans, A. S., and Kantor, F. S., 1974, Depression of cell mediated immunity during acute infectious mononucleosis, *N. Engl. J. Med.* **291**:1149.

Ochs, H. D., Sullivan, J. L., Wedgwood, R. W., Seeley, J. K., Sakamoto, K., and Purtilo, D. T., 1983, X-linked lymphoproliferative syndrome: Abnormal antibody responses to bacteriophages ϕx174, *Birth Defects Orig. Artic. Ser.* **19**:321.

Provisor, A. J., Iacuone, J. J., Chilcote, R. R., Neiburger, R. G., Crussi, F. G., and Baehner, R. L., 1975, Acquired agammaglobulinemia after a life-threatening illness with clinical and laboratory features of infectious mononucleosis in three related male children, *N. Engl. J. Med.* **293**:62.

Purtilo, D. T., 1976, Pathogenesis and phenotypes of an X-linked recessive lymphoproliferative syndrome, *Lancet* **2**:882.

Purtilo, D. T., Yang, J. P., Cassel, C. K., Harper, R., Stephenson, S. R., Landing, B. H., and Vawter, G. F., 1975, X-linked recessive progressive combined variable immunodeficiency, *Lancet* **1**:935.

Purtilo, D. T., DeFloria, D., Jr., Hutt, L., Bhawan, J., Yang, J., Otto, R., and Edwards, W., 1977, Variable phenotypic expression of an X-linked recessive lymphoproliferative syndrome, N. Engl. J. Med. **297**:1077.

Purtilo, D. T., Sakamoto, K., Barnabei, V., Seeley, J., Bechtold, T., Rogers, G., Yetz, J., and Harada, S., 1982, Epstein–Barr virus-induced diseases in boys with the X-linked lymphoproliferative syndrome (XLP), Am. J. Med. **73**:49.

Quinnan, G. V., Kirmani, N., Rook, A. H., Manischewitz, J. F., Jackson, L., Moreschi, G., Santos, G. W., Saral, R., and Burns, W., 1982 Cytotoxic T cells in cytomegalovirus infection, N. Engl. J. Med. **307**:6–13.

Reinherz, E. L., O'Brien, C., Rosenthal, P., and Schlossman, S. F., 1980, Cellular basis for viral-induced immunodeficiency: Analysis by monoclonal antibodies, J. Immunol. **125**:1269–1274.

Rickinson, A. B., Wallace, L. E., and Epstein, M. A., 1980, HLA-restricted T-cell recognition of Epstein–Barr virus-infected B cells, Nature **283**:865.

Robinson, J., Smith, D., and Niederman, J., 1980, Mitotic EBNA-positive lymphocytes in peripheral blood during infectious mononucleosis, Nature **287**:334.

Royston, I., Sullivan, J. L., Periman, P. O., and Perlin, E., 1975, Cell-mediated immunity to Epstein–Barr virus-transformed lymphoblastoid cells in acute infectious mononucleosis, N. Engl. J. Med. **293**:1159.

Sakamoto, K., Freed, H. J., and Purtilo, D. T., 1980, Antibody responses to Epstein–Barr virus in families with the X-linked lymphoproliferative syndrome, J. Immunol. **125**:921.

Sakamoto, K., Seeley, J. K., Lindsten, T., Sexton, J., Yetz, J., Ballow, M., and Purtilo, D. T., 1982, Abnormal anti-Epstein–Barr virus antibodies in carriers of the X-linked lymphoproliferative syndrome and in females at risk, J. Immunol. **128**:904.

Seeley, J. K., and Golub, S. H., 1978, Studies on cytotoxicity generated in human mixed lymphocyte cultures. I. Time course and target spectrum of several distinct concomitant cytotoxic activities, J. Immunol. **120**:1415.

Seeley, J., Sakamoto, K., Ip, S. H., Hansen, P. W., and Purtilo, D. T., 1981, Abnormal lymphocyte subsets in X-linked lymphoproliferative syndrome, J. Immunol. **127**:2618.

Sixbey, J. W., Vesterinen, E. H., Nedrud, J. G., Raab-Traub, N., Walton, J. A., Pagano, W. S., 1983a, Replication of Epstein-Barr virus in human epithelial cells infected in vitro, Nature **306**:480.

Sixbey, J. W., Pagano, J. S., Sullivan, J. L., Gurwith, M., Pleisher, G., and Clemons, R. H., 1983b, Treatment of infectious mononucleosis with intravenous acyclovir, Clin. Res. **31**:542A.

Sullivan, J. L., Byron, K. S., Brewster, F. E., and Purtilo, D. T., 1980, Deficient natural killer cell activity in the X-linked lymphoproliferative syndrome, Science **210**:543.

Sullivan, J. L., Byron, K. S., Brewster, F. E., Sakamoto, K., Shaw, J. E., and Pagano, J. S., 1982, Treatment of life threatening Epstein–Barr virus infections with acyclovir, Am. J. Med. **73**:262.

Sullivan, J. L., Byron, K. S., Brewster, F. E., Baker, S. M., and Ochs, H. D., 1983, X-linked lymphoproliferative syndrome, J. Clin. Invest. **71**:1765.

Svedmyr, E., and Jondal, M., 1975, Cytotoxic effector cells specific for B cell lines transformed by Epstein-Barr virus are present in patients with infectious mononucleosis, Proc. Natl. Acad. Sci. USA **72**:1622.

Thorley-Lawson, D. A., 1979, A virus-free immunogen effective against Epstein-Barr virus, Nature **281**:486.

Thorley-Lawson, D., 1981, The transformation of adult but not newborn human lymphocytes by Epstein–Barr virus and phytohemagglutinin is inhibited by interferon: The early suppression by T cells of Epstein–Barr infection is mediated by interferon, J. Immunol. **126**:829.

Tobi, M., Ravid, Z., Feldman-Weiss, V., Ben-Chetrit, E., Mong, A., Chowers, I., Michaeli, Y., Shalit, M., and Knobler, H., 1982, Prolonged atypical illness associated with serological evidence of persistent Epstein–Barr virus infection, Lancet **1**:61.

Tosato, G., Magrath, I. T., Koski, I., Dooley, N., and Blaese, M., 1979, Activation of suppressor T cells during Epstein–Barr-virus-induced infectious mononucleosis, *N. Eng. J. Med.* **301**:1133.

Vankey, F., Gorsky, T., Gorsky, Y., Masucci, M. G., and Klein, E., 1982, Lysis of tumor biopsy cells by autologous T lymphocytes activated in mixed cultures and propagated with T cell growth factor, *J. Exp. Med.* **155**:83.

CHAPTER 12

Abnormal Responses to EBV Infection in Patients with Impairment of the Interferon System

Jean-Louis Virelizier

I. INTRODUCTION

Infectious mononucleosis (IM) is typically a self-limited lymphoprolifer-ation, i.e., an array of host responses normally brings to a natural halt the intense proliferation of B lymphocytes induced by Epstein–Barr virus (EBV) infection (Carter, 1975). That immune mechanisms are responsible for the control of EBV infection is clearly deduced from the observation that immunosuppressed patients, such as renal transplant recipients, manifest severe forms of IM as well as a spectrum of lymphoproliferative disorders associated with EBV (Hanto et al., 1981). Similarly, an X-linked deficiency that appears to be specific for EBV underlies the unusually severe forms of IM observed in patients with X-linked lymphoprolifera-tive (XLP) syndrome (Purtilo, 1976; Sullivan, this volume). Finally, ab-normal antibody responses to EBV are observed in immunologically com-promised individuals (Henle and Henle, 1981). Thus, the finding of an abnormal response to EBV, manifested by an unusually severe disease caused by EBV infection, should induce a search for an immunological defect. We have been looking systematically at EBV serology in a large series of patients with well-defined congenital immune deficiencies such

JEAN-LOUIS VIRELIZIER • Groupe d'Immunologie et de Rhumatologie Pediariques (IN-SERM U 132), Hopital Necker-Enfants Malades, 75743 Paris Cedex 15, France.

as Chediak–Higashi or Wiskott–Aldrich syndrome. As discussed below, we found abnormal anti-EBV serological profiles in a number of patients, but, much to our surprise, we did not observe any severe form of IM in these patients. In contrast, we found severe IM in patients with no known primary or secondary immune deficiency detectable by classical tests, but with clear, permanent abnormalities of the interferon (IFN) system. In one case, IFN production by leukocytes was profoundly diminished. In other cases, IFN secretion was normal, but natural killer (NK) activity could not be enhanced by addition of IFN *in vitro* or IFN administration *in vivo*. Such observations are compatible with one hypothesis that IFN plays an important role in host defense against EBV infection. The aim of the present chapter is to describe clinical and immunological observations on patients with abnormal responses to EBV infection, and to discuss the possible role of endogenous production of IFN and of cytotoxic effectors of the IFN system in the control of EBV infection.

II. IMMUNOLOGICAL CONTROL OF EBV INFECTION

One of the peculiarities of IM is that EBV infects only B lymphocytes; the only cell type shown to have receptors for this virus (Pattengale *et al.*, 1973; Jondal and Klein, 1973). EBV is able to "immortalize" B lymphocytes by transforming them into permanently proliferating lymphoblastoid cells *in vitro*. Such lymphoblastoid cell lines derived from spontaneously transformed peripheral blood lymphocytes of patients with IM contain the EBV genome (Nonoyama and Pagano, 1971). Thus, an efficient host immune response must both inhibit viral replication, as is the case with any acute viral infection, and, in addition, restrict proliferation of nonproductively infected B lymphocytes. The latter aspect may require different mechanisms of host defense, not necessarily involved in other acute viral infections. As is the case with other herpesvirus infections, there is also a need to postulate efficient specific surveillance mechanisms to control the lifelong EBV latency. How exactly immune mechanisms control EBV infection and restrict EBV-induced B-cell proliferation is far from being elucidated. A diagram representing the immune mechanisms that might control EBV-induced B-cell proliferation is shown in Fig. 1. It is generally assumed that the production of specific antibody plays little (if any) role in host recovery, since reactivation of EBV infection occurs despite the presence of serum antibody to EBV. However, it is possible to envisage a preventative role for specific antibody. Thus, specific antibody to EBV membrane antigens may participate in the eradication of EBV-infected B cells through antibody-dependent cell-mediated cytolysis (ADCC) (Pearson and Orr, 1976; Jondal, 1976). The main antigen recognized by antibody in this *in vitro* system appears to be the late membrane antigen (LMA) (Takaki *et al.*, 1980). *In vivo*, LMA-expressing cells appear to represent only a small proportion of the

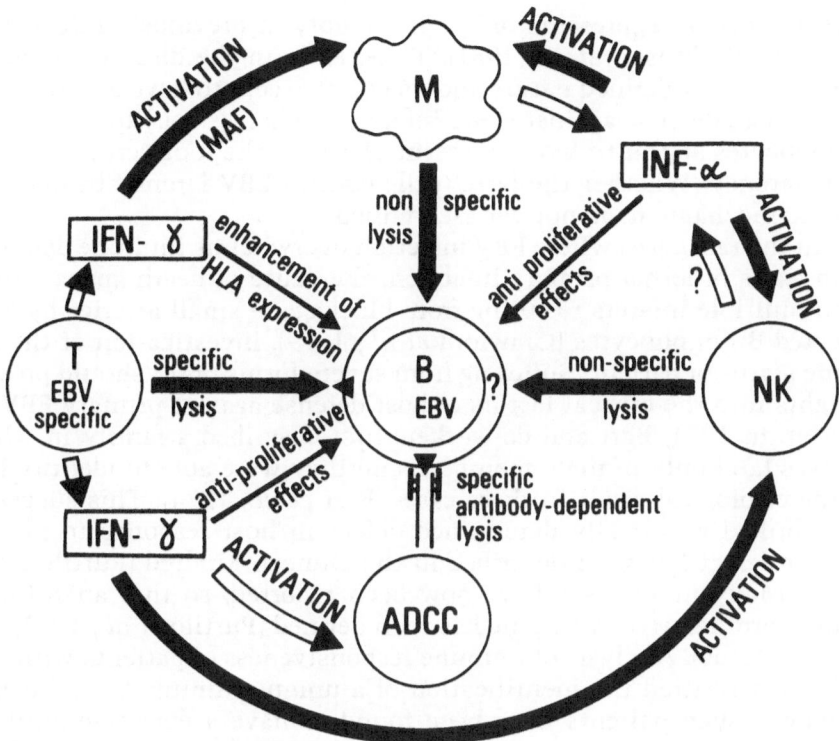

FIGURE 1. Schematic representation of the main possible effector mechanisms controlling EBV-induced B-lymphocyte proliferation. (M) macrophages; (NK) natural killer leukocytes; (ADCC) effector cells of the antibody-dependent cell-mediated cytotoxicity; (T) T lymphocytes; (MAF) macrophage-activating factor.

total EBV-transformed cell population, a finding that limits the importance of ADCC in host defense against EBV infection.

Cell-mediated immunity is more likely to have a critical role in recovery from EBV infection. The exact nature and mechanisms of action of cytotoxic cells found in the blood of patients with acute IM have been the subject of numerous and apparently contradictory reports. During the early phase of infection, the cytotoxic effector cells capable of destroying EBV-infected cells were found not to be HLA-restricted (Svedmyr and Jondal, 1975; Seeley et al., 1981). Such effector cells, however, seem to be different from NK cells in that they appear to be EBV-specific. The "activated lymphocyte killers" (ALK) induced by lymphokines (presumably IFN) or mixed leukocyte cultures (Masucci et al., 1980) also appear to be T lymphocytes. On the other hand, HLA-restricted, EBV-specific cytotoxic T lymphocytes can be found in the blood of seropositive individuals (Misko et al., 1980; Rickinson et al., 1980). Similarly, specific T lymphocytes capable of inducing the regression of EBV-transformed B cells are also an expression of T-cell memory in previously infected individuals (Rickinson et al., 1981). Thus, early in the disease, cytotoxic cells of a poorly defined nature and specificity seem to have an important role in host defense against acute infection, whereas specific T-memory lymphocytes appear to have an essential role in the long-term control of virus latency. Whether the latter cells control EBV latency through cytotoxic mechanisms is not yet ascertained.

In the rare cases where EBV infection overwhelms immune defenses, as in cases of lethal primary infection, the cause of death appears to be from multiple infarcts resulting from blockage of small arteries by EBV-infected B lymphocytes (Crawford et al., 1979). Investigation of the immune status of patients suffering from severe forms of IM should provide insights into the critical factors of host defense against primary EBV infection. In 1974, Barr and co-workers first described a family in which IM was fatal only in male members, and they were able to identify EBV as the etiological agent in these cases (Barr et al., 1974). This suggested an X-linked genetically determined defect in host response to EBV. A similar defect has been described in the Duncan kindred (Purtilo et al., 1975). Many more cases have now been reported, so that an X-linked lymphoproliferative syndrome has been defined (Purtilo et al., 1982; Sullivan, volume). Analysis of immune responsiveness of patients with XLP has not permitted the identification of a unique immunological defect. However, such patients have been found to have a defective antibody response to EBV (Sakamoto et al., 1980), deficient NK activity (Sullivan et al., 1980), and abnormal cell-mediated immunity to EBV-infected autologous B cells (Harada et al., 1982). To further complicate the issue, it should be stressed that familial susceptibility to EBV is not always X-linked, as a family has been described where both males and females were affected (Fleisher et al., 1982). Also, a fatal case of EBV infection has been observed in a girl despite demonstrable humoral and cellular immunity (including cytotoxicity) to EBV (Britton et al., 1978). We have reported the case of a girl with persistent EBV infection progressively leading to polyclonal activation of immunoblasts and death (Virelizier et al., 1978). In the study of this patient, the only detected immunological defect was a selective deficiency of immune IFN production, as described below. As IFN is able to regulate immune responses and to activate both nonspecific and specific cytotoxic cells, the possibility that defective production or abnormal sensitivity to IFN underlies some cases of abnormal responses to EBV infection has to be considered.

III. THE IFN SYSTEM AND ITS PLACE IN ANTIVIRAL IMMUNITY

What is now globally known as the "interferon system" includes the various mechanisms of induction and production of three major classes of IFNs in a variety of cell types, and the antiviral, antiproliferative, and immunoregulatory effects exerted by IFNs on a series of target cells (Stewart, 1979). IFN-β is mainly produced by nonimmunocompetent cells infected by a virus. IFN-α (previously known as "leukocyte IFN") is produced by macrophages and "null" (non-T, non-B) leukocytes stimulated by virus particles or tumors. IFN-γ (or "immune" IFN) is a lymphokine secreted by sensitized T lymphocytes stimulated by a specific antigen. The secretion of IFN-γ can also be induced *in vitro* by nonspecific mitogens. Specific IFN-γ induction is HLA-restricted and macrophage-dependent. In the mouse, it is clear that immune reactions restricted by either class I or class II histocompatibility antigens lead to IFN-γ secretion, and this is likely also to be the case in humans. Compared to IFN-α or IFN-β IFN-γ has very clear (of the order of 100-fold) preferential effects on immunocompetent cell functions. The preferential effects of IFN-γ, and the respective role of nonspecific α/β IFN production versus specific induction of IFN-γ in antiviral immunity have recently been reviewed (Virelizier, 1984a). That early production of IFN-α/β has a critical role in host defense during acute viral infections is clearly shown by the effects of antiserum to these types of IFN in experimental models. Natural, genetically determined resistance to viruses can be abolished by such antisera, with increased virus replication and death (Lopez, this volume). Whether IFN acts directly (through its antiviral effects in all cells of the body) or through immune mechanisms (such as activated NK, K, ALK, ADCC, macrophages, or cytotoxic T lymphocytes) is not known. Later in the course of viral infections, IFN-γ can be induced by viral antigens. Indeed, the production of this lymphokine is a clear expression of specific T-cell memory, borne by long-lived, recirculating lymphocytes able to be recruited into inflammatory lesions (Virelizier and Guy-Grand, 1980). The role of IFN-γ secretion in antiviral immunity has not yet been determined, but this lymphokine is likely to contribute to immune surveillance mechanisms against virus latency and abnormal cell proliferation. Again, these effects may be mediated directly (through induction of the antiviral state or direct antiproliferative effects) or indirectly (through the preferential activating effects of IFN-γ on cytotoxic leukocytes). Another intriguing property of IFN-γ is its ability to enhance the membrane expression of HLA-A-B in Burkitt cells (Fellous *et al.*, 1981) and HLA-DR on human monocytes (Virelizier *et al.*, 1984). IFN-γ, through its preferential effects on membrane expression of both class I and II antigens, may thus facilitate virus antigen presentation by monocytes to both class I-restricted T8 and class II-restricted T4 lymphocytes and, therefore, the destruction of virus-infected cells by HLA-restricted, specific cytotoxic T lymphocytes.

IV. EVIDENCE FOR A ROLE OF IFN IN THE CONTROL OF EBV INFECTION *IN VITRO*

EBV infection of human leukocytes *in vitro* results in the transformation and the proliferation of B lymphocytes. However, a regression of this outgrowth of infected B cells is usually observed after a few days in unseparated leukocytes. The study of this "regression phenomenon" may provide insights into the physiological mechanisms controlling EBV infection *in vitro*. Thorley-Lawson *et al.* (1977) have shown that this regression is mediated by T lymphocytes, as B-cell outgrowth is not controlled when T cells are removed from the culture. The T lymphocytes mediating this control must be specifically sensitized, for only leukocytes from seropositive subjects are able to provide this effect (Moss *et al.*, 1977). From these observations it is obvious that a T-dependent, EBV-specific mechanism(s) is able to inhibit transformation and/or outgrowth of EBV-infected B lymphocytes *in vitro*. The mechanism by which this inhibition is mediated is unknown. It is possible that specific T lymphocytes may limit EBV replication or proliferation of infected B lymphocytes. Although some authors have reported that soluble mediators do not play any major role in the regression phenomenon (Rickinson *et al.*, 1979), it has been reported that sensitized leukocytes inhibit EBV-induced transformation through the secretion of mediators with IFN-like properties (Lai *et al.*, 1977). The mediator(s) was shown to be labile at pH 2, a property of IFN-γ and of a newly described type of IFN-α molecules. The critical role of IFN secretion in the regression of EBV-induced transformation has been conclusively shown by the abrogation of the phenomenon with the addition of an antiserum to IFN-α/β to the culture (Thorley-Lawson, 1981). Conversely, the regression phenomenon could be mimicked by the addition of human leukocyte IFN. It was shown that IFN is released in response to virus-infected B cells, not the virus alone (Thorley-Lawson, 1981). More recent work has provided evidence for a role for IFN-γ in the regression phenomenon (Hasler *et al.*, 1983). Thus, normal T cells stimulated in mixed leukocyte reactions inhibited EBV-induced DNA synthesis, and this inhibitory activity was blocked by the addition of monoclonal antibody specific for human IFN-γ. Altogether, these data indicate that the secretion of IFN has a critical role in the control of EBV infection *in vitro*. Some confusion remains about which type of IFN is required, as antisera to either IFN-α or IFN-γ abolished the regression. It may be that both IFN classes are needed, or that IFN-γ secretion by T cells is a prerequisite for IFN-α production by another cell type. The latter hypothesis is compatible with our recent finding that the production of IFN-α by human monocytes critically depends on the previous activation of monocytes by IFN, especially IFN-γ (Arenzana-Seisdedos and Virelizier, in preparation), suggesting that a cascade of IFN production can occur *in vitro*. Whatever the exact type of IFN is responsible, the question arises of whether IFN acts directly or indirectly in the regression phenomenon. There is some indication that IFN does not protect EBV genome-negative lymphoblastoid cell lines infected with EBV *in vitro* from transformation and nuclear antigen production (Menezes *et al.*, 1976). If this inability of IFN to directly inhibit transformation is confirmed with normal adult B lymphocytes, it would suggest that IFN operates indirectly, presumably through its ability to activate cytotoxic cells (T, NK, and macrophages) and/or to enhance HLA antigens on the membrane of infected B lymphocytes.

V. EVIDENCE FOR A ROLE OF IFN IN THE CONTROL OF EBV INFECTION IN PATIENTS

A. IFN Administration in Immunosuppressed Patients

IFN administration is often of little benefit in controlling severe viral infections in normal subjects. This may be due to the fact that administration of exogenous IFN only modestly increases the serum titers of endogenous IFN induced by the viral infection itself. However, the situation appears to be different in immunosuppressed patients, probably because endogenous production of IFN is impaired. Indeed, leukocytes from immunosuppressed patients have been shown to have a diminished ability to produce IFN-α and IFN-γ (Rytel and Balay, 1973; Levin *et al.*, 1978). As IFN-α is produced early during infection in normal subjects, one might predict that IFN-α administration will have a protective role if given early after exogenous infection or endogenous reactivation; or, better still, if it is administered prophylactically. Indeed, IFN-α administered twice a week beginning just before renal transplantation in a placebo-controlled trial clearly diminished EBV virus excretion in the saliva, especially in the group of patients profoundly immunosuppressed by antithymocyte globulin (Cheeseman *et al.*, 1980). The observation that administration of IFN reconstitutes the ability of immunosuppressed patients to control EBV reactivation provides indirect evidence for a role of IFN secretion in the physiological control of EBV latency.

B. Selective Defects of IFN Secretion

During the past years, we have been systematically investigating IFN production by leukocytes from patients with various types of immune disorders. IFN secretion was frequently found to be normal. When IFN secretion was impaired, this defect was usually not selective. For example, leukocytes from patients with severe combined immunodeficiency due to the complete absence of T lymphocytes do not secrete IFN-γ after stimulation with mitogens or antigens, an expected finding as T cells are known producers of this lymphokine. In contrast, in some rare patients suffering from severe or persistent infections, we found a profound defect of IFN secretion, whereas all other immune functions tested were found to be normal, except for a secondary defect of NK activity (Table I). Three observations of such patients with a "selective" defect of leukocyte IFN production have been reported elsewhere (Virelizier, 1984b).

TABLE I. Immune Status in Three Patients with Selective Defects of IFN
Secretion

Immune function tested	Patient tested		
	SG	NS	DF
Serum immunoglobulin	Increased	Increased	Normal
Antibody production	Normal	Normal	Normal
% B cells	Normal	Elevated	Diminished
% T cells	Normal	Subnormal	Normal
Skin tests	Normal	Normal	Normal
T-cell proliferation	Normal	Normal	Subnormal
Allogeneic T-cell cytotoxicity	Normal	Normal	Normal
IFN-γ secretion	Abolished	Abolished	Abolished
IFN-α (tumors)	Abolished	Abolished	Abolished
IFN-α (viruses)	Normal	Subnormal	Subnormal
NK (spontaneous	Diminished	Diminished	Diminished
NK (IFN-boosted)	NT	Normal	Normal

Production of serum immunoglobulins, specific antibodies, delayed hypersensitivity tests, markers and functions of T and B lymphocytes, and phagocyte functions were found to be normal in the three patients. Secretion of IFN-γ in response to stimulation with mitogens or antigens was abolished, despite normal proliferation of T lymphocytes in these tests. Secretion of IFN-α was found to be normal after infection with Sendai or Newcastle disease virus, but was profoundly diminished in coculture with EBV genome-positive Burkitt cell lines (such as Raji cells) or EBV-negative tumors (such as K562 erythroleukemia cells). In none of these patients could the IFN secretion defect be demonstrated as being of genetic origin, i.e., no other family cases could be documented. The only other immunological abnormality detected was a profound defect of NK activity. In two patients tested, this NK defect could be reversed *in vitro* by preincubation of effector cells with IFN, and *in vivo* by the administration of IFN-α indicating that the NK defect was secondary to a defect in endogenous production of IFN, as discussed elsewhere (Virelizier and Griscelli, 1980, 1981). In the three patients, an abnormal serological response to EBV was observed, with high antibody titers to VCA and Ea and low antibody titers to EBNA. In two patients, we could not find signs of active EBV infection. In one case, however, we found clear evidence of persistent primary EBV infection, as reported previously (Virelizier *et al.*, 1978). This 5-year-old girl had a chronic disease with fever, lymphoid hyperplasia due to polyclonal immunoblastic proliferation, interstitial pneumonitis, moderate thrombocytopenia, and diffuse hypergammaglobulinemia. Evidence of severe, persistent EBV infection included extremely high titers of IgG anti-VCA and anti-EA, persistent IgM anti-VCA, detection of EBNA in about 1% of lymphocytes in peripheral blood and lymph node, and spontaneous immortalization of peripheral leukocytes into EBNA-positive lymphoblastoid cell lines during a 1-year follow-up. The child died 19 months after the onset of the disease with a dissemination of the polyclonal immunoblastic proliferation to the bone marrow. The selectivity of the IFN deficiency associated with this unusually severe form of IM is compatible with the hypothesis that endogenous IFN production has an important role in recovery from acute IM and in long-term control of EBV latency.

VI. CHRONIC NEUTROPENIA TRIGGERED BY EBV INFECTION AND ASSOCIATED WITH IMPAIRED ACTIVATION OF NK CELLS BY IFN

In the past few years we have studied, in collaboration with G. Tchernia and G. Lenoir, a series of five male patients with chronic neutropenia of central origin. In three patients, a history of very severe EBV infection could be recorded. Polymorphonuclear counts were below 500 per μl in three patients, whereas variable, subnormal counts were observed in the other two patients. Prednisone therapy was able to reverse the hematological abnormality in three patients, but neutropenia relapsed when administration of the drug was stopped. As shown in Table II, an abnormal antibody response to EBV was observed in all patients, with very low anti-EBNA titers contrasting with persistently high anti-VCA and anti-EA titers. No evidence of abnormal humoral response to other microorganisms could be found, and no gross immune dysfunction could be detected by the use of classical tests. IFN production was normal after stimulation of leukocytes *in vitro* with either EBV-positive Burkitt cells or nonspecific mitogens. Two immunological abnormalities, however, were consistently observed in this series of patients. The percentage of peripheral lymphocytes with the T8 (suppressor) phenotype was increased, so that the T8/T4 ratio was abnormal. As seen in Table II, spontaneous NK activity was normal in four patients and profoundly diminished in one. The role of EBV in these observations, although not decisively demonstrated, is likely in view of the severity of the initial IM recorded and the persistent abnormality of the antibody response to EBV antigens. The elevated number of T8 lymphocytes is reminiscent of that observed transiently in the course of normal, acute IM (Tosato *et al.*, 1979). The striking lack of sensitivity of NK cells to the activating effects of IFN observed in the five patients *in vitro*, and in one patient *in vivo* (Virelizier and Griscelli, 1980) suggests an active inhibition of NK cells, possibly mediated by EBV-activated T8 lymphocytes. Although the relationship between resistance of NK cells to the activating effects of IFN and decreased host resistance to EBV is not understood, it is interesting to note that such a resistance has also been observed in other

TABLE II. Immunological Data in Patients with EBV-Induced Chronic Neutropenia

Subjects tested	EBV serology			IFN production		% T-cell subpopulations				NK activity	
	VCA	EA	EBNA	α (tumor-induced)	γ (mitogen)	T3	T4	T8	(8/4)	Spont.	+IFN
Controls	285.8	3.8	120.6	92–2304	96–6144	77.5	49.3	31.4	(0.63)	51 ± 17	73 ± 16
FA	1280	640	5	144	288	93	10	64	(6.4)	42	43
PC	>1280	>1280	20	576	144	64	1	67	(67)	31	31
DB	>1280	1280	20	144	6144	28	7	27	(2.4)	1.5	1.5
EH	>1280	>1280	80		168	73	23	58	(2.5)	44	48
JB	>1280	>1280	5	576	288	67	28	30	(1.07)	28	29

patients with abnormal response to EBV infection (Thestrup-Pedersen *et al.*, 1980; Fleisher *et al.*, 1982).

VII. CONCLUSION

Much remains to be learned about the mechanisms responsible for controlling EBV infection. Investigation of the immunological status of patients with unusually severe IM or abnormal antibody responses to EBV should help to better define the respective *in vivo* roles of the various effector functions described *in vitro*. Unfortunately, no single mechanism appears to be solely responsible for recovery from EBV infection. Our studies had intended to correlate EBV serology with selective defects of immune functions (Vilmer *et al.*, 1984). For example, patients with the Chediak–Higashi syndrome show a selective defect of NK activity and persistently elevated titers of anti-EA antibodies. Also, in nine patients with the Wiskott–Aldrich syndrome (apart from a defective IgM production), the only cellular immunological defect that we have been able to detect is a profound impairment of allogeneic cytotoxicity, which correlated with a lack of antibody to EBNA. In patients with these two syndromes, there is thus serological evidence of previous IM with abnormal antibody responses, but no evidence of unusual susceptibility to EBV as no severe form of IM was found in these patients. Thus, a preferential defect of either NK activity of allogeneic T-cell cytotoxicity appears not to be a sufficient condition to severely impair host defense against EBV. Similarly, a deficiency of endogenous secretion of IFN may not be sufficient to expose patients to severe IM, for only one patient in our series of three children with a selective defect of IFN developed a severe form of IM, whereas the two other patients showed serological evidence of previous contact with EBV but no unusually severe disease. Sensitivity of cytotoxic effector cells to IFN, and possibly to B lymphocytes themselves, is an intriguing possibility that deserves further evaluation. EBV-specific memory T cells, able to induce regression of EBV-induced B-cell proliferation *in vitro*, have clearly an important role in the long-term surveillance of EBV latency and IFN-γ may be an important mediator of these T lymphocytes. IFN may act either through its direct effects on infected B lymphocytes or through its ability to modulate immune responses, enhance membrane HLA expression, and activate cytotoxic cells. As is the case with other virus models, host defense against EBV is likely to be mediated by a series of interwoven mechanisms of which IFN is only one of the many facets.

REFERENCES

Barr, R. S., Delow, J. C., Clausen, K. P., Hurtubise, P., Henle, W., and Hewetson, J. F., 1974, Fatal infectious mononucleosis in a family, *N. Engl. J. Med.* **290**:363.

Britton, S., Andersson-Anvret, M., Gergely, P., Henle, W., Jondal, M., Klein, G., Sandstedt, B., and Svedmyr, E., 1978, Epstein–Barr virus immunity and tissue distribution in a fatal case of infectious mononucleosis, N. Engl. J. Med. 298:89.

Carter, R. L., 1975, Infectious mononucleosis: Model for self-limiting lymphoproliferation, Lancet 1:846.

Cheeseman, S. H., Henle, W., Rubin, R. H., Tolkoff, N. E., Cosimi, B., Cantell, K., Winkle, S., Herrin, J. T., Black, P. H., Russell, P. S., and Hirsch, N. S., 1980, Epstein–Barr virus infection in renal transplant recipients: Effects of antithymocyte globulin and interferon, Ann. Intern. Med. 93:39.

Crawford, D. H., Epstein, M. A., Achong, B. G., Finerty, S., Newman, J., Liversedge, S., Tedder, R. S., and Stewart, J. W., 1979, Virological and immunological studies on a fatal case of infectious mononucleosis, J. Infect. 1:37.

Fellous, M., Bono, R., Hyafil, F., and Gresser, I., 1981, Interferon enhances the amount of membrane-bound β_2 microglobulin and its release from human Burkitt cells, Eur. J. Immunol. 11:524.

Fleisher, M. D., Starr, S., Koven, N., Kamiya, H., Douglas, S. D., and Henle, W., 1982, A non-X linked syndrome with susceptibility to severe Epstein–Barr virus infections, J. Pediatr. 100:727.

Hanto, D. W., Frizzera, G., Purtilo, D. T., Sakamoto, K., Sullivan, J. L., Saemundsen, A. K., Klein, G., Simmons, R. L., and Najarian, J. S., 1981, Clinical spectrum of lymphoproliferative disorders in renal transplant recipients and evidence for the role of Epstein–Barr virus, Cancer Res. 41:4253.

Harada, S., Bechtold, T., Seeley, J. K., and Purtilo, D. T., 1982, Cell-mediated immunity to Epstein–Barr virus (EBV) and natural killer (NK) cell activity in the X-linked lymphoproliferative syndrome, Int. J. Cancer 30:739.

Hasler, F., Bluestein, H. G., Zvaifler, N. J., and Epstein, L. B., 1983, Analysis of the defects responsible for the impaired regulation of Epstein–Barr virus-induced B cell proliferation by rheumatoid arthritis lymphocytes. I. Diminished gamma interferon production in response to autologous stimulation, J. Exp. Med. 157:173.

Henle, W., and Henle, G., 1981, Epstein–Barr virus-specific serology in immunologically compromised individuals, Cancer Res. 41:4222.

Jondal, M., 1976, Antibody-dependent cellular cytotoxicity (ADCC) against Epstein–Barr virus determined membrane antigens. I. Reactivity in sera from normal persons and from patients with acute infectious mononucleosis, Clin. Exp. Immunol. 25:1.

Jondal, M., and Klein, G., 1973, Surface markers on human B and T lymphocytes. II. Presence of Epstein–Barr virus receptors on B lymphocytes, J. Exp. Med. 138:1365.

Lai, P. K., Alpers, M. P., and MacKay-Scollay, E. M., 1977, Epstein–Barr virus infection: Inhibition by immunologically induced mediators with interferon-like properties, Int. J. Cancer 20:21.

Levin, M. J., Parkman, R., Oxman, M. N., Rappaport, J. M., Simpson, M., and Leary, D. L., 1978, Proliferative and interferon responses by peripheral blood mononuclear cells after bone marrow transplantation in humans, Infect. Immun. 20:678.

Masucci, M. G., Klein, E., and Argov, S., 1980, Disappearance of the NK effect after explantation of lymphocytes and generation of similar nonspecific cytotoxicity correlated to the level of blastogenesis in activated cultures, J. Immunol. 124:2458.

Menazes, J., Patel, P., Dussault, H., Joncas, J., and Leibold, W., 1976, Effect of interferon on lymphocyte transformation and nuclear antigen production by Epstein–Barr virus, Nature 260:430.

Misko, I. S., Moss, D. J., and Pope, J. H., 1980, HLA-antigen related restriction of T lymphocyte cytotoxicity to Epstein–Barr virus, Proc. Natl. Acad. Sci. USA 77:4247.

Moss, D. J., Scott, W., and Pope, J. H., 1977, An immunological basis for inhibition of transformation of human lymphocytes by EB virus, Nature 268:735.

Nonoyama, M., and Pagano, J. S., 1971, Detection of Epstein–Barr viral genome in nonproductive cells, Nature New Biol. 233:103.

Pattengale, P. K., Smith, R. W., and Gerber, P., 1973, Selective transformation of B lymphocytes by EB virus, *Lancet* **2**:93.

Pearson, G. R., and Orr, T. W., 1976, Antibody-dependent lymphocyte cytotoxicity against cells expressing Epstein–Barr virus antigen, *J. Natl. Cancer Inst.* **56**:485.

Purtilo, D. T., 1976, Pathogenesis and phenotypes of an X-linked recessive lymphoproliferative syndrome, *Lancet* **2**:882.

Purtilo, D. T., Cassel, C., Yang, J. P. S., Stephenson, S. R., Harper, R., Landing, G. H., and Vawter, G. F., 1975, X-linked recessive progressive combined variable immunodeficiency (Duncan's disease), *Lancet* **1**:935.

Purtilo, D. T., Sakamoto, K., Barnabel, K., Seeley, V., Bechtold, J., Rogers, G., Yetz, J., and Harald, S., 1982, Epstein–Barr virus induced diseases in boys with the X-linked lymphoproliferative syndrome (XLP), *Am. J. Med.* **73**:49.

Rickinson, A. B., Moss, D. J., and Pope, J. H., 1979, Long-term T cell-mediated immunity to Epstein–Barr virus in man. II. Components necessary for regression in virus-infected leucocyte cultures, *Int. J. Cancer* **23**:610.

Rickinson, A. B., Wallace, L. E., and Pope, J. H., 1980, HLA-restricted T cell recognition of Epstein–Barr virus-infected B cells, *Nature* **283**:865.

Rickinson, A. B., Moss, D. J., Wallace, L. E., Rowe, M., Misko, J. S., Epstein, M. A., and Pope, J. H., 1981, Long-term T cell-mediated immunity to Epstein–Barr virus, *Cancer Res.* **41**:4216.

Rytel, M. W., and Balay, J., 1973, Impaired production of interferon in lymphocytes from immunosuppressed patients, *J. Infect. Dis.* **127**:445.

Sakamoto, K., Freed, H. J., and Purtilo, D. T., 1980, Antibody response to Epstein–Barr virus in families with the X-linked lymphoproliferative syndrome, *J. Immunol.* **125**:921.

Seeley, J., Svedmyr, E., Weiland, O., Klein, G., Muller, E., Erichsson, E., Andersson, K., and Van der Waal, L., 1981, Epstein–Barr virus selective T cells in infectious mononucleosis are not restricted to HLA-A and B antigens, *J. Immunol.* **127**:293.

Stewart W. E., II (ed.), 1979, *The Interferon System*, Springer-Verlag, Berlin.

Sullivan, J. L., Byron, K. S., Brewster, F. E., and Purtilo, D. T., 1980, Deficient natural killer activity in X-linked lymphoproliferative syndrome, **210**:543.

Svedmyr, E., and Jondal, M., 1975, Cytotoxic effector cells specific for B cell lines transformed by Epstein–Barr virus are present in patients with infectious mononucleosis, *Proc. Natl. Acad. Sci. USA* **72**:1622.

Takaki, K., Horado, M., Sairenji, T., and Hinuma, Y., 1980, Identification of target antigen for antibody-dependent cellular cytotoxicity on cells carrying Epstein–Barr virus genome, *J. Immunol.* **125**:2112.

Thestrup-Pedersen, K., Esmann, V., Jensen, J. R., Hastrup, J., Thorling, K., Saemundsen, A. K., Bisballe, S., Pallesen, G., Madsen, M., Masucci, M. G., and Ernberg, I., 1980, Epstein–Barr virus-induced lymphoproliferative disorder converting to fatal Burkitt-like lymphoma in a boy with interferon-inducible chromosomal defect, *Lancet* **2**:997.

Thorley-Lawson, D. A., 1981, The transformation of adult but not newborn human lymphocytes by Epstein–Barr virus and phyto-hemagglutinin is inhibited by interferon: The early suppression by T cells of Epstein–Barr infection is mediated by interferon, *J. Immunol.* **126**:829.

Thorley-Lawson, D. A., Chess, L., and Strominger, J. L., 1977, Suppression of *in vitro* Epstein–Barr virus infection: A new role for adult human T lymphocytes, *J. Exp. Med.* **146**:495.

Tosato, G., Magrath, I., Koski, I., Dooley, N., and Blaese, M., 1979, Activation of suppressor T cells during Epstein–Barr virus-induced infectious mononucleosis, *N. Engl. J. Med.* **301**:1133.

Vilmer, E., Lenoir, G. M. Virelizier, J. L., and Griscelli, C., 1984, Epstein–Barr serology in immunodeficiencies: An attempt to correlate to immune abnormalities in Wiskott–Aldrich and Chediak–Higashi syndromes and ataxia-telengiectasia, *Clin. Exp. Immunol.* **55**:249.

Virelizier, J. L., 1984a, Immunoregulatory effects of interferons and their therapeutic im-
    plications, in: *Approaches to Antiviral Agents*, Macmillan and Co., London, in press.
Virelizier, J. L., 1984b, Deficiencies of interferon production in children, *Presse Med.* **13**:495.
Virelizier, J. L., and Griscelli, C., 1980, Interferon administration as an immunoregulatory
    and antimicrobial treatment in children with defective interferon secretion, in: *Primary
    Immunodeficiencies* (M. Seligmann and W. H. Hitzig, eds.), pp. 231–234, Elsev-
    ier/North-Holland, Amsterdam.
Virelizier, J. L., and Griscelli, C., 1981, Defaut selectif de secretion d'interferon associe a
    un deficit d'activite cytotoxique naturelle, *Arch. Fr. Pediatr.* **38**:77.
Virelizier, J. L., and Guy-Grand, D., 1980, Immune interferon secretion as an expression of
    immunological memory to transplantation antigens: *In vivo* generatin of long-lived,
    recirculating memory cells, *Eur. J. Immunol.* **10**:375.
Virelizier, J. L., Lenoir, G., and Griscelli, C., 1978, Persistent Epstein–Barr virus infection
    in a child with hypergammaglobulinemia and immunoblastic proliferation associated
    with a selective defect in immune interferon secretion, *Lancet* **2**:231.
Virelizier, J. L., Perez, N., Arenzana-Seisdedos, F., and Devos, R., 1984, Pure IFN- enhances
    class II HLA-antigens on human monocyte cell lines, *Eur. J. Immunol.* **14**:106.

CHAPTER 13

Vaccination against Herpes Simplex Virus Infections

BERNARD MEIGNIER

I. INTRODUCTION

Many viral diseases have been effectively prevented by the use of vaccines, and these accomplishments make it desirable and logical to develop a vaccine that will prevent the diseases caused by herpes simplex virus (HSV), a requirement made urgent by the increased incidence of genital herpes infections in recent years.

Studies on antiherpetic immunity in animals began shortly after recognition of the viral nature of the agent responsible for herpes (Levaditi and Harvier, 1920; Lipschütz, 1921). However, notwithstanding generations of experiments and trials, a satisfactory vaccine that would "tame maverick virus" (Gunby, 1983) has not been developed. Approaches to prophylaxis of HSV infections have been hindered by several difficulties: (1) There is a lack of sufficient knowledge regarding immunity to HSV and an absence of "markers" of immunity comparable to neutralizing antibodies, which were so helpful in the development of the polio vaccines. Evaluation of the herpes vaccines must, therefore, be done through clinical trials. (2) In most vaccine studies, efforts have been made to prevent the recurrent manifestations of the disease in infected patients rather than to prevent or at least mitigate the primary infections in herpes-free individuals. As it is difficult to demonstrate convincingly that recurrences have really been reduced or abolished in the absence of placebo-controlled studies, little usable conclusions emerged from these studies. (3) A major problem stems from the characteristics of HSV itself. On the

BERNARD MEIGNIER • Institut Mérieux, Marcy l'Etoile, Charbonnières les Bains, France.

one hand, the protective immunity it induces as measured by recurrences and superinfections seems weak, while, on the other hand, HSV can establish a latent infection and may be involved in the cause of cervical cancers in women, raising concerns about the possible safety of vaccines.

Together, these difficulties may explain why, among the viral vaccines under development, "probably the toughest problem of all comes with Herpes" (Osborn, 1979).

The general concepts of antiherpetic vaccines and vaccination as well as benefit risk considerations have been discussed in several papers (Parks and Rapp, 1975; Wise et al., 1977; Allen and Rapp, 1982; Moreschi and Ennis, 1982). It is the purpose of this chapter to review these concepts and their application to the development of HSV vaccines. Some importance has been given to historical data as the modern concepts regarding HSV vaccines have developed from past experience and also to illustrate how this development has been dependent on the general advances of "vaccinology."

It is certainly worth mentioning that, although an effective vaccine would certainly be a major tool for the control of herpes infections, it is doubtful that by itself, it would solve all of the herpes-related problems. It may be expected that, in conjunction with education and the development of antiviral drugs (Roizman et al., 1982), ways will be found to subjugate the disease.

II. THE USE OF WILD-TYPE VIRUS

In 1919, Loewenstein demonstrated that several clinical forms of herpes (ocular, cutaneous, or mucosal) were caused by an invisible virus that induced a typical keratitis in rabbits.

Shortly thereafter, successful experimental infections could be obtained in humans by autoinoculation; scarifying the forearm of patients with vesicle fluid obtained from their vesicular lesions, as well as by heteroinoculation from one individual to another one (Teissier et al., 1922, 1923; Tureau, 1924), or from experimentally infected rabbit to human (Nicolau and Poincloux, 1923). Typical lesions were obtained at the site of inoculation in 40–80% of the attempts and were sometimes followed by authentic recrudescent herpes (Nicolau and Banciu, 1924; Paulian, 1932).

Because these experiments were designed either to explore the virulence of the virus or the possible immunity to herpes of patients suffering other diseases, the authors did not focus on the possible beneficial influence of experimental herpes infection on the course of their recurrent herpetic disease.

Nothwithstanding the well-established risks of creating new sites of recurrent infection, live HSV has been repeatedly used in attempts to treat patients with recurrent diseases. Autoinoculation was claimed to be successful by Macher (1957) in one case of genital herpes and by Goldman (1961) in another. The latter's patient experienced recurrences at the site of inoculation. Good results were also obtained by Panscherewski and Rhode (1957) in 10 of 17 patients receiving repeated injections of embryonated egg-grown virus and these patients did not develop complications or side effects after inoculation.

Autoinoculation, however, is definitely not an innocuous procedure: Lazar (1956) and Blank and Haines (1973) observed new herpetic lesions at the site of autoinoculation (abdomen, forearm, or thigh) and, subsequently, found simultaneous recurrences at both the original and the "therapeutic" sites.

By modern standards, the lack of efficacy and the risk of recurrent infections preclude the use of live virulent virus as possible vaccines. However, experiments with virulent virus have highlighted two points that have been major pitfalls in the development of antiherpetic vaccines. (1) Stimulation of the immune systems, which was expected from autoinoculation, does not significantly affect the course of recurrent herpetic diseases; and (2) patients with recurrent herpes infection are still susceptible to infection by autoinoculation or superinfection with a heterologous strain, indicating that the immunity induced by spontaneous infections is only relative.

III. THE USE OF INACTIVATED (KILLED) VIRUS

Divergent reasons led researchers to look for an alternative to autoinoculation: one was the natural reluctance of some practitioners to use a living virus (Biberstein and Jessner, 1958); the other was observations that a significant number of patients did not develop lesions by autoinoculation (Brain, 1936; Frank, 1938). Since then, safety and efficacy have been the major goals of vaccine developers.

A comment is necessary before reviewing the studies with inactivated vaccines. In all but one study, the purpose of vaccination was to reduce or prevent recurrences or recurrent episodes in patients already suffering from recrudescent herpetic disease. Therefore, vaccination was considered to be a treatment (vaccinotherapy), rather than a prophylaxis of disease. Consequently, the efficacy of vaccination was usually assessed by comparing the frequency and severity of recurrences before and after vaccination, an evaluation known to be biased by the placebo effect.

A. The Placebo Effect

The placebo effect is the effect produced by any therapeutic procedure on a patient, a symptom, syndrome, or disease but which is objectively without specific activity for the condition being treated. It is a neglected and berated aspect of patient care and may involve practically any organ in the body (Benson and Epstein, 1975).

The possible role of the placebo effect in the improvement of the condition of herpetic patients was recognized early. "The injection of herpes viruses or their antigens might only provide a non specific stimulus, comparable to any foreign protein injection, coupled with being a suitable vehicle for supportive reassurance by the physician. Many persons with recurrent herpes establish very rapidly a strong positive transference to the physician. They are anxious to please, are cooperative, eager and submissive. Provided the physician gives them strong encouragement, aided by a new 'method treatment' they often respond with a disappearance of lesions. When therapy is withdrawn, and with it the support of the physician, lesions may recur." (Jawetz et al., 1955). Estimated as the percentage of control subjects showing improvement in the severity and/or the frequency of the attacks, the placebo effect varies from 30% (Steppert, 1956) or 40% (Bierman, 1976) to 52% (Kern and Schiff, 1959) and may be as high as 75% (Kern and Schiff, 1964; Vallée, 1980).

It follows that, unless an appropriate placebo group is included in the study design, one may not be able to distinguish the relative roles of the placebo effect and specific activity of the vaccine in the improvement of patients. That is the justification for presenting separately the studies without placebo controls.

B. Studies without a Placebo Group

Most studies evaluating the efficacy of inactivated vaccines have been carried out without a placebo control group. The resulting publications have been frequently quoted as indicating the efficacy of vaccinotherapy in recurrent herpes. However, because of their flawed design, those studies cannot be considered conclusive. Whether "perhaps there is some fire behind all the smoke" (Hilleman et al., 1982) remains speculative.

The very first clinical attempts used serial injections of preparations derived from animal tissues infected with the virus and inactivated with various chemical or physical agents: Brain emulsion inactivated with phenol, formalin (Frank, 1938), or heat (Biberstein and Jessner, 1958) as well as formalin-treated suspension of guinea pig pads (Brain, 1936) have been used. Because of the methods used to prepare them, the difficulty in scaling up production and because of concerns about the possible complications such as demyelination due to organ-specific antibodies in response to animal antigens, vaccines made from tissue suspension did not reach a wide application. However, they established the concept that inactivated vaccines could be a cure for recurring herpes. They also raised the question of which mechanisms mediated the beneficial effects of the vaccine, for in studies evaluating the neutralizing antibodies in the serum of four patients, Frank (1938) could not detect changes of titers in spite of clinical improvement experienced by patients who had received 10–50 ml of infected brain suspension.

The development of herpes vaccines benefited from the general impetus derived from the development of modern techniques of cell culture and virus propagation and from the success of the polio vaccination in the middle and late 1950s. Successful immunization of laboratory animals (Anderson and Kilbourne, 1961) also prompted interest in antiherpetic vaccines and several laboratories developed vaccines prepared either in embryonated eggs or in primary cell cultures and inactivated by either physical or chemical agents. The choice of ultraviolet irradiation was based on the premise that heat or chemical agents might denature the antigens (Lepine and De Rudder, 1964). In several countries mass production was achieved either by commercial companies (Diamant in France, Hermal in Germany, producing Lupidon, Eli Lilly in the United States) or by public institutions (Bulgaria, USSR).

A large number of papers have been published during the past 20 years concerning the results of clinical trials with these vaccines. Many of them are redundant and/or more worshipful than technically documented. Papers quoted in Table I have been selected on the criterion that they give enough numerical data to fill in the columns of the table for comparative purposes.

The stage was set rather negatively by Nagler (1946) who used formalinized amniotic fluid to vaccinate six herpetic patients, four of whom experienced recurrences within the 6-month period of follow-up. Later, Jawetz et al. (1955) reported their observations on a series of 50 patients with recurrent herpes of skin, mucous membranes, or cornea and—included as a control—patients with recurrent aphthous stomatitis, a nonherpetic condition. These individuals, who had failed to improve with other conventional forms of treatment, received a course of three or four weekly injections of a vaccine prepared from allantoic fluid of infected embryonated eggs. Definite improvement could be seen in 47% of the patients in the herpetic group and in 25% of the patients with recurrent aphthous stomatitis. None of the patients showed a significant change in their complement fixing or neutralizing antibody levels. Interestingly enough, the authors carefully discussed the possibility that "the limited beneficial effect of the preparation used might be, in part due to desensitization with a specific antigen and in part due to the strong psychogenic impact of a 'new' treatment on suggestible individuals," a reserve in the conclusion that was forgotten or neglected by their followers.

The possibility that repeated vaccination produced a desensitization of the patients was the most commonly cited rationale for using vaccinotherapy. Henocq (1967, 1972a) based his approach on the fact that almost all patients had neutralizing antibodies in their sera and 80% of them showed positive skin test responses with herpes antigen during recurrences. It was therefore thought logical to apply courses of serially

TABLE I. Clinical Efficacy in Man of Inactivated Vaccines—Studies Lacking a Placebo Group

| Study | Antigen | | Number of patients | Type of patients[a] | % of patients improved |
	Substrate	Inactivation			
Nagler et al. (1946)	Amniotic fluid	Formalin	6	NS	0
Jawetz et al. (1955)	Allantoic fluid	UV	73	K, CM	47
Sóltz-Szóts (1960)	Chorioallantoic membrane	Ultracentrifugation	46	CM	78
Fanta et al. (1974)	Chicken cell culture	Not specified	621	K, CM	77
Henocq et al. (1964)	Sheep kidney cells	UV	20	K, CM	90
Degos and Touraine (1964)			50	NS	76
Offret et al. (1965)			20	K	80
Santoianni (1966)			30	CM	70
Henocq (1972b)			30	K	80
Macotela-Ruiz (1973)			20	CM	45
Baron et al. (1966)			28	K	75
Brückner (1970)			13	K	92
Bubola and Mancosu (1967)			30	CM	83
Pouliquen et al. (1966)			48	K	87
Chapin et al. (1962)	Rabbit kidney cells	Formalin	50	K	72
Hull and Peck (1966)			333	K, CM	72–79
Nasemann (1965, 1970)	Chorioallantoic membrane	Heat	170	CM	90–96
Nasemann and Schaeg (1973)			1200	NS	70–96
Schmersahl and Rüdiger (1975)			1059	K, CM	81
Kitagawa (1973)			4	CM	100
Schneider and Rhode (1972)			188	K, CM	81
Weitgasser (1973)			76	NS	63
von Rodovsky et al. (1971)	Rabbit kidney cells	Formalin	15	CM	100
Andonov et al. (1979)			477	CM	85
Dundarov et al. (1982)			2350	K, CM	60–80
Schubladze et al. (1978)			114	K	89

[a] NS, not specified; K, herpetic keratitis; CM, cutaneo-mucous herpes.

increasing doses of vaccine; a treatment used to desensitize allergic patients. Although less categorical, Nasemann (1965, 1973) referred to the same explanations. A series of 6–10 injections was usually a minimum and as many as 20–40 vaccinations was not uncommon. As shown in Table I, the published results were uniformly good to excellent with only one exception (Nagler, 1946). The follow-up periods were frequently not mentioned and, when indicated, varied from a few months to 1 year with few exceptions (Dundarov et al., 1982; Pouliquen et al., 1966; von Rodovsky et al., 1971).

When evaluated, despite repeated inoculations, antibody titers, as measured by neutralization or complement fixation, remained unchanged (Henocq et al., 1964; Remy et al., 1975; Bubola and Mancosu, 1967) or underwent only a limited rise (von Rodovsky et ál., 1971; Söltz-Szöts, 1960). Studying selected cases, Jarisch and Sandor (1977) observed increased activity in the macrophage inhibition test in three out of four patients.

Almost all publications insisted upon the absence of side effects following the vaccination. Some authors, however, reported rare focal or general side effects and/or called for caution when using the vaccine, particularly in keratitis patients (Henocq, 1967; Lepine and de Rudder, 1964; Pouliquen et al., 1966; Schneider and Rhode, 1972; Söltz-Szöts, 1971).

C. Studies Including a Placebo Group

As opposed to uncontrolled trials, studies that included a placebo group are rather scarce: Some studies were performed on patients with recurrent disease and were carried out on children without a past history of herpes.

In herpetic patients, the results (Table II) are dependent on the way data were collected: Whereas the treatment produced a significant improvement in studies where the follow-up period was short (Mullin, 1966) or not specified (Söltz-Szöts, 1971; Weitgasser, 1977), no difference was detected between treated and placebo group in a double-blind trial with a follow-up of 36 months (Kern and Schiff, 1964). It is noteworthy that the studies by Mullin and by Kern and Schiff were performed with the same vaccine and the same protocol. Also, the vaccine used in these studies induced neutralizing antibodies in rabbits and guinea pigs and produced resistance of immunized mice to direct challenge with live HSV (Hull and Peck, 1966).

The trial in uninfected children was done in an orphanage with a vaccine prepared from amniotic fluid, killed by heat and formalin (Anderson et al., 1950). Ten children in each group, placebo and vaccinated, were immunized between the ages of 7 and 10 months and were followed over 1 year. Among the vaccinees, four of nine entered the study with

TABLE II. Clinical Efficacy in Man of Inactivated Vaccines—Studies with a Placebo Group

Study	Antigen		Type of study	Follow-up period	Patients improved in	
	Substrate	Inactivation			Treated group	Placebo group
Kern and Schiff (1964)	Rabbit kidney cells	Formalin	Double blind	36 months	16/23 (70%)	22/29 (76%)
Mullin (1966)			Double blind	6 months	10/12 (83%)	2/11 (18%)
Weitgasser (1977)	CAM[a]	Heat	Double blind	NS[b]	56/68 (82%)	36/120 (30%)
Söltz-Szöts (1971)	CAM	NS	NS	NS	285/366 (78%)	10/66 (15%)

Studies in noninfected children

					Vaccinees	Control
Anderson et al. (1950)	Amniotic fluid	Formalin	Open	1 year	Postinfection antibodies	
					9/10	8/10
		Heat			Herpes stoma-titis	
					6/10	5/10

[a] CAM, chicken chorioallantoic membrane.
[b] NS, not specified.

no herpetic antibodies and seroconverted, presumably due to the vaccination. Nine of ten and eight of ten in the vaccinated and unvaccinated groups, respectively, experienced an antibody rise after clinical (6/9 and 5/8) or presumptive infections, a result that speaks clearly against the efficacy of the vaccination in the study. According to the authors, the failure of the vaccine to protect may have been due to the inadequate nature of the antigenic stimulus.

Briefly, the clinical trials with inactivated vaccines showed that:

1. Results of uncontrolled studies were uniformly good but cannot be critically appraised.
2. When double-blind, controlled trials were undertaken, vaccinotherapy demonstrated efficacy in herpetic patients on a short-term but not on a long-term basis.
3. In one pulished trial, vaccination of babies did not prevent subsequent clinical herpetic infections in a significant number of cases.

Although no formal evidence proved that such vaccines are ineffective, a feeling developed in the Western medical community that the vaccines did not meet the expectations generated at the beginning and they are no longer in use, except in Eastern Europe (Dundarov et al., 1983; Barinsky et al., 1983). However, the major reason for their withdrawal comes from the finding that HSV might be involved in the etiology of cervical cancer, that HSV can transform cells in vitro, and that other members of the herpes group are oncogenic (Parks and Rapp, 1975; Wise et al., 1977). Although one experiment failed to detect oncogenic properties of inactivated vaccines in hamsters (Petersen, 1977), the consensus was that vaccines containing herpes DNA were too risky and that their use should be discontinued. At that time, the progress made in mass production of virus and in the technology of protein purification opened the possibility for subunit vaccines.

IV. SUBUNIT VACCINES

The rationale for trying subunit vaccines for the immunoprophylaxis of herpes infection relies on several lines of considerations. First, the lack of efficacy in previous studies might be due to the poor quality of the antigens used or to the flaws in the design of most studies. Moreover, the fact that in controlled trials there was no case control difference indicates a brief but real response to vaccination (Skinner et al., 1982a). Therefore, purified concentrated antigen preparations may induce a stronger and a longer-lasting immunity than that that could be obtained with inactivated virus. Second, the procedures used in the preparation of subunit vaccines require the destruction of virus particles, obviating concerns about the possibility of there being live residual virus in the inactivated material (Fenyves and Strupp, 1982). Third, the various steps involved in their preparation usually result in getting rid of most or all of the herpes simplex DNA, so that such vaccines are not subject to the criticism that they might be oncogenic. Consequently, they would be easier to develop and more acceptable to the scientific and medical community.

A. Preparation

Subunit vaccines, sometimes called envelope antigens or subunit antigens, contain a mixture of viral polypeptides and glycoproteins, ranging from 9 to 10 or 12 species (Cappel *et al.*, 1980; Larson and Lehman, 1979). The common basic step in their preparation consists of detergent extraction of the antigens with NP-40, Triton X-100, or SDS. Extraction can be applied to a purified virus suspension obtained by ultracentrifugation (Cappel *et al.*, 1980; Hilfenhaus *et al.*, 1981; Klein *et al.*, 1981) or directly to the infected cells (Larson and Lehman, 1979), in some cases concurrently with (Skinner *et al.*, 1978, 1980a) or after (Kitces *et al.*, 1977) formalin inactivation. Capsids are usually eliminated by ultracentrifugation and DNase treatment may be added to digest residual viral DNA (Larson and Lehman, 1979; Kitces *et al.*, 1977). Further purification may also be achieved by ion-exchange or affinity chromatography using lectin-Sepharose columns (Larson and Lehman, 1979; Zweerink *et al.*, 1981). Bertland and Lampson (1982) developed a variant where antigens are extracted with 4 M urea applied directly to the infected cell layers, which avoided their collection by scraping and centrifugation. In another variant, Thornton *et al.* (1982) prepared vesicles of plasma membranes by treatment of infected cells with dithiotreitol and formaldehyde. On the whole, subunit vaccines represent preparations that differ from one laboratory to another by their degree of purification and characterization. This heterogeneity as well as the diversity of immunization schedules and animal models used, explains the heterogeneity of results observed.

Although no figure is currently available, one must expect subunit vaccines to be expensive to prepare and use. The cost of purification, especially considering the losses obtained during the process, and the cost of propagation of the virus will be high. In addition, because of the inert nature of the antigen, immunization will require larger amount of immunogen than live virus vaccine. According to the data from Cappel *et al.* (1982a), one dose for an adult, about 100μg, would require the cultivation of 500 cm^2 of chick embryo fibroblasts. One 1200-ml roller bottle yields one dose for rabbit immunization (Rajcani *et al.*, 1980) or 25–40 doses for mouse immunization (Slichtova *et al.*, 1982).

B. Experience in Animals

With the exception of one failure in mice (Zweerink *et al.*, 1981), subunit vaccines have been found to induce various levels of neutralizing antibodies in mice (Kitces *et al.*, 1977; Kutinova *et al.*, 1980; Slichtova *et al.*, 1980; Klein *et al.*, 1981), guinea pigs (Thornton *et al.*, 1982; Kutinova *et al.*, 1982; Yoshino *et al.*, 1982), and rabbits (Cappel *et al.*, 1979; Carter *et al.*, 1981). Following immunization, antibodies were also detected after immunization by ELISA in mice (Hilfenhaus *et al.*, 1981; Thomson *et al.*, 1983) or complement fixation and complement-dependent cytolysis in guinea pigs (Thornton *et al.*, 1982). Cell-mediated immune response has also been reported in mice (Klein *et al.*, 1981) and in rabbits (Cappel, 1976). Generally speaking, in spite of the diversity of the adjuvants and immunization schedules, there is agreement that subunit preparations are antigenic in laboratory animals.

A correlation was observed by Hilfenhaus *et al.* (1981) in hairless mice between the intensity of HSV antibodies induced by immunization and the level of protection achieved; but Klein *et al.* (1981) could not find a similar correlation. Instead they observed a correlation with cell-mediated immune stimulation.

The immunogenicity of subunit vaccines has been successfully assessed by virulent challenge of immunized animals. In mice, the experimental models used included skin infection in various sites of hairless mice (Zweerink *et al.*, 1981; Klein *et al.*, 1981; Hilfenhaus *et al.*, 1981), lip abrasion (Kitces *et al.*, 1977; Thomson *et al.*, 1983), intravaginal inoculation (Skinner *et al.*, 1978, 1980b), intradermal injection in the ear pinna (Slichtova *et al.*, 1980) or the foot pad (Hilleman *et al.*, 1982), and injection by intraperitoneal route (Kutinova *et al.*, 1980). Rabbits were challenged by ocular route (Cappel, 1976; Rajcani *et al.*, 1980; Carter *et al.*, 1981). In guinea pigs, Scriba (1981) used subcutaneous, intradermal, and intravaginal routes and the latter was also used by Thornton *et al.* (1982).

The variety of experimental conditions used makes it difficult to compare in detail the published results. General features, however, can be presented: Subunit vaccines are capable of protecting vaccinated animals against morbidity and/or mortality, usually reducing both the incidence and the severity of lesions produced by virulent challenge. The effect on the establishment of latency or on reactivation of latent virus is less clear-cut: positive in mice according to Thomson *et al.* (1983), Kutinova *et al.* (1980), and Hilfenhaus and Moser (1981) and in rabbits (Rajcani *et al.*, 1980), but negative in mouse studies by Cappel *et al.* (1980) and Klein *et al.* (1981). When compared with inactivated virus, subunit vaccines regularly produced the best protection (Cappel, 1976; Skinner *et al.*, 1978), but weaker immunity or the necessity of higher dosages to obtain the same level of protection have also been reported (Hilfenhaus *et al.*, 1981; Carter *et al.*, 1981; Rajcani *et al.*, 1980; Schneweis *et al.*, 1981) and the ratio can be as high as 10- to 100-fold (Hilfenhaus and Moser, 1981). Kitces *et al.* (1977) found that in mice the purification of antigens from viral suspension reduced their potency compared to crude detergent extract. Finally, achieving satisfactory protection usually requires several injections combined with the use of adjuvants such as complete Freund's adjuvant, PICLC (polyinosinic–polycytidylic acid complexed with poly L-lysine and carboxymethylcellulose), or alum.

In summary, experiments in rodents definitely established the ability of subunit vaccines to trigger an immune response and to induce protection of the animals against challenges with live, virulent virus. One must keep in mind, however, that these results were obtained in animal models with a virus that is not indigenous to the host and results may not be entirely relevant to man. Moreover, "caution must be applied to the data obtained from prevention of HSV disease with vaccines in animal models" (Moreschi and Ennis, 1982).

Paradoxically, animal models are indispensable to HSV vaccine development, but their validity will only be assessed when conclusive trials in man allow the comparison of animal data to protection in the field. The necessity to carefully define the models used is underlined by the experiments of Schneweis et al. (1981) who compared different challenge systems in mice (homologous versus heterologous as well as homotopic versus heterotopic) and showed that protection data are to some extent dependent on the models used.

Experiments in monkeys are generally considered of special interest because of their closer phylogenetic relationship to man. In fact, they are not exceptions to the previous reservation, for the different species display a great range of susceptibility and reactivity to HSV. Thus, although they provide interesting data about seroconversion, the experiments done with rhesus monkeys (Skinner et al., 1982c) are difficult to interpret with regard to immunogenicity and protection, as this species has been found to be very difficult to infect (London et al., 1971).

Neutralizing antibodies have been detected following vaccination of rhesus (Skinner et al., 1982c), chimpanzees (Cappel et al., 1979, 1980), and Cebus (Hilleman et al., 1982) monkeys. Antibody level peaked a few weeks after vaccination and lasted for 2–8 months. A later booster injection (Cappel et al., 1979) or challenge (Skinner et al., 1982c) triggered an anamnestic response. In Cebus monkeys, the seroconversion was found to depend on the dosage and the repetition of the injection: With 2 μg of antigen, three inoculations were necessary to obtain transient seropositivity whereas 50 μg induced antibodies in all eight animals tested after one immunization; subsequent repeats maintained antibodies in all animals over a 4-month follow-up period after the last injection.

C. Experience in Man

Studies in man (Table III) confirmed the ability of subunit vaccines to produce either seroconversion in seronegative individuals or to raise antibodies in seropositive subjects as assessed by seroneutralization, complement fixation, ELISA, complement-dependent cytotoxicity, or im-

TABLE III. Clinical Efficacy in Man of Subunit Vaccines

Study	Antigen		Criterion	Results	
	Substrate	Virus		Vaccinees	Control
Skinner et al. (1982a)	Human diploid cells	Type 1	Improvement of recurrences	17/24 improved	None
Skinner et al. (1982b)[a]			Protection of consorts	0/42 acquired genital herpes	10/40 acquired genital herpes
Woodman et al. (1983)[a]			Rate of recurrences after first episode	7/22 experienced recurrences	17/20 experienced recurrences
Cappel et al. (1982b)[b]	CEF	Type 2	Delay between consecutive relapses	on 38 patients: 72 ± 17 days	on 30 patients: 29 ± 16 days
Ashley and Corey (1983)	NS	Type 2	Protection of consorts	2/12 acquired genital herpes	None

[a] Control group: historical compilation of patients seen before the start of the clinical trial.
[b] Control group was not placebo vaccinated.

munoprecipitation (Hilleman *et al.*, 1982; Skinner *et al.*, 1982a,b; Cappel *et al.*, 1982a; Ashley and Corey, 1983).

These experiments also confirmed the need for repeated injections in order to obtain antibodies in a high proportion of the vaccinees (Hilleman *et al.*, 1982; Cappel *et al.*, 1982a). The data of Cappel *et al.* indicate that, in recipients seronegative prior to immunization, neutralizing antibodies remained stable over a 5-month period following the last injection but that complement fixing and complement-dependent cytolytic activity decreased.

Unfortunately, it is impossible to determine whether subunit vaccines are effective in humans because, here again, only open trials without a placebo group—*stricto sensu*—were used to evaluate efficacy. Control patients in the Cappel *et al.* (1982b) study were not mock vaccinated and the Birmingham group compared observations in vaccinees to data collected from a similar population of patients who were seen before the start of the vaccination program.

In the absence of double-blind placebo-controlled studies, therapeutic effects of subunit vaccines in herpetic patients need no further comment.

More information and work will be necessary to reconcile the striking differences observed between the excellent results of Skinner *et al.* (no case of herpes infection in 42 vaccinated consorts with a follow-up period of 3–18 months) and the two disappointing breakthroughs, one clinical and the other one subclinical, among 12 vaccinees, observed in Seattle within 3 months after the second immunization (Ashley and Corey, 1983).

D. Future Trends

A frequent criticism addressed to conventional subunit preparations is that they contain a mixture of antigens that varies with the method of preparation and may also vary from batch to batch. Standardization of a vaccine would, therefore, be difficult to establish.

Monoclonal antibodies might help solve this problem as it is possible to couple them to substrates and prepare immunosorbent columns that specifically extract the desired immunogen from crude cell lysates (Eisenberg *et al.*, 1982). Such a procedure has been used to purify glycoproteins gD and gB from HSV-1 for immunization of immunized mice, which were then challenged with HSV-1 and HSV-2 (Chan, 1983). Administered without adjuvant, gD conferred excellent and gB partial homologous protection against intraperitoneal or subcutaneous challenge. However, both preparations failed to give cross-protection against type 2 intraperitoneal infection. It would be interesting to determine the duration of immunity, for in this experiment, protection was only tested 1 week after the third and last immunization.

Although the use of monoclonal antibodies may solve the problem of purity of antigens, they certainly do not help with the cost of production and the necessity of generating large quantities of virus during the preparation process.

It is the purpose of molecular cloning to overcome these limitations by providing means whereby gene products from pathogenic agents can be expressed in virtually unlimited quantities in a nonpathogenic form. In addition to safety and economy, it is also believed that genetic engineering would make it possible to compensate for the weak immunogenicity of subunit vaccines by allowing the use of larger dosages of antigens, produced more easily and cheaply.

In *Escherichia coli*, gD of HSV-1 has been cloned and expressed using plasmid construction (Watson *et al.*, 1983; Weis *et al.*, 1983) or using phage M13 as a vector (Eisenberg *et al.*, 1983).

The gene for gD of HSV-1 has also been expressed in a stable mammalian cell line, which gets around the problem that membrane proteins—including HSV envelope antigens—tend to aggregate and become insoluble when expressed in *E. coli* (Berman *et al.*, 1983). According to the same authors, expression of gD1 was also obtained in *Saccharomyces cerevisiae*.

Lastly, an interesting approach has been initiated with the synthesis of an antigenic determinant of gD1, which stimulated neutralizing antibodies in mice and rabbits (Cohen *et al.*, 1983) and conferred protection in mice against the lethal (heterologous) type 2 challenge (Dietzschold and Cohen, 1984).

There is no doubt that different experiments are being carried out in animals to test the antigenicity and immunogenicity of new preparations and to determine whether such antigens live up to their promise in providing improved subunit vaccines. These studies should also provide a better understanding of viral antigenicity and of the immune response in humans and animals infected with HSV.

V. LIVE VACCINES

The arguments in favor of the concept of live antiviral vaccines have been known for decades: Because of their ability to replicate in the recipient, they require smaller doses, which makes them more economical. In addition, they produce a stronger and longer-lasting immunity, which avoids frequent boosters and makes them more practical when implementing vaccine programs. These points are of particular importance for the prevention of herpes infection for the following reasons. First, although natural infection induces some immunity, as indicated by the diminution of clinical and virological signs in subsequent episodes (Corey *et al.*, 1983), the existence of recurrences and of documented superinfection indicates that the wild-type infections yield only relative protection, at least in some individuals. Therefore, vaccines should at the least match the protective effect of natural infections if they are expected to prevent or ameliorate primary infections and subsequent recurrences. Second, epidemiological studies have shown that genital herpes, a concern for public health, is transmitted as a venereal disease and although its pattern of incidence is strongly influenced by socioeconomic factors, the major determinant of the risk is obviously the sexual behavior and the greatest age-specific incidence is found among young adults who are sexually active. The estimated cumulative age-specific incidence of HSV-2 shows that 60% of primary infections are contracted between the ages of 15–35 (Rawls and Campione-Piccardo, 1982). This means that a vaccine, in order to protect efficiently the population at risk, should be given before the beginning of sexual activity (to children or young students) and it should induce a long-lasting immunity. Interestingly enough, the prevention of type 1 infection would also benefit from such a schedule as the same authors have estimated that after the wave of primary infections in early childhood, a second peak of infections occurs in adolescence and early adult life among people living in industrialized societies.

Several laboratories are investigating the potential of developing live vaccines against HSV; at this point, all of them are still at the laboratory stage.

A. HSV Mutants

The first experimenters recognized very early that the virulence of HSV in animals may differ enormously from one isolate to another (Levaditi, 1926) but the suggestion that spontaneous low-virulence strains might be useful in immunization procedures (Wheeler, 1960) was not—to our knowledge—put into application.

It is not uncommon that mutants, whatever marker was used for their selection, are less pathogenic than wild-type parents and proposals have been made to take advantage of that situation to prepare vaccine strains by selection for thermosensitivity (Zygraich and Huyegelen, 1974) or in two steps for double mutation for resistance to 5-ethyl-2'-desoxyuridine and phosphonoformic acid (Gauri et al., 1981).

Whether obtained spontaneously or induced, mutants cannot be considered a promising way of obtaining vaccine candidates because they occur as a result of what Chanock (1982) referred to as "genetic roulette" and it is impossible to anticipate to what extent virulence, latency, and other biological properties of the mutants will be affected. Furthermore, there is no assurance of their genetic stability if they happen to be passed among susceptible individuals, and there are examples of the reversion to pathogenicity of nonpathogenic strains by serial passage in mice (Thompson and Stevens, 1983; Kaerner et al., 1983).

B. Heterologous Herpesviruses

The use in man of heterologous herpesviruses, which worked well with Mareks disease, has been evoked by Parks and Rapp (1975) as a frightening prospect in view of the virulence of different primate herpesviruses in the heterologous host. It is likely that safety concerns would actually limit the application in man of the thymidine kinase-negative marmoset herpesvirus deletion mutant isolated by Kit *et al.* (1983) although it was found to be attenuated when tested for lethality in mice.

C. Antigens Expressed by Non-HSV Viral Vectors

Vaccinia virus was recently suggested as a possible vector for the expression of antigens of unrelated pathogens (Smith *et al.,*, 1983). A chimera was constructed where the HBs Ag gene of B hepatitis virus was inserted within the vaccinia virus genome. The technique has been further exploited to construct engineered vaccinia viruses expressing foreign genes from a variety of sources, including the gD gene of HSV (Paoletti and Panicalli, 1984; Smith *et al.*, 1984). The recombinant prepared by Paoletti and Panicalli was able to elicit neutralizing antibodies in rabbits and to protect mice against homotypic type 1 intraperitoneal challenge. However, this very interesting approach raises several difficult problems (Beale, 1983). One is that the take of "vaccinia carrier" inoculation might be impaired by the immune memory of the world population because a high proportion have been vaccinated with vaccinia virus. Also, the general usefulness of the method might be limited by the fact that the first campaign with one vaccinia virus vehicle would create immunity detrimental to the efficacy of further vaccinations with other immunogens. Finally, vaccinia virus has been responsible for rare but dramatic postvaccination accidents, which led the World Health Organization to recommend discontinuance of its use after the achievement of smallpox eradication.

D. Genetically Engineered HSV

The principal objections to the use of live attenuated HSV as a vaccine stem from some of their unwanted biological properties: ability to establish latency, ability to undergo reactivation, transforming properties, and oncogenic potential. It has already been noted that the classical procedures of selection of either spontaneous or induced attenuated mutants usually leave these properties unaffected. The basic principle of genetic engineering of an HSV virus for use as a vaccine is that the unwanted properties are dependent on the function of one, or a few, specific gene products of the herpes genome and they do not belong to the minimal set of functions necessary for replication and packaging of the genome. Rather, they represent functions developed by the virus to survive in nature. It is therefore conceivable to construct viruses that would maintain the ability to replicate but in which sequences responsible for neurovirulence, latency, or oncogenicity would be inactivated or deleted (Roizman *et al.*, 1982).

Two procedures might be expected to yield strains suitable for testing as vaccines.

1. Intertypic Recombination

The first one involves the selection of HSV-1 × HSV-2 recombinants obtained by double infection of cultured cells. Recombination is possible because of the colinearity of the genomes of HSV-1 and HSV-2; however, because the genomes differ significantly in nucleotide sequences and in the properties of the proteins they specify, mismatches are expected to happen at the site of recombination, which would yield viruses crippled in some of their functions. Representatives of four series of recombinants (Morse *et al.*, 1977) have been found to be able to grow at least as well as the wild-type viruses in HEp-2 and Vero cells but they showed reduced pathogenicity when inoculated intracerebrally route in mice; although they showed an impaired ability to replicate in the brain of mice, they were still able to protect mice against lethal injection with virulent HSV-1 and HSV-2 (Roizman *et al.*, 1982). Data shown in Fig. 1 confirm both the attenuation and the immunogenicity of several recombinants.

Recombinants have the advantages that they can be selected for optimal growth in culture and that they may be selected to specify both serotypes. The recombination procedure, however, has two limitations: (1) it occurs randomly, and it does not allow precise intervention in selected genes; and (2) in spite of the very low probability of reversion, not all recombinants are genetically stable (see below), which limits the general usefulness of the approach.

2. Construction of Specific Deletion Mutants

The second approach has been technically made possible by the development of procedures that permit the deletion of specific genes as well as the construction of rearrangements in specific locations on the genome (Post and Roizman, 1981; Post *et al.*, 1981). The advantages of this approach are the following:

1. The mutants are selected for ability to grow.
2. Because of the deletion the virus cannot revert.
3. It opens the possibility of constructing recombinants carrying determinant sites of both HSV-1 and HSV-2.

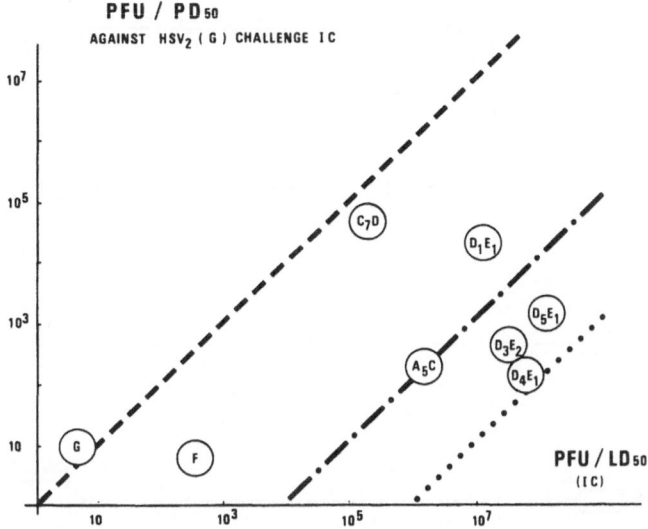

FIGURE 1. Attenuation and immunogenicity of several intertypic HSV-1 × HSV-2 recombinants. *Attenuation:* Groups of 20 BALB/c mice were injected intracerebrally with serial 10-fold dilutions of titered suspension of intertypic recombinants. Based on the pattern of death, lethal doses 50% were calculated. The number of plaque-forming units that would kill 50% of the mice (PFU/LD$_{50}$) is used as an index of attenuation and the strains have been plotted on the abscissa according to that index. *Immunogenicity:* Survivors of the attenuation tests were challenged 7 weeks later intracerebrally with 3000 LD$_{50}$ of HSV-2 genome. Based on the pattern of survival and referring to the amount of recombinant virus received initially, the number of PFU necessary to protect 50% of the mice against the challenge was determined (PFU/PD$_{50}$). The PFU/PD$_{50}$ index was used to plot the strains on the ordinate. The oblique lines represent the LD$_{50}$/PD$_{50}$ ratios: (----) 1; (—·—) 10^4; (·····) 10^6. F and G stand respectively for wild-type viruses HSV-1(F) and HSV-2(G). The first letter of the recombinant denomination indicates different crosses used to produce the recombinants (Morse *et al.*, 1977). All recombinants had a *ts*$^+$ phenotype.

4. It might be possible to construct recombinants in which increased production of antigen could be obtained by substituting more effective promoters for the natural ones.

5. It might be possible to alter the order of expression of some genes in order to change the appearance of their products in the infected cells (Roizman *et al.*, 1982).

Several constructs have been prepared so far, all of which are derived from HSV-1. Details of their construction and properties are reported elsewhere (Mocarski: *et al.*, 1980; Poffenberger, *et al.*, 1980; Post *et al.*, 1980; Post and Roizman, 1981). They have been used to test the practicability of genetic engineering for the design of vaccines. For that purpose they have been tested in animals to determine whether they meet the criteria that a vaccine strain would be expected to meet: attenuation, latency, and immunogenicity. Schematic representations of the genomes of the viruses tested to date are given in Fig. 2.

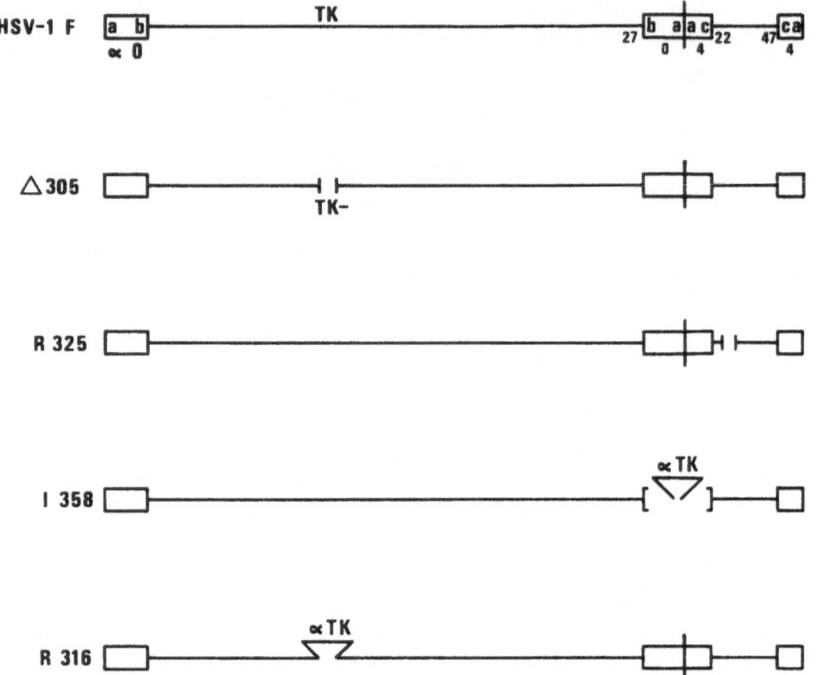

FIGURE 2. Schematic structure of the genome of constructs tested in animals. The top line illustrates the prototype arrangement of the sequences of wild-type HSV-1 DNA. The rectangles represent the inverted repeat sequences flanking the long (left) and short (right) sequences. TK refers to the location of the thymidine kinase gene. The numbers 0, 4, 22, 47 indicate the location of the α genes. Δ305 is deleted in the TK gene, R325 is deleted in the α22 gene. I358 is deleted in the junction region and a TK gene has been inserted in place of the deletion, under an α promoter. In R316, TK gene under α regulation has been inserted at the deleted site of the normal TK gene.

a. Studies in Mice

Table IV summarizes the data obtained in mice. Genetically engineered viruses were found to be markedly attenuated, yielding plaque forming units per lethal dose 50% ratio (PFU/LD$_{50}$) greater than or equal to 6×10^5 by the intracerebral route whereas wild-type strains (subsequently used for challenges) have a PFU/LD$_{50}$ ratio of less than 10.

None of the attenuated viruses were able to establish latent infections when inoculated into the pinna of the ear and their ability to establish latent infection in the trigeminal ganglia when inoculated into the eye varied from being unable (Δ305) to establishment of latency in 50% of the animals (R325), while, in contrast, 100% of the mice inoculated with wild-type HSV-1(F) developed latent infections.

Immunogenicity was tested by intracerebral challenge with 3000 LD$_{50}$, a severe test, 6–7 weeks after immunization. The potency of the

TABLE IV. Properties of Genetically Engineered HSV in Mice

Virus strain[a]	Phenotype	Virulence[b] i.c. PFU/LD$_{50}$	Latency[c] Dose	Latency[c] Eye route	Latency[c] Ear route	Route of immunization	Protection[d] PFU/PD$_{50}$ against HSV-1(MGH$_{10}$)	Protection[d] PFU/PD$_{50}$ against HSV-2(G)
HSV-1(F)	Wild	3×10^2	10^7	10/10	9/10	i.c.	16	5
						i.p.	1×10^5	4×10^4
						s.c.	8×10^3	6×10^6
HSV-2(G)	Wild	3	10^4	$1/3^e$	5/10		ND^h	
			10^6	dead	9/9			
HSV-1(Δ305)	TK$^-$	3×10^6	$>10^8$	0/10	0/10			
HSV-1(R325)	ICP22$^-$	10^7	10^7	5/10	0/9	i.c.	80	5×10^2
						i.p.	5×10^7	$>10^8$
						s.c.	5×10^6	3×10^7
HSV-1(R328)	ICP22$^-$	4×10^6		ND		i.c.	2×10^2	6×10^2
						i.p.	4×10^6	3×10^7
						s.c.	6×10^5	10^6
HSV-1(l358)	Noninverting	6×10^5	10^7	$1/20^f$	0/10	i.c.	10^2	4×10^3
						i.p.	8×10^4	6×10^4
						s.c.	8×10^3	10^5
HSV-1(R316)	αTK	ND	10^8	$6/20^g$	0/10		ND	

[a] HSV-1(F) and HSV-2(G) are wild-type viruses used as references. The structures of the genetically engineered viruses tested are given in Fig. 2.

[b] Virulence was determined in 6-week-old BALB/c mice by the intracerebral route. Results are expressed as the number of plaque-forming units (PFU) that kills 50% of the mice (PFU/LD$_{50}$) as calculated from mortality patterns among groups of 20 mice inoculated with serial dilutions of virus.

[c] Latency: BALB/c mice were inoculated either by scarification of the cornea or by intradermal infection in the ear pinna. Appropriate trigeminal or cervical dorsal root ganglia were taken 4 weeks later and assayed for the presence of latent virus by incubation for 4–5 days at 37°C in cell culture medium followed by homogenization and plating on Vero cells. Attempts were made in parallel to detect infectious virus in the ganglia; all were negative.

[d] Protection: Protective doses, expressed as number of PFU necessary to protect 50% of the mice (PFU/PD$_{50}$), were calculated from the pattern of survival after challenge among groups of 20 mice immunized with serial 10-fold dilutions of viruses by intracerebral (i.c.), intraperitoneal (i.p.), or subcutaneous (s.c.) routes. Challenges were done i.c. with 3000 LD$_{50}$ of either HSV-1(MGH$_{10}$) or HSV-2(G).

[e] Dosage-dependent mortality; survivors only are shown in denominator.

[f] Virus isolated from ganglia was identical to input.

[g] All isolates from ganglia contained nonidentical deletions.

[h] ND, not done.

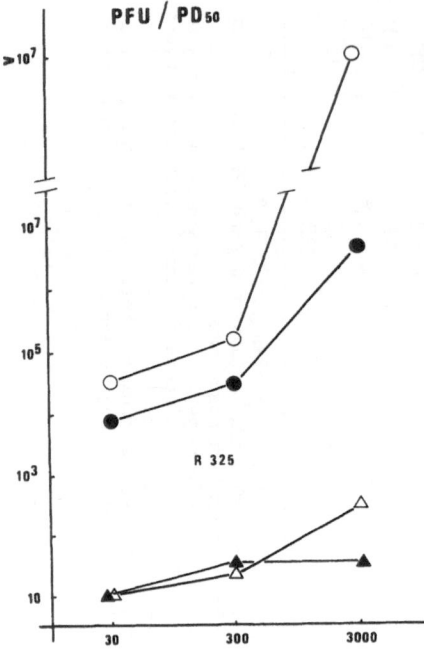

FIGURE 3. Effect of the challenge doses on the evaluation of immunogenicity. Groups of 10 BALB/c mice were immunized with 10-fold serial dilutions of a titered suspension of R325 virus by intracerebral (i.c.) or intraperitoneal (i.p.) routes. They were challenged 7 weeks later i.c. with 30, 300, or 3000 LD_{50} of wild-type viruses, either HSV-1(MGH_{10}) (solid symbols) or HSV-2(G) (open symbols). Based on the pattern of survival the amount of R325 expressed in plaque-forming units necessary to protect 50% of the mice against challenge was calculated (PFU/PD_{50}). This experiment also confirmed that the i.p. route (circles) was less effective in mice than the i.c. route (triangles) and that protection against type 2 requires higher immunization doses than protection against type 1.

viruses was assessed by determining the number of PFU necessary to protect 50% of the animals (PFU/PD_{50}) against the challenge. The ratio was found to be minimal in mice immunized intracerebrally (survivors from virulence test) against homotypic type 1 challenge where values were on the order of magnitude of 10^2. Intraperitoneal and subcutaneous routes of immunization required higher doses to achieve the same level of protection (from 10^5 to > 10^8 $PFLU/LD_{50}$). As would be expected, protection against heterotypic challenge required an even larger dose for a given route of immunization. The dependence of the results on size of the challenge dose has been explored with one virus (R325): groups of immunized mice, challenged with 10-fold serial dilutions of the same wild-type virus suspension, were protected with reduced amounts of immunizing virus (Fig. 3). The effect was more pronounced when immunization was done intraperitoneally, probably because of the even poorer multiplication of the vaccine by this route.

Genetic stability was tested by serial passages in mice intracerebrally. Because of the poor replication of the virus in the brain, virus recovered from one animal had to be grown in culture in Vero cells before further inoculation. R325 was found stable over nine passages; that is, no significant change was observed in the ratios of virus recovered in mouse brain per amount of virus inoculated and, moreover, restriction enzyme patterns of both the ninth passage isolate and the original strain were identical. By comparison, the HSV-1 × HSV-2 recombinant A_5C

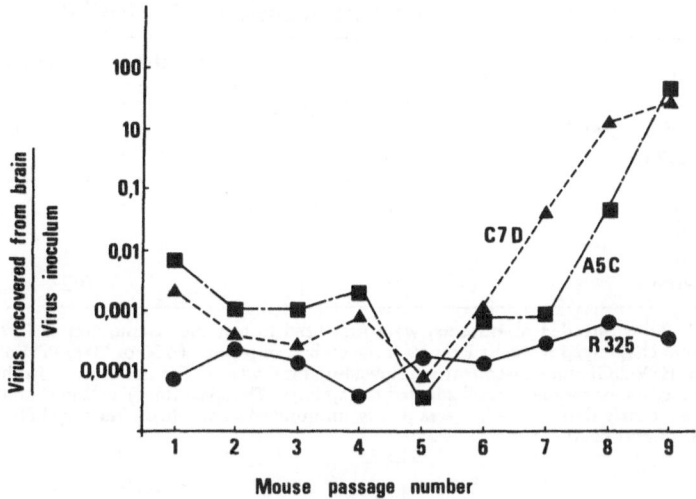

FIGURE 4. Genetic stability of virus R325 through serial passages in mice. BALB/c mice were injected intracerebrally with 10^5–10^6 PFU of each virus. They were killed 3–4 days later and the homogenate of a pool of three brains was plated on Vero cells in order to produce virus suspensions that served to inoculate the next group of mice. The alternation in mice and cells was necessary owing to the poor replication of the virus in the mouse brain. In one preliminary experiment, direct inoculation from brain to brain could not be pursued beyond the second passage. At each mouse passage, virus titers of both the inoculum and the brain suspension were determined on Vero cells to calculate the ratios plotted on the ordinate. R325 derives from HSV-1(F) by deletion of the α22 gene. A_5C and C_7D are intertypic recombinants. The sensitivity of the procedure is attested by the fact that it detected recovery of both ability to grow in mouse brain and neurovirulence with the two recombinants. Restriction enzyme analysis showed that A_5C was a mutant that has regained virulence, but C_7D was found to be contaminated with an HSV-1 wild-type virus. R325 remained genetically stable over the ninth passage as shown by the regularly low levels of multiplication in brain. No difference could be found between the ninth passage isolate and the initial strain in the restriction endonuclease pattern or in the PFU/LD$_{50}$ ratio.

yielded a virulent mutant between the sixth and the ninth passage, which was not a contaminant as the mutant and A_5C shared similar restriction endonuclease patterns (Fig. 4).

b. Studies in Monkeys

Experiments were done in two species. *Saguinus fuscicollis* (brown-headed marmoset) is a tamarin, which preliminary experiments showed susceptible to small amounts of wild-type HSV-1 (MCH$_{10}$) or HSV-2(G) by intracerebral route but not by eye or vaginal inoculation. Monkeys died 6–8 days after being injected into the brain with 10^2 PFU of either type of virus. The intracerebral route was, therefore, used to test the attenuation of the deleted virus R325. The survival rates were 6/8, 4/4, and 7/8 in animals injected with 10^2, 10^5, and 10^8 PFU, respectively.

TABLE V. Protection Test in *Saguinus fuscicollis*[a]

Immunizing dose (PFU) of HSV-1(R325) TK$^+$	Survivors/total after i.c. challenge			
	HSV-1(MGH$_{10}$) PFU		HSV-1(G) PFU	
	200	2000	200	2000
10^2	1/1	1/1		1/2
10^4		1/1		0/2
10^6	1/1		0/1	2/2
None	0/2		0/2	0/2

[a] *S. fuscicollis* (brown-headed marmosets) were first used to test the attenuation of R325 virus. The survivors were challenged 6 weeks later by intracerebral injection of 200 or 2000 PFU of either HSV-1(MGH$_{10}$) or HSV-2(G) and subsequently followed-up for 6 weeks. The data shown in this table were obtained in a group of monkeys well adapted to captivity. The experiment included another group of six monkeys recently delivered who were poorly immunized as 5/6 died after the challenge with 200 or 2000 PFU of HSV-2(G).

Almost all *Saguinus* tamarins were heavily parasitized with *Filaria* and the relative roles of parasitism, stress of intracerebral inoculation, and virus infection in the cause of death are difficult to determine. It is significant, however, that deaths did not follow a dose–effect relationship.

Challenge of the survivors from attenuation studies with either HSV-1 or HSV-2 virus by the intracerebral route showed better protection against HSV-1(MGH$_{10}$) than against HSV-2(G) (Table V).

Monkeys that survived challenge with virulent virus were tested for the presence of latent virus in the trigeminal ganglia 6 weeks later. Interestingly, no virus could be isolated from type 1-challenged tamarins whereas all three survivors of type 2 challenges carried latent virus that was identified as type 2 by restriction enzyme analysis.

The second species of monkeys, *Aotus trivirgatus* or owl monkey, was chosen because of its known susceptibility to HSV infection by peripheral routes (Katzin *et al.*, 1967; Melendez *et al.*, 1969). In fact, in preliminary experiments, an inoculum as small as 10^2 PFU of either HSV-1(MGH$_{10}$) or HSV-2(G) killed the monkeys. Vaginal inoculation of monkeys with 10^8 PFU of R325 resulted in two monkeys shedding virus for a few days, but both survived infection with no symptoms or signs other than a discrete and transient vulvar inflammation. Following infection with R325, both monkeys developed antibodies against a range of viral polypeptides, as detected by immune precipitation (B. Norrild, personal communication).

Confirmation of the take of the infection by R325 *Aotus* monkeys, despite the paucity of clinical signs, came from the isolation of latent R325 virus from lumbosacral ganglia 2 months after infection.

In summary, genetically engineered HSV strains exhibited some of the features required for vaccine strains. They are attenuated, antigenic, and immunogenic. Depending on the animal species tested, they were also able to establish latent infections. The ability to establish latent infection is diversely affected in mice.

VI. CONCLUSION

The field of antiherpetic vaccines is equally exciting and despairing. One is tempted to say that with the contribution of the new disciplines of virology, such as molecular biology, immunochemistry, and genetic engineering, unprecedented possibilities are open for the development of vaccines. But on the other hand, one is tempted to be extremely circumspect for too many enthusiastic predictions, based on progress in virology, have later turned into frustrations.

Several important points that deal with the general approach to evaluation of vaccines against herpetic disease have emerged from past studies. The lack of evidence of efficacy of the vaccines in patients with recurrent herpesvirus has several implications. One implication is that such patients can no longer be considered as a suitable population for testing new vaccines, except in the very first stage, to assess their safety and antigenicity. Efficacy testing should be performed in individuals without past history of herpetic infection and because they are at greater risk of infection, consorts of patients with genital herpes represent the most appropriate group for field trials. It is unlikely that vaccination will be beneficial for already infected individuals. Therefore, vaccination should be scheduled in infancy or adolescence prior to the time of maximum risks of infection and immunity induced so that it covers several years. As it is difficult to implement vaccination programs that include booster injections (e.g., tetanus), the number of repeated administrations will be an important determinant of success.

Because of the diversity of clinical forms of herpetic diseases, the unpredictability of the patterns of recurrences, combined with the impact of the placebo effect, the evaluation of the vaccines will require studies with a careful definition of objectives (prevention of primary infection and/or prevention of subsequent recurrences, etc.). Careful experimental design will be necessary, in particular, including the presence of placebo subjects in appropriate numbers and the double-blind collection of data. It is also essential that biological measurements (viral expression, antibody-and cell-mediated responses, etc.) complement the clinical observations.

Animal models may not have the predictive value of the efficacy in man with which they have been credited. The diversity of susceptibility of the different species as well as the effect of doses and routes of either immunization or challenge makes it necessary to define the systems used for testing new vaccines and some standardization would be useful to allow comparisons. It is noteworthy that all the vaccines have been found to be effective in animals, no matter what the results were in man.

One consequence might be that animals should primarily serve to assess the safety of vaccines, preparing the way for efficacy assessment in humans.

Within the past few years the conceptual design of vaccine has moved from global to analytical approaches.

In the past, either live virulent or killed vaccines were used in attempts to evoke immune responses mimicking the ones observed in spontaneous infections. The present orientation of research on subunit vaccines is based on the idea that a limited number of viral antigenic structures—among which glycoproteins come in first place—are responsible for immunity. Those structures could therefore be prepared under purified form either by extraction from HSV-infected cells or by cloning and expression in any of the systems used by genetic engineering—eukaryotic cells, yeasts, or *E. coli*. A better understanding of the immunobiology of HSV is required to establish the validity of this approach; particularly, to determine the respective roles played by each glycoprotein in the induction of the immunity and whether one or a few of them would induce strong and durable immunity. The degree of protection against both types will be of particular importance to cope with a recent tendency of finding increased proportions of type 1 isolates in genital infections and type 2 in infections other than genital.

The same remarks apply to vaccinia recombinants as the appropriate antigenic determinants will have to be selected in order to be introduced into the construct. Vaccinia virus as a carrier offers space for many additional genes and if it were necessary it could serve as a vector to express several antigenic determinants. The usefulness of this approach will greatly depend on the intensity of the replication of recombinants in the host.

All arguments in favor of live vaccines apply to live genetically engineered HSV vaccines. One may expect them to induce immune responses at least equal to those evoked by spontaneous infections, or even stronger and broader responses because of the incorporation of both types of immunogenic determinants and by the enhancement of their expression. Preliminary experiments in animals have indicated the feasibility of this approach but its achievement depends on the knowledge still to be developed of the genes involved in virulence, latency, and ability to transform, to enable a rational design of the construction of the vaccine virus. Of particular importance will be the answers to the safety concerns, and specifically answers regarding the oncogenic potential of a vaccine that would contain HSV DNA.

REFERENCES

Allen, W. P., and Rapp, F., 1982, Concept review of genital herpes vaccines, *J. Infect. Dis.* **145**:413–421.

Anderson, S. G., Hamilton, J., and Williams, S., 1950, An attempt to vaccinate against herpes simplex, *Aust. J. Exp. Biol. Med. Sci.* **28**:579–584.

Anderson, W. A., and Kilbourne, E. D., 1961, Immunization of mice with inactivated herpes simplex virus, *Proc. Soc. Exp. Biol. Med.* **107**:518–520.

Andonov, P., Dundarov, S., and Bakalov, B., 1979, Preparation and prophylactic therapeutical effect of Bulgarian inactivated polyvalent herpes vaccines, *Acta Med. Bulg. Eng.* **6**:55–64.

Ashley, R. L., and Corey, L., 1983, Development of antibodies to a herpes simplex virus type 2 (HSV-2) subunit vaccine in seronegative and seropositive volunteers, 8th International Herpes Virus Workshop, Oxford, England.

Barinsky, I. F., Shubladze, A. K., Kasparov, A. A., Zaitseva, N. S., and Zdanov, V. M., 1983, Inactivated herpes virus vaccine and cell immunity reactions in the study of its efficacy, 8th International Herpes Virus Workshop, Oxford, England.

Baron, A., Rozan, Fouanon, and Guillot, 1966, Herpes cornéen, traitement par vaccin spécifique, *Bull. Soc. Ophthalmol. Fr.* **66**:137–142.

Beale, J., 1983, Hybrid vaccinia virus for mass hepatitis immunization?, *Nature (London)* **302**:476–477.

Benson, H., and Epstein, M. D., 1975, The placebo effect: A neglected asset in the care of patients, *J. Am. Med. Assoc.* **232**:1227.

Berman, P. W., Dowbenco, D., Lasky, L. A., and Simonsen, C. C., 1983, Detection of antibodies to herpes simplex virus with a continuous cell line expressing cloned glycoprotein D, *Science* **222**:524–527.

Bertland, A. U., and Lampson, G. P., 1982, Herpes simplex type 1 subunit vaccine and process for its preparation, European Patent Application No. 81401407.2.

Biberstein, H., and Jessner, M., 1958, Experiences with herpin in recurrent herpes simplex, together with a review and analysis of the literature on the use of CNS as a virus antigen carrier, *Dermatologica* **117**:267–287.

Bierman, S. M., 1976, BCG immunoprophylaxis of recurrent herpes progenitalis, *Arch. Dermatol.* **112**:1410–1415.

Blank, H., and Haines, H. G., 1973, Experimental human reinfection with herpes simplex virus, *J. Invest. Dermatol.* **61**:223–225.

Brain, R. T., 1936, Biological therapy in virus diseases, *Br. J. Dermatol. Syph.* **48**:21–26.

Brückner, R., 1970, Erste persönliche Erfahrungen mit dem Vaccin antiherpétique "Diamant," *Ophthalmologica* **161**:104–107.

Bubola, D., and Mancosu, A., 1967, Indagine clinica ed immunologica di 30 malati di erpete recidivante dopo vaccinazione specifica, *G. Ital. Dermatol.* **108**:107.

Cappel, R., 1976, Comparison of the humoral and cellular immune responses after immunization with live UV inactivated herpes simplex virus and a subunit vaccine and efficacy of these immunizations, *Arch. Virol.* **32**:29–35.

Cappel, R., de Cuyper, F., and de Braekeler, J., 1979, Antibody and cell mediated immunity to a DNA free herpes simplex subunit vaccine, *Dev. Biol. Stand.* **43**:381–385.

Cappel, R., de Cuyper, F., and Rickaert, F., 1980, Efficacy of a nucleic acid free herpetic subunit vaccine, *Arch. Virol.* **65**:15–23.

Cappel, R., Sprecher, S., and de Cuyper, F., 1982a, Immune responses to DNA free herpes simplex proteins in man, *Dev. Biol. Stand.* **52**:345–350.

Cappel, R., Sprecher, S., Rickaert, F., and de Cuyper, F., 1982b, Immune response to a DNA free herpes simplex vaccine in man, *Arch. Virol.* **73**:61–67.

Carter, C. A., Hartley, C. E., Skinner, G. R. B., Turner, S. P., and Easty, D. L., 1981, Experimental ulcerative herpetic keratitis. IV. Preliminary observations on the efficacy of a herpes simplex subunit vaccine, *Br. J. Ophthalmol.* **65**:679–682.

Chan, W. L., 1983, Protective immunization of mice with specific HSV-1 glycoproteins, *Immunology* **49**:343–352.

Chanock, R. M., 1983, Vaccines—New developments, *Arch. Virol.* **76**:163–177.

Chapin, H. B., Wong, S. C., and Reapsome, J., 1962, The value of tissue culture vaccine in the prophylaxis of recurrent attacks of herpetic keratitis, *Am. J. Ophthalmol.* **54**:255–265.

Cohen, G. H., Dietzschold, B., Ponce de Leon, M., Long, D., Golub, E., Varichio, A., and Eisenberg, R. J., 1983, Construction of an antigenic determinant of herpes simplex virus glycoprotein D which stimulates production of neutralizing antibody, 8th International Herpes Virus Workshop, Oxford, England.

Corey, L., Adams, H. G., Brown, Z. A., and Holmes, K. K., 1983, Genital herpes simplex virus infections: Clinical manifestations, course and complications, *Ann. Intern. Med.* **98:**958–972.

Degos, R., and Touraine, R., 1964, Traitement de l'herpes récidivant par un vaccin spécifique, *Bull. Soc. Fr. Dermatol. Syphiligr.* **71:**161–166.

Dietzschold, B., and Cohen, G. H., 1984, Synthesis of an antigenic determinant of the herpes simplex virus (HSV) glycoprotein D which stimulates the production of virus neutralizing antibody and which confers protection against a lethal challenge infection of HSV in: *Modern Approaches to Vaccines—Molecular and Chemical Basis of Virus Virulence and Immunospecificity*, Cold Spring Harbor Laboratory, Cold Spring Harbor, New York.

Dundarov, S., Andonov, P., Bakalov, B., Nechev, K., and Tomov, C., 1982, Immunotherapy with inactivated polyvalent herpes vaccines, *Dev. Biol. Stand.* **52:**351–358.

Dundarov, S., Andonov, P., Bakalov, B., Tomov, S., Nechev, K., Kazandjiev, S., Antonov, L., Dundarova, D., and Todorova, R., 1983, Therapy with herpes simplex vaccines, 8th International Herpes Virus Workshop, Oxford, England.

Eisenberg, R. J., Wilcox, W., Long, D., Ponce de Leon, M., Pereira, L., Long, D., and Cohen, G. H., 1982, Purification of glycoprotein D of herpes simplex virus types 1 and 2 by use of monoclonal antibodies, *J. Virol.* **41:**1099–1104.

Eisenberg, R. J., Wilcox, W., Long, D., Ponce de Leon, M., and Cohen, G. H., 1983, Expression of herpes simplex virus glycoprotein D in *E. coli* transformed with bacteriophage M 13, 8th International Herpes Virus Workshop, Oxford, England.

Fanta, D., Kraft, D., and Söltz-Szöts, J., 1974, Problematik und Therapie des rezivierenden Herpes Simplex, *Z. Hautkr.* **49:**597–606.

Fenyves, A., and Strupp, L., 1982, Heat resistant infectivity of herpes simplex virus revealed by viral transfection, *Intervirology* **17:**228–239.

Frank, D. B., 1938, Formalized herpes virus therapy and the neutralizing substance in herpes simplex, *J. Invest. Dermatol.* **1:**267–282.

Gauri, K. K., Pressler, K., and Chenk, K. D., 1981, Procédé pour la préparation de nouveaux mutants de virus de l'herpes type 1 et type 2, mutants obtenus et leur application à la fabrication de vaccins contre les infections herpétiques, Demande de Brevet d'Invention No. 81 11576, Institut National de la Propriété Industrielle, Paris.

Goldman, L., 1961, Reactions of an auto inoculation for recurrent herpes simplex, *Arch. Dermatol.* **84:**1025–1026.

Gunby, P., 1983, Genital herpes research: Many aim to tame maverick virus, *J. Am. Med. Assoc.* **250:**2417–2427.

Henocq, E., 1967, Vaccin anti-herpétique, *Rev. Med. (Paris)* numéro spécial **8:**695–708.

Henocq, E., 1972a, Hypersensibilités cutanées mixtes avec présence d'anticorps précipitants dans le sérum, *Rev. Fr. Allergol.* **12:**35–41.

Henocq, E., 1972b, Actualité du vaccin anti-herpétique, *Therapeutique* **48:**485–488.

Henocq, E., de Rudder, J., Maurin, J., and Lepine, P., 1964, Essai de thérapeutique de l'herpès récidivant par un vaccin préparé en culture cellulaire et inactivé par les rayons ultraviolets. II. Essais cliniques, *Sem. Hop.* **40:**1474–1480.

Hilfenhaus, J., and Moser, H., 1981, Prospects for a subunit vaccine against herpes simplex virus infections, *Behring Inst. Mitt.* **69:**45–56.

Hilfenhaus, J., Christ, H., Köhler, R., Moser, H., Kirchner, H., and Levy, H. B., 1981, Protectivity of herpes simplex virus antigens: Studies in mice on the adjuvant effect of PICLC and on the dependence of protection on T cell competence, *Med. Microbiol. Immunol.* **169:**225–235.

Hilleman, M. R., Larson, V. M., Lehman, E. D., Salerno, R. A., Conard, R. A., Conard, P., and McLean, A., 1982, Subunit herpes simplex virus vaccine, in: *The Human Herpesviruses: An Interdisciplinary Perspective* (A J. Nahmias, W. R. Dowdle, and R. F. Schinazi, eds.), pp. 503–506, Elsevier/North-Holland, Amsterdam.

Hull, R. N., and Peck, F. B., 1966, Vaccination against herpes virus infections, *Pan Am. Health Org. Sci. PUbl.* **147:**266–275.

Jarisch, R., and Sandor, I., 1977, MIF in der Therapiekontrolle von Herpes simplex recidivans: Behandlung mit Levamisole, BCG, Urushiol und Herpes Antigen Vaccine, *Arch. Dermatol. Res.* **258:**1151–159.

Jawetz, E., Allende, M. E., and Coleman, V. R., 1955, Studies on herpes simplex virus. VI. Observations of patients with recurrent herpetic lesions infected with herpes simplex virus or their antigens, *Am. J. Med. Sci.* **229:**477.

Kaerner, H. C., Schröder, C. H., Ott-Hartmann, A., Kümmel, G., and Kirchner, H., 1983, Genetic variability of herpes simplex virus: Development of a pathogenic variant during passaging of a non pathogenic herpes simplex virus type 1 virus strain in mouse brain, *J. Virol.* **46:**83–93.

Katzin, D. S., Connor, J. D., Wilson, L. A., and Sexton, R. S., 1967, Experimental herpes simplex infection in the owl monkey, *Proc. Soc. Exp. Biol. Med.* **125:**391–398.

Kern, A. B., and Schiff, B. L., 1959, Small pox vaccinations in the management of recurrent herpes simplex: A controlled evaluation, *J. Invest. Dermatol.* **33:**99.

Kern, A. B., and Schiff, B. L., 1964, Vaccine therapy in recurrent herpes simplex, *Arch. Dermatol.* **89:**844–845.

Kit, S., Quavi, H., Dubbs, D. R., and Otsuka, H., 1983, Attenuated marmoset herpes virus isolated from recombinants of virulent marmoset herpes virus and hybrid plasmids, *J. Med. Virol.* **12:**25–36.

Kitagawa, K., 1973, Therapy of herpes simplex with heat inactivated herpes virus *hominis* type 1 and type 2, *Z. Hautkr.* **48:**533–535.

Kitces, E. N., Morahan, P. S., Tew, J. G., and Murray, B. K., 1977, Protection from oral herpes simplex virus infection by a nucleic acid-free virus vaccine, *Infect. Immun.* **16:**955–960.

Klein, R. J., Buimovici-Klein, E., Moser, H., Moucha, R., and Hilfenhaus, J., 1981, Efficacy of a virion envelope herpes simplex virus vaccine against experimental skin infections in hairless mice, *Arch. Virol.* **68:**73–80.

Kutinova, L., Vonka, V., and Slichtova, V., 1980, Immunogenicity of subviral herpes simplex virus preparations: Protection of mice against intraperitoneal infection with live virus, *Acta Virol.* (*Engl. Ed.*) **24:**391–398.

Kutinova, L., Slichtova, V., and Vonka, V., 1982, Subviral herpes simplex vaccine, *Dev. Biol. Stand.* **52:**313–319.

Larson, V., and Lehman, E. D., 1979, Antigenic, immunogenic herpes virus subunit and process for preparing it, European Patent Application No. 78400098.6.

Lazar, M. P., 1956, Vaccination for recurrent herpes simplex infection: Initiation of a new disease site following the use of unmodified material containing the live virus, *Arch. Dermatol.* **73:**70–71.

Lepine, P., and de Rudder, J., 1964, Bilan des résultats obtenus dans le traitement de l'herpès au moyen d'un nouveau vaccin inactivé: Les Entretiens de Bichat, *Sem. Ther.* **40:**469–474.

Levaditi, C., 1926, *L'herpès et le zona*, Masson, Paris.

Levaditi, C., and Harvier, P., 1920, Etude expérimentale de l'encéphalite léthargique, *Ann. Inst. Pasteur* (*Paris*) **34:**911–972.

Lipschütz, B., 1921, Untersuchungen über die Etiologie der Krankenheiten des Herpesgruppe, *Arch. Dermatol. Syphilol.* **136:**428–482.

Löwenstein, A., 1919, Aetiologische Untersuchungen über den Fieberhaften Herpes, *Med. Wschr.* **28:**769–770.

London, W. T., Catalano, L. W., Nahmias, A. J., Fucillo, D. A., and Sever, J. L., 1971, Genital herpes virus hominis type 2 infection of monkeys, *Obstet. Gynecol.* **37:**501–509.

Macher, E., 1957, Zur Behandlung des chronisch-rezivierenden Herpes simplex, *Z. Haut Geschlechtskr.* **23:**18–22.

Macotela-Ruiz, E., 1973, Vacuna anti-herpetica y unguento con 5 Yodo-2-desoxyuridine en el herpes simple recidivante cutaneo mucoso, *Prensa Med. Mex.* **38:**362–367.

Melendez, L. V., Espana, C., Hunt, R. D., Daniel, M. D., and Garcia, F. G., 1969, Natural herpes simplex infection in owl monkey (Aotus trivirgatus), *Lab. Anim. Care* **19:**38–45.

Mocarski, E. S., Post, L. E., and Roizman, B., 1980, Molecular engineering of the herpes simplex virus genome: Insertion of a second L-S junction into the genome causes additional genome inversions, *Cell* **22**:243–255.

Moreschi, G. I., and Ennis, F. A., 1982, Prevention and treatment of HSV infections, in: *The Human Herpesviruses: An Interdisciplinary Perspective,* (A. J. Nahmias, W. R. Dowdle, and R. F. Schinazi, eds.), pp. 441–446, Elsevier/North-Holland, Amsterdam.

Morse, L. S., Buchman, R. G., Roizman, B., and Schaffer, P. A., 1977, Anatomy of herpes simplex virus DNA. IX. Apparent exclusion of some parental DNA arrangements in the generation of intertypic (HSV 1 × HSV 2) recombinants, *J. Virol.* **24**:231–248.

Mullin, W. B., 1966, Unpublished data cited in Hull and Peck (1966).

Nagler, F. P. O., 1946, A herpes skin test reagent from amniotic fluid, *Aust. J. Exp. Biol. Med. Sci.* **24**:103–105.

Nasemann, T., 1965, Die Behandlung der Infektionen durch das Herpes simplex virus, *Ther. Ggw.* **104**:294–309.

Nasemann, T., 1970, Neuere Behandlungsmethoden unterschiedlicher Herpes simplex Infektionen, *Arch. Klin. Exp. Dermatol.* **237**:234–237.

Nasemann, T., 1973, Herpes progenitalis durch Herpes simplex virus type 1?, *Hautarzt* **24**:367–368.

Nasemann, T., and Schaeg, G., 1973, Herpes simplex virus, type II: Mikrobiologie und klinische Erfahrungen mit einer abgetöten Vaccine, *Hautarzt* **24**:133–139.

Nicolau, S., and Banciu, A., 1924, Herpès récidivant expérimental chez l'homme, *C.R. Soc. Biol.* **90**:138–140.

Nicolau, S., and Poincloux, P., 1923, Quelques caractères de l'herpès expérimental chez l'homme, *C.R. Soc. Biol.* **89**:779–781.

Offret, G., Payrau, P., de Rudder, J., Pouliquen, Y., Faure, J. P., and Cuq, G., 1965, De l'application du vaccin inactivé dans la kératite herpétique, *Arch. Ophthalmol.* **25**:287–300.

Osborn, J. E., 1979, Viral vaccines under development: A third generation, *Adv. Exp. Med. Biol.* **118**:61–82.

Panscherewski, D., and Rhode, B., 1957, Zur Serologie und Therapie des Herpes simplex recidivans, *Hautarzt* **13**:275–278.

Paoletti, E., and Panicali, D., 1984, Genetically engineered pox viruses as live recombinant vaccines, in: *Modern Approaches to Vaccines—Molecular and Chemical Basis of Virus Virulence and Immunospecificity,* Cold Spring Harbor Laboratory, Cold Spring Harbor, N.Y., pp. 295–299.

Parks, W. P., and Rapp, F., 1975, Prospects for herpes virus vaccination: Safety and efficacy considerations, *Prog. Med. Virol.* **21**:188–206.

Paulian, D., 1932, Le virus herpétique et la sclérose latérale amyotropique, *Bull. Acad. Med. Roum.* **107**:1462–1467.

Petersen, E. E., 1977, Onkogenicität von Herpes simplex Viren and Lupidon, *Z. Hautkr.* **52**:417–426.

Poffenberger, K. L., Tabares, E., and Roizman, B., 1980, Characterization of a viable, non-inverting herpes simplex virus 1 genome derived by insertion and deletion of sequences at the junction of components L and S, *Proc. Natl. Acad. Sci. USA* **80**:2690–2694.

Post, L. E., and Roizman, B., 1981, A generalized technique for deletion of specific genes in large genomes: α gene 22 of herpes simplex virus 1 is not essential for growth, *Cell,* **25**:227–232.

Post, L. E., and Roizman, V., 1981, A generalized technique for deletion of specific genes in large genomes: α gene 22 of herpes simplex virus 1 is not essential for growth in cell culture, *Cell* **25**:227–232.

Post, L. E., Mackem, S., and Roizman, B., 1981, The regulation of alpha genes of herpes simplex virus: Expression of chimeric genes produced by fusion of thymidine kinase with α gene promoters, *Cell* **34**:555–556.

Pouliquen, Y., Jollivet, J., and Argenton, C., 1966, Mode d'utilisation du vaccin anti-herpétique en pratique ophtalmologique, *Arch. Ophthalmol.* **26**:565–568.

Rajcani, J., Kutinova, L., and Vonka, V., 1980, Restriction of latent herpes virus infection in rabbits immunized with subviral herpes simplex virus vaccine, *Acta Virol.* (*Engl. Ed.*) **24**:183–193.

Rawls, W. E., and Campione-Piccardo, J., 1982, Epidemiology of herpes simplex virus type 1 and type 2 infections, in: *The Human Herpesviruses: An Interdisciplinary Perspective* (A. J. Nahmias, W. R. Dowdle, and R. F. Schinazi, eds.), pp. 137–152, Elsevier/North-Holland, Amsterdam.

Remy, W., Antoniadis, G., Bockendahl, H., and Remy, B., 1975, Antikörpertiter-Verläufe bei Patienten mit rezivierenden Herpes simplex virus Infektionen unter Vakzinationen mit Hitzinaktivierten Herpes Viren, *Z. Hautkr.* **51**:103–107.

Roizman, B., Warren, J., Thuning, C. A., Fanshaw, M. S., Norrild, B., and Meignier, B., 1982, Application of molecular genetics to the design of live herpes simplex virus vaccines, *Dev. Biol. Stand.* **53**:287–304.

Santoianni, P., 1966, La nostra esperienza sullo terapia dell-erpete semplice recidivante con un vaccino specifico, *Minerva Dermatol.* **41**:30–35.

Schmersahl, P., and Rüdiger, G., 1975, Behandlungsergebnisse mit dem Herpes simplex Antigen Lupidon, H., bzw. Lupidon G., *Z. Hautkr.* **50**:105–112.

Schneider, J., and Rhode, B., 1972, Zur Antigen Therapie des rezidivierenden Herpes simplex mit dem Herpes simplex Impfstoff Lupidon H und G, *Z. Haut Geschlechtskr.* **47**:973–980.

Schneweis, K. E., Gruber, J., Hilfenhaus, J., Möslein, A., Kayser, M., and Wolff, M. H., 1981, The influence of different modes of immunization on the experimental genital herpes simplex virus infection of mice, *Med. Microbiol. Immunol.* **169**:269–279.

Schubladze, A. K., Zaitseva, N. S., Maevskaya, T. M., Cheglakov, Y. A., Muravieve, T. V., Slepova, O. S., Kasparov, A. A., and Barinsky, I. F., 1978, (On the mechanism of specific vaccine therapy in ocular herpes simplex), *Vopr. Virusol.* **1**:63–68.

Scriba, M., 1981, Vaccination against herpes simplex virus: Animal studies on the efficacy against acute, latent and recurrent infections, in: *Herpetische Augenkrankungen* (R. Sundmacher, ed.), Springer-Verlag (Bergmann), Berlin.

Skinner, G., Williams, D. R., Buchan, A., Whitney, J., Harding, M., and Bodfisch, K., 1978, Preparation and efficacy of an inactivated subunit vaccine (NFU1 BHK) against type 2 herpes simplex virus infection, *Med. Microbiol. Immunol.* **166**:119–132.

Skinner, G., Buchan, A., Hartley, C. E., Turner, S. P., and Williams, D. R., 1980a, The preparation, efficacy and safety of "antigenoid" vaccine NFU1 (S⁻L⁺) MRC towards prevention of herpes simplex virus infections in human subjects, *Med. Microbiol. Immunol.* **169**:39–51.

Skinner, G. R. B., Williams, D. R., Moles, A. W., and Sargent, A., 1980b, Prepubertal vaccination of mice against experimental infection of the genital tract with type 2 herpes simplex virus, *Arch. Virol.* **64**:329–338.

Skinner, G. R. B., Woodman, G. B., Hartley, C. E., Buchan, A., Fuller, A., Durham, J., Synnot, M., Clay, J. C., Melling, J., Wiblin, C., and Wilkins, G., 1982a, Preparation and immunogenicity of vaccine Ac NFU1 (S⁻) MRC towards the prevention of herpes genitalis, *Br. J. Vener. Dis.* **58**:381–386.

Skinner, G. R. B., Woodman, G., Hartley, C., Buchan, A., Fuller, A., Wiblin, C., Wilkins, G., and Melling, J., 1982b, Early experience with "antigenoid" vaccine Ac NFU1 (S⁻) MRC towards prevention or modification of herpes genitalis, *Dev. Biol. Stand.* **52**:333–344.

Skinner, G. R. B., Buchan, A., Williams, D., Marsden, J., Hartley, C., Wilbanks, G., Turyk, M., and Namkoong, E. S., 1982c, Immunogenicity and protective efficacy in a rhesus monkey model of vaccine Ac NFU1 (S⁻) MRC against primary type 2 herpes simplex virus infection, *Br. J. Exp. Pathol.* **63**:378–387.

Slichtova, V., Kutinova, L., and Vonka, V., 1980, Immunogenicity of subviral herpes simplex virus type 1 preparation: Protection of mice against intradermal challenge with type 1 and type 2 viruses, *Arch. Virol.* **66**:207–214.

Slichtova, V., Kutinova, L., and Vonka, V., 1982, Immunogenicity of a subviral herpes simplex type 1 preparation: Reduction of recurrent disease in mice, *Arch. Virol.* **71**:75–78.

Smith, G. L., Mackett, M., and Moss, B., 1983a, Infectious vaccinia virus recombinants that express hepatitis B virus surface antigen, *Nature* (*London*) **302**:490–495.

Smith, G. L., Mackett, M., Murphy, B., and Moss, B., 1984, Vaccinia virus recombinants expressing genes from pathogenic agents have potential as live vaccines, in: *Modern Approaches to Vaccines—Molecular and Chemical Basis of Virus Virulence and Immunospecificity*, Cold Spring Harbor Laboratory, Cold Spring Harbor, N.Y., pp. 313–317.

Söltz-Szöts, J., 1960, Neue Methode einer spezifischen Vaccination bei rezivierendem Herpes simplex, *Hautartz* **11**:465–467.

Söltz-Szöts, J., 1971, Die Behandlung des rezivierenden Herpes simplex, *Z. Haut Geschlechtskr.* **46**:267–272.

Steppert, A., 1956, Zur Behandlung des rezivierenden Herpes simplex, *Wien. Klin. Wochenschr.* **68**:452–453.

Teissier, P., Gastinel, P., and Reilly, J., 1922, L'inoculabilité de l'herpes; présence du virus kératogène dans les lésions, *C.R. Soc. Biol.* **87**:638–649.

Teissier, P., Gastinel, P., et Reilly, J., 1923, L'inoculabilité de l'herpès chez les encéphalitiques, *C.R. Soc. Biol.* **88**:255–256.

Thompson, R. L., and Stevens, J., 1983, Replication at body temperature selects a neurovirulent herpes simplex virus type 2, *Infect. Immun.* **41**:855–857.

Thomson, T. A., Hilfenhaus, J., Moser, H., and Morahan, P. S., 1983, Comparison of effects of adjuvants on efficacy of virion envelope herpes simplex virus vaccine against labial infection of Balb/c mice, *Infect. Immun.* **41**:556–562.

Thornton, B., Griffiths, J. B., and Walkland, A., 1982, Herpes simplex vaccine using cell membrane associated antigen in animal model, *Dev. Biol. Stand.* **50**:201–206.

Tureau, 1924, L'herpès expérimental de l'homme, Thèse No. 364, Legrand, Paris.

Vallée, G., 1980, Traitement de l'herpès récurrent par le vaccin antipoliomyélitique Sabin, Thèse Médecine, Université de Paris VII.

von Rodovsky, J., Dbaly, V., and Benda, R., 1971, Präventivbehandlung des rezivierenden Herpes mit einer aus Kaninchennierenzellen hergestellt Formolvakzine, *Dermatol. Monatsschr.* **157**:701–708.

Watson, R. J., Weis, J. H., Salstrom, J. S., and Endquist, L. W., 1983, Herpes simplex virus type 1 glycoprotein D gene: Nucleotide sequence and expression in *Escherichia coli*, *Science* **218**:381–384.

Weis, J. H., Endquist, L. W., Salstrom, J. S., and Watson, R. J., 1983, An immunologically active chimaeric protein containing herpes simplex virus type 1 glycoprotein D, *Nature* (*London*) **302**:72–74.

Weitgasser, H., 1973, Neue Behandlungsmöglichkeiten des Herpes simplex in der dermatologischen Praxis, *Hautarzt* **24**:298–301.

Weitgasser, H., 1977, Kontrollierte klinische Studie mit den Herpes Antigenen Lupidon H und Lupidon G, *Z. Hautkr.* **52**:625–628.

Wheeler, C. E., 1960, Herpes simplex virus, *Arch. Dermatol.* **82**;141–149.

Wise, T. G., Pavan, P. R., and Ennis, F. A., 1977, Herpes simplex virus vaccines, *J. Infect. Dis.* **136**:706–711.

Woodman, C. B. J., Buchan, A., Fuller, A., Hartley, C., Skinner, G. R. B., Stocker, D., Sugrue, D., Clay, J. C., Wilkins, G., Wiblin, C., and Melling, J., 1983, Efficacy of vaccine Ac NFU (S⁻) MRC5 given after an initial clinical episode in the prevention of herpes genitalis, *Br. J. Vener. Dis.* **59**:311–313.

Yoshino, K., Yanagi, K., and Abe, K., 1982, Efficacy of intradermal administration of herpes simplex virus subunit vaccine, *Microbiol. Immunol.* **26**:753–757.

Zweerink, H. J., Martinez, D., Lynch, R. J., and Stanton, L. W., 1981, Immune responses in mice against herpes simplex virus: Mechanisms of protection against facial and ganglionic infections, *Infect. Immun.* **31**:267–275.

Zygraich, N., and Huygelen, C., 1974, Vaccin contre le virus Herpès simplex type 2, procédé pour sa préparation et ses applications, Brevet d'invention No. 74.10112, Institut National de la Propriété Industrielle, Paris.

CHAPTER 14

CMV Vaccines

STANLEY A. PLOTKIN

I. INTRODUCTION

The prevention of infections by human CMV presents a number of vexing problems common to the herpesviruses. Latency and reactivation, the possible oncogenicity of the virus, and the imperfection of even natural immunity with respect to CMV conspire together to make prospects for vaccination complex and daunting. Nevertheless, a good deal of progress has been made, if we define progress as movement in a forward direction, without drawing any inference as to proximity to the goal.

It is unnecessary to go into detail about the complicated natural history of CMV (reviewed by Meyers, this volume); however, for the purposes of this chapter, we do have to ask: at what steps in viral transmission is prevention possible? Table I lists the principal means by which transmission of CMV occurs, and the means by which prevention might be effected.

Vaccination would appear to be the only feasible strategy for prevention of transplacental infection of the fetus following primary infection of the mother. There is no way to prevent transplacental or perinatal infection of the fetus of immune mothers, but it may be unnecessary to do so, as the epidemiologic evidence thus far suggests that primary infection is much more likely to cause disease (Stagno *et al.*, 1982, 1984).

Similarly, prevention of transmission by breast milk and respiratory secretions may be unnecessary, as little or no disease is produced (Stagno *et al.*, 1980). Sexual transmission may occasionally result in an infectious mononucleosis (IM)-type syndrome (Chretien *et al.*, 1977), and vaccination of seronegative sexual partners might be feasible, but is of doubtful necessity. Most homosexual men are already seropositive (Drew *et al.*,

STANLEY A. PLOTKIN • The Children's Hospital of Philadelphia, The University of Pennsylvania, and the Wistar Institute, Philadelphia, Pennsylvania 19104.

TABLE I. Strategies for Prevention of HCMV Disease

Type of infection	Means of prevention
Transplacental	Vaccination of mother before pregnancy (primary infection only)
Perinatal	?
Breast milk	Prevention may be unnecessary
Respiratory	Prevention may be unnecessary
Sexual	Vaccination of seronegative partners (may be too late in homosexuals)
Blood transfusion	Selection of donors Hyperimmune globulin
Allograft transplantation	Donor selection (seronegative kidney recipients) Hyperimmune globulin (bone marrow) Vaccination (seronegative kidney recipients) Interferon (seropositive kidney recipients)

1981), but it may be mentioned parenthetically that vaccination of seronegatives might reduce the immune suppression caused by natural CMV infection.

Transfusion-transmitted CMV might be prevented by selection of donors (Monif *et al.*, 1976) or by administration of hyperimmune globulin to the recipients (Condie and O'Reilly, 1984). Finally, transmission of CMV by organ grafts is susceptible to prevention by several of the strategies listed in Table I.

The methods of CMV prevention currently available to us are four: donor selection, administration of passive antibody, vaccination, and prophylactic antivirals.

II. DONOR SELECTION

Donor selection is useful in reducing infection due to blood transfusion, particularly in premature newborns; in renal transplant recipients; and perhaps in bone marrow transplant recipients (reviewed by Meyers, this volume). Yeager and her colleagues demonstrated originally that transfusion-acquired CMV is a significant threat to premature infants, who develop pneumonia, hepatitis, thrombocytopenia, and hemolytic anemia in response to infection (Yeager, 1974; Yeager *et al.*, 1981). This syndrome could be prevented by transfusion with blood from donors who were antibody-negative, as shown in Table II.

III. PASSIVE CMV ANTIBODY

At first thought it seems strange that a pathogen generally considered to be one of the most cell-associated could be prevented by passive an-

TABLE II. CMV Infection in Infants According to
Antibody Status of Their Mothers and Blood
Donors[a]

Mother's antibody	Donors' antibody	
	Positive	Negative
Positive	9/60 (15%)	23/131 (18%)
Negative	10/74 (14%)	0/9 (0%)

[a] From Yeager et al. (1981).

tibody, yet that is the case. Evidence for the efficacy of passive CMV antibody has been elicited in transfusion disease of newborn and in post bone marrow transplant pneumonia.

In the studies from the Stanford group, Yeager et al. (1981) showed that premature infants who acquired CMV antibodies from their mothers did not develop illness despite having later received blood from seropositive donors. Thus, their maternal antibody protected them from the CMV syndrome to which seronegative-transfused infants are subject (Table III). The Stanford group (Yeager et al., 1983) also showed recently that maternally derived perinatal infection could result in severe disease in prematures, but that this disease occurred when the passive antibody had declined to negligible levels.

As previously stated, bone marrow transplant patients also appear to acquire significant disease from blood transfusion (Meyers, this volume). Condie and O'Reilly (1984) prepared CMV hyperimmune human IgG for intravenous use, and antibody-deficient human IgG as a control. Three infusions of 200 mg/kg were administered at 25, 50, and 75 days posttransplant. During the first 120 days posttransplant the patients who received CMV hyperimmune globulin were completely protected against CMV infection *and* disease (Table IV), indeed a remarkable result.

The group at UCLA (Winston et al., 1982) had previously used both CMV immune plasma and ordinary intravenous γ-globulin in bone marrow transplant recipients. They found that CMV infection was not inhibited by either form of therapy, but the clinical outcome of infection was better in the treated patients.

TABLE III. Rate of Serious
Illness in CMV-Infected
Prematures According to Prior
Serologic Status[a]

Antibody negative	5/10
Antibody positive	0/32

[a] From Yeager et al. (1981).

TABLE IV. Incidence of CMV Infection and Interstitial
Pneumonia in Prophylactic Globulin Trial[a]

	CMV hyperimmune globulin	CMV-negative globulin	No therapy control
CMV infection	0/17	6/18	10/20
Pneumonia	0	3	6

[a] From Condie and O'Reilly (1984).

Meyers *et al.* (1983) have tested intramuscular CMV immune globulin on a weekly schedule. Patients who received only whole blood had
a significantly lower CMV infection rate if they also received prophylactic
globulin. On the other hand, this effect was abolished in patients who
received granulocyte transfusions.

It is interesting to speculate about the meaning of these positive
results. The simplest explanation is that CMV carried in blood must go
through an extracellular phase before it reaches a susceptible tissue cell.
If there is enough CMV antibody this passage is blocked, but the larger
the virus dose, the less likely the protection. CMV is carried in several
types of white blood cell, the granulocyte and probably both T and B
lymphocytes and macrophages (Winston *et al.*, 1980; Wu and Ho, 1979).
Apparently, genomic material is not transferred by cell-to-cell contact,
but rather complete virions must be synthesized and pass out of the cell
before infection is established.

The implication for active vaccination is important: namely, that
induction of an antibody response may be sufficient to protect against
certain types of primary infection, or at least to cut down the dissemination of CMV within the body.

IV. VACCINATION

All through the history of vaccination there has been the dichotomy
between live and killed vaccines, between the disciples of Jenner and
Pasteur on the one hand and of Theobald Smith and Almroth Wright on
the other. CMV has been no exception. The protagonists of live vaccines
cry efficacy while the protagonists of killed vaccines cry safety. No futile
attempt to settle this issue will be made here but before reviewing data
on vaccination in man, it is profitable to examine the experience with
animal models.

Two animal CMVs have been extensively studied as models for
human infection. Mouse CMV has received the most attention, despite
the fact that it does not cross the placenta. Attenuated viruses developed
for experimental work have in general only been slightly more attenuated

TABLE V. Protection of Mice by TS Variants of
MCMV[a]

Immunizing agent	Morbidity	Mortality	Virus-positive tissues
Wild	0/9	0/9	5/22
ts 21	2/8	0/8	0/22
ts 29	0/9	0/9	1/22
Control	8/8	8/8	20/22

[a] From Minamishima and Tonari (1984).

than wild virus. Recently, however, Minamishima and Tonari (1984) have developed a further attenuated MCMV that protects against wild virus challenge without becoming latent in the mouse (Table V).

Guinea pig CMV is even more interesting. Bia, Hsiung, and their colleagues have demonstrated the GP CMV simulates human infection in many ways, including transplacental infection (Bia et al., 1983). They have compared attenuated virus and inactivated virus for protective effect, with results summarized in Table VI. Whereas both live and killed GP CMV showed significant protection of mother and fetus, live vaccine was significantly better in reducing maternal viremia.

The start of our work on live attenuated CMV vaccine was the isolation in WI-38 cultures of a virus from the urine of an infant named Towne (Plotkin et al., 1975). The original strain was passaged in WI-38 at various temperatures, but eventually it was decided to use an ordinary 37°C passaged virus that had been cloned three times by plaquing. Pools were made at the 125th to 133rd passages for testing in man (Plotkin et al., 1976).

Normal volunteers have been inoculated by us and by others (Starr et al., 1981; Plotkin, 1979; Friedman et al., 1982) with consistent results. Towne vaccine elicits CMV antibodies and cellular responses that are qualitatively similar to those following natural infection, but without clinical or laboratory evidence of systemic reaction or of virus excretion. There is a local reaction to subcutaneous inoculation.

TABLE VI. Effect of Vaccination on Guinea Pig CMV Infection in Pregnant
Animals[a]

Vaccination	Maternal viremia	Fetal death	Fetal virus
None	9/10	13/26	7/26
Passive Ab	5/9		0/25
Envelope Ag	9/10	5/16	0/16
Live attenuated virus	2/9	1/25	1/25

[a] From Bia et al. (1983).

TABLE VII. Comparison of Infection Due to Natural Virus and Towne Live Vaccine

	Natural	Live vacine
Febrile illness	+	0
Lymphocytosis	+	0
Increase in transaminases	+	0
Local reaction	0	+
Reversed T-cell helper/suppressor ratio	+	0
Virus excretion	+	0
CF antibody	+	+(1 year)
ACIF antibody	+	+
Neut. antibody	+	+
IgM antibody	+	+
Early antigen antibody	+	+
Lymphocyte proliferation (CMV Ag)	± → +	+
T-cell cytotoxicity	+	+ (shorter duration)

Table VII compares the results of vaccination with events following natural infection in normal individuals. Vaccine-induced infection is extremely mild, but nevertheless many immune responses common to natural infection are stimulated.

A question that frequently arises with respect to CMV is the breadth of immunity. We have not seen strain-specificity with respect to lymphocyte proliferation or to fluorescent antibodies when various different strains have been used to prepare antigens. However, in testing for neutralizing antibodies a higher response has been noted to the Towne virus than to heterologous strains.

Thus, the studies in normal volunteers give evidence for short-term innocuity and immunogenicity of Towne. How do we prove that a live vaccine virus does not become latent in the vaccinee? And how do we provide evidence for efficacy in preventing what is largely a subclinical infection? The renal transplant candidate (RTC) seemed to provide a population in which to answer these questions for the following reasons:

1. RTC who are seronegative and who receive a kidney from a seropositive donor almost always become infected.
2. RTC who sustain primary infection frequently manifest significant serious CMV disease.
3. Seropositive RTC who have had previous natural infection will often reactivate the latent virus after transplant immunosuppression.

Accordingly, we believed that an attempt to convert seronegative RTC to seropositive by vaccination was justified. Unfortunately, one problem we might have anticipated was that the RTC are immunodebilitated even before transplant. As shown in Table VIII, antibody re-

TABLE VIII. Serologic Responses to Towne at 8 Weeks Postvaccination[a]

Subjects		Antibodies		
		CF	ACIF	Neut.
Normals	N	41	41	10
	% positive	95	100	100
	GMT	25	54	28
RTC	N	37	37	15
	% positive	76	83	87
	GMT	9	19	12

[a] CF, complement fixation; ACIF, anticomplement immunofluorescence; Neut., neutralizing antibodies; GMT, geometric mean titer; RTC, renal transplant candidate.

sponses in RTC were considerably lower than in normals. Table VIII also shows the rather uniform responses of normal volunteers.

Similarly, CMV-specific lymphocyte proliferation responses in RTC are considerably less striking in comparison to normals (Fig. 1).

Although the immune stimulus generated by vaccine is blunted in RTC, we did go on to evaluate the effect of such immune responses on

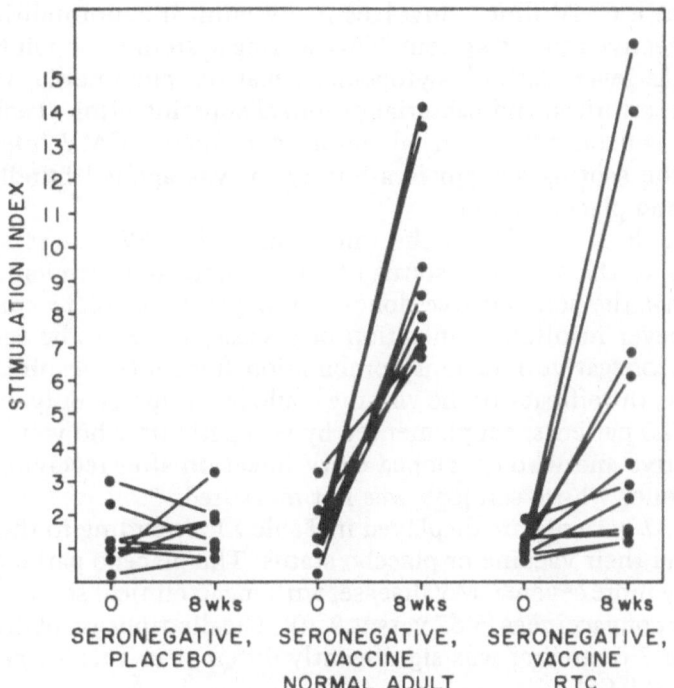

FIGURE 1. Stimulation indices against CMV antigen at the time of vaccination and 8 weeks postvaccination.

TABLE IX. Scoring System for CMV Disease[a]

Manifestation	Points	Manifestation	Points
Fever	1–3	Glomerulonephritis	1–3
Leukopenia	1	Arthritis	2
Thrombocytopenia	1	Superinfection	3
Hepatitis	1–3	GI bleed	3
Pneumonia	1–3	Death	4
Encephalitis	1–3		

[a] Defined as febrile illness occurring contemporaneous with laboratory evidence of CMV infection.

subsequent challenge by CMV (Glazer *et al.*, 1979; Plotkin *et al.*, 1984). The trial was conducted by a double-blind randomized protocol in which patients received either vaccine or placebo. Transplantation was then done as soon as a kidney became available, providing 8 weeks had passed since vaccination, and the patients were then observed for CMV infection and disease. Infection was defined as virus excretion of fourfold antibody rise, while disease consisted of a febrile illness occurring in time relation to laboratory evidence of infection. The CMV syndrome in renal transplant patients is well described, but it is often difficult to distinguish it from other syndromes unless virologic data are considered.

Because CMV illness involves many clinical abnormalities, we set up an objective scoring system. This scoring system gave points for such changes as fever, thrombocytopenia, hepatitis, pneumonia, glomerulonephritis, arthritis, and bacterial or fungal superinfection (Table IX) that occurred concomitant with laboratory evidence of CMV infection. Although the scoring system is arbitrary, it was applied blindly to both vaccine and placebo groups.

In Table X are shown the infections and CMV illnesses analyzed according to the serologic status of both donors and recipients. It is apparent that the seronegative donor–seronegative recipient combination almost never resulted in infection or disease, whereas the seropositive donor–seronegative recipient combination frequently resulted in both. Therefore, the efficacy of the vaccine could be evaluated only in the latter group of 30 patients, supplemented by two patients who were originally seronegative and who developed CMV infection after receiving a kidney from a donor whose serology was not measured.

The 32 patients are displayed in Table XI, according to their clinical scores and their vaccine or placebo status. The placebo patients had significantly more severe CMV disease, with mean clinical scores more than twice that of vaccinees (5.67 versus 2.70). The distribution of disease with a score of 7 or greater was significantly different in vaccine and placebo group ($p < 0.05$).

A study with Towne strain vaccine similar to ours has been under way at the University of Minnesota, under the leadership of Dr. Henry

TABLE X. CMV Illness and Infection in Renal Transplant Patients

Recipients	Vaccinees[a]								Placebo[a]							
	Donor +		Donor −						Donor +		Donor −					
	Sick	Inf[b]		Sick	Inf				Sick	Inf			Sick	Inf		
Seronegative[c]	9/16	15/16		0/20	0/20				10/14	11/14			1/12	1/12		
Seropositive[c]	2/8	6/7		1/9	3/9				2/7	6/7			0/5	2/5		

[a] (+), Seropositive; (−), seronegative.
[b] Inf, infected.
[c] Original serologic status before vaccination.

TABLE XI. Clinical Scores in R-D⁺ Group by Vaccine Status

Group	Score															
	0	1	2	3	4	5	6	7	8	9	10	11	12	13	14	15
Vaccinees ($N = 17$)	7	1	1	2	2	1	2						1			
Placebo recipients ($N = 15$)	4				3	1		2		3			1			1

TABLE XII. Nonliving CMV Vaccine Preparations

Inactivated virions
Detergent-separated envelope
Dense bodies
Noninfectious enveloped particles
Glycoproteins purified by affinity chromatography
Cloned genes with expression of immunogenic proteins
Synthetic peptides corresponding to neutralizing epitopes

Balfour. The Minnesota results have been entirely comparable to our own (Balfour *et al.*, 1984). Combining the results of the two studies the risk of severe disease was 46% in placebo recipients and 43% in vaccinees, a protective effect of 93%.

Thus, the vaccine seems to have been moderately effective in preventing severe disease, though ineffective in preventing infection of renal transplant patients.

What about long-term safety? This may be divided into two questions: Is latent infection established with the vaccine virus, and are there long-term ill effects of the virus? The second question may be answered simply, at least for the time being. To date no vaccinee has developed a tumor or other manifestation that could be attributed to latent or recrudescent CMV.

We have two pieces of information that bear on the question of latency. First, there are the observations on seronegative recipients who were vaccinated but who received a kidney from a seronegative donor. Their only CMV experience was with vaccine virus. Twenty of those have been observed with no evidence of reactivation despite immunosuppression for transplant.

Second, there were 22 seronegative vaccinees who received a kidney from a seropositive donor and who excreted virus. We tested 44 isolates from these patients for their similarity to vaccine virus by restriction endonuclease assays (Huang *et al.*, 1976). All 44 were different from vaccine virus. Therefore, the evidence for latency of vaccine virus is negative.

Recently, we have been attempting to study vaccine efficacy by artificial challenge. A new isolate from an infant was prepared as a pool at the fifth tissue culture passage. This virus, called Toledo-1, is being used as a challenge to both natural and vaccine-induced immunity, to determine if they are the same or different.

V. SUBUNIT VACCINES

It is obvious that a subunit vaccine, if efficacious, would bypass concerns about latency and oncogenicity of a living preparation. At least seven types of nonliving preparations are possible, listed in Table XII.

TABLE XIII. Responses of Guinea Pigs to HCMV Antigens

Test[a]	Immunizing antigen		
	Virion	Nucleocapsid	Envelope
IFA			
LA-NUCL	50	400	<10
CYTO	1000	1600	1600
EA-NUCL	<10	200	<10
CYTO	400	200	400
NI-NUCL	<10	<10	<10
CYTO	400	200	200
CF			
CMV	128	128	16
NI	64	<8	8
NEUT	320	40	128
LPR	+	+	+

[a] IFA, indirect fluorescence; LA, CMV-infected cells expressing late antigens; NUCL, nuclear staining; CYTO, cytoplasmic staining; EA, CMV-infected cells expressing early antigens; NI, noninfected cells; CF, complement fixation; CMV, CF antigen made from infected cells; NEUT, neutralizing antibodies; LPR, lymphocyte proliferation response to CMV antigen.

The first two possibilities are not really feasible, for killed virions still contain DNA and viral envelope is too expensive to prepare directly from virions for general use. However, it is useful to show that virions and envelope are immunogenic, as lesser units of the virus are unlikely to work if these two preparations do not.

We have made both formalin-inactivated virus and detergent-freed envelope vaccines for tests in guinea pigs, shown in Table XIII (Furukawa et al., 1984). It was possible to induce immune responses with both types of preparation. The results with the viral envelope and with the whole virion vaccines show the induction of anticellular antibody, suggesting that cellular proteins are part of the viral envelope. Nevertheless, it was certainly of interest that antibody that neutralizes HCMV could be induced in guinea pigs, at least with the use of Freund's adjuvant. Lymphocyte sensitization to HCMV antigens was also evoked.

The next two possibilities are naturally synthesized particles that are largely free of viral DNA: dense bodies and noninfectious enveloped particles (NIEP). Dense bodies are spherical bodies larger in diameter than virions and surrounded by an outer envelope containing the viral glycoproteins (Craighead et al., 1972). Dense bodies are produced in quantity in infected cells, and can be separated from virions by centrifugation. They contain relatively large amounts of CMV matrix protein (69K) (Sarov and Abady, 1975; Stinski, 1976; Kim et al., 1976).

NIEP are similar in diameter to infectious virions and are synthesized in similar amounts by strain AD-169 (Kanich and Craighead, 1972; Irmiere and Gibson, 1983). They lack DNA cores and indeed have less matrix protein, but they have all the envelope glycoproteins. Gibson and

TABLE XIV. Major Proteins of HCMV

M_r ($\times$ 10^{-3})	Location	Functional data
153–155	Capsid	Abundant
150	Tegument	Phosphoprotein
130–145	Envelope	Glycoprotein with neut. specificity
72		Immediate—early, size variable
68–69	Matrix	Abundant
64–66	Envelope	Glycoprotein with neut. specificity
55–57	Envelope	Glycoprotein
?	Capsid?	DNA polymerase–early
?	Tegument	Protein kinase

Irmiere (1984) report that inhibitors of viral DNA synthesis do not prevent synthesis of NIEP. Both dense bodies and NIEP contain viral protein kinase activity.

At least in principle it would be possible to concentrate and treat preparations of dense bodies or NIEP to inactivate all DNA and to use these preparations as immunogens.

More esthetically attractive approaches are based on single polypeptides: either by affinity chromatography purification of immunogenic glycoproteins; by cloning of the genes for these proteins and obtaining expression; or by chemical synthesis of peptides carrying neutralizing specificities.

For any of these approaches to succeed, one must know the antigenic structure of the virion. Our current information is summarized in Table XIV, which represents an interpretation of the notable work performed by Pereira, Stinski, and Gibson, among others (Stinski, 1977; Gibson, 1981; Pereira et al., 1982). From the point of view of vaccination, the two glycoproteins with neutralizing specificities are the most significant proteins. However, it is possible that immunization against other important functional proteins would be necessary.

Zaia et al. (1984a) have started to isolate and characterize the 64–66K glycoprotein. They find that it can induce neutralizing antibodies in guinea pigs if injected with complete Freund's adjuvant. More recently they have obtained some data on the nucleotide sequence of the gene coding for that protein (Pande et al., 1984b).

One must of course point out that any peptide to be used as a vaccine will have to be capable of inducing cellular as well as humoral antibody.

VI. INTERFERON

The prophylactic use of IFN in renal transplant patients was recently studied by Hirsch et al. (1983). Based on earlier work (Cheeseman et al., 1979), they realized that IFN might have an effect on reactivation of CMV occurring in patients who were seropositive before transplant. They administered 3×10^6 units of IFN-α, thrice weekly for 6 weeks, and then twice weekly for 8 weeks. Although the CMV infection rates were not significantly different in IFN-treated or placebo recipients, disease was less in the former: 7 of 22 placebo recipients developed CMV illness, while only 1 of 20 IFN recipients did so.

VII. SUMMARY

Prevention of HCMV disease is possible through donor selection in certain blood transfusion and organ transplant situations. When infection is by blood products, γ-globulin containing high-titer CMV antibodies can exert significant protection.

Active immunization is also practicable. The Towne live vaccine produces an essentially asymptomatic infection with induction of humoral and cellular immune responses. Efficacy against severe CMV disease has been demonstrated under adverse immunologic conditions in renal transplant patients.

Subunit vaccines containing noninfectious particles appear to be theoretically possible, and two major glycoproteins have been identified with properties that suggest their potential utility for immunization.

ACKNOWLEDGMENTS. The author's research was supported by the National Institutes of Health (Grant AI-14927) and The Hassel Foundation. The author wishes to acknowledge the contributions of Clyde F. Barker, Donald C. Dafoe, Gary R. Fleisher, Allan D. Friedman, Harvey M. Friedman, Robert A. Grossman, M. Lynn Smiley, Stuart E. Starr, and Cliff Wlodaver to the work described herein.

REFERENCES

Balfour, H. H., Jr., Sachs, G. W., Welo, P., Gehrz, R. C., Simmons, R., and Najarian, J. S., 1984, Cytomegalovirus vaccine in renal transplant candidates: Progress report of a randomized, placebo-controlled, double-blind trial, in: *CMV: Pathogenesis and Prevention of Human Infection* (S. A. Plotkin, S. Michelson, J. Pagano, and F. Rapp, eds.), Liss, New York.

Bia, F. J., Griffith, B. P., Fong, C. K. Y., and Hsiung, G. D., 1983, Cytomegaloviral infections in the guinea pig: Experimental models for human disease, *Rev. Infect. Dis.* 5:177.

Cheeseman, S. H., Rubin, R. H., Stewart, J. A., Tolkoff-Rubin, N. E., Cosimi, A. B., Cantell, K., Gilbert, J., Winkle, S., Herrin, J. T., Black, P. H., Russell, P. S., and Hirsch, M. S., 1979, Controlled clinical trial of prophylactic human-leukocyte interferon in renal transplantation: Effects on cytomegalovirus and herpes simplex virus infections, *N. Engl. J. Med.* 300:1345.

Chretien, J. H., McGinniss, C. G., and Muller, A., 1977, Venereal causes of cytomegalovirus replication, *J. Am. Med. Assoc.* 238:1644.

Condie, R. M., and O'Reilly, R. J., 1984, Prevention of CMV infection in bone marrow transplant recipients by prophylaxis with an intravenous, hyperimmune CMV globulin, in: *CMV: Pathogenesis and Prevention of Human Infection* (S. A. Plotkin, S. Michelson, J. Pagano, and F. Rapp, eds.), Liss, New York.

Craighead, J. E., Kanich, R. E., and Almeida, J. D., 1972, Nonviral microbodies with viral antigenicity produced in cytomegalovirus-infected cells, *J. Virol.* **10:**766.

Drew, W. L., Mintz, L., Miner, R. C., Sands, M., and Ketterer, B., 1981, Prevalence of cytomegalovirus infection in homosexual men. *J. Infect. Dis.* **143:**188.

Friedman, A. D., Furukawa, T., and Plotkin, S. A., 1982, Detection of antibody to cytomegalovirus early antigen in vaccinated, normal volunteers and renal transplant candidates, *J. Infect. Dis.* **146:**255.

Furukawa, T., Gonczol, E., Starr, S., Tolpin, M. D., Arbeter, A., and Plotkin, S. A., 1984, HCMV envelope antigens induce both humoral and cellular immunity in guinea pigs, *Proc. Soc. Exp. Biol. Med.* **175:**243–250.

Gibson, W., 1981, Structural and nonstructural proteins of strain Colburn cytomegalovirus, *Virology* **15:**516.

Gibson, W., and Irmiere, A., 1984, Selection of particles and proteins for use as human cytomegalovirus subunit vaccines, in: *CMV: Pathogenesis and Prevention of Human Infection* (S. A. Plotkin, S. Michelson, J. Pagano, and F. Rapp, eds.), Liss, New York.

Glazer, J. P., Friedman, H. M., Grossman, R. A. Starr, S. E., Barker, C. F., Perloff, L. J., Huang, E.-S., and Plotkin, S. A., 1979, Live cytomegalovirus vaccination of renal transplant candidates: A preliminary trial, *Ann. Intern. Med.* **91:**676.

Hirsch, M. S., Schooley, R. T., Cosimi, A. B., Russell, P. S., Delmonico, F. L., Tolkoff-Rubin, N. E., Herrin, J. T., Cantell, K., Farrell, M.-L., Rota, T. R., and Rubin, R. H., 1983, Effects of interferon-alpha on cytomegalovirus reactivation syndromes in renal-transplant recipients, *N. Engl. J. Med.* **308:**1489.

Huang, E.-S., Kilpatrick, B. A., Huang, Y. T., and Pagano, J. S., 1976, Antisera to human cytomegaloviruses prepared in the guinea pig: Specific immunofluorescence and complement fixation test, *J. Immunol.* **112:**528.

Irmiere, A. F., and Gibson, W., 1983, Isolation and characterization of a noninfectious virion-like particle released from cells infected with human strains of cytomegalovirus, *Virology* **130:**118.

Kanich, R. E., and Craighead, J. E., 1972, Human cytomegalovirus infection of cultured fibroblasts. II. Viral replicative sequence of a wild and an adapted strain, *Lab. Invest.* **27:**273.

Kim, K. S., Sapienza, R. I., Carp, R. I., and Moon, H. M., 1976, Analysis of structural polypeptides of purified human cytomegalovirus, *J. Virol.* **20:**604.

Meyers, J. D., Leszczynski, J., Zaia, J. A., Flournoy, N., Newton, B., Snydman, D. R., Wright, G. G., Levin, M. J., and Thomas, E. D., 1983, Prevention of cytomegalovirus infection by cytomegalovirus immune globulin after marrow transplantation, *Ann. Intern. Med.* **98:**442.

Minamishima, Y., and Tonari, Y., 1984, A murine model for immunoprophylaxis of cytomegalovirus infection, in: *CMV: Pathogenesis and Prevention of Human Infection* (S. A. Plotkin, S. Michelson, J. Pagano, and F. Rapp, eds.), Liss, New York.

Monif, G. R. G., Daicoff, G. I., and Flory, L. F., 1976, Blood as a potential vehicle for the cytomegaloviruses, *Am. J. Obstet. Gynecol.* **126:**445.

Pande, H., Baak, S., Zaia, J. A., Clark, B. R., Shively, J. E., and Riggs, A. D., 1983, Isolation and characterization of a gene fragment coding for a 64K dalton glycoprotein of human cytomegalovirus (HCMV) using synthetic gene-specific oligonucleotide probe(s), Eighth International Herpesvirus Workshop, Oxford, England, p. 272.

Pereira, L., Hoffman, M., Gallo, D., and Cremer, N., 1982, Monoclonal antibodies to human cytomegalovirus: Three surface membrane proteins with unique immunological and electrophoretic properties specify cross-reactive determinants, *Infect. Immun.* **36:**924.

Plotkin, S. A., 1979, Is it possible to vaccinate against herpes virus infections and their effects?, *Behring Inst. Mitt.* **63:**123.

Plotkin, S. A., Furukawa, T., Zygraich, N., and Huygelen, C., 1975, Candidate cytomegalovirus strain for human vaccination, *Infect. Immun.* **12:**521.

Plotkin, S. A., Farquhar, J., and Hornberger, E., 1976, Clinical trials of immunization with the Towne 125 strain of human cytomegalovirus, *J. Infect. Dis.* **134**:470.

Plotkin, S. A., Smiley, M. L., Friedman, H. M., Starr, S. E., Fleisher, G. R., Wlodaver, C., Dafoe, D. C., Friedman, A. D., Grossman, R. A., and Barker, C. F., 1984, Prevention of cytomegalovirus disease by Towne strain live attenuated vaccine, in: *CMV: Pathogenesis and Prevention of Human Infection* (S. A. Plotkin, S. Michelson, J. Pagano, and F. Rapp, eds.), Liss, New York.

Sarov, I., and Abady, I., 1975, The morphogenesis of human cytomegalovirus: Isolation and polypeptide characterization of cytomegalovirions and dense bodies, *Virology* **19**:243.

Stagno, S., Reynolds, D. W., Pass, R. F., and Alford, C. A., 1980, Breast milk and the risk of cytomegalovirus infection, *N. Engl. J. Med.* **302**:1073.

Stagno, S., Pass, R. F., Dworsky, M. E., Henderson, R. E., Moore, E. G., Walton, P. D., and Alford, C. A., 1982, Congenital cytomegalovirus infection: The relative importance of primary and recurrent maternal infection, *N. Engl. J. Med.* **306**:945.

Stagno, S., Pass, R. F., Dworsky, M. E., Britt, W. J., and Alford, C. A., 1984, Congenital and perinatal CMV infections: Clinical characteristics and pathogenic factors, in: *CMV: Pathogenesis and Prevention of Human Infection* (S. A. Plotkin, S. Michelson, J. Pagano, and F. Rapp, eds.), Liss, New York.

Starr, S. E., Galzer, J. P., Freidman, H. M., and Plotkin, S. A., 1981, Specific cellular and humoral immunity following immunization with live Towne strain cytomegalovirus vaccine, *J. Infect. Dis.* **143**:585.

Stinski, M. F., 1976, Human cytomegalovirus: Glycoproteins associated with virions and dense bodies, *J. Virol.* **19**:594.

Stinski, M., 1977, Synthesis of proteins and glycoproteins in cells infected with human cytomegalovirus, *J. Virol.* **23**:751.

Winston, D. J., Ho, W. G., Howell, C. L., Miller, M. J., Mickey, R., Martin, W. J., Lin, C.-H., and Gale, R. P., 1980, Cytomegalovirus infections associated with leukocyte transfusions, *Ann. Intern. Med.* **93**:671.

Winston, D. J., Pollard, R. B., Ho, W. G., Gallagher, J. G., Rasmussen, L. E., Huang, S. N. Y., Lin, C.-H., Gossett, T. G., Merigan, T. C., and Gale, R. P., 1982, CMV immune plasma in bone marrow transplant recipients, *Ann. Intern. Med.* **97**:11.

Wu, G. C., and Ho, M., 1979, Characteristics of infection of B and T lymphocytes from mice after inoculation with cytomegalovirus, *Infect. Immun.* **24**:856.

Yeager, A. S., 1974, Transfusion-acquired cytomegalovirus infection in newborn infants, *Am. J. Dis. Child.* **128**:478.

Yeager, A. S., Grumet, F. C., Hafleigh, E. B., Arvin, A. M., Bradley, J. S. and Prober, C. G., 1981, Prevention of transfusion-acquired cytomegalovirus infections in newborn infants, *J. Pediatr.* **98**:281.

Yeager, A. S., Palumbo, P. E., Malachowski, N., Ariagno, R. L., and Stevenson, D. K., 1983, Sequelae of maternally derived cytomegalovirus infections in premature infants, *J. Pediatr.* **102**:918.

Zaia, J. A., Clark, B. R., Ting, Y.-P., Vanderwal-Urbina, E., Troiannello, L. J., Balce-Director, L., and Eberle, R., 1984, Production, detection, and characterization of antibody to a 64–66,000 dalton glycoprotein of human cytomegalovirus, in: *CMV: Pathogenesis and Prevention of Human Infection* (S. A. Plotkin, S. Michelson, J. Pagano, and F. Rapp, eds.), Liss, New York.

CHAPTER 15

Live Attenuated Varicella Vaccine

ANNE A. GERSHON

I. HISTORICAL PERSPECTIVE

Natural History of the Clinical Illness

1. Varicella

Most physicians regard varicella (chicken pox), primary infection with varicella–zoster virus (VZV), as merely a mild infection of childhood. It might even be viewed by some as a prerequisite of normal growth and development, a traditional milestone that almost all individuals have endured and that new generations ought to endure as well. Clearly, however, this is an old-fashioned view. Recent advances in medical treatment such as chemotherapy of childhood cancer and organ transplantation require clinicians and virologists to reassess the impact of this disease today.

It is now known that varicella may be severe or even fatal in children with a deficit in cell-mediated immunity (CMI), due to treatment of an underlying disease or on a congenital basis, and in newborn infants and adults. In actual numbers of patients, those with iatrogenic depression of CMI due to treatment of cancer constitute the largest group at highest risk.

The propensity for children with an underlying malignancy to develop severe varicella was first noted by Cheatham *et al.* (1956). The exact mortality rate for children with malignant disease who contract varicella

---

ANNE A. GERSHON • New York University School of Medicine, New York, New York 10016.

remains unknown, and certainly it varies with the type of disease and its form of treatment. It is known that the majority of immunocompromised patients who experience varicella will survive. One retrospective study of 77 patients with varicella, most of whom had underlying leukemia, followed at St. Jude Children's Research Hospital from 1962 through 1973 (Feldman *et al.*, 1975), demonstrated a 32% incidence of severe disseminated varicella with spread to the lungs and/or CNS, and a 7% mortality rate. It is possible that more recently, as chemotherapeutic regimens for cancer have become more immunosuppressive, the mortality rate from varicella in these high-risk children may now be greater. In some series of patients the mortality rate has approached 50% (Gershon, 1980), but these studies have included only small numbers of patients, and the populations were often selected to include only children who were sick enough to require hospitalization for varicella. In a recent study of the efficacy of vidarabine for treatment of varicella in immunosuppressed patients, most of whom had underlying cancer, the mortality rate for varicella in the group of 15 patients who received placebo was 13% (Whitley *et al.*, 1982). However, even if the mortality rate from varicella in children with cancer is only 7%, obviously something must be done about it.

Varicella in normal children is still thought to be a relatively benign illness. Most individuals acquire varicella before 10 years of age, and they experience an illness that lasts for 5–7 days. The incidence of complications in normal children appears to be very low, but a wide range of complications have been reported. These include bacterial superinfection of skin and lungs, encephalitis, arthritis, glomerulonephritis, Reye's syndrome, purpura fulminans, and gangrene (Gershon, 1980).

Following an attack of varicella, both antibody and CMI to VZV develop and often persist for life. Because most persons acquire varicella as children, it might be predicted that the incidence of varicella in adults would be rather low, which it is. Unfortunately, however, those adults who have escaped varicella as children and then later contract it, often from their own children, are at risk to develop severe disease. This risk appears to be lower than it is for children with an underlying malignancy, but its exact magnitude is unknown. Occasional fatalities due to chicken pox in adults have been reported (Gershon, 1980).

2. Zoster

Individuals with a prior history of varicella are potentially at risk to develop zoster. This localized rash illness is a secondary infection with VZV. It is believed to be due to reactivation of latent VZV, acquired during the attack of chicken pox. The reasons why latent VZV reactivates and causes zoster are still unclear, but zoster is more common in immunocompromised persons than in those with normally functioning immune systems. The incidence of zoster also increases with age, reaching a rate of 1% in persons in the 10th decade of life (Hope-Simpson, 1965). This may be due to loss of VZV CMI secondary to aging. The occurrence of zoster clearly does not seem to be related to a loss of specific antibody (Gershon, 1980).

II. DEVELOPMENT OF VARICELLA VACCINE

A. First Studies in Japan

In 1974, Takahashi and his colleagues reported the development of a live attenuated varicella vaccine. This vaccine had been prepared by passage of a VZV isolate, obtained from a 3-year-old normal boy with varicella, 11 times in human embryonic lung cells at 34°C, 12 times in guinea pig embryo cell cultures at 37°C, and 1 to 21 times in WI-38 cells at 37°C. As there is no practical animal model of varicella infection, all studies had to be performed in humans. At first, 71 normal children were inoculated, and when it was clear that there were only mild side effects, 39 chronically ill children were vaccinated on a hospital ward, in an attempt to terminate a nosocomial outbreak of chicken pox. The only side effect noted in these vaccinees was an occasional mild rash.

Takahashi's early studies were hampered by the lack of availability of a sensitive test for susceptibility to varicella. Therefore, some of his vaccinees may have been immune before immunization. In addition, it was impossible to distinguish between rashes secondary to vaccination and mild natural varicella that might have been caused by the wild virus. His overall experience, however, strongly suggested that vaccination against varicella could be accomplished. This publication also served to stimulate strong opinions both for (Sabin, 1977; Plotkin, 1977; Kempe and Gershon, 1977) and against (Brunell, 1975, 1977a,b) further use and testing of the vaccine. Fortunately, this controversy in America did not inhibit the Japanese who continued to pursue their studies vigorously.

Alternatives to varicella vaccine for prevention of severe varicella in high-risk children were not available in Japan in 1974. In the United States, it had been shown that passive immunization of immunocompromised patients could be successfully accomplished by the use of zoster immune globulin (ZIG) (Gershon et al., 1974), but this drug was not licensed even in the United States until 1981. Antiviral therapy for varicella was also very much in the experimental stage in 1974, so that it seems quite justifiable for the Japanese to have approached the problem of prevention of severe varicella by immunization. It appears that the rationale was to develop the vaccine mainly for high-risk varicella susceptibles but that eventually the vaccine might be used in normal children.

B. Subsequent Studies in Japan

Following the initial publication concerning varicella vaccine (Takahashi et al., 1974), a spate of articles from these investigators appeared. The rationale appeared to be, as there was no appropriate animal model, to immunize first those individuals at little or no risk from any potential side effects of the vaccine, and then to proceed gradually to immunize those hosts at greater potential risk, assuming that side effects were minimal at each stage of the investigation, and that immune responses to VZV developed. These subsequent studies were also performed by utilizing neutralization and other sensitive VZV antibody assays such as the fluorescent antibody to membrane antigen assay (FAMA) (Williams et al., 1974) and immune adherence to test for susceptibility to chicken pox prior to vaccination (Gershon, 1980).

In one important study that followed, not only safety of the vaccine but also its protective effects were demonstrated (Asano et al., 1977). Eighteen normal varicella-susceptible children were immunized within 3 days of the occurrence of varicella in a sibling. There was a control group of 19 similar children who were not immunized and who also had a sibling who had contracted varicella. All of the controls became ill, while none of the vaccinees developed varicella. In addition, all of the 18 vaccinees developed detectable antibody to VZV within 4–8 weeks after vaccination.

The vaccine had been given to children on steroids for renal disease without untoward effects. These children were at moderate risk to develop severe varicella (Takahashi et al., 1974). Subsequently, it was reported (Izawa et al., 1977; Asano and Takahashi, 1977) that children with an underlying malignancy could also be safely immunized against varicella. Eleven children with acute leukemia in remission from 1 to 43 months were given 200 to 1500 PFU of varicella vaccine virus. Apparently on an arbitrary basis, their maintenance cancer chemotherapy, consisting of daily 6-mercaptopurine, weekly methotrexate, and intermittent vincristine and steroids, was suspended for 1 week before and 1 week following immunization. All vaccinees developed VZV-neutralizing antibody within 4 weeks after immunization. Side effects of vaccination were minimal. Two vaccinees (18%) developed extremely mild papulovesicular rashes that did not seem to correlate with either a larger dose of immunizing virus or a shorter interval between the time of remission and vaccination. One of these children subsequently had a household exposure to varicella 1½ years later, and he did not contract varicella after the exposure. In this same study, six children with neuroblastoma or retinoblastoma were also immunized with 200 PFU of VZV vaccine, with no significant side effects. All had suspension of cancer chemotherapy for 2 weeks as did the leukemics, and all developed VZV antibody (Izawa et al., 1977). At about the same time, at another institution in

Japan, four additional leukemic children and two additional children with solid tumors were safely and successfully immunized with varicella vaccine (Asano and Takahashi, 1977).

Because it seemed preferable from the point of view of treatment of the underlying malignancy to avoid suspension of chemotherapy if possible, this was then tried (Ha *et al.*, 1980). Of 15 children with acute leukemia in remission for 1 to 53 months who were vaccinated with 500 PFU of Oka varicella vaccine without suspending their chemotherapy, 10 (67%) developed mild papular rashes with 3–32 lesions, occurring 3–4 weeks after vaccination. Eight children (53%) also had fever in conjunction with rash. Of 13 children with solid tumors whose chemotherapy was also not suspended, only 1 developed a rash, but it was extensive, lasted 4 days, and was accompanied by fever to 39°C (Ha *et al.*, 1980). The virus isolated from this patient was recently demonstrated to be vaccine virus on the basis if its DNA and growth in guinea pig tissue culture cells (M. Takahashi personal communication). It was pointed out (Ha *et al.*, 1980) that while vaccination without suspending chemotherapy was in all likelihood not particularly dangerous, it would probably be preferable to suspend chemotherapy in most instances, to decrease the possibility of complications from the vaccine. It was further stated that there was no relationship between relapse of leukemia and suspension of chemotherapy for 2 weeks for vaccination against varicella, although no data were given to support this statement (Ha *et al.*, 1980).

III. STUDIES IN THE UNITED STATES ON VARICELLA VACCINE

A. Studies in Normal Children

When it became clear from studies in Japan that live attenuated varicella vaccine appeared to be safe and effective, investigators in this country began clinical trials. While most studies of varicella vaccine have employed the Oka strain, a few studies have been performed with a vaccine strain developed in the United States, the KMcC strain (Neff *et al.*, 1981; Arbeter *et al.*, 1983). In contrast to the Oka strain, KMcC was passaged exclusively in WI-38 cells. In the first published clinical study (Neff *et al.*, 1981), doses of 190–9500 PFU of passage levels 10, 30, 40, 45, 50, and 60 were administered. In these studies a dose of 1150 PFU of passage 40 virus caused 100% seroconversion, and 6500 PFU of passage 50 virus provided 83% seroconversion, as determined by the immune adherence assay for VZV antibody. It was felt that passage 50 gave the best degree of immunogenicity without significant toxicity (Neff *et al.*, 1981).

In the subsequent study of KMcC vaccine, the responses of 26 normal children to passage 40 vaccine virus and of 17 to passage 50 vaccine virus were examined in greater detail (Arbeter *et al.*, 1983). Passage 40 was used at a dose of 1150 PFU, and passage 50 was used at a dose of 6500 PFU. There was 100% seroconversion to VZV induced by both vaccines, determined by the FAMA method, and every child also manifested evidence of VZV CMI measured by lymphocyte stimulation. Mild papular skin reactions were noted in 31% of vaccinees who received passage 40 but in only 6% of those who received passage 50 virus, although the dose was 10 times higher. The difference in seroconversion rates for the two study groups at passage 50, 83 versus 100%, may probably be explained by the fact that FAMA is slightly more sensitive for measurement of VZV antibody than is immune adherence (Gershon, 1980). It was proposed that passage 50 KMcC strain would be safe to administer to leukemic children (Arbeter *et al.*, 1983), but this has not yet been done.

Arbeter *et al.* (1982) also administered the Oka strain to 91 healthy children in order to gain information on the safety of this vaccine in American children and thereby determine if it would be safe to administer this vaccine to children with leukemia in the United States. These studies served to confirm those performed in normal Japanese children. They also indicated that American children reacted similarly to the Oka strain of vaccine as did Japanese children. There was a 100% conversion rate in vaccinees who received 540 PFU of vaccine prepared in Japan, and a 94% seroconversion rate in vaccinees who received 600 PFU of a commercially prepared Oka strain vaccine made in Europe from a Japanese seed lot; the latter was used at passage 36. Side effects of vaccination were unusual and consisted only of minor papular rashes, and there was no evidence of transmission of virus to others.

B. Studies in Children with an Underlying Malignancy

Brunell *et al.* (1982) immunized 23 children with leukemia with approximately 500 PFU of Oka strain, passage 29 or 33. Twelve of the vaccinees were no longer receiving chemotherapy, and in the other 11 chemotherapy was suspended for 2 weeks. The results of this study essentially confirmed those of Japanese investigators in that none of the immunocompromised vaccinees experienced adverse reactions from the vaccine, and all developed immune responses to VZV. As in data on immunocompromised children vaccinated in Japan, there was little or no information concerning the protective efficacy of the vaccine in these patients. From this study, however, came the first evidence that vaccinees could be contagious to others. One varicella-susceptible sibling of a vaccinee developed VZV antibody several weeks after the vaccinee had manifested a mild rash. Formerly, there had been no evidence of spread of VZV to contacts of vaccinees (Gershon, 1980). In addition, one vaccinee in this study developed a mild case of varicella after a household exposure to the disease; it occurred 2 years after vaccination. Breakthrough cases of varicella had also been reported in some children vaccinated in Japan (Sakurai et al., 1982).

The significance of these breakthrough cases of varicella in vaccinees has become clearer in studies conducted by Gershon and colleagues in the Collaborative Varicella Vaccine Study sponsored by the National Institute of Allergy and Infectious Diseases (Gershon et al., 1984a,b). In these studies, over 200 children with leukemia in remission have been vaccinated. Thus far, over 20 of these vaccinees have subsequently had household expsoures to VZV. One would expect an incidence of varicella in 80–90% of susceptible children after this type of exposure (Ross et al., 1962). In contrast, the attack rate in leukemic vaccinees was significantly lower, about 20%. In addition, all of the cases of varicella were extremely mild, even though the children who developed chicken pox were receiving chemotherapy for their underlying disease.

Mild cases of varicella have occurred, however, not only in some of the vaccinees with household exposures, but also in several children who have not had household exposures. For example, one child with leukemia in remission, off chemotherapy, developed a mild varicelliform rash 10 months after immunization. She had no known exposure to VZV. Two weeks prior to the onset of her illness her FAMA titer was $1:8$ ($> 1:4$ is positive) and her VZV CMI lymphocyte stimulation index was 7 (> 3 is positive). VZV was isolated from her rash, and when it was analyzed by restriction endonucleases (see below) it was found to be wild virus (Gershon et al., 1984a). These observations, and additional similar ones, require a reevaluation of the factors that provide immunity to VZV. Apparently having demonstrable specific antibody and CMI in the peripheral blood does not guarantee that an individual will be completely protected from varicella, particularly if that person is immunocompromised.

On a practical level, however, this experience has indicated that while varicella may occur in immunocompromised vaccinees, they are clearly protected from severe disease. First, the attack rate of vaccinees with household exposures, about 20%, is significantly lower than the attack rate for normal susceptible children with household exposures, 80–90%, reported by Ross et al. (1962), (Gershon et al., 1984b). Second, based on the severity of disease in immunocompromised children reported by Feldman et al. (1975), vaccinated children receiving chemotherapy at the time of exposure to varicella have a signifcantly milder disease, if they do become ill, than unvaccinated ones. Immunized children off chemotherapy at the time of exposure are likely to develop subclinical varicella manifested only by a significant increase in antibody titer (Gershon et al., 1984b).

Immunologically normal children, in contrast to those with leukemia, probably achieve more complete protection from vaccination against varicella. In a study by R. Weibel and colleagues (1984), two groups of normal children were given either varicella vaccine or placebo in a double-blind placebo-controlled randomized study. During the course of a subsequent outbreak of varicella in the community, 39 cases of varicella occurred, but only in those who had received placebo. Not a single case of varicella occurred in children who received vaccine. Thus, 100% protection was seen in normal children, even after household exposure to varicella. Two issues concerning this study must be raised. First, the immunizing dose of virus was higher than that used in leukemic children, 8700 PFU of VZV. Second, the children were vaccinated within a few months to a few weeks before they were exposed to varicella. It will be very important, therefore, to follow the vaccinees who did not yet have household exposures for additional exposures to varicella in ensuing years to obtain data concerning the duration of protection against varicella and whether it remains complete in normal children.

One of the most helpful advances in the analysis of data from varicella vaccine studies has been the development of a marker for the vaccine virus. Based on restriction endonuclease studies, the DNA of VZV vaccine has been found to be clearly different from that of wild virus (Martin et al., 1982). In studies by Gershon et al. (1984a,b), it has been possible to demonstrate that vaccinees with mild breakthrough varicella have been infected with wild virus. Therefore, the possibility that they had experienced disseminated zoster due to reactivation of latent vaccine virus was ruled out. Vaccine-type virus, however, has been isolated from patients with vaccine-associated rashes, on occasion.

IV. PROBLEMS REQUIRING FURTHER RESOLUTION

A. Persistence of Immunity after Vaccination

1. Reinfection with VZV

A recent follow-up study of 26 healthy children immunized in Japan, revealed that 100% had VZV-neutralizing antibody, and 96% had VZV FAMA antibody detectable 5 years after immunization (Asano et al., 1983). None of these vaccinees developed varicella during that interval although many had been exposed to the virus. Because no decrease in antibody titer was observed during this time, and because of the many exposures, the authors postulate that subclinical reinfections may have occurred, causing boosts in VZV antibody titer. Other investigators have also observed the phenomenon of subclinical reinfection in recipients of VZV vaccine (Brunell et al., 1982; Gershon et al., 1984a,b). Subclinical reinfection with VZV has also been observed after natural varicella (Brunell et al., 1975; Gershon et al., 1982; Arvin et al., 1983). In addition, occasional mild clinical reinfections with VZV have been observed in normal persons after natural varicella (Gershon et al., 1984a). Whether these secondary booster responses are important in maintaining long-term immunity to VZV is an extremely important subject for further research and may have significant implications for the future use of varicella vaccine. The relative importance of local immunity in the respiratory tract also deserves further evaluation. In one such study, it was found that while secretory IgA could be detected after natural varicella, it could not be found after varicella vaccination of normal children, even if the vaccine was given by inhalation (Bogger-Goren et al., 1982).

2. Loss of Detectable VZV Antibody in Vaccinees

Loss of detectable VZV FAMA antibody has also been observed in leukemic vaccinees 1–2 years after immunization (Gershon et al., 1984b). The significance of this phenomenon is unknown, as a booster antibody response was usually found after a second dose of vaccine had been given, suggesting that the child was immune to varicella despite the lack of detectable antibody. In addition, as has been mentioned, the presence of detectable antibody in leukemics does not necessarily indicate complete protection against clinical varicella but at times only partial protection. Our studies strongly suggest the necessity for a second dose of vaccine in leukemic children for yet another reason. Only 82% of leukemic vaccinees have been found to seroconvert after 1 dose of 1000 PFU of Oka commercially prepared vaccine (Gershon et al., 1984b). Thus, further observations concerning the duration of protection in immunocompromised vaccinees, the specific factors responsible for protection, and whether booster injections are required, are crucial.

A 6-year follow-up of leukemic children immunized in Japan indicates persistence of VZV-neutralizing antibody in those tested (Sakurai et al., 1982). A major difference between studies carried out in Japan and in the United States, however, has been that in Japan routine skin testing every 6–12 months with an intradermal injection of VZV antigen is performed. We have found that skin testing boosts VZV antibody (La Russa et al., 1984), which may account for the better persistence of VZV antibody in Japanese vaccinees compared with American vaccinees. Our experience with the skin test is that it is too cumbersome for routine use in young children. Therefore, we have been approaching the problem of apparent declining immunity in vaccinees by routinely giving two doses of vaccine 3 months apart rather than by skin testing.

B. Dose of Vaccine Virus Necessary to Induce Immunity

Doses employed by various investigators have ranged from 20 to 8700 PFU of VZV vaccine. For example, in the study of Weibel and colleagues (1984) in which normal vaccinees were totally protected against varicella, a dose of 8700 PFU was given. In contrast, we have employed 1000 PFU, but our leukemic patients have not been completely protected against varicella, and their immune responses have waned (Gershon et al., 1984b).

Whether this is a reflection of the host or of the dose of immunizing virus deserves further study.

C. Safety

1. Short Term

In the short run, varicella vaccine is clearly safe, in both normal and certain immunocompromised patients. It is likely, based on the studies of M. Takahashi (personal communication), that severely immunocompromised patients, such as those with lymphosarcoma and Hodgkin's disease, will not necessarily tolerate the vaccine well, and extreme caution must be exercised if such patients are to be immunized. Certainly for children with leukemia in remission on maintenance therapy, the vaccine has proved to be safe. Because of the possibility of spread of vaccine virus to other high-risk children, especially if the vaccinee develops a rash, follow-up examinations of vaccinees with rash should not be performed in a clinic setting. In our study (Gershon et al., 1984b) the incidence of spread of vaccine virus to varicella-susceptible household members was 3% overall and 10% in children with rash. Other serious complications of varicella, such as Reye's syndrome, have not been observed in vaccine recipients, although this merits further study, as large numbers of children would have to be immunized to rule out this possibility as a complication because its incidence is very low even after natural varicella.

2. Long Term

Concern for long-term safety of the vaccine involves two issues, zoster and malignancy. As is true for any phenomenon, it is easier to rule in a causal factor than to rule one out. It is difficult to rule out VZV as a cause of malignancy, but this particular herpesvirus has never been linked to any form of cancer. It has been found that infected cells possess transforming ability in vitro although cell-free material (e.g., the vaccine) does not have this property (Yamanishi et al., 1981). In addition, the clinical relevance of this experimental observation remains unknown.

Another issue with regard to malignancy is whether there is an increase in the relapse rate of leukemia after varicella vaccination. It is especially important to look for this because chemotherapy is stopped for at least 2 weeks in vaccinees. There has been no reported increase in relapse of leukemia in vaccinees in Japan (Ha et al., 1980). In the NIAID Collaborative Varicella Vaccine Study there has also been no difference in relapse rate between leukemic children who have been vaccinated (13%) and those who have experienced natural varicella (8%) (Gershon et al., 1984b).

It was feared that zoster might be a serious complication of varicella vaccination (Brunell, 1975). A 7-year follow-up of leukemic children in Japan has revealed that although some vaccinees have developed zoster, the incidence, 10–15%, and severity are no different than that seen in leukemics who have had natural varicella (Sakurai *et al.*, 1982). It is unfortunate that no virus isolates have been obtained from any of these cases of zoster. It would be of great interest to determine whether they were caused by vaccine virus or by wild virus. One of our 240 vaccinees, followed for as long as 3 years in some cases, has developed zoster (Gershon *et al.*, 1984b), and further long-term follow-up of these vaccinees will be important.

V. LOGISTICS OF FUTURE VACCINE USE

A. Potential Vaccine Candidates

It seem fairly obvious that children with leukemia in remission and probably children with solid tumors ought to be offered protection from severe varicella by immunization. Other groups that would potentially benefit from the vaccine include varicella-susceptible adults, especially those who work in hospitals and/or those who have close contact with young children. We have immunized 35 normal adults with no adverse effects in the Collaborative Varicella Vaccine Study. One of the 35 has failed to seroconvert, even after three doses of vaccine. Immune responses to VZV in adults as well as leukemic children are improved by giving two doses of vaccine. Loss of detectable VZV antibody has also occurred in adult vaccinees after 1–2 years. Of three adults who have had household exposures to varicella, none has developed clinical chicken pox.

Varicella vaccine has been employed in Japan to terminate hospital outbreaks of varicella (Takahashi *et al.*, 1974; Baba *et al.*, 1978). This is another potentially useful approach to nosocomial varicella, as it may be extremely difficult to terminate and may impose great danger on hospitalized immunocompromised patients. It has been found that one simple means to identify varicella-susceptible normal adults is the intradermal skin test for varicella (Steele *et al.*, 1982; LaRussa *et al.*, 1984). One can determine whether an individual is susceptible to varicella in 48 hr, and susceptibles may be immunized immediately. Unfortunately, however, the skin test appears to be accurate only in immunologically normal persons.

Whether varicella vaccine will be used in normal children on a routine basis is, at this time, difficult to predict. One of the greatest reasons for caution at this time is that the duration of immunity to varicella after vaccination is still unknown. One would not want to convert clinical varicella from a childhood disease into an illness of adulthood by administering a vaccine that induces only temporary immunity. In addition, as has been mentioned, the importance of reexposure to varicella or zoster with boosts in immune responses in maintaining immunity is unknown. Finally, it is unknown how often VZV vaccine virus becomes latent after its administration. It is possible that some of the cases of zoster observed in Japan were due to wild VZV. The role that silent reactivation of latent infection might play in maintaining immunity to VZV is also unknown. If the vaccine virus does not become latent, it may not provide long-lasting immunity to VZV. Subclinical boosts in VZV antibody titer due to silent reactivation of latent VZV after infection with wild virus have been described (Gershon *et al.*, 1982). It is obvious that we have much to learn about the factors that contribute to immunity to VZV. Until more is known about them, it is difficult to recommend widespread use of varicella vaccine in normal children.

B. Future Planning

The major decision as to whether to use varicella vaccine routinely in normal children is not one that can be made now. It will have to be clear that potential benefits outweigh potential risks before the vaccine can be used extensively in normal children. Decisions as to the future use of this vaccine will require stepwise analysis of the data obtained from various groups. Gradually a reasoned, practical approach to the use of the vaccine in various populations should emerge.

ACKNOWLEDGMENT. The author's research was supported by Grant AI-12814 and Contract AI-02639 from the National Institutes of Health.

REFERENCES

Arbeter, A. M., Starr, S. E., Weibel, R. E., and Plotkin, S. A., 1982, Live attenuated varicella vaccine: Immunization of healthy children with the OKA strain, *J. Pediatr.* **100**:886–893.

Arbeter, A. M., Starr, S. E., Weibel, R. E., Neff, B. J., and Plotkin, S. A., 1983, Live attenuated varicella vaccine: The KMcC strain in healthy children, *Pediatrics* **71**:307–312.

Arvin, A., Koropchak, C. M., and Wittek, A. E., 1983, Immunologic evidence of reinfection with varicella–zoster virus, *J. Infect. Dis.* **148**:200–205.

Asano, Y., and Takahashi, M., 1977, Clinical and serologic testing of a live varicella vaccine and two-year follow-up for immunity of the vaccinated children, *Pediatrics* **60**:810–814.

Asano, Y., Nakayama, H., Yazaki, T., Kato, R., Hirose, S., Tsuzuki, K., Ito, S., Isomura, S., and Takahashi, M., 1977, Protection against varicella in family contacts by immediate inoculation with live varicella vaccine, *Pediatrics* **59**:3–7.

Asano, Y., Albrecht, P., Vujcic, L. K., Quinnan, G. V., Kawakami, K., and Takahashi, M., 1983, Five-year follow-up study of recipients of live varicella vaccine using enhanced neutralization and fluorescent antibody membrane antigen assays, *Pediatrics* **72**:291–294.

Baba, K., Yabuuchi, H., Okuni, H., and Takahashi, M., 1978, Studies with live varicella vaccine and inactivated skin test antigen: Protective effect of the vaccine and clinical application of the skin test, *Pediatrics* **61**:550–555.

Bogger-Goren, S., Baba, K., Hurly, P., Yabuuchi, H., Takahashi, M., and Ogra, P. L., 1982, Antibody response to varicella–zoster virus after natural or vaccine-induced infection, *J. Infect. Dis.* **146**:260–265.

Brunell, P., 1975, Live varicella vaccine [letter], *Lancet* **1**:98.

Brunell, P., 1977a, Protection against varicella: Commentary, *Pediatrics* **59**:1–2.

Brunell, P., 1977b, Brunell's brush off [letter], *Pediatrics* **59**:954.

Brunell, P., Gershon, A., Uduman, S. A., and Steinberg, S., 1975, Varicella–zoster immunoglobulins during varicella, latency, and zoster, *J. Infect. Dis.* **132**:49–54.

Brunell, P., Geiser, C., Shehab, Z., and Waugh, J. E., 1983, Administration of live varicella vaccine to children with leukemia, *Lancet* **2**:1069–1072.

Cheatham, W. J., Weller, T. H., Dolan, T. F., and Dower, J. C., 1956, Varicella: Report of two fatal cases with necropsy, virus isolation, and serologic studies, *Am. J. Pathol.* **32**:1015–1028.

Feldman, S., Hughes, W. T., and Daniel, C. B., 1975, Varicella in children with cancer: 77 cases, *Pediatrics* **56**:388–397.

Gershon, A., 1980, Live attenuated varicella–zoster vaccine, *Rev. Infect. Dis.* **2**:393–405.

Gershon, A., Steinberg, S., and Brunell, P. A., 1974, Zoster immune globulin: A further assessment, *N. Engl. J. Med.* **290**:243–245.

Gershon, A., Steinberg, S., Borkowsky, W., Lennette, D., and Lennette, E., 1982, IgM to varicella–zoster virus: Demonstration in patients with and without clinical zoster, *Pediatr. Infect. Dis.* **1**:164–167.

Gershon, A., Steinberg, S., Gelb, L., Galasso, G., Borkowsky, W., LaRussa, P., Ferrara, A., and the National Institute of Allergy and Infectious Diseases Varicella Vaccine Collaborative Study Group, 1984a, Live attenuated varicella vaccine: Efficacy for children with leukemia in remission, *J. Am. Med. Assn.* **252**:355–362.

Gershon, A., Steinberg, S., and Gelb, L., 1984b, Clinical reinfection due to varicella–zoster virus, *J. Infect. Dis.* **149**:137–142.

Ha, K., Baba, K., Ikeda, T., Nishida, M., Yabuuchi, H., and Takahashi, M., 1980, Application of live varicella vaccine to children with acute leukemia or other malignancies without suspension of anticancer therapy, *Pediatrics* **65**:346–350.

Hope-Simpson, R. E., 1965, The nature of herpes zoster: A long-term study and a new hypothesis, *Proc. R. Soc. Med.* **58**:1–12.

Izawa, T., Ihara, T., Hattori, A., Iwasa, T., Kamiya, H., Sakurai, M., and Takahashi, M., 1977, Application of a live varicella vaccine in children with acute leukemia or other malignant diseases, *Pediatrics* **60**:805–809.

Kempe, H., and Gershon, A., 1977, Varicella vaccine at the crossroads, *Pediatrics* **60**:930–931.

Martin, J. H., Dohner, D. E., Wellinghoff, W. J., and Gelb, L. D., 1982, Restriction endonuclease analysis of varicella–zoster vaccine virus and wild-type DNAs, *J. Med. Virol.* **9**:69–76.

Neff, B. J., Weibel, R. E., Villarejos, V. M., Buynak, E. B., McLean, A. A., Morton, D. H., Wolanski, B. S., and Hilleman, M. R., 1981, Clinical and laboratory studies of KMcC strain live attenuated varicella virus, *Proc. Soc. Exp. Biol. Med.* **166**:339–347.

Plotkin, S. A., 1977, Varicella vaccine: Plotkin's plug [letter], *Pediatrics* **59**:954.

Ross, A. H., Lencher, E., and Reitman, G., 1962, Modification of chickenpox in family contacts by administration of gamma globulin, *N. Engl. J. Med.* **267**:369–376.

Sabin, A. B., 1977, Varicella–zoster virus vaccine: Commentary, *J. Am. Med. Assoc.* 1731–1733.

Sakurai, M., Ihara, T., Ito, M., Hirai, S., Iwasa, T., Oitani, K., Kamiya, H., Izawa, T., and Takahashi, M., 1982, Application of a live varicella vaccine in children with acute leukemia, in: *Herpesvirus: Clinical, Pharmacological, and Basic Aspects* (H. Shiota, Y.-C. Cheng, and W. H. Prusoff, eds.), pp. 87–93, Excerpta Medica, Amsterdam.

Steele, R. W., Coleman, M. A., Fiser, M., and Bradsher, R. W., 1982, Varicella zoster in hospital personnel: Skin test reactivity to monitor susceptibility, *Pediatrics* **70:**604–608.

Takahashi, M., Otsuka, T., Okuno, Y., Asano, Y., Yazaki, T., and Isomura, S., 1974, Live varicella vaccine used to prevent the spread of varicella in children in hospital, *Lancet* **2:**1288–1290.

Weibel, R. E., Neff, B. J., Kuter, B. J., Guess, H. A., Rothenberger, C. A., Fitzgerald, A. J., Connor, K. A., McLean, A. A., Hilleman, M. R., Buynak, E. B., and Skolnick, E. M., 1984, Live attenuated varicella virus vaccine. Efficacy trial in healthy children. *N. Engl. J. Med.* **310:**1409–1415.

Whitley, R., Hilty, M., Haynes, R., Bryson, Y., Connor, J. D., Soong, S.-J., Alford, C. A., and the National Institute of Allergy and Infectious Diseases Collaborative Antiviral Study Group, 1982, Vidarabine therapy of varicella in immunosuppressed patients, *J. Pediatr.* **101:**125–131.

Williams, V., Gershon, A., and Brunell, P. A., 1974, Serologic response to varicella–zoster membrane antigens measured by indirect immunofluorescence, *J. Infect. Dis.* **130:**669–672.

Yamanishi, H., Matsunaga, Y., Ogino, T., and Lopetegui, P., 1981, Biochemical transformation of mouse cells by varicella–zoster virus, *J. Gen. Virol.* **56:**421–430.

CHAPTER 16

A Subunit Vaccine against Epstein–Barr Virus

M. A. Epstein

I. INTRODUCTION

Epstein–Barr virus (EBV) was discovered in 1964 (Epstein *et al.*, 1964) and since that time an enormous amount has been learned about the biological and molecular attributes of this agent (Epstein and Achong, 1979a). In consequence, it is no longer difficult to explain how a ubiquitous virus that infects huge numbers of normal individuals in all human populations (W. Henle and Henle, 1979) is also both the cause of infectious mononucleosis (IM) in developed countries with high living standards (G. Henle and Henle, 1979), and involved in the induction, together with essential cofactors, of endemic Burkitt's lymphoma (BL) (Epstein, 1978) and undifferentiated nasopharyngeal carcinoma (NPC) (Klein, 1979).

A. Immunological Control of EBV Infection

It has been clear for some years that IM accompanies primary EBV infection in young adults (G. Henle and Henle, 1979) and the role of kissing in this context was recognized (Hoagland, 1955) long before the mechanisms were understood (Epstein and Achong, 1977). Once initiated, the disease runs its course until the patient develops virus-neutralizing antibodies which arrest the spread of infection in the body, and a full cellular immunological response which eliminates most virus-infected cells. This cellular reaction includes multiple T and NK responses together with the crucial, HLA-restricted, specific cytotoxic T-cell re-

M. A. EPSTEIN • Department of Pathology, University of Bristol Medical School, Bristol BS8 1TD, England.

sponse, which maintains surveillance of the lifelong carrier state that always follows primary EBV infection (Rickinson *et al.*, 1980, 1981). As regards the carrier state, the fact that latently infected B cells are liberated into the circulation throughout life while being steadily eliminated by specific cytotoxic T cells (Rickinson *et al.*, 1981) indicates that there must be an immunologically privileged site where infection of target cells continues at a low level. Shedding of virus particles into the buccal fluid, the source of person-to-person infection, is another indication of continuing infection of cells but in this case the result is virus replication (see Epstein and Achong, 1979b). It would appear that exactly similar control mechanisms operate in those whose primary infection is silent, i.e., without the clinical manifestations of IM.

Where immunological controls fail, other clinical conditions arise. Thus, an X-linked recessive, fatal lymphoproliferative syndrome (XLP syndrome or Duncan's disease) has been recognized in which males in affected families die, following primary EBV infection, from IM or even lymphomas owing to a genetically determined lack of EBV specific cytotoxic T cells (Bar *et al.*, 1974; Purtilo *et al.*, 1975, 1982). Lesser degrees of failure are responsible for chronic IM, while immunosuppression induced therapeutically in organ graft recipients gives rise to EBV carrying lymphomas seemingly as a result of impaired specific cytotoxic T-cell function (Crawford *et al.*, 1981; Gaston *et al.*, 1982).

B. EBV and Human Tumors

1. Burkitt's Lymphoma

The strikingly close association of EBV with endemic BL has been fully documented and reviewed (Epstein and Achong, 1979c) and quite recent work on cellular oncogene activation in BL suggests possible explanations for it. Thus, as discussed elsewhere (Epstein and Morgan, 1983), EBV appears to act on a target B-lymphocyte population altered by hyperendemic malaria, an essential cofactor in endemic BL (Burkitt, 1969), to give either specially large numbers of transformed B cells or B cells transformed in an unusual way, such that repeated cell divisions increase the likelihood of one or other of three specific chromosomal translocations (Lenoir *et al.*, 1982) occurring. It is now thought that these translocations ensure that the cellular *myc* oncogene is moved from its normal site on chromosome 8 to the immediate vicinity of one of the Ig genes active in the lymphoid cells (Dalla-Favera *et al.*, 1982; Taub *et al.*, 1982) where it could be affected by the Ig gene promoter, with subsequent selection of an *myc* oncogene-driven clone of malignant BL cells. Other new findings also implicate the cellular HuBlym-1 oncogene (Diamond *et al.*, 1983) but how these contribute has not yet been determined. Nevertheless, it is evident that progression to the full malignancy of endemic

BL involves numerous interdependent steps for one of which at least, EBV seems to be essential.

EBV is also apparently involved in the causation of about 20% of cases of sporadic BL (Philip et al., 1982) but in the absence of such co-factors as hyperendemic malaria this is a relatively rare disease.

2. Nasopharyngeal Carcinoma

As regards NPC, EBV is associated only with the undifferentiated form and this holds good for tumors from all parts of the world; 100% of this type of NPC carry EBV DNA in the malignant epithelial cells (Wolf et al., 1973; Pagano et al., 1975; Andersson-Anvret et al., 1977, 1979). A full account of this and all other aspects of the relationship of the virus to the tumor has been published (Klein, 1979).

C. Comment

The foregoing sections give a brief outline of those aspects of EBV that are relevant to a consideration of vaccination against the agent. It should be noted that other features of EBV are dealt with in depth elsewhere in this volume (Chapters 8, 9, 11, 12).

II. REASONS FOR AN ANTIVIRAL VACCINE

The weight of evidence implicating EBV in the causation of endemic BL and undifferentiated NPC when considered in conjunction with the numerical importance of these tumors in world cancer terms first prompted proposals for a vaccine against the virus (Epstein, 1976). For although endemic BL is not of particular significance, undifferentiated NPC is, being the most common cancer of men and the second most common of women among southern Chinese (Shanmugaratnam, 1971) and Eskimos (Lanier et al., 1980), and having a moderately high incidence in north Africa (Cammoun et al., 1974), east Africa (Clifford, 1970), and through most of southeast Asia (Shanmugaratnam, 1971). Arguments for the development of such a vaccine to prevent or modify EBV infection and thereby reduce the occurrence of the malignancies have become more cogent, as current studies continue to clarify the agent's role (Epstein and Morgan, 1983).

Furthermore, herpesvirus-induced animal cancers provide encouraging precedents. The lymphoma of chickens caused by Marek's disease herpesvirus has been largely eliminated by inoculation of apathogenic virus (Churchill et al., 1969; Okazaki et al., 1970), and is thus the first example of antiviral vaccination reducing the frequency of a naturally occurring cancer. Similarly, in the case of the malignant lymphoma that can be induced experimentally with *Herpesvirus saimiri* in subhuman primates (Meléndez *et al.*, 1969), animals given heat- and formalin-treated virus were protected against infection and the consequential development of tumors (Laufs and Steinke, 1975). Other work with these systems has shown that injection of antigen-containing membranes from cells infected with Marek's disease herpesvirus markedly reduces the incidence of lymphomas in chickens (Kaaden and Dietzschold, 1974), while membrane preparations from *H saimiri*-infected cells induce virus-neutralizing antibodies when inoculated into marmosets (Pearson and Scott, 1977). In the context of a vaccine against EBV these results with membrane preparations have focused attention on the EBV-determined membrane antigen (MA) (Klein *et al.*, 1966) and the importance of the latter has been further emphasized by well-established observations showing that antibodies from EBV-infected individuals that neutralize the virus are those directed against MA (Pearson *et al.*, 1970, 1971; Gergely *et al.*, 1971; de Schryver *et al.*, 1974).

III. EBV MEMBRANE ANTIGEN

The EBV MA complex consists largely of two high-molecular-weight components of 340,000 and 270,000 (gp340 and gp270) (Qualtière and Pearson, 1979; Thorley-Lawson and Edson, 1979; Strnad *et al.*, 1979), which are also present on the viral envelope (North *et al.*, 1980); these molecules elicit virus-neutralizing antibodies (North *et al.*, 1982a) and monoclonal antibodies that react with both components are likewise virus-neutralizing (Hoffman *et al.*, 1980; Thorley-Lawson and Geilinger, 1980; Thorley-Lawson and Poodry, 1982).

A. Quantification of MA gp340

In order to work out a satisfactory method for the preparation of MA molecules, it was necessary to be able to measure the product so as to introduce modifications giving the most efficient yields. A quantitative radioimmunoassay (RIA) was therefore elaborated based on small amounts of highly purified [^{125}I]gp340 and a full account of this procedure has been given elsewhere (North *et al.*, 1982b).

B. A Preparation Method for MA gp340

Most EBV-carrying, virus-producing, lymphoid cell lines synthesize roughly equal amounts of gp340 and gp270. However, certain lines are anomalous in this respect, and the B95-8 marmoset lymphoblastoid line (Miller *et al.*, 1972) presents special advantages for molecular-weight-based isolation techniques because it expresses almost exclusively the larger molecule, gp340.

With the indispensable help of the RIA, a preparative SDS-PAGE procedure was developed for the isolation of gp340 in tractable amounts from B95-8 cell membranes. The procedure included essential steps to renature the product and ensure that it retained antigenicity; these consisted of separation of the gp340 from SDS in the presence of urea to prevent protein refolding, followed by removal of urea during dialysis against buffer containing nonionic detergents. Details of this preparation method have been reported (Morgan et al., 1983).

C. Immunization with MA gp340 Using Novel Adjuvants

Preliminary experiments with gp340 prepared by the above technique showed it to be only rather weakly immunogenic when given to mice and rabbits in microgram amounts in Freund's adjuvant, and virus-neutralizing antibody could only be obtained after multiple injections.

In order to overcome the disadvantages of repeated immunization with scarce, purified glycoprotein and the need for Freund's adjuvant which could never be suitable for administration to man, alternative methods were sought. In addition, comparative immunogenicity studies were undertaken in mice and rabbits using various adjuvants, routes of injection, and dose schedules. The only animals known to be fully susceptible to experimental infection with EBV are the owl monkey (Aotus) (Epstein et al., 1973a,b, 1975) and the cotton-top tamarin (Saguinus oedipus oedipus) (Shope et al., 1973) (see Miller, 1979, for review). The former "species" has recently been found to have considerable karyotypic heterogeneity (Ma et al., 1976, 1978; Ma, 1981) and shows much variation in disease susceptibility, so that the cotton-top tamarin is preferable for in vivo studies with EBV. Although this endangered species can now be bred very successfully in captivity (Brand, 1981; Kirkwood et al., 1983; Kirkwood, 1983), it is nevertheless necessary to work out methodologies with banal laboratory animals before applying them to the rare and costly tamarins. Reports of highly effective immunization with viral antigens incorporated in artificial liposomes (Morein et al., 1978; Manesis et al., 1979) and of the benefit of adding lipid A (Naylor et al., 1982) suggested a fruitful approach.

Any method using liposomes gave better results as assessed by antibody titers than antigen presented with Freund's adjuvant. Where mice were immunized i.p., antigen in liposomes with killed B. pertussis organisms was more effective than without, although antigen in liposomes alone was highly immunogenic when given i.v. More important, irrespective of the mode of administration, gp340 in liposomes with lipid A always engendered superior responses. This last finding also held good for the cotton-top tamarins where antibodies were seen earlier after immunization with gp340 in liposomes incorporating lipid A than without lipid A.

All the sera reacted specifically with MA gp340 from B95-8 cells and failed to recognize other molecules from either the surface or the interior of such cells. However, when tested against cells that expressed both gp340 and gp270, the latter antigen was recognized as well, indicating once again the immunological relatedness of the two molecules. Finally, the sera obtained after presenting gp340 in these novel ways were virus-neutralizing, and for vaccine studies, this is of particular importance in the case of the tamarins. These experiments have been described in full (North *et al.*, 1982b; Morgan *et al.*, 1984a).

D. Structure of MA gp340

Although purified gp340 has been prepared and shown to be both antigenic and immunogenic, it seems likely that any EBV MA subunit vaccine for human use will ultimately be made by recombinant DNA technology or chemical synthesis of peptides. Understanding of the structure of gp340 and the role, if any, of the sugar moiety is obviously a prerequisite for either of these procedures and the molecule has therefore been studied by analyzing the effects on it of various glycosidases and V8 protease.

After treatment with glycosidases or exposure to tunicamycin during synthesis, the carbohydrate moiety has been found to constitute about 50% of the molecular mass and to consist of both O-linked and N-linked sugars. This high carbohydrate content of gp340 conferred resistance to proteolysis so that V8 protease was only effective at concentrations above 1 mg/ml. Removal of sialic acid before V8 protease digestion did not alter the pattern or antigenicity of digestion fragments. Antigenicity of the intact molecule was likewise unaffected by removal of sialic acid nor were the O-linked or N-linked carbohydrate moieties necessary for this property. The binding of virus-neutralizing human sera and monoclonal antibody by gp340 from which either O-linked or N-linked sugars had been removed indicates that the sites on the molecule that generate neutralizing antibodies are present in the protein component.

It would appear, therefore, that only the polypeptide of gp340 is relevant for use as an immunogen and further vaccine studies need only concentrate on this portion of the molecule (for details, see Morgan *et al.*, 1984b).

IV. DISCUSSION

Having successfully induced EBV-neutralizing antibodies in cotton-top tamarins by immunization with gp340, it is clear that the protective effects of this procedure must be investigated *in vivo* as the next step.

These key experiments are no longer inherently difficult now that breeding colonies of the necessary animals are available (Brand, 1981; Kirkwood et al., 1983; Kirkwood, 1983) and it is likely that results will be obtained in the near future. If a gp340 vaccine can indeed be shown to protect cotton-top tamarins against challenge with disease-inducing doses of virulent EBV, the way will be open for the evaluation of vaccination in man.

The efficacy of subunit antiviral vaccines has been demonstrated in both human and veterinary diseases (Szmuness et al., 1981; Bittle et al., 1982) and development of an EBV vaccine must surely proceed in this direction. While recombinant DNA experiments, and sequencing and synthesis studies, are going forward, the most satisfactory human system in which to assess gp340 would be in the context of IM. It has long been possible to determine who among a group of young adults in Western countries has escaped silent childhood infection with EBV (Pereira et al., 1969) and is therefore at risk for delayed infection with its accompanying clinical IM in 50% of cases (Niederman et al., 1970; University Health Physicians and PHLS Laboratories, 1971). Groups of such populations at risk come into universities and other training establishments each year and the effectiveness of immunization with gp340 in preventing infection and reducing the expected incidence of IM would rapidly be evident using quite small numbers. Thereafter, the effect on tumor incidence of an efficacious vaccine against EBV could be tested in one of the well-recognized high-incidence areas of endemic BL, and because this type of BL has its peak incidence around the age of 6 years (Burkitt, 1963), the results of a blanket vaccination program covering all 3- to 12-month-old children could be judged in less than 10 years. Such a program would involve no greater logistic problems than the large-scale WHO prospective seroepidemiological study of EBV and BL carried out over 7 years in Uganda (de Thé et al., 1978) and would share many features with hepatitis B vaccine trials already under way in several Third World countries (Coursaget et al., 1980; Anonymous, 1983).

Success in reducing the incidence of endemic BL analogous to the antiviral vaccine control of Marek's disease lymphomas (Biggs, 1974) would then provide inescapable arguments for addressing the much more difficult task of intervention against undifferentiated NPC. This is a disease of middle and later life in the high-incidence areas (Shanmugaratnam, 1971) and thus calls for the maintenance of immunity over a great many years. However, the huge numbers of cases in high-incidence populations must surely serve as sufficient incentive for such an undertaking.

ACKNOWLEDGMENTS. Work from the author's laboratory was supported by the Medical Research Council, London (Special Project Grant SPG/978/32/5) together with the Cancer Research Campaign, London (out of funds donated by the Bradbury Investment Company of Hong Kong).

REFERENCES

Andersson-Anvret, M., Forsby, N., Klein, G., and Henle, W., 1977, Relationship between the Epstein—Barr virus and undifferentiated nasopharyngeal carcinoma: Correlated nucleic acid hybridization and histopathological examination, *Int. J. Cancer* **20**:486.

Andersson-Anvret, M., Forsby, N., Klein, G., Henle, W., and Biörklund, A., 1979, Relationship between the Epstein—Barr virus genome and nasopharyngeal carcinoma in Caucasian patients, *Int. J. Cancer* **23**:762.

Anonymous, 1983, Prevention of primary liver cancer, *Lancet* **1**:463.

Bar, R. S., Delor, C. J., Clausen, K. P., Hurtubise, P., Henle, W., and Hewetson, J. F., 1974, Fatal infectious mononucleosis in a family, *N. Engl. J. Med.* **290**:363.

Biggs, P. M., 1974, The control of Marek's disease with special reference to the United Kindgon, *Acta Ve. (Brno)* **43**:153.

Bittle, J. L., Houghten, R. A., Alexander, H., Shinnick, T. M., Sutcliffe, J. G., Lerner, R. A., Rowlands, D. J., and Brown, F., 1982, Protection against foot and mouth disease by immunization with a chemically synthesized peptide predicted from the viral nucleotide sequence, *Nature* **298**:30.

Brand, H. M., 1981, Husbandry and breeding of a newly established colony of cotton-topped tamarins (*Saguinus oedipus*), *Lab. Anim.* **15**:7.

Burkitt, D., 1963, A lymphoma syndrome in tropical Africa, in: *International Review of Experimental Pathology* (G. W. Richter and M. A. Epstein, eds.), Vol. 2, pp. 67—138, Academic Press, New York.

Burkitt, D. P., 1969, Etiology of Burkitt's lymphoma—An alternative hypothesis to a vectored virus, *J. Natl. Cancer Inst.* **42**:19.

Cammoun, M., Hoerner, G. V., and Mourali, N., 1974, Tumors of the nasopharynx in Tunisia: An anatomic and clinical study based on 143 cases, *Cancer* **33**:184.

Churchill, A. E., Payne, L. N., and Chubb, R. C., 1969, Immunization against Marek's disease using a live attenuated virus, *Nature* **221**:774.

Clifford, P., 1970, A review: On the epidemiology of nasopharyngeal carcinoma, *Int. J. Cancer* **5**:287.

Coursaget, P., Maupas, P., Goudeau, A., Chiron, J.-P., Drucker, J., Denis, F., and Diop-Mar, I., 1980, Primary hepatocellular carcinoma in intertropical Africa: Relationship between age and hepatitis B virus etiology, *J. Natl. Cancer Inst.* **65**:687.

Crawford, D. H., Sweny, P., Edwards, J., Janossy, G., and Hoffbrand, A. V., 1981, Long-term T-cell-mediated immunity to Epstein—Barr virus in renal-allograft recipients receiving cyclosporin A, *Lancet* **1**:10.

Dalla-Favera, R., Bregni, M., Erikson, J., Patterson, D., Gallo, R. C., and Croce, C. M., 1982, Human c-*myc onc* gene is locate on the region of chromosome 8 that is translocated in Burkitt lymphoma cells, *Proc. Natl. Acad. Sci. USA* **79**:7824.

de Schryver, A., Klein, G., Hewetson, J., Rocchi, G., Henle, W., Henle, G., Moss, D. J., and Pope, J. H., 1974, Comparison of EBV neutralization tests based on abortive infection or transformation of lymphoid cells and their relation to membrane reactive antibodies (anti MA), *Int. J. Cancer* **13**:353.

de Thé, G., Geser, A., Day, N. E., Tukei, P. M., Williams, E. H., Beri, D. P., Smith, P. G., Dean, A. G., Bornkamm, G. W., Feorino, P., and Henle, W., 1978, Epidemiological evidence for causal relationship between Epstein—Barr virus and Burkitt's lymphoma: Results of the Ugandan prospective study, *Nature* **274**:756.

Diamond, A., Cooper, G. M., Ritz, J., and Lane, M.-A., 1983, Identification and molecular cloning of the human Blym transforming gene activated in Burkitt's lymphomas, *Nature* **305**:112.

Epstein, M. A., 1976, Epstein—Barr virus—Is it time to develop a vaccine program?, *J. Natl. Cancer Inst.* **56**:697.

Epstein, M. A., 1978, An assessment of the possible role of viruses in the aetiology of Burkitt's lymphoma, in: *Viruses and Human Cancer* (Y. Ito, ed.), pp. 72—99, Karger, Basel.

Epstein, M. A., and Achong, B. G., 1977, Pathogenesis of infectious mononucleosis, *Lancet* **2**:1270.

Epstein, M. A., and Achong, B. G., (eds.), 1979a, *The Epstein–Barr Virus*, Springer-Verlag, Berlin.

Epstein, M. A., and Achong, B. G., 1979b, Introduction and general biology of the virus, in: *The Epstein–Barr Virus*, (M. A. Epstein and B. G. Achong, eds.), pp. 1–22, Springer-Verlag, Berlin.

Epstein, M. A., and Achong, B. G., 1979c, The relationship of the virus to Burkitt's lymphoma, in: *The Epstein–Barr Virus* (M. A. Epstein and B. G. Achong, eds.), pp. 321–337, Springer-Verlag, Berlin.

Epstein, M. A., and Morgan, A. J., 1983, Clinical consequences of Epstein–Barr virus infection and possible control by an antiviral vaccine, *Clin. Exp. Immunol.* **53**:257.

Epstein, M. A., Achong, B. G., and Barr, Y. M., 1964, Virus particles in cultured lymphoblasts from Burkitt's lymphoma, *Lancet* **1**:702.

Epstein, M. A., Hunt, R. D., and Rabin, H., 1973a, Pilot experiments with EB virus in owl monkeys (*Aotus trivirgatus*). I. Reticuloproliferative disease in an inoculated animal, *Int. J. Cancer* **12**:309.

Epstein, M. A., Rabin, H., Ball, G., Rickinson, A. B., Jarvis, J., and Meléndez, L. V., 1973b, Pilot experiments with EB virus in owl monkeys (*Aotus trivirgatus*). II. EB virus in a cell line from an animal with reticuloproliferative disease, *Int. J. Cancer* **12**:319.

Epstein, M. A., zur Hausen, H., Ball, G., and Rabin, H., 1975, Pilot experiments with EB virus in owl monkeys (*Aotus trivirgatus*). III. Serological and biochemical findings in an animal with reticuloproliferative disease, *Int. J. Cancer* **15**:17.

Gaston J. S. H., Rickinson, A. B., and Epstein, M. A., 1982, Epstein–Barr-virus-specific T-cell memory in renal-allograft recipients under long-term immunosuppression, *Lancet* **1**:923.

Gergely, L., Klein, G., and Ernberg, I., 1971, Appearance of Epstein–Barr virus-associated antigens in infected Raji cells, *Virology* **45**:10.

Henle, G., and Henle, W., 1979, The virus as the etiologic agent of infectious mononucleosis, in: *The Epstein–Barr Virus* (M. A. Epstein and B. G. Achong, eds.), pp. 297–300, Springer-Verlag, Berlin.

Henle, W., and Henle, G., 1979, Seroepidemiology of the virus, in: *The Epstein–Barr Virus* (M. A. Epstein and B. G. Achong, eds.), pp. 61–78, Springer-Verlag, Berlin.

Hoagland, R. J., 1955, The transmission of infectious mononucleosis, *Am. J. Med. Sci.* **229**:262.

Hoffman, G. J., Lazarowitz, S. G., and Hayward, S. D., 1980, Monoclonal antibody against a 250,000-dalton glycoprotein of Epstein–Barr virus identifies a membrane antigen and a neutralizing antigen, *Proc. Natl. Acad. Sci. USA* **77**:2979.

Kaaden, O. R., and Dietzschold, B., 1974, Alterations of the immunological specificity of plasma membranes of cells infected with Marek's disease and turkey herpes viruses, *J. Gen. Virol.* **25**:1.

Kirkwood, J. K., 1983, Effects of diet on health, weight and litter size in captive cotton-top tamarins *Saguinus oedipus oedipus*, *Primates* **24**:515.

Kirkwood, J. K., Epstein, M. A., and Terlecki, A. J., 1983, Factors influencing population growth of a colony of cotton-top tamarins, *Lab. Anim.* **17**:35.

Klein, G., 1979, The relationship of the virus to nasopharyngeal carcinoma, in: *The Epstein–Barr Virus* (M. A. Epstein and B. G. Achong, eds.), pp. 339–350, Springer-Verlag, Berlin.

Klein, G., Clifford, P., Klein, E., and Stjernswärd, J., 1966, Search for tumor-specific immune reactions in Burkitt lymphoma patients by the membrane immunofluorescence reaction, *Proc. Natl. Acad. Sci. USA* **55**:1628.

Lanier, A., Bender, T., Talbot, M., Wilmeth, S., Tschopp, C., Henle, W., Henle, G., Ritter, D., and Terasaki, P., 1980, Nasopharyngeal carcinoma in Alaskan Eskimos, Indians and Aleuts: A review of cases and study of Epstein–Barr virus, HLA and environmental risk factors, *Cancer* **46**:2100.

Laufs, R., and Steinke, H., 1975, Vaccination of non-human primates against malignant lymphoma, *Nature* **253:**71.

Lenoir, G. M., Preud'homme, J. L., Bernheim, A., and Berger, R., 1982, Correlation between immunoglobulin light chain expression and variant translocation in Burkitt's lymphoma, *Nature* **298:**474.

Ma, N. S. F., 1981, Chromosome evolution in the owl monkey, *Aotus*, *Am. J. Phys. Anthropol.* **54:**293.

Ma, N. S. F., Jones, T. C., Miller, A. C., Morgan, L. M., and Adams, E. A., 1976, Chromosome polymorphism and banding patterns in the owl monkey (*Aotus*), *Lab. Anim. Sci.* **26:**1022.

Ma, N. S. F., Rossan, R. N., Kelley, S. T., Harper, J. S., Bedard, M. T., and Jones, T. C., 1978, Banding patterns of the chromosomes of two new karyotypes of the owl monkey, *Aotus*, captured in Panama, *J. Med. Primatol.* **7:**146.

Manesis, E. K., Cameron, C. H., and Gregoriadis, G., 1979, Hepatitis B surface antigen-containing liposomes enhance humoral and cell-mediated immunity to the antigen, *FEBS Lett.* **102:**107.

Meléndez, L. V., Hunt, R. D., Daniel, M. D., Garcia, F. G., and Fraser, C. E. O., 1969, Herpesvirus saimiri. II. An experimentally induced primate disease resembling reticulum cell sarcoma, *Lab. Anim. Care* **19:**378.

Miller, G., 1979, Experimental carcinogenicity by the virus *in vivo*, in: *The Epstein–Barr Virus* (M. A. Epstein and B. G. Achong, eds.), pp. 351–372, Springer-Verlag, Berlin.

Miller, G., Shope, T., Lisco, H., Stitt, D., and Lipman, M., 1972, Epstein–Barr virus: Transformation, cytopathic changes, and viral antigens in squirrel monkey and marmoset leukocytes, *Proc. Natl. Acad. Sci. USA* **69:**383.

Morein, B., Helenius, A., Simons, K., Pettersson, R., Kääriäinen, L., and Schirrmacher, V., 1978, Effective subunit vaccines against an enveloped animal virus, *Nature* **276:**715.

Morgan, A. J., North, J. R., and Epstein, M. A., 1983, Purification and properties of the gp340 component of Epstein–Barr (EB) virus membrane antigen (MA) in an immunogenic form, *J. Gen. Virol.* **64:**455.

Morgan, A. J., Epstein, M. A., and North, J. R., 1984a, Comparative immunogenicity studies on Epstein–Barr (EB) virus membrane antigen (MA) with novel adjuvants in mice, rabbits and cotton-top tamarins, *J. Med. Virol.* **13:**281.

Morgan, A. J., Smith, A. R., Barker, R. N., and Epstein, M. A., 1984b, A structural investigation of the Epstein–Barr (EB) virus membrane antigen glycoprotein, gp340, *J. Gen. Virol.* **65:**397.

Naylor, P. T., Larsen, H. L., Huang, L., and Rouse, B. T., 1982, *In vivo* induction of anti-herpes simplex virus immune response by type 1 antigens and lipid A incorporated into liposomes, *Infect. Immun.* **36:**1209.

Niederman, J. C., Evans, A. S., Subrahmanyan, L., and McCollum, R. W., 1970, Prevalence, incidence and persistence of EB virus antibody in young adults, *N. Engl. J. Med.* **282:**361.

North, J. R., Morgan, A. J., and Epstein, M. A., 1980, Observations on the EB virus envelope and virus-determined membrane antigen (MA) polypeptides, *Int. J. Cancer* **26:**231.

North, J. R., Morgan, A. J., Thompson, J. L., and Epstein, M. A., 1982a, Purified EB virus gp340 induces potent virus-neutralizing antibodies when incorporated in liposomes, *Proc. Natl. Acad. Sci. USA* **79:**7504.

North, J. R., Morgan, A. J., Thompson, J. L., and Epstein, M. A., 1982b, Quantification of an EB virus-associated membrane antigen (MA) component, *J. Virol. Methods* **5:**55.

Okazaki, W., Purchase, H. G., and Burmester, B. R., 1970, Protection against Marek's disease by vaccination with herpesvirus of turkeys, *Avian Dis.* **14:**413.

Pagano, J. S., Huang, C. H., Klein, G., de Thé, G., Shanmugaratnam, K., and Yang, C. S., 1975, Homology of Epstein–Barr virus DNA in nasopharyngeal carcinomas from Kenya, Taiwan, Singapore and Tunisia, in: *Oncogenesis and Herpesviruses II* (G. de Thé, M. A. Epstein, and H. zur Hausen, eds.), Part 2, pp. 179–190, IARC, Lyon, France.

Pearson, G. R., and Scott, R. E., 1977, Isolation of virus-free herpesvirus saimiri antigen-positive plasma membrane vesicles, *Proc. Natl. Acad. Sci. USA* **74:**2546.

Pearson, G., Dewey, F., Klein, G., Henle, G., and Henle, W., 1970, Relation between neutralization of Epstein–Barr virus and antibodies to cell-membrane antigens induced by the virus, *J. Natl. Cancer Inst.* **45:**989.

Pearson, G., Henle, G., and Henle, W., 1971, Production of antigens associated with Epstein–Barr virus in experimentally infected lymphoblastoid cell lines, *J. Natl. Cancer Inst.* **46:**1243.

Pereira, M. S., Blake, J. M., and Macrae, A. D., 1969, EB virus antibody at different ages, *Br. Med. J.* **4:**526.

Philip, T., Lenoir, G. M., Bryon, P. A., Gérard-Marchant, R., Souillet, G., Philippe, N., Freycon, F., and Brunat-Mentigny, M., 1982, Burkitt-type lymphoma in France among non-Hodgkin malignant lymphomas in Caucasian children, *Br. J. Cancer* **45:**670.

Purtilo, D. T., Cassel, C., Yang, J. P. S., Stephenson, S. R., Harper, R., Landing, G. H., and Vawter, G. F., 1975, X-linked recessive progressive combined variable immunodeficiency (Duncan's disease), *Lancet* **1:**935.

Purtilo, D. T., Sakamoto, K., Barnabei, V., Seeley, J., Bechtold, T., Rogers, G., Yetz, J., and Harada, S., 1982, Epstein–Barr virus-induced diseases in boys with the X-linked lymphoproliferative syndrome (XLP), *Am. J. Med.* **73:**49.

Qualtière, L. F., and Pearson, G. R., 1979, Epstein–Barr virus-induced membrane antigens: Immunochemical characterisation of Triton X100 solubilized viral membrane antigens from EBV-superinfected Raji cells, *Int. J. Cancer* **23:**808.

Rickinson, A. B., Wallace, L. E., and Epstein, M. A., 1980, HLA-restricted T cell recognition of Epstein–Barr virus-infected B cells, *Nature* **283:**865.

Rickinson, A. B., Moss, D. J., Wallace, L. E., Rowe, M., Misko, I. S., Epstein, M. A., and Pope, J. M., 1981, Long term T-cell-mediated immunity to Epstein–Barr virus, *Cancer Res.* **41:**4216.

Shanmugaratnam, K., 1971, Studies on the etiology of nasopharyngeal carcinoma, in: *International Review of Experimental Pathology*, Vol. 10 (G. W. Richter and M. A. Epstein, eds.), pp. 361–413, Academic Press, New York.

Shope, T., Dechairo, D., and Miller, G., 1973, Malignant lymphoma in cotton-top marmosets after inoculation with Epstein–Barr virus, *Proc. Natl. Acad. Sci. USA* **70:**2487.

Strnad, B. C., Newbauer, R. H., Rabin, H., and Mazur, R. A., 1979, Correlation between Epstein–Barr virus membrane antigen and three large cell surface glycoproteins, *J. Virol.* **32:**885.

Szmuness, W., Stevens, C. E., Zang, E. A., Harley, E. J., and Kellner, A., 1981, A controlled clinical trial of the efficacy of the hepatitis B vaccine (Heptavax B): A final report, *Hepatology* **1:**377.

Taub, R., Kirsch, I., Morton, C., Lenoir, G., Swan, D., Tronick, S., Aaronson, S., and Leder, P., 1982, Translocation of the c-*myc* gene into the immunoglobulin heavy chain locus in human Burkitt lymphoma and murine plasmacytoma cells, *Proc. Natl. Acad. Sci. USA* **79:**7837.

Thorley-Lawson, D. A., and Edson, C. M., 1979, The polypeptides of the Epstein–Barr virus membrane antigen complex, *J. Virol.* **32:**458.

Thorley-Lawson, D. A., and Geilinger, K., 1980, Monoclonal antibodies against the major glycoprotein (gp350/220) of Epstein–Barr virus neutralize infectivity, *Proc. Natl. Acad. Sci. USA* **77:**5307.

Thorley-Lawson, D. A., and Poodry, C. A., 1982, Identification and isolation of the main component (gp350-gp220) of Epstein–Barr virus responsible for generating neutralizing antibodies in vivo, *J. Virol.* **43:**730.

University Health Physicians and PHLS Laboratories, 1971, Infectious mononucleosis and its relationship to EB virus antibody, *Br. Med. J.* **4:**643.

Wolf, H., zur Hausen, H., and Becker, V., 1973, EB viral genomes in epithelial nasopharyngeal carcinoma cells, *Nature New Biol.* **244:**245.

CHAPTER 17

A Perspective on the Therapy of Human Herpesvirus Infections

Richard J. Whitley

I. INTRODUCTION

In less than a decade, significant advances have been made in developing antiviral drugs for the therapy of human viral infections. In large part, a foundation for these advances was set in the late 1950s when Prusoff (1959) synthesized the pyrimidine analog 5-iodo-2'-deoxyuridine (idox-uridine), which was shown to have both antitumor and antiviral properties. Subsequently, the *in vitro* antiviral activity of idoxuridine against several DNA viruses was elucidated (Herrmann, 1961). Kaufman reported the activity of this drug in both animal (Kaufman *et al.*, 1962a,b) and human studies of herpes simplex keratoconjunctivitis (Kaufman, 1962).

These studies led to idoxuridine ophthalmic ointment becoming the first licensed antiviral compound in the United States. Toxicity prevented this useful drug from being extensively utilized for parenteral therapy. The recognition of the potential toxicity both *in vitro* and *in vivo* of the first-generation antiviral drug likely slowed enthusiasm for the development of new antiviral agents. Thus, it was not until the 1970s that the realistic possibilities for therapy of herpesvirus infections began to attract the attention of biochemists, pharmacologists, virologists, and biomedical investigators. As the field of antiviral therapy has evolved, the family of herpesviruses, among the most common of viral pathogens of humans, became the most likely targets for controlled trials and, ultimately, the

RICHARD J. WHITLEY • Department of Pediatrics and Microbiology, The University of Alabama, Birmingham, Alabama 45294.

most amenable to therapeutic success. In part, these successes have been the consequence of prompt and specific clinical diagnosis, resulting from the cutaneous manifestations of varicella–zoster (VZV) and herpes simplex virus (HSV) infections. These clinical manifestations lead to the early introduction of therapy, a necessity for chemotherapeutic successes.

The propensity of these infections to be of enhanced severity during symptomatic primary infections or when occurring in the immunocompromised host offered a broader therapeutic window for the demonstration of clinical effectiveness. This review will focus on the prospects and problems encountered in achieving successful therapy for human herpesvirus infections, particularly as it relates to establishing a therapeutic index or ratio of efficacy to toxicity, predicated upon successfully altering the natural history of disease. Ultimately, the success of any therapeutic program must be weighed against the enhanced benefits that might be achieved if the disease could be either prevented or ameliorated by vaccination. Some background, however, is essential before considering specific therapeutic approaches to different diseases.

A. Development of Antivirals for Clinical Evaluation

Two approaches have led to the introduction of antiviral drugs into evaluation programs: the random screening of newly synthesized compounds to determine antiviral activity and the synthesis of compounds targeted specifically to inhibit enzymes unique for viral replication. The former approach has generally been an outgrowth of screening trials for anticancer drugs, as was the case with synthesis of idoxuridine. Compounds with known cellular inhibitory activity underwent structural alterations and, then, were tested for both anticancer and antiviral activity.

This approach has led to compounds that have been defined as first-generation antiviral drugs because, for the most part, they are nonselective inhibitors of both host-cell and viral replication. It is conceivable that a selective inhibitor of only viral replication might be developed in such a random fashion. Examples of these compounds other than idoxuridine include vidarabine (9-β-D-arabinofuranosyl adenine) and cytarabine (1-β-D-arabinofuranosyl cytosine). With the exception of vidarabine, these two drugs demonstrated significant evidence of toxicity both on a cellular level as well as in clinical trials. More recently, the structure of nucleoside analogs has been specifically altered in order to create antiviral drugs that act at targets unique for viral replication. These specific inhibitors of viral replication have become identified as second-generation antiviral drugs and include acyclovir, bromovinyl deoxyuridine, and phosphonoformate. In fact, the development of acyclovir was in part fortuitous, arising from a synthetic program designed to develop inhibitors of adenosine deaminase.

After structural identification, new compounds are screened *in vitro* for antiviral activity against a variety of RNA and DNA viruses. Any indication of drug activity is confirmed in a variety of cell culture systems and further defined for the specific site of action on a molecular basis. Of those compounds considered for human trials, acyclovir and phosphonoformate are prototypic models, as will be discussed below.

Compounds that survive *in vitro* screening programs may still fail when efficacy is assessed in the intact host, be it animal or, ultimately, humans. Discrepancies between the *in vitro* single-cell systems and those of intact multiorgan systems should be identified prior to controlled trials in humans. Cellular considerations of importance, aside from the issue of specificity of drug action, include the need to determine the ability of a compound and its metabolites to traverse the cell membrane and attain an optimal intracellular concentration, the site where the active compound(s) must be present to effectively inhibit viral replication. This is a particularly important problem with nucleoside analogs, especially when applied topically to the skin. In addition, the metabolic fate of the drug and its derivatives within the cell must be defined. This knowledge would allow for some judgment of either potential toxicity or efficacy prior to the initiation of expensive and difficult human trials.

Translating cellular events to an intact host introduces new, more complex variables that must be assessed before proceeding to controlled trials designed to determine efficacy. A useful intermediary stage of development has been the utilization of animal models for the purpose of testing potential usefulness. Depending upon the animal species selected for the model, drug metabolism may well vary, resulting in metabolites capable of mediating either beneficial or harmful effects. A drug that exemplifies this principle is vidarabine, a compound discussed in detail below. Additional considerations when evaluating animal model data include the strain of virus, animal species selected, quantity and route of virus inoculated, and host manipulation with drugs. Importantly, animal models of human disease remain imperfect as they do not truly mimic the human illness as demonstrated by no animal model reflecting the true course of herpes simplex encephalitis.

Human trials of antiviral agents magnify the difficulties encountered in assessing improvement because of variables such as the natural history of the infection and peculiar propensities of the infecting virus. In contrast to bacterial infections where signs and symptoms of disease parallel microbial replication, clinical illness associated with viral infections occurs only after an appropriate incubation period during which time viral replication peaks. Thus, therapy initiated at the appearance of clinical illness will actually be late in the course of infection at a time when productive viral replication is waning. Extending these observations to human illness introduces fascinating new variables. Namely, there is a broad spectrum of illness with any viral infection, ranging from the totally asymptomatic, the most usual occurrence, to symptomatic illness with its obvious clinical manifestations. Because of the difficulties of laboratory diagnosis of most viral infections, only a few, at least at the present, are targets for antiviral chemotherapy; such targets include those previously noted as herpes simplex and herpes zoster, where skin manifestations can allow for early clinical diagnosis.

The unique characteristics of the infecting organism must be taken into consideration, as well. Herpesviruses provide excellent examples of organisms with such unique characteristics, as they are subject to latency, chronicity, and recurrences. Whatever the therapeutic approach, active or productive infections must be controlled and, ideally, the state of latency must be abated or blocked so as to significantly reduce recurrent attacks. This must be accomplished without subjecting the patient to undue acute or chronic toxic effects such as a propensity for developing cancer or harboring a drug-resistant virus with unknown pathogenic potential. This is especially true for herpetic infections as the natural history is most often clinically silent or, with notable exceptions, results in annoying medical problems rather than life-threatening disease. Thus, therapeutic measures, including prevention, must be devoid of adverse acute or chronic reactions, particularly for nuisance herpesvirus infections.

To determine the potential effects, good or bad, following administration of any presumed antiviral compound, evaluation must proceed through blind and controlled trials that employ a placebo, at least at the outset. Such an approach allows for the definition of the natural history of the disease in the patient population evaluated and can provide a critical baseline for the evaluation of any future experimental compounds. In so doing, the therapeutic index or ratio of efficacy to toxicity can be established. In diseases with predictably high mortality rates or severe permanent morbidity (e.g., herpes simplex encephalitis), the allowable index can be much lower than when the disease is less severe (e.g., herpes labialis). This is especially important when distinguishing the effect of a drug on outcome from that of a natural host response; thus, each disorder and drug must be evaluated separately. For those diseases where an appropriate therapeutic index is achieved, dosage duration of therapy, and route of administration will be further predicated on such critical variables as age, severity of the underlying disease process, concomitant medications administered, site of involvement, and intactness of the immune system, to mention but a few. Purine and pyrimidine derivatives have the potential for toxicity; therefore, it is important that proof of diagnosis be mandatory.

With these considerations in mind, topical and systemic therapy of herpesvirus infections are discussed from the standpoint of the mechanism of drug action, development for potential clinical usefulness, and clinical relevancy. The focus will be on herpes simplex and herpes zoster viral infections as treated with idoxuridine, vidarabine, trifluorothymidine, acyclovir, phosphonoformate, and bromovinyl deoxyuridine.

B. Antiviral Drugs

Idoxuridine was the first of the antiviral agents tested on both *in vitro* and *in vivo* levels of antiviral activity. The replacement of the methyl group at the 5' position of the pyrimidine ring with a halogen results in alterations of the electron configuration of thymidine. Although the pK_a becomes more acidic, the van der Waal radii remain nearly identical, allowing the halogenated compounds and subsequent derivatives to enter similar reactions as thymidine and its intermediaries. The mechanism of action of this drug focuses on its steric similarities to thymidine. The three sites of action by idoxuridine are: (1) competitive inhibition of the phosphorylating system leading to incorporation of thymidine into DNA, (2) feedback inhibition by idoxuridine triphosphate of thymidine kinase, dCMP deaminase, and ribonucleotide reductase, and (3) alteration of DNA expression after incorporation of idoxuridine. Of note, trifluorothymidine acts in a similar fashion. Notably, the activity of this drug, as is the case for other antiviral agents currently available, is at the level of β gene expression during the replicative cycle of HSV. It is during this phase of viral replication that enzymes essential for the production of progeny virus are synthesized. Although usefulness of idoxuridine for therapy of herpes simplex keratoconjunctivitis is established, toxicity in the form of bone marrow suppression precludes parenteral therapy.

Vidarabine was first synthesized in the early 1960s, as a potential anticancer agent, and was found to have antiviral activity against HSV and vaccinia virus in cell culture (Privat de Garilhe and De Rudder, 1964). These findings were extended at approximately the same time by investigators at Parke–Davis and Company (Ann Arbor, Mich.) and Southern Research Institute (Birmingham, Ala.) who found similar significant antiviral activity of an antibiotic concentrate derived from fermentation with *Streptomyces antibioticus* (Lee *et al.*, 1964; Dixon *et al.*, 1968). However, this compound was soon identified as vidarabine, which could be prepared inexpensively from fermentation or laboriously isolated along with other arabinosyl nucleosides, such as cytosine arabinoside, from the Caribbean sponge *Cryptotethia crypta*. Detailed reviews of these compounds appeared soon after their discovery (Cohen, 1966; Schabel, 1968; Suhadolnik, 1970).

Following the initial *in vitro* reports of Privat de Garilhe and De-Rudder (1964) of the antiviral activity of adenine arabinoside against HSV, Miller *et al.* (1968) demonstrated that other DNA viruses, particularly two other members of the herpesvirus family, VZV and cytomegalovirus (CMV), were sensitive to this compound when studied in other cell culture systems. *In vitro* assessment of this drug when compared with cytarabine, idoxuridine, and trifluorothymidine indicated a higher therapeutic index than the others, primarily because of lower cellular toxicity within an apparent efficacious dose range (Schabel, 1968). Each of these drugs acts to varying degrees on the incorporation of its triphosphate derivative into host-cell DNA, as noted previously.

The mechanism of action of the arabinosyl nucleosides has recently been reviewed by Cohen (1966). Vidarabine, in spite of deamination to arabinosyl hypoxanthine (Ara-Hx), is transported into cells where phosphorylation to the mono-, di-, and triphosphate derivatives occurs. The triphosphate derivative acts as a competitive inhibitor of the DNA polymerase, perhaps even more so for the virus-induced than for the host-cell α- or β-DNA polymerases. It is also a potent inhibitor of ribonucleoside diphosphoreductase. In addition, vidarabine is incorporated into the growing DNA molecule, with the 3'-exonuclease of HSV DNA polymerase acting as a pseudo-chain terminator. Vidarabine itself is capable of binding to 5-adenosyl homocysteine hydrolase, leading to the accumulation of 5-adenosyl homocysteine, an inhibitor of methylation reactions such as the capping of mRNA. Even though vidarabine acts at many other sites, its essential mechanism of action, although not clearly defined, appears to be similar to idoxuridine. Because of the lack of specificity, toxicity is potentially inherent in any of these first-generation antiviral agents.

Second-generation antiviral compounds, namely specific inhibitors of viral replication, have recently been introduced into clinical trials. One drug attracting significant attention in the biomedical community is acyclovir, 9-(2-hydroxyethoxymethyl)guanine. This compound was synthesized by Schaeffer at Burroughs Wellcome Company Research Laboratories in the late 1960s (Schaeffer et al., 1971, 1978). The compound was fortuitously discovered through a biosynthetic program designed to synthesize inhibitors of adenosine deaminase and was found to have excellent antiviral activity against herpesviruses and, in particular, HSV. Further laboratory studies comparing acyclovir to other antivirals demonstrated a significantly higher relative potency when compared to vidarabine, cytosine arabinoside, and idoxuridine. More importantly, however, the mechanism of action of acyclovir was found to be unique when compared to the already existing agents. Elion et al. (1977) determined that acyclovir is selectively phosphorylated by HSV-specified thymidine kinase to acyclovir monophosphate, which is then converted by cellular kinases to di- and triphosphate derivatives. Acyclovir triphosphate can be found in HSV-infected cells in concentrations 40–100 times higher than that found in uninfected cells. The triphosphate derivative of acyclovir then acts as a competitive inhibitor of HSV DNA polymerase by competing with deoxyguanosine triphosphate for viral DNA polymerase (Elion et al., 1977). HSV DNA polymerases exhibit a 10- to 30-fold greater affinity for acyclovir triphosphate than do cellular DNA polymerases. Small quantities of acyclovir triphosphate, however, are incorporated into host-cell DNA. Furthermore, the bioactivation of acyclovir can occur with other herpesviruses, namely Epstein–Barr virus (EBV), which does not produce its own thymidine kinase. Thus, vagaries remain both about its activity (EBV) and incorporation into DNA (host cell). These studies, nevertheless, have provided a landmark foundation for the development of specific inhibitors of viral replication, recognizing again that HSV inhibition is at the level of β gene expression.

Two other specific inhibitors have approached clinical trials for which data *in vitro* indicate specificity of action. The first, phosphonoformate, was synthesized 60 years ago by Nylen (1924) but was not recognized to have activity as an antiviral agent until recently when selective inhibition of cell-free DNA polymerase activity induced by HSV was demonstrated. This drug can be considered an analog of pyrophosphate. The mechanism of action is similar to the structurally related compound, phosphonoacetate, a drug not considered for clinical trials because of potential toxicity. The activity of the compound is attributed to its inhibitory effect on herpesvirus-induced DNA-dependent DNA polymerase while sparing host-cell polymerases. Inhibition of the polymerase is likely the result of direct interaction at the pyrophosphate-binding site (Helegstrand *et al.*, 1978). It is unclear whether drug binds at or overlaps the pyrophosphate-binding site; yet, both HSV DNA polymerase activity and the associated 3'-exonuclease activity are inhibited.

Several other compounds are nearing clinical evaluation for therapy of herpesvirus infections. One of the most promising of the compounds to approach clinical trials is (E)-5-(2-bromovinyl)-2'-deoxyuridine, reportedly a specific inhibitor of HSV replication and extremely potent against HSV-1. This compound can be utilized to distinguish between HSV-1 and HSV-2 by differences in the concentration of bromovinyl deoxyuridine required to inhibit replication (DeClercq *et al.*, 1979). Compounds of similar activity exist in a class of deoxycytidine derivatives of which 1-(2-deoxy-2-fluoro-β-D-arabinofuranosyl)-5-iodocytosine (FIAC) is the progenitor. The complicated metabolism of FIAC leads to 5-iodouracil (FIAU) and 5-methyluracil (FMAU) derivatives, which also have significant antiviral activity. However, because of only limited amounts of these compounds being available, clinical studies remain very preliminary and inconclusive. Numerous other compounds, perhaps with less likelihood for reaching clinical trials in therapy of herpesvirus infections, include ribavarin, amino and ethyl derivatives of idoxuridine, AIU, and EtdUrd, and 1-β-D-arabinofuranosyl thymine, among others.

II. THE ROAD TO CONTROLLED TRIALS

Studies of the clinical pharmacology, safety, and tolerance of idoxuridine, trifluorothymidine, phosphonoformate, and bromovinyl deoxyuridine are limited. Idoxuridine is rapidly excreted and metabolized so that nearly all drug and metabolites are recovered from the urine within 24 hr after single-dose administration (Calabresi *et al.*, 1961). Drug does not enter the CNS after intravenous administration. The remainder of this section will focus upon the currently popular antivirals vidarabine and acyclovir.

A. Vidarabine

Studies of the pharmacokinetics, safety, and potential clinical usefulness of vidarabine have been detailed (Whitley *et al.*, 1980c). Early studies show that when man and higher animal species, such as simians, are given vidarabine, they deaminate the parent compound to Ara-Hx, a derivative that has antiviral activity, although less than the primary compound. Lower animal forms primarily oxidize vidarabine to xanthine, which is freely excreted in the urine (Glazko *et al.*, 1975). This latter observation implies the need for caution when interpreting experimental data generated from animals that oxidize rather than deaminate vidarabine. Second, results from radioactive tracer studies indicate that the plasma half-life of vidarabine approximates 4 hr. These studies have been performed primarily in adults; thus, age variation may well occur. Third, similar radioactive tracer studies demonstrated the accumulation of active drug and metabolites within erythrocytes over 5–7 days, with persistence as long as 3 weeks. These observations were extended to several terminal leukemia patients whose tissues were examined several weeks after dosing for the presence of active compound. The highest drug levels were found in kidney, liver, and spleen; somewhat lower levels were obtained from skeletal muscle and, importantly, brain. Fourth, during and as long as 10 days after therapy, vidarabine and its metabolites can be found in the cerebrospinal fluid at approximately one-half the serum levels. Fifth, multidose pharmacokinetic studies in humans demonstrated that the primary extracellular metabolite was Ara-Hx. Because of its poor solubility, the drug was administered by constant infusion over 12 hr. During such infusions of dosages of 10–15 mg/kg per day, Ara-Hx peaks at termination of administration at 4–8 µg/ml. Vidarabine is present but only at barely detectable levels of approximately 0.2 µg/ml, the lower limit of detection. These data imply no extracellular accumulation following multiple-dose administration. Finally, vidarabine is cleared from the body by the kidney with nearly 60% appearing in the urine primarily as Ara-Hx over the first 24 hr after administration. All of these data provided the foundation for field trials in man designed to test clinical usefulness.

This basic understanding of drug disposition in man led to uncontrolled field trials designed to determine a safe dosage range or human tolerance for targeted herpesvirus infections. The targeted diseases included only those with potentially significant morbidity and/or mortality such as herpes simplex encephalitis, neonatal HSV infection, progressive mucocutaneous HSV infections, and VZV infections in the immunocompromised host. Numerous investigators in this country and abroad contributed to these endeavors by performing open drug evaluations. These studies provide a logical sequential step in the development of any antiviral drug for clinical trials and have helped in the determination of a safe dose range to be used in later controlled studies.

These early phase II trials of vidarabine can be summarized as follows. First, vidarabine therapy of progressive and prolonged (longer than 21 days) mucocutaneous HSV-1 infections of the immunosuppressed patient resulted in decreased virus shedding paralleled by clinical resolution of lesions. These patients demonstrated a 90% or greater reduction in the quantity of virus excreted from the involved site within 72 hr of initiation of therapy. The response to genital HSV, due to type 2, was less remarkable than that to mucocutaneous HSV, due to type 1, a finding not completely surprising based upon *in vitro* sensitivity data (Chien *et al.*, 1973). Second, a similar acceleration of cutaneous healing in VZV infections was noted as evidenced by decreased new vesicle formation and elimination of virus from vesicles. For both of these studies, vidarabine at dosages of 5–15 mg/kg per day over 5–10 days suppressed or reduced viral replication and allowed the altered host to control the infection without signs of toxicity (Johnson *et al.*, 1975). It appears as though some degree as of yet undefined host response was necessary to bring the infection under control. Third, vidarabine therapy of newborns infected with HSV appeared to decrease mortality if instituted early in the course of infection (Chien *et al.*, 1975). Fourth, systemic administration of this compound to patients with deep herpetic uveitis and topical administration to patients with herpes keratoconjunctivitis resulted in prompt healing without adverse reactions (Kaufman *et al.*, 1972). Fifth, data have become available that demonstrate suppression of the virus-induced DNA polymerase in patients with chronic hepatitis B infection who received therapy with vidarabine (Hafkin *et al.*, 1979). In this patient population, unexplained toxicity has appeared at seemingly appropriate dosages; however, mainly in combination with interferon. This finding may be related to the diseased liver, a site necessary for deamination of vidarabine. From these trials, a safe dose range was established at 5–15 mg/kg per day given as a continuous intravenous infusion over 12 hr. At these dosages, no clinically significant laboratory toxicity was detected. The only adverse clinical effects were nausea with or without vomiting and diarrhea in small percentages of drug recipients. Exceeding these doses, at least in adults, can result in evidence of overt toxicity. At 20 mg/kg per day, evidence of weight loss, tremors, and megaloblastosis of the erythroid series occurred (Ross *et al.*, 1976). If the dosage exceeded 30 mg/kg per day, severe thrombocytopenia and leukopenia were detected after 10–14 days of therapy (Bodey *et al.*, 1975).

B. Acyclovir

Pharmacokinetic studies to establish the appropriate dose and frequency of administration have been conducted with acyclovir for each of the three major routes of drug administration: topical, oral, and intravenous. These studies have recently been reviewed along with therapeutic trials (Gnann *et al.*, 1983). It is important to recognize that, at this stage in the development of antiviral drugs, many questions regarding pharmacokinetics remain unresolved. For example, plasma concentrations of antiviral drugs may not reflect drug concentrations within host cells, the site where viral replication occurs. Similarly, peak and trough plasma drug concentrations are of unknown importance in the therapy of viral infections. Nevertheless, the systemic and thorough evaluation of the clinical pharmacology of acyclovir in a variety of patient populations provides a model for the development of future antiviral drugs. It is worth briefly reviewing these developments.

1. Plasma Concentration

Single-dose pharmacokinetic studies were performed in human volunteers by infusing 0.5, 1.0, 2.5, or 5.0 mg/kg of acyclovir intravenously over 1 hr and serially measuring plasma and urine acyclovir concentrations. These studies demonstrated that the postinfusion plasma concentrations of acyclovir declined biphasically such that plasma concentration–time data fit a two-compartment open model with zero-order input (deMiranda *et al.*, 1979a; Laskin *et al.*, 1982a; Spector *et al.*, 1981). The mean peak plasma concentrations at each of the four doses studied was 6.4 ± 1.2, 12.1 ± 4.7, 14.9 ± 3.8, and 33.7 ± 15.8 μM, respectively (deMiranda *et al.*, 1979). Each of these peak plasma acyclovir levels exceeds concentrations required to inhibit replication of HSV-1, HSV-2, and VZV *in vitro*. The mean half-life of infused acyclovir was 3.16 ± 0.20 hr and did not vary significantly with dosage. There was proportionality between the area under the plasma concentration–time curve from zero to infinity (AUC) and the dose.

The single-dose study was followed by a single-day, multidose study utilizing dosages of acyclovir of 2.5, 5.0, and 10 mg/kg per day divided into three 1-hr infusions 8 hr apart (deMiranda *et al.*, 1979b). These data also demonstrated biphasic clearance of acyclovir. Trough plasma concentrations were achieved at 1.0 to 4.1 μM for the dosages studied, while peak plasma concentrations ranged from 10 to 80 μM (Whitley *et al.*, 1982a).

Multiday, multidose studies were conducted in which peak plasma concentrations could be measured after repetitive dosing and steady-state plasma concentrations and disposition kinetics determined. The mean peak plasma drug levels after the first dose of 2.5, 5.0, 10.0, or 15.0 mg/kg were 22.5 ± 12.0, 38.2 ± 5.5, 86.8 ± 18.1, and 94.8 ± 30 μM, followed by steady-state plasma levels of 30.1 ± 12.8, 43.2 ± 4.3, 88.9 ± 5.5, and 91.7 ± 11.1 μM, respectively. There was a linear correlation between steady-state peak concentrations and AUC with dosage when data from two patients with unusual creatinine clearance values were excluded (Whitley *et al.*, 1982a).

The half-life data indicate that an appropriate dosing interval for acyclovir is every 8 hr and that there is no accumulation of drug with repetitive dosing. Similar plasma levels are achieved for adults and children when equivalent doses are calculated on the basis of body surface area.

2. Urinary Clearance of Acyclovir

Cumulative urinary recovery of unchanged drug ranged from 30 to 69% of the dose in the single-dose study and from 38 to 76% in the single-day, multidose studies. The mean urinary recovery in the multiday, multidose study was 68 ± 14% with a range of 33–84%. Patients with the highest serum creatinines and lowest creatinine clearances had the lowest total urinary recovery of acyclovir. The mean renal clearance of acyclovir for each of the three studies was 125 ± 54, 177 ± 87, and 184 ± 79 ml/min per 1.73 m^2, each being significantly higher than the corresponding creatinine clearance (57 ± 13, 62.9 ± 3.3, and 70.2 ± 40.7 ml/min per 1.73 m^2) and indicating tubular secretion (Whitley *et al.*, 1982a).

Analyses of selected urine samples by high-performance liquid chromatography indicated the presence of a metabolite identified as 9-carboxymethoxymethylguanine (deMiranda *et al.*, 1981). Quantitative assays for this compound showed that 2.4–13.6% of the acyclovir dose was present in the urine as this metabolite in patients with normal renal function. These data indicate that the kidney is the major route for clearance of acyclovir and that both glomerular filtration and tubular secretion are important mechanisms of elimination. Tubular secretion of acyclovir can be partially blocked with probenecid, though any clinical applicability of this observation remains to be defined (Laskin *et al.*, 1982c).

3. Acyclovir Concentrations in Cerebrospinal Fluid

Samples of cerebrospinal fluid from two patients, obtained 2 hr after infusion of 5 mg/kg doses of acyclovir, showed drug levels of 16.6 and 17.8 μM. Model-predicated plasma levels at the times of cerebrospinal fluid collection were 35.0 and 36.1 μM, respectively. These data are important if therapy of HSV infections of the brain with acyclovir are anticipated (Whitley *et al.*, 1982c).

4. Acyclovir Pharmacokinetics in Renal Failure

Data have been generated for clearance of acyclovir in end-stage renal disease (Laskin *et al.*, 1982b). From these studies the following relationships between drug half-life and clearance and renal function have been established, indicating that with decreasing creatinine clearance the half-life of acyclovir in the plasma increases significantly. In the anuric patient, hemodialysis will eliminate drug as indicated by an extraction coefficient of 0.45 ± 12 with a fourfold enhancement in the elimination of acyclovir during dialysis (Laskin *et al.*, 1982b). A 6-hr dialysis period will reduce the plasma acyclovir concentration by 60%.

5. Alternative Routes of Drug Delivery

Pharmacokinetic properties of acyclovir following oral administration and topical application have been determined. Peak plasma levels after oral administration of 200 mg of acyclovir were found at 1.5–1.75 hr postdose and ranged from 1.4 to 4.0 μM (Van Dyke *et al.*, 1982; deMiranda *et al.*, 1982). Further studies are currently under way utilizing 400- and 800-mg doses. Studies of topical application indicate that cutaneous absorption is minimal through intact skin. Application of acyclovir 5% ointment in a polyethylene glycol base to intact skin (surface area of 40 cm²) four times daily for 7 days did not result in detectable drug levels in either plasma or urine. However, utilizing a similar regimen of application on lesions of herpes zoster, acyclovir was detected in the plasma at very low concentrations (1.20–3.35 μM) (M. J. Levin, Sidney Farber Cancer Institute, personal communication). Similar absorption studies performed in patients with initial genital HSV infections have not yet been reported; however, it is unlikely that systemic absorption would exceed that found in therapy of localized zoster where the surface area of damaged skin involved is greater.

III. CLINICAL TRIALS

A. Herpes Simplex Keratoconjunctivitis

Herpetic infection of the eye has become recognized as one of the most common infectious causes of blindness in the United States. Of all modalities of antiviral chemotherapy available to date, the earliest advances were in the area of topical therapy of herpes keratoconjunctivitis. Interestingly, these infections are most commonly treated as recurrences. Indoxuridine, trifluorothymidine, and vidarabine all can abort individual attacks, each with varying but acceptable degrees of toxicity, the most pronounced being the allergic reactions witnessed with idoxuridine therapy. The genesis of this allergic raction remains undefined. All three of these drugs are licensed by the Food and Drug Administration for therapy of herpes simplex infection of the eye.

It is important to recognize that topical therapy of HSV infections of the eye simply accelerates the time to total healing, hopefully improving visual acuity in the process. Therapy does not influence latency and, therefore, recurrences are problematic. The eye represents a unique organ of the body for which therapy of a recurrent herpes simplex infection is not only possible but feasible, a rare demonstration indeed.

B. Cutaneous HSV Infections

A major focus of antiviral research in the United States has been attempts to develop therapeutically useful and simple modalities of therapy for herpes labialis and genital HSV infections. As reviewed recently by Overall (1980), numerous compounds have been tested for these diseases with limited evidence of success. No doubt these failures are in large part related to the inability of these nucleoside derivatives to penetrate the skin and reach the site of viral replication.

In marked contrast to the availability of topical therapy for herpetic eye infections, evidence of efficacy for therapy of individual attacks has only recently been demonstrated for local treatment of mucocutaneous disease due to HSV-2 but not HSV-1. The compounds successful for therapy of eye infections were similarly tested on patients with genital herpes and recurrent nuisance fever blisters. To date, only the British experience with 5% idoxuridine in dimethylsulfoxide (DMSO) applied topically or administered by air gun injection has been useful for therapy of acute disease (Juel-Jensen and MacCallum, 1965). In this country, the Food and Drug Administration has only recently allowed trials utilizing the vehicle DMSO; however, no data have been reported from American trials.

The introduction of acyclovir into clinical trials in the formulation of polyethylene glycol or modified aqueous cream has spawned many studies of primary and recurrent HSV infections of the skin. It is essential to distinguish between the type of skin infections and the site of involvement as outcome expectations will be markedly different.

No studies, to our knowledge, have utilized topically applied acyclovir for primary herpes simplex gingivostomatitis, an extremely common clinical problem encountered in the practice of pediatrics. Recognizing that most HSV-1 infections are asymptomatic, latency with the subsequent potential for recurrences is established. It is recognized that the prevalence of antibodies to HSV is from 50 to 80%, depending upon socioeconomic status. As many as 70 million Americans suffer from recurrent herpes labialis, a disease of 7 days' duration and one that uncommonly leads to facial disfigurement in the normal host. In acyclovir treatment of active recurrent herpes labialis, no benefit could be demonstrated for time to total healing, crusting, and loss of pain; however, there was a trend to accelerated loss of virus from lesions (Spruance and Crumpacker, 1982). A follow-up study to determine if therapy instituted during the prodomal phase of illness similarly failed to show evidence of a drug effect (Spruance et al., 1982).

Before leaving the subject of orolabial HSV infections, it is worth briefly noting findings of two series of studies to determine both prophylactic and therapeutic effects of systemic drug for these infections in the immunocompromised host. A unique study performed by Saral *et al.* (1982) demonstrated that acyclovir could be successfully utilized to prevent severe HSV infections after bone marrow transplantation. Similarly, Wade *et al.* (1982) were able to demonstrate the clinical usefulness of intravenous therapy for disease in bone marrow transplant recipients. These studies as well as others indicate that acyclovir is a useful antiviral for immunocompromised patients at high risk for developing severe and progressive HSV infections. Clearly, in contrast to normal hosts with genital HSV infections, the risks of the infection, particularly the morbidity from continuous viral excretion and expanding lesions, far outweigh those of therapy. Likely, oral therapy will be useful in these patients as well.

Genital HSV infections are of increasing public health significance. It has been estimated that in the United States between 300,000 and 500,000 new cases occur annually and as many as 10–20 million individuals have recurrent infections. The actual prevalence of genital HSV-1 and HSV-2 remains unknown because of the difficulty distinguishing between antibodies to these two viruses. Primary genital infection is defined as first exposure to either type of HSV. It is associated with systemic symptoms in over 50% of patients and between 16 and 26% develop meningitis (Corey *et al.*, 1983a). The mean duration of viral shedding and pain is 11–12 days. Lesions persist on the average of 3 weeks. Recurrent disease is much less severe such that pain persists 4–6 days, viral shedding ceases in approximately 4 days, and lesions are healed in 10 days. Disease of intermediary form is nonprimary but initial genital HSV infection. In this situation, preexisting antibodies and/or cell-mediated immune responses to labial or cutaneous herpes simplex infection ameliorate the clinical course such that systemic symptoms are decreased to 16% and pain and viral shedding are decreased to 9 and 7 days, respectively (Corey *et al.*, 1983a).

Recently, data have been presented to show clinical usefulness for topical therapy of primary genital HSV infections. In a study of 77 patients with first-episode genital herpetic infection, topical acyclovir applied six times daily reduced the duration of viral shedding, alleviated local symptoms, and accelerated time to crusting by mean times of approximately 3, 2, and 4 days, respectively. Topical therapy, however, did not significantly shorten the duration of total disease or decrease the appearance of new lesions. These data did not appear to be mediated by sex or preexisting antibodies to HSV. Furthermore, therapy did not alter the frequency or severity of recurrences (Corey *et al.*, 1982).

In the same study, 111 patients were similarly evaluated for the therapeutic effectiveness of acyclovir on the management of recurrent genital herpes infections. The data suggest an effect on shedding of virus from lesions and, perhaps, the time to crusting of lesions in men but not women. Overall, topical application of acyclovir to existing lesions had no effect on the symptoms of disease or healing time in either sex, with the exception of crusting in women. Similarly, no effect on recurrences was apparent. A follow-up study to determine therapeutic effectiveness when application of medicine is introduced during the prodrome has been completed and similarly appears to show no benefit (Luby et al., 1983).

These trials encapsulate the difficulties encountered in topical studies of antiviral agents. Specifically, issues regarding penetration of nucleoside derivatives to the site of viral replication and the clinical value of therapy become foremost questions. In that recurrences are not prevented following therapy of either primary or established latent infections, improved therapeutic modalities must be developed. Furthermore, the potential for development of drug resistance is apparent in a disease such as genital HSV infection with a known propensity for recurrence. The probability of developing a resistant virus increases with repeated exposure to medications following each episode (Crumpacker et al., 1982). Such resistant viruses have been documented, following repeated systemic therapy with acyclovir. The frequency of development of resistance and the potential pathogenicity of these isolates will require further extensive investigation. Nevertheless, one must weigh the value of topical therapy of HSV infections against the need to reserve such potentially useful drugs for truly life-threatening disease lest problems of drug resistance encountered in the management of bacterial disease occur also with viral infections.

Only recently have field trials been initiated with phosphonoformate, a compound developed in Sweden, for topical therapy of both herpes labialis and genital HSV infections. These trials have employed crossover study designs and can only be considered preliminary because of the limited numbers of patients studied and, consequently, an inability to appropriately stratify analyses for demographic characteristics as well as for treatment responses. Regardless, these initial studies demonstrate a trend toward accelerated healing, particularly loss of virus from lesions and time to total crusting. A larger, collaborative trial has recently been initiated to assess clinical usefulness in Scandinavia. With phosphonoformate, as with acyclovir, resistant strains of virus have been encountered and, clearly, this possibility must lead to careful definition of frequency of occurrence. Should this occur even as frequently as 1–2% following a course of therapy, serious consideration must be given to the possible resultant harm that could result from drug application to a usually benign illness.

With the apparent difficulties of skin penetration of nucleoside analogs when applied topically, two alternate systemic routes of therapy have been studied; these are oral and intravenous administration of acyclovir. Preliminary findings from these studies are encouraging for acceleration of skin healing. Findings of a study of intravenous administration of acyclovir at 15 mg/kg per day for 5 days to 31 hospitalized patients with initial disease demonstrated a reduction of viral shedding by a mean of 11 days ($p = 0.001$), in symptoms by a mean of 5 days, and acclerated time to total healing (Corey *et al.*, 1983b). Furthermore, complications of infection were eliminated in those who received drug. Placebo recipients had complications that included extragenital disseminated lesions (2 of 16 patients) and urinary retention (2 of 16 patients). These data clearly indicate enhanced healing when compared to data derived from the topical trial; however, cost-effectiveness of intravenous therapy, which requires 5 days of hospitalization, can only be justified if recurrences are prevented or for therapy of severe complications. In the follow-up of these 31 patients, 60% of acyclovir recipients versus 88% of placebo recipients experienced a recurrence within 10 months of infection. These data can be transposed to a recurrence frequency per month of 0.26 and 0.50 for acyclovir and placebo recipients, respectively. A larger number of patients is being assessed to determine the effect of intravenous therapy on recurrences. In this trial there was no evidence of significant clinical or laboratory toxicity nor the appearance of resistant virus in either randomization group.

A double-blind placebo-controlled study has been completed for oral therapy of initial genital HSV infections. This study involved 48 patients who received 200 mg of acyclovir or placebo five times daily for 10 days (Bryson *et al.*, 1983). The data from this trial are similar to those found following intravenous administration of drug; thereby, a much less costly modality of therapy is available. A beneficial effect on the frequency of recurrences was seen but this remains to be verified. Here also, however, the physician must be concerned with the possibility of drug resistance following repeated exposures to medication.

Unique study designs have examined the potential beneficial effect of continuous acyclovir suppressive therapy on the frequency of recurrences of genital herpes. These trials (Larson *et al.*, 1983; Corey *et al.*, 1983a; Straus *et al.*, 1983) all suggest benefit by continuous therapy such that recurrences are significantly diminished in number. These beneficial effects must be weighed against the possible harmful effects of continuous administration of a nucleoside analog capable of becoming incorporated in its normal host-cell DNA.

Although the results of several trials substantiate the antiviral effects of acyclovir for genital HSV infections, the effect focuses on acceleration of healing of acute disease. These findings alone represent major advances in therapy of viral disease. Yet, clinical usefulness may be short-lived if recurrences are not prevented and resistant virus develops.

C. Herpes Simplex Encephalitis

1. Background

Herpes simplex encephalitis is the most common cause of sporadic fatal encephalitis in the United States today. It is estimated to occur in

1 in 750,000 to 1,000,000 individuals annually; and as many as one-third of individuals with biopsy-proven disease are children. Patients with herpes simplex encephalitis have clinical evidence of an acute febrile encephalopathy, temporal lobe signs with disordered mentation, evidence of localized CNS disease by diagnostic studies (brain scan, computerized axial tomography, and/or electroencephalography), and CSF findings compatible with viral encephalitis. In the presence of any evidence of localization on neurodiagnostic assessment and excluding alternative diagnoses, these patients should be referred for diagnostic brain biopsy in order to confirm the etiology of the focal CNS process and for treatment by experienced physicians. No other clinical or laboratory method currently exists to allow the physician to establish the diagnosis unequivocally. Brain biopsy specimens should be examined by culture and for histologic evidence of perivascular cuffing, hemorrhagic necrosis, and intranuclear inclusions. Each of these pathologic findings is suggestive but not pathognomonic of HSV infection (Whitley et al., 1982c). Other rapid diagnostic techniques include direct fluorescence microscopy and electron microscopy performed by the pseudoreplica method.

2. Therapy

The treatment of herpes simplex encephalitis introduced new variables into the clinical assessment of antivirals. Because of flawed study designs and inadequate diagnostic methods, idoxuridine had become the standard therapeutic agent for the disease in the late 1960s and early 1970s. To some extent, this level of acceptance was by default in that no other antiviral drug was available for treatment. However, by 1974 the cooperative efforts of the National Institute of Allergy and Infectious Diseases Antiviral Study Group and the Boston Interhospital Study Group (1975) had shown that idoxuridine was both ineffective and profoundly toxic for biopsy-proven disease. Clearly, the therapeutic index was negative for utilization of this drug to treat herpes simplex encephalitis.

These double-blind placebo-controlled studies were continued utilizing vidarabine and were reported (Whitley et al., 1977). In these studies, mortality was decreased from 70% in placebo recipients to 28% in drug recipients with biopsy-proven disease. These findings were statistically significant 30 days after the onset of therapy ($p = 0.03$). Quality of life of survivors in both groups varied strikingly. Overall, 20% of placebo patients had an acceptable outcome, returning to normal function or suffering from moderate debility (returned to functional existence but with some impairment) compared to 39% of drug recipients. Nevertheless, many questions remained, including verification of mortality with therapy, precise definition of morbidity to determine clinical usefulness, elucidation of diseases that mimic herpes simplex encephalitis in those biopsy-negative, and the value of brain biopsy of ineffective or placebo therapy could be verified from other studies employing identical criteria.

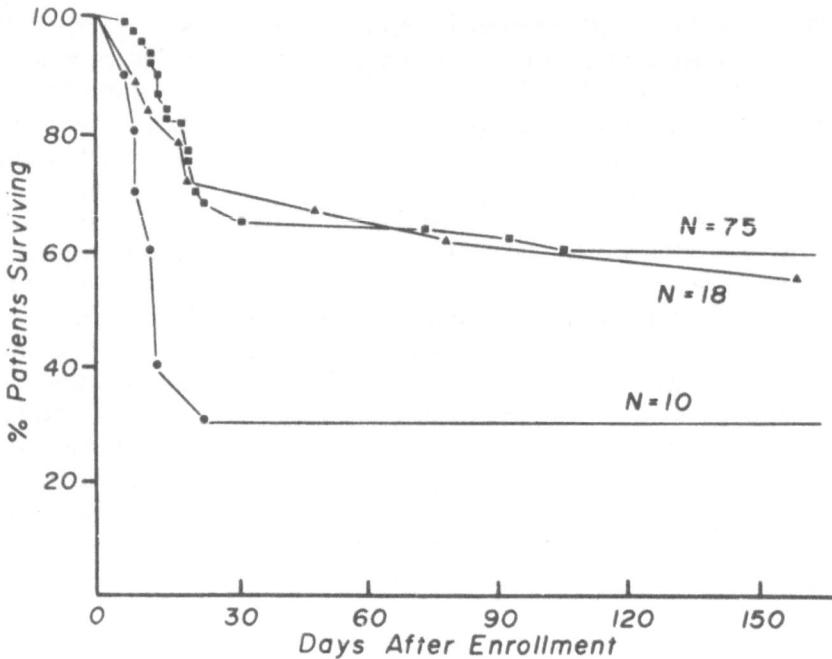

FIGURE 1. Mortality in biopsy-proved herpes simplex encephalitis. Comparisons of results from double-blind study with those from open therapeutic trial. Reprinted with permission from *N. Engl. J. Med.* **304**:313–318, 1981.

These findings indicate that without efficacious therapy, mortality will exceed 70%. Overall, 13% will return to normal function or have moderate neurologic impairment following placebo or ineffective drug treatment (Whitley *et al.*, 1981). These results further support the placebo data from the original vidarabine controlled trial.

The NIAID Collaborative Antiviral Study Group pursued these observations in order to answer several unresolved questions by performing an open and uncontrolled assessment of therapy in all biopsy-proven patients treated with vidarabine (Whitley *et al.*, 1981). The findings can be summarized as follows. First, the reduction of mortality in 75 biopsy-proved patients was virtually identical to that of the first trial as displayed in Figure 1. With this larger series of patients, it was possible to combine data from the first and second trials to assess factors that influenced outcome, namely level of consciousness and age, as summarized in Fig. 2. Individuals lethargic at the time of diagnosis and institution of therapy and those less than 30 years of age have the best prognosis, as only 10% die and more than 60% return to a prior level of function. Lethargic individuals, irrespective of age, return to the same level of function as prior to the onset of the encephalitis in 50% of all cases studied. Delaying diagnosis and institution of therapy until semicoma or coma ensues decreases the number of normal individuals at 1 year or older to approxi-

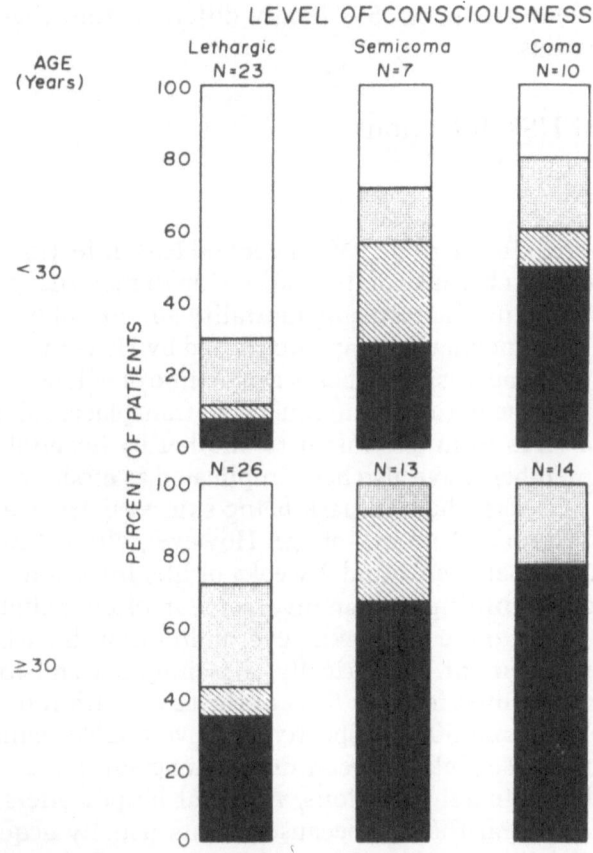

FIGURE 2. Influence of level of consciousness and age on outcome following Ara-A therapy. (□) Returned to normal function. (▦) Moderate debility. (▩) Severe impairment. (■) Death. Reprinted with permission from *N. Engl. J. Med.* **304:**313–318, 1981.

mately 20% overall. On long-term follow-up, significant neurologic improvement is evident in all survivors but particularly so in those who are young. For those less than 18 years, improvement continues over a 2-year period following therapy.

Therapy of herpes simplex encephalitis with vidarabine results in decreased mortality and improved morbidity. Nevertheless, mortality and morbidity remain significant; therefore, second-generation antiviral agents, such as acyclovir, are currently in clinical trials. It is hoped that the advent of newer therapeutic agents for this disease process and the development of noninvasive diagnostic procedures will further improve the outcome of this devastating infection. In the ongoing trials of confirmed herpes simplex encephalitis, a comparison of vidarabine to acyclovir has led to the enrollment of 52 patients with proven disease. Adequate numbers of patients have not been entered in order to allow for breaking the code; however, the 6-month follow-up mortality, irrespective of treatment group, is 35% hardly different from that in the prior vidarabine studies.

D. Neonatal HSV Infections

1. Background

First reported in the mid 1930s, neonatal HSV infection has attracted particular attention because of its relationship to maternal genital herpes, the patient's age, and the resulting mortality and morbidity. The finding that most cases of neonatal herpes are caused by HSV-2 provided support for the theory that acquisition occurs at the time of delivery from infected maternal genital secretions. Infrequently, transplacental transmission and spread of virus from one infant to another by hospital personnel or from family members have also been implicated as modes of spread. Manifestations of disease, the hallmark being skin vesicles, can be noted at birth or delayed up to 1 month of age. However, clinical manifestations occur most often between 1 and 2 weeks of life. Infection may be either disseminated with multiple organ involvement, often including the CNS, or can be localized to the CNS, skin, eye, and/or mouth without evidence of visceral involvement. Historically, disseminated and localized CNS disease have accounted for over 80% of the cases with mortality rates of approximately 80 and 50%, respectively. Severe CNS damage and, less frequently, ocular sequelae are common among survivors.

Of all the perinatal infections, neonatal herpes offers the greatest prospect for antiviral therapy because infants usually acquire infection at the time of delivery rather than early gestation. This is in contrast to the situation with congenital rubella and cytomegalovirus infection. Hence, it is conceivable that effective therapy can be delivered early in the course of disease rather than long after infection.

2. Prevention

Administration of biologic materials (interferon or antibodies) and chemotherapeutic drugs to the pregnant woman with genital herpes at the time of delivery has not been studied as a means of preventing neonatal HSV infection. The major concern in this situation is the toxic effects on the developing fetus either acutely or chronically. Cesarean section has been recommended as an effective measure for preventing neonatal HSV infection when performed within 4 hr after the onset of labor. These recommendations remain to be verified in larger studies. However, the management of each pregnant woman with genital HSV infection must be individualized in order to avoid unnecessary abortions or cesarean sections.

Recommendations for these situations have recently been reviewed (Nahmias et al., 1983). It is important to note that the great majority of these mothers had clinically recognizable infections. The question whether or not to intervene surgically when asymptomatic genital infection is diagnosed fortuitously by virologic or cytologic methods near term is unresolved. That clinically silent infection can be transmitted to the newborn infant is suggested by occurrences of HSV-2 infections in many newborn infants whose mothers denied genital lesions; the frequency of such occurrences is unknown. These findings suggest that cesarean section is warranted if genital lesions, especially internal lesions, are present at the onset of labor. Intervention in silent maternal infection can be considered only in the presence of cytologic or virologic evidence of infection. It is clear that information is needed concerning the persistence of virus in silent HSV maternal infection detected near term in order to make an informed decision about the need for cesarean section. Virus isolation should be used to assist in making this decision. However, once the decision is made to intervene, cesarean section should be accomplished as soon after rupture of membranes as possible.

3. Therapeutic Considerations

In contrast to other perinatally acquired infections, the chances for successful therapy should be greater in neonatal HSV infection simply because the disease is generally acquired intrapartum. Therefore, treatment can theoretically be instituted early in the course of disease, provided the infection is diagnosed promptly. Herein lies a problem of considerable importance. In a recent study of the natural history of maternal and newborn infection, an assessment of population data revealed that most women (70%) who delivered babies developing neonatal HSV infections were completely without signs or symptoms of infection. Admittedly, a few of these mothers noted a prior history of herpetic infection as well as presumed penile herpetic lesions of a sexual partner. Nevertheless, over half of the population was devoid of history, signs, or symptoms of genital herpetic infection surrounding the time of delivery. Thus, identification of babies born to women with genital infection will be difficult at best (Whitley et al., 1982b).

A compounding problem is the presentation of disease in the newborn infant. Specifically, skin vesicles, the hallmark of infection, never develop in approximately 10–20% of babies with HSV infection. These babies have evidence of encephalitis or visceral disease. As a consequence, diagnosis and appropriate institution of therapy are frequently delayed. An aggressive approach to diagnosis of herpetic infection in the newborn infant is essential. Aside from searching for virus in cultures of vesicular fluid, CSF, mouthwash, urine, conjunctival swab, and stool, a maternal history of genital infection should be sought and maternal genital cultures obtained. Once the diagnosis is confirmed by culture or implicated by more rapid means (electron miscroscopy or Papanicolaou smear of the lesion base), therapy must be instituted immediately.

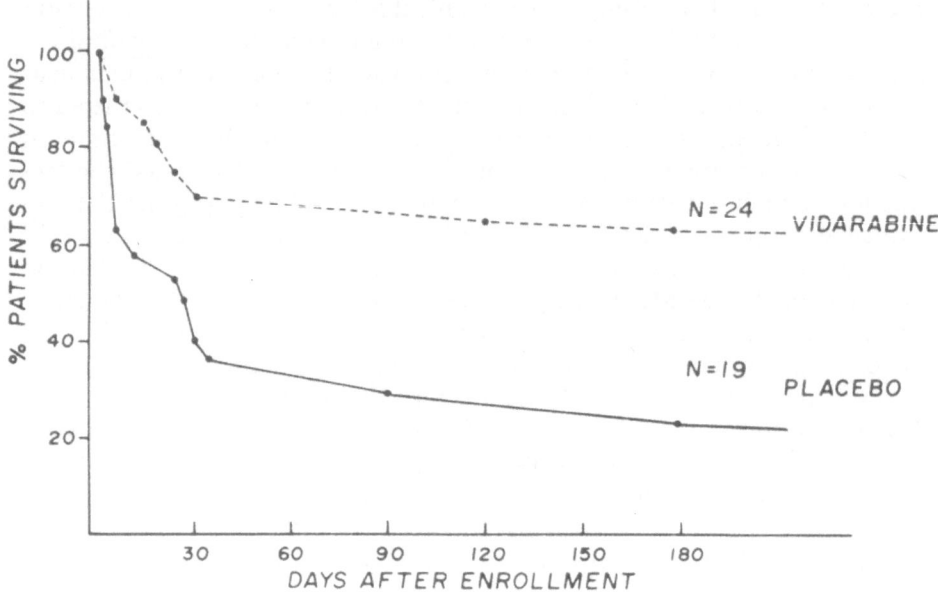

FIGURE 3. Outcome of HSV infection in neonates according to type of disease and therapy. Points represent last death(s). Reprinted with permission from *N. Engl. J. Med.* **304:**313–318, 1981.

4. Therapy

Many drugs have been attempted for experimental therapy of neonatal HSV infections. However, only one drug, vidarabine, has shown an indication of effectiveness as a therapeutic. Mortality was reduced from 74% to 38% (Fig. 3). For this study, drug was given intravenously at a dosage of 15 mg/kg per day over 12 hr for 10 days. A recent study was performed to verify effectiveness at this dosage and to determine if a higher dose, 30 mg/kg per day, further influenced mortality. The analyses of these data reinforce the observation of decreased mortality (Whitley *et al.*, 1984). However, therapy at this higher dose did not have a further beneficial effect on decreasing mortality or improving morbidity. The type of neonatal herpetic infection was the major determinant of mortality as displayed in Fig. 4. Recognizing the limited number of patients in each category, mortality rates were reduced with therapy from 50% to 10% in infants who had localized CNS disease and 85% to 57% in those with disseminated infection. These observations have also been verified in a larger series of treated newborns (Whitley *et al.*, 1984). Drug did not appear to prevent the development of more severe disease, as an equal number of drug and placebo recipients progressed during the study.

The true usefulness of a therapeutic agent can only be gauged by the quality of life of the survivors. Babies with disseminated and CNS infection are reported for normalcy according to treatment regimen. Those babies assessed as normal from both trials (Whitley *et al.*, 1980a, 1984) occurred approximately 2.5 times more frequently following therapy than did their placebo counterparts. Of the severe groups, treated babies with localized CNS disease appeared to derive the greatest benefit. Vidarabine therapy appears to reduce viral replication but obviously cannot reverse existing damage; thus, survivors had, as would be expected, a broad range of neurologic impairment.

Outcome is significantly improved if therapy is instituted prior to the appearance of brain or disseminated disease. When the disease remains limited to the skin, eye, or mouth, death is most uncommon, as revealed in a recent study. In this study, one-third of the placebo recipients had severe developmental impairment (microcephaly and spasticity) on long-term assessment. In contrast, only one of the drug recipients had evidence of minimal chorioretinitis, which did not appear until nearly 2 years of age. For therapy to be effective, it must be instituted with the earliest evidence of disease, especially as 70% of patients who present with skin vesicles alone progress to exhibit more severe forms of disease (disseminated and CNS infections). As HSV infections of the newborn infant are fulminant, it is essential to establish the extent of involvement and progression with appropriate monitors in order to define the prognosis following treatment.

At the dosages employed, evidence of adverse clinical reactions and laboratory toxicity did not occur. Bone marrow and liver function did not show evidence of impairment with therapy and, in fact, improved more rapidly in those babies receiving therapy for disseminated disease.

Vidarabine is beneficial for treatment of neonatal HSV infection, but therapy must be improved. Whether increasing the dose of drug or using other potentially effective drugs, such as acyclovir, might improve the outcome is the aim of future trials. Similarly, usefulness of administering HSV antibodies to infected newborn infants is yet to be determined. Prospective studies are under way to determine the frequency and risk of transmission of virus from mother to baby in relationship to transplacental antibodies and development of cell-mediated immunity. Management of the newborn infant with HSV infection is most complicated. Identification of babies with herpetic infection should lead to prompt transfer to regional centers that have experience in evaluation and treatment of these high-risk infants.

E. VZV Infections

1. Chicken Pox

At this time the only experience available on treatment of VZV infections is in the immunocompromised host and with the drug vidarabine. The most common infection caused by this virus is chicken pox, a benign and self-limited disease of short duration in normal children in whom serious complications are rare. In immunocompromised children, chicken pox can be severe and even life-threatening. Of particular concern is the prolonged state of the acute disease, which is associated with visceral spread, especially pneumonitis. In leukemia and other childhood malignancies, exaggerated disease is particularly likely; and there is a substantial case fatality rate. Preventive and therapeutic approaches are currently being assessed.

Presently the most acceptable preventive approach is administration of zoster immune globulin. Several groups are currently evaluating prophylaxis by active immunization with attenuated live virus or by administration of transfer factor, interferon, or vidarabine.

Data on vidarabine treatment of chicken pox in immunocompromised children are now available through the NIAID Collaborative Study Group (Whitley *et al.*, 1982d). In this study of 34 patients, vidarabine was administered intravenously at a dosage of 10 mg/kg per day for 5 days in a fashion similar to that for herpes simplex encephalitis. Significant improvement in outcome was demonstrated when therapy was administered within 72 hr of the onset of chicken pox. New vesicle formation, the hallmark of continued viral replication, ceased significantly sooner in drug-treated patients than in placebo-treated controls. Additionally, the duration of fever was decreased from a median of 6 days in the control patients to 3 days in those who received therapy. Finally and most importantly, early institution of therapy significantly diminished the frequency of visceral complications. Specifically, pneumonitis, which is reported to occur in at least 30% of immunocompromised children with chicken pox, was reduced from 27% to 5% with vidarabine therapy. There were no deaths when treatment was instituted early in the course of disease in spite of an expected mortality rate of 10–15% in untreated patients. The mortality rate in placebo recipients was 13%. A similar study reports the benefits of acyclovir for therapy of varicella in a trial involving a more limited number of patients (Prober *et al.*, 1982).

Ideally, our best approach to the control of this disease is prevention. Even so, however, cases will likely occur. It is unlikely that data on any other antiviral will generate better outcome data. Nevertheless, an oral medication, such as acyclovir or bromovinyl deoxyuridine, with established usefulness for outpatient therapy would be desirable.

2. Herpes Zoster

Herpes zoster, a common infection in immunocompromised patients, particularly those with lymphoproliferative malignancies, can be life-threatening if the viscera become involved. As a consequence, these infections have become a prime target for experimental antiviral drug trials. In 1976, results of a blind, controlled, crossover study of vidarabine therapy for herpes zoster infections in the immunocompromised host demonstrated accelerated cutaneous healing, as measured by loss of virus from lesions, cessation of new vesicle formation, time to total pustulation, and reduction of the pain of acute neuritis. In addition, the greatest benefit for cutaneous healing was derived by those patients who were $\leq$ 38 years of age, with lymphoproliferative malignancies and, importantly, received therapy within 6 days of disease onset. Because of the crossover study design (mandated by ethical concerns at the time), the value of vidarabine for the prevention and treatment of complications (postherpetic neuralgia, skin and visceral dissemination), the major medical problems associated with herpes zoster in the immunosuppressed host, could not be evaluated. The true clinical usefulness of vidarabine consequently remained poorly defined (Whitley et al., 1976).

A follow-up study was performed to specifically determine if treatment instituted within 72 hr of disease onset: (1) prevented cutaneous dissemination, (2) decreased the frequency and severity of visceral complications, (3) decreased the frequency or ameliorated the severity of postherpetic neuralgia, and (4) had additive toxic effects when administered with concurrent chemotherapeutic regimens. In addition, risk factors that might alter the course of herpes zoster in the immunocompromised host with or without treatment were also sought. Results from this study established clinical usefulness of early vidarabine therapy when administered early in the course of localized herpes zoster infections in the immunocompromised patient (Whitley et al., 1982b). In this study of 121 patients with localized herpes zoster $\leq$ 72 hr in duration, 63 received vidarabine (10mg/kg per day for 5 days) and 58 got placebo intravenously for 5 days. Populations were matched for pertinent characteristics. Therapy accelerated cutaneous healing as evidenced by more rapid cessation of lesion formation and reduction in time to both total pustulation and scabbing. Furthermore, and more importantly, it decreased cutaneous dissemination from 24% to 8% ($p = 0.014$), zoster-related visceral complications from 19% (11 patients) to 5% ($p = 0.015$) (Fig. 4) and the duration of postherpetic neuralgia ($p = 0.047$). In this study, patients with lymphoproliferative cancers and those $\geq$ 38 years of age were at greatest risk for complications and benefited most from therapy. Significant drug toxicity was not encountered. Vidarabine therapy is valuable for reduction of herpes zoster-related complications in immunocompromised patients when given early (< 72 hr) in the course of the infection; however, because natural healing is common, an oral medication for outpatient therapy is essential.

Utilizing a similar study design, the Burroughs Wellcome Cooperative Study Group has demonstrated that acyclovir administration will decrease progression to visceral complications in a similar high-risk population (Balfour et al., 1983). Nevertheless, therapy did not accelerate the events of cutaneous healing, a major component of this disease. Likely,

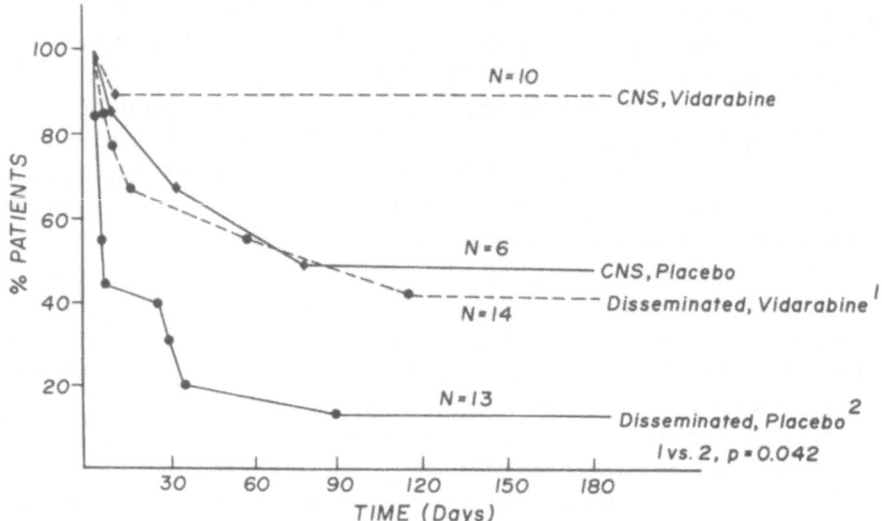

FIGURE 4. Outcome of HSV infection in neonates with disseminated and localized CNS disease—effect of vidarabine therapy. Vidarabine recipients (-----); placebo recipients (——). Statistical assessment by generalized Wilcoxon's test shows a *p* value of 0.014 (two-tailed). Points represent study death(s). Reprinted with permission from *N. Engl. J. Med.* **304:**313–318, 1981.

it will be essential to accelerate the time to total healing in order to impact on the expression of pain.

IV. CONCLUSION

The introduction of vidarabine for intravenous therapy and acyclovir for topical, oral, and intravenous therapy of herpesvirus infections has created a wave of enthusiasm for the development of newer and more specific antiviral agents. The data available for the aforementioned drugs clearly delineate the progress achieved in the therapy of these diseases. In the published literature, significant progress has only been achieved in the most severe of herpesvirus infections, namely herpes simplex encephalitis, neonatal HSV infections, primary genital herpes simplex, and VZV infections in the immunocompromised host. Obviously, with each of these diseases, their severity or protracted nature provides "time-to-event" outcomes that can be significantly altered. Less severe diseases, as recurrent genital or labial HSV infections, have not been particularly responsive to therapy. Whether the failure to impact on these clinical manifestations of HSV is a function of the propensity of virus to become latent, the short duration of disease, the failure of drug delivery to the involved site of viral replication, or the fact that each of these drugs acts at an intermediary site of viral replication (β gene expression) remains unknown. This latter point is not trivial as the newer drugs, including bromovinyl deoxyuridine, DHPG, and fluoroarabinosylpyrimidine derivatives, all act at similar sites. Whether these newer antivirals can decrease the mortality and improve morbidity for severe herpesvirus infections remains unanswered at the present time; but, for some diseases (e.g., herpes simplex encephalitis), it is unlikely. Perhaps, the future of the control of herpesvirus infections rests in a better understanding of the mechanism of latency as well as those portions of the viral genome that control this occurrence. Then, and only then, can true prevention via vaccination or other methods be undertaken.

A great deal has been learned with respect to the application of chemotherapeutic agents in man. Such lessons include the need for an understanding of the natural history of the disease, host factors that influence pathogenesis and, therefore, outcome, the necessity for definitive diagnosis, and the establishment of a therapeutic index through blind and controlled efforts. Most importantly, therapy must be targeted for specific disease states. Only by so doing can an appropriate risk–benefit ratio be determined. At least now, the recognition of inherent difficulties encountered in evaluating these compounds as well as the specific areas of present achievement should lead to refinements both of agents and of designs for their study.

ACKNOWLEDGMENTS. The author's research was initiated supported by a contract from the Developmental and Applications Branch, National Institute of Allergy and Infectious Diseases (NO-1-AI-12667), and grants from the National Cancer Institute (NCI-13148) and the Division of Research Resources (RR-032).

REFERENCES

Balfour, H. H., Bean, B., Laskin, O. L., Ambinder, R. F., Meyers, J. D., Wade, J. C., Zaia, J. A., Aeppli, D., Kirk, L. E., Segreti, A. C., Keeney, R. E., and the Burroughs Wellcome Collaborative Acyclovir Study Group, 1983, Acyclovir halts profession of herpes zoster in immunocompromised patients, *N. Engl. J. Med.* **308**:1448–1453.

Bodey, G. P., Gottlieb, J., McCredie, K. B., Freireich, E., 1975, Adenine arabinoside in cancer chemotherapy in: *Adenine Arabinoside: An Antiviral Agent*, pp.182–285 (D. Pavan-Langston, R. Buchanan, and C. Alford, Jr., eds.), Raven, New York.

Boston Interhospital Collaborative and NIAID Sponsored Cooperative Antiviral Clinical Study, 1975, Failure of high dose 5-iodo-2'-deoxyuridine in the therapy of herpes simplex virus encephalitis: evidence of unacceptable toxicity, *N. Engl. J. Med.* **292**:600–603.

Bryson, Y. J., Dillon, M., Lovett, M., Acuna, G., Taylor, S., Cherry, J. D., Johnson, B. L., Wiesmeier, E., Growdow, W., Creagh-Kirk, T., and Keeney, R., 1983, Successful treatment of initial genital herpes simplex virus infections with oral acyclovir, *N. Engl. J. Med.* **308**:916–921.

Calabresi, P., Carnosos, S. S., Finch, S. C., Kligerman, M. M., Von Essen, C. F., Chu, M. Y., and Welch, A. D., 1961, Initial clinical studies with 5-iodo-2'-deoxyuridine, *Cancer Res.* **21**:550–559.

Chien, L. T. Cannon, N. J., Charamella, L. J., Dismukes, W. E., Whitley, R. J., Buchanan, R. A., and Alford, C. A., Jr., 1973, Effect of adenine arabinoside on severe herpesvirus hominis infections in man. Preliminary Report, *J. Infect. Dis.* **128**:658–663.

Chien, L. T., Whitley, R. J., Nahmias, A. J., Lewin, E. B., Linnemann, C. C., Jr., Frenkel, L. D., Bellanti, J. A., Buchanan, R. A., and Alford, C. A., Jr., 1975, Antiviral chemotherapy and neonatal herpes simplex virus infection, a pilot study: experience with adenine arabinoside (ara-A), *Pediatrics* **55**:678–685.

Cohen, S. S., 1966, Introduction to the biochemistry of D-arabinosylnucleosides, *Prog. Nucl. Acid. Res. Mol. Biol.* **5**:1–88.

Corey, L., Adams, H. G., Brown, Z. A., and Holmes, K. K., 1983a, Genital herpes simplex virus infections: clinical manifestations, course, and complications, *Ann. of Intern. Med.* **98**:958–972.

Corey, L., Fife, K. H., Benedetti, J. K., Winter, C., Fahnlander, A., Connor, J. D., Hintz, M. A., and Holmes, K. K., 1983b, Intravenous acyclovir in the treatment of primary genital herpes, *Ann. Int. Med.* **98**:914–921.

Corey, L., Nahmias, A. J., Guinan, M. E., Benedetti, J. K., Critchlow, C. W., and Holmes, K. K., 1982, A trial of topical acyclovir in genital herpes simplex virus infections *N. Engl. J. Med.* **306**:1313–1319.

Crumpacker, C. S., Schnipper, L. E., Marlowe, S. K., Kowalsky, P. N., Hershey, B. J., and Levin, M. J., 1982, Resistance to antiviral drugs of herpes simplex virus isolated from a patient treated with acyclovir, *N. Engl. J. Med.* **305**:343.

DeClercq, E., Descamps, J., DeSomer, P., Barr, P., Jones, A. S., and Walker, R. T., 1979, (E)5-(2-bromovinyl) 2'-deoxyuridine: a potential and selective antiherpes agent, *Proc. Natl. Acad. Sci. USA* **76**:2947–2951.

DeRudder, J., Andreeff, F., and Privat de Garilhe, M., 1967, Action inhibitric de la 9-beta-D-xylofuranosyladenine sur la multiplication du virus de l'herpes en culture cellulaire, *C.R. Acad. Sci.* **264**:677–680.

deMiranda, P. D., Whitley, R. J., Blum, M. R., Keeney, R. E., Barton, N., Cochetto, B. S., Goode, S., Hemstreet, G. P., Kirk, L. E., Pager, D. A., and Elion, G. B., 1979a, The pharmacokinetics after intravenous infusion, *Clin. Pharmacol. Ther.* **26**:718–728.

deMiranda, P., Whitley, R. J., Blum, M. R., Krasney, H. C., Connor, J. D., and Lietman, P. S., 1979b, Pharmacokinetics of the antiviral drug acyclovir (Zovirax®) in man, Programs and Abstracts, 19th Interscience Conference on Antimicrobial Agents and Chemotherapy, Abstract No. 255.

deMiranda, P., Good, S. S., Laskin, O. L., Krasney, H. C., Connor, J. D., and Lietman, P. S., 1981, Deposition of intravenous radioactive acyclovir, *Clin. Pharmacol. Ther.* **30**:662–672.

deMiranda, P., Whitley, R. J., Barton, N., Page, D., Creagh-Kirk, T., Liao, S., and Blum, R., 1982, Systemic absorption and pharmacokinetics of acyclovir (ACV) (Zovirax®) capsules in immunocompromised patients with herpesvirus infections, Proceedings of the 22nd Interscience Conference on Antimicrobial Agents and Chemotherapy, Miami Beach, Abstract No. 418.

Dixon, G. J., Sidwell, R. W., Miller, F. A., Sloan, B. J., 1968, Antiviral activity of 9-beta-D-arabinofuranosyladenine V. Activity against intracerebral vaccinia virus infections in mice, *Antimicrob. Agents Chemother.* **8**:172–179.

Elion, G. B., Furman, P. A., Fyfe, J. A., deMiranda, P., Beaucham, P. L., and Schaeffer, R., 1977, Selectivity of action of an anti-herpetic agent, 9(2-hydroxyethoxymethyl) guanine, *Proc. Natl. Acad. Sci. USA* **74**:5716–5720.

Glazko, A. J., Chang, T., Drach, J. C., Mourer, D. R., Borondy, P. E., Schneider, H., Croskey, L., and Maschernske, E., 1975, Species differences in the metabolic disposition of adenine arabinoside in: *Adenine Arabinoside: An Antiviral Agent*, (D. Pavan-Langston, R. Buchanan, and C. Alford, Jr., eds.), pp. 111–133, Raven, New York.

Gnann, J. W., Jr., M. D., Whitley, R. J., and Barton, N., 1983, Acyclovir, *Pharmacotherapy* **73**:275–283.

Hafkin, B., Pollard, R. B., Tiku, M. L., Robinson, W. S., and Merigan, T. C., 1979, Effects of interferon and adenine arabinoside treatment of hepatitis B virus infection on cellular immune responses, *Antimicrob. Agents Chemother.* **16**:781–787.

Helegstrand, E., Eriksson, B., Johansson, N. G., Lannero, B., Larson, A., Misiorny, A., Noren, J. P., Fjoberg, B., Stenberg, K., Stridh, S., and Oberg, B., 1978, Trisodium phosphonoformate, a new antiviral compound, *Science* **201**:819–821.

Herrmann, E. C., Jr., 1961, Plaque inhibition test for detection of specific inhibitors of DNA containing viruses, *Proc. Soc. Exp. Biol. Med.* **107**:142–145.

Johnson, M. T., Luby, J. P., Buchanan, R. A., and Mikulec, D., 1975, Varicella-zoster virus infections treated with adenine arabinoside (ara-A), *J. Infect. Dis.* **131**:225–229.

Juel-Jensen, B. E., MacCallum, F. O., 1965, Herpes simplex lesions of face treated with idoxuridine applied by spray gun: results of a double-blind controlled trial, *Biol. Med. J.* **5439**:901–903.

Kaufman, H. E., 1962, Clinical cure of herpes simplex keratitis by 5-iodo-2-deoxyuridine, *Proc. Soc. Exp. Biol. Med.* **109**:251–252.

Kaufman, H. E., Marola, E., Dohlman, C., 1962a, Use of 5-iodo-2'-deoxyuridine. (IDU) in treatment of herpes simplex keratitis, *Arch. Ophthalmol.* **68**:235–239.

Kaufman, H. E., Nesburn, A. B., Maloney, E. D., 1962b, IDU: therapy of herpes simplex, *Arch. Ophthalmol.* **67**:583–591.

Larson, T., Dillon, M., Goldman, L., Connor, J., Bryson, Y., 1983, Double-blind, placebo-controlled study of acyclovir (ACV) prophylaxis in frequently recurrent genital herpes simplex virus (HSV) infections, Proceedings of the 23rd Interscience Conference on Antimicrobial Agents and Chemotherapy, Las Vegas, Abstract No. 562.

Laskin, O. L., deMiranda, P., King, D., Page, D. A., Lonstreth, J. A., Rocco, L., and Lietman, P. S., 1982a, Effects of probenecid on the pharmacokinetics and elmination of acyclovir in humans, *Antimicrob. Agents Chemother.*, **21**:804–807.

Laskin, O. L., Lonstreth, J. A., Saral, R., deMiranda, P., Keeney, R. E., and Lietman, P. S., 1982c, Pharmacokinetics and tolerance of acyclovir, a new anti-herpesvirus agent in humans, *Antimicrob. Agents Chemother.* **21**:393–398.

Laskin, O. L., Lonstreth, J. A., Whelton, A., Krasney, H. C., Keeney, R. E., Rocco, L., and Lietman, P. S., 1982b, Effect of renal failure on the pharmacokinetics of acyclovir. *Am. J. Med.* **73**:197–200.

Lee, W. W., Benitez, A., Goodman, L., and Baker, B. R., 1960, Potential anti-cancer agents. XL. Synthesis of beta-anomer of 9-(D-arabino-duranosyl) adenine, *J. Am. Chem. Soc.* **83**:2648–2649.

Luby, J., Douglas, J., Friedman-Klien, A., Gnann, J., Mills, J., Nahmias, A., and Sacks, S., 1983, A patient-initiated, double-blind trail of topical acyclovir (ACV) in recurrent genital herpes, Proceedings of the 23rd Interscience Conference on Antimicrobial Agents and Chemotherapy, Las Vegas, Abstract No. 560.

Miller, F. A., Dixon, G. J., Ehrlich, J., Solan, B. J., and McLean, I. W., Jr., 1980, Antiviral activity of 9-beta-D-arabionfuranosyladenine. I. Cell culture studies, *Antimicrob. Agents Chemother.* **8**:155–160.

Nahmias, A. J., Keyserling, H. L., and Kerrick, G. M., 1983, Herpes simplex, in: *Infectious Diseases of the Fetus and Newborn Infant*, (J. S. Remington, M. D. Klein, and J. O. Klein, eds.), pp. 636–678, W. B. Saunders Co., Philadelphia.

Nylen, P., 1924, Beitrag zur Kenntnis der organischer phosphor—verbindungen. *Chem. Ber.* **57**:1023–1038.

Overall, J. C., Jr., 1980, Antiviral chemotherapy of oral and genital herpes simplex virus infections, in: *The Human Herpesviruses*, (A. J. Nahmias, W. R. Dowdle, and R. E. Schinazi, eds.), pp. 447–465, Elsevier/North-Holland, New York.

Prober, C. G., Kirk, L. E., Keeney, R. E., 1982, Acyclovir therapy of chickenpox in immunosuppressed children—a collaborative study. *J. Pediatr.* **101**:622–625.

Prusoff, W. H., 1959, Synthesis and biological activities of iododeoxyuridine, an analogue of thymidine, *Biochem. Biophys. Acta.* **32**:295–296.

Ross, A. J., Julia, A., Balakrishnan, C., 1976, Toxicity of adenine arabinoside in humans, *J. Infect. Dis.* **133**:192–198.

Saral, R., Burns, W. H., Laskin, O. L., Santos, G. W., and Lietman, P. S., 1982, Acyclovir prophylaxis of herpes simplex virus infections, *N. Engl. J. Med.* **305**:63–67.

Schabel, F. M., Jr., 1968, The antiviral activity of 9-beta-D-arabinofuranosyl adenine (ara-A), *Chemotherapy* **13**:321–338.

Schaeffer, H. J., Gurwara, S., Vince, R., and Bittner, S., 1971, Novel substrate of adenosine deaminase, *J. Med. Chem.* **4**:367–369.

Schaeffer, H. J., Beauchamp, L., deMiranda, P., Elion, G. B., Bayer, D. J., and Collins, P., 1978, 9-(2-hydroxyethoxymethyl) guanine activity against viruses of the herpes group, *Nature* **272**:583–585.

Spector, S. A. Connor, J. D., Hintz, M., Quinn, R. P., Blum, M. R., and Keeney, R. E., 1981, Single-dose, pharmacokinetics of acyclovir, *Antimicrob. Agents Chemother.* **19**:608–612.

Spruance, S. L., and Crumpacker, C. S., 1982, Topical 5% acyclovir in polythylene glycol for herpes simplex labialis: antiviral effect without clinical benefit, *Am. J. Med.* **73**:315–319.

Spruance, S. L., Schnipper, L. E., Overall, J. C., Kern, E. R., Webster, B., Modlin, J., Wernerstrom, G., Burton, C., Arndt, K. A., Chiu, G., and Crumpacker, C. S., 1982, Treatment of herpes simplex labialis with topical acyclovir in polyethylene glycol, *J. Infect. Dis.* **146**:85–91.

Struas, S., Seidlin, M., Takiff, H., Bachrach, S., DiGiovanna, J., Western, K., Creagh-Kirk, T., Lininger, L., and Alling, D., 1983, Suppression of recurrent genital herpes with oral acyclovir, *Clin. Res.* **31**:543A.

Suhadolnik, R. J., 1970, *Nucleoside antibiotics*, Wiley Interscience, New York.

Van Dyke, D. B., Connor, J. D., Wyborney, C., 1982, Pharmacokinetics of orally administered acyclovir in patients with herpes progenitalis, *Am. J. Med.* **73**:172–175.

Wade, J. C., Newton, B., McLaren, C., Flournoy, N., Keeney, R. E., and Meyers, J. D., 1982, Intravenous acyclovir to treat mucocutaneous herpes simplex virus infection after marrow transplantation, *Ann. Int. Med.* **96**:265–269.

Whitley, R. J., Chien, L. T., Dolin, R., Galasso, G. J., Alford, C. A., Jr., and the Collaborative Antiviral Study Group, 1976, Adenine arabinoside therapy of herpes zoster in the immunosuppressed: NIAID Collaborative Antiviral Study, *N. Engl. J. Med.* **294**:1193–1199.

Whitley, R. J., Soong, S. J., Dolin, R., Galasso, G. J., Chien, L. T., Alford, C. A., Jr., and the NIAID Collaborative Antiviral Study Group, 1977, Adenine arabinoside therapy of biopsy-proved herpes simplex encephalitis: National Institute of Allergy and Infectious Diseases Collaborative Antiviral Study Group, *N. Engl. J. Med.* **297**:289–294.

Whitley, R. J., Nahmias, A. J., Soong, S. J., Galasso, G. J., Fleming, C. L., Alford, C. A., Jr., and the NIAID Collaborative Antiviral Study Group with special assistance from J. Connor, Y. Bryson, and C. Linneman, 1980a, Vidarabine therapy of neonatal herpes simplex virus infextion, *Pediatrics* **66**:495–501.

Whitley, R. J., Nahmias, A. J., Visintine, A. M., Fleming, C. L., Alford, C. A., Jr., and the NIAID Collaborative Antiviral Study Group, 1980b, The natural history of herpes simplex virus infection of mother and newborn, *Pediatrics* **66**:489–494.

Whitley, R. J., Alford, C. A., Jr., Hess, F., and Buchanan, R. A., 1980c, Vidarabine: a preliminary review of its pharmacological properties and therapeutic use, *Drugs* **20**:267–282.

Whitley, R. J., Soong, S. J., Hirsch, M. S., Karchmer, A. W., Dolin, R., Galasso, G. J., Dunnick, J. K., Alford, C. A., Jr., and the NIAID Collaborative Antiviral Study Group, 1981, Herpes simplex encephalitis: vidarabine therapy and diagnosti problems, *N. Engl. J. Med.* **304**:313–318.

Whitley, R. J., Blum, M. R., Barton, N., and deMiranda, P., 1982a, Pharmacokinetics of acyclovir in humans following intravenous administration, *Am. J. Med.* **73**:165–171.

Whitley, R. J., Hilty, M., Haynes, R., Bryson, Y., Connor, J. D., Soong, S. J., Alford, C. A., Jr., and the NIAID Collaborative Antiviral Study Group, 1982b, Vidarabine therapy of varicella in immunosuppressed patients. *J. Pediatr.* **101**:125–131.

Whitley, R. J., Tilles, J., Linnemann, C., Liu, C., Pazin, G., Hilty, M., Overall, J., Visintine, A. M., Soong, S. J., Alford, C. A., Jr., and the NIAID Collaborative Antiviral Study Group, 1982c, Herpes simplex encephalitis: clinical assessment, *J. Am. Med. Assn.* **247**:317–320.

Whitley, R. J., Hilty, M., Haynes, R., Bryson, Y., Connor, J. D., Soong, S. J., Alford, C. A., Jr., and the NIAID Collaborative Antiviral Study Group, 1982d, Vidarabine therapy of varicella in immunosuppressed patients. *J. Pediatr.* **101**:125–131.

Whitley, R. J., Yeager, A., Kartus, P., Bryson, Y., Connor, J. D., Nahmias, A. J., Soong, S. J., and the NIAID Collaborative Antiviral Study Group, 1984, Neonatal herpes simplex virus infection: follow-up evaluation of vidarabine therapy, *Pediatrics*, **72**:778–785.

CHAPTER 18

The Role of Interferon in Immunity and Prophylaxis

FRANÇOIS LEBEL AND MARTIN S. HIRSCH

I. INTRODUCTION

The term interferon (IFN) was first used by Isaacs and Lindenmann in 1957 to describe a factor, present in the supernatant of cultured chick chorioallantoic membranes previously injected with heat-inactivated influenza virus, that could render fresh chick membranes resistant to challenge with live virus. Since that time, studies performed in many laboratories have not only confirmed this phenomenon, but have greatly expanded our understanding of the IFN family of proteins and their role in viral infections. Moreover, a number of clinical trials have suggested that IFNs may be of use in the prophylaxis and therapy of herpesvirus infections. We will discuss the role of IFNs in the biological response to herpesviruses and review the trials addressing their clinical utility.

II. THE INTERFERONS

IFNs are presently defined as proteins that exert virus-nonspecific antiviral activity in homologous cells through cellular metabolic processes involving synthesis of both RNA and protein. They are classified into three types of functionally related proteins, α, β, and γ, largely on the basis of their antigenic specificities. This classification was adopted in 1980 by an international committee (NIAID et al., 1980) to replace

FRANÇOIS LEBEL • Department of Medicine, Massachusetts General Hospital and New England Deaconess Hospital, and Harvard Medical School, Boston, Massachusetts 02114. MARTIN S. HIRSCH • Department of Medicine, Massachusetts General Hospital and Harvard Medical School, Boston, Massachusetts 02114.

the old nomenclature of Type I leukocyte, Type I fibroblast, and Type II immune IFN, interferon, which was based on cell of origin and physiochemical properties.

IFN-α are produced mainly by leukocytes and to a lesser extent by fibroblasts. Human leukocyte-derived IFN-α were initially shown on SDS-PAGE to consist of a population that migrated at 20,000–25,000 daltons and another at 15,000–18,000 daltons (Stewart and Desmyter, 1975). Since then, IFN-α have been further resolved into over a dozen distinct isoelectric forms, partially attributable to their differing carbohydrate moieties (Stewart *et al.*, 1977). This heterogeneity of human IFN-α has been confirmed by recombinant DNA cloning techniques. Over a dozen different human IFN-α genes have been cloned (Nagata *et al.*, 1981), and have been found to have unique nucleotide sequences. Most studies on the biological activity of human IFN-α have employed relatively crude preparations with specific activities of 10^4 to 10^6 U/mg protein, whereas pure IFN-α can have a specific activity of the order of 10^9 U/mg protein (Berg and Heron, 1980). Most clinical studies of IFN-α were done with preparations containing less than 1% IFN, the small amount present being a mixture of subtypes of IFN-α, some of which have already been shown to have different biological properties (Weck *et al.*, 1981). Consequently, before ascribing any particular property or side effect to IFN with confidence, one must ensure that the preparation used is homogeneous and has a high specific activity.

IFN-β is a glycoprotein of 22,000 daltons (Heine *et al.*, 1980) produced by fibroblasts when stimulated by viruses or nucleic acids. It has been purified to homogeneity, but subtypes probably exist, based on analysis of mRNA coding for IFN-β (Sehgal and Sagar, 1980). IFN-α and IFN-β have similar molecular weights and stability at pH 2, but they differ in their host range and their antigenicity. In contrast to IFN-β, human IFN-α has significant antiviral activity on bovine cells, and this property can be used in assays to differentiate the two types (Gresser *et al.*, 1974; Hayes *et al.*, 1979). IFN-α and IFN-β can also be distinguished by the use of non-cross-reacting neutralizing antibodies (Havell, 1975). To date only one gene for human IFN-β has been cloned (Taniguchi *et al.*, 1980). Both IFN-α and IFN-β genes are located on human chromosome 9 (Owerbach *et al.*, 1981). Until very recently, supplies of IFN-β have been scarce and fewer clinical trials have been performed with this IFN type. Recombinant DNA technology has eliminated this problem, and many trials are currently under way.

IFN-γ, formerly known as Type II or immune IFN, is a glycoprotein that is sensitive to pH 2. Recent studies suggest that human IFN-γ may be composed of dimers of 20,000- and 25,000-dalton subunits (Yip *et al.*, 1982). IFN-γ is produced by T lymphocytes following nonspecific stimulation by mitogens *in vitro* or by sensitized lymphocytes stimulated with specific antigen *in vitro* or *in vivo* (Epstein, 1979). IFN-γ can be considered a member of the large family of more than 60 lymphokines, as it is a nonantibody soluble mediator of cellular immunity produced by lymphocytes as a result of immune recognition (Epstein, 1979). IFN-γ may possess greater immunomodulatory and antiproliferative effects than IFN-α or IFN-β (Epstein, 1979; Rubin and Gupta, 1980a). A single gene for human IFN-γ has been identified and cloned (Gray et al., 1982).

Recently, and unusual acid-labile human IFN-α has been described in the blood of certain patients with connective tissue diseases (e.g., systemic lupus erythematosis) and in patients with AIDS (DeStefano et al., 1982; Eyster et al., 1983). A similar IFN was also found in many patients with chronic active infections with Epstein–Barr virus but without other clinical manifestations of AIDS (Preble and Friedman, 1983). This acid-labile IFN is produced by immune lymphocytes in vitro (Balkwill et al., 1983). Whether it is normally present as a minor component of human IFN-α or is a specific marker if immune dysfunction is unknown at the present time.

III. ANTIVIRAL MECHANISMS OF INTERFERON

RNA viruses were originally shown to induce IFN (Isaacs and Lindenmann, 1957); subsequently, IFN-α and IFN-β have been induced by double-stranded (ds) RNAs in the form of synthetic polyribonucleotides (Field, 1967). It has been suggested that ds RNA might be the common link among all IFN inducers, as even a ds viral RNA could be extracted from cells infected with vaccinia, a DNA virus. Indeed, in the case of DNA viruses, the inducer may be ds RNA resulting from symmetrical transcription. No ds RNA has been extracted from cells infected with herpesviruses. However, in some viral infections, although ds RNA is below threshold detection levels, it can induce IFN (Marcus, 1982); this possibility cannot be excluded with the herpesviruses. Once induced, IFNs are secreted or shed by virus-infected cells into the extracellular environment and bind to specific cell surface receptors composed of gangliosides and/or glycoproteins (Vengris et al., 1976). Development of the antiviral state is dependent upon intact ATPase functions (Lebon et al., 1975). IFN-α and IFN-β appear to compete for the same receptor, while IFN-γ does not compete with IFN-α (Branca and Baglioni, 1981). In human cells, the gene for the IFN receptor resides or chromosome 21 (Tan, 1975). Following the binding of IFN to the receptor, an undefined cellular message leads to a new mRNA and protein synthesis, which in turn induces an antiviral state. Several new proteins are synthesized (Rubin and Gupta, 1980b) and begin to appear after 2–4 hr, peaking at 18–24 hr (Revel et al., 1980). At least three of these proteins have been shown to be associated with antiviral activity: a protein kinase, a 2′,5′-oligoadenylate (2-5A) synthetase, and a phosphodiesterase. The protein kinase and 2-5A synthetase are both activated by ds RNA. The 2-5A synthetase catalyzes the synthesis of 2-5A, an oligonucleotide made of ATP. The 2-5A in turn activates an endogenous ribonuclease that can degrade the viral mRNA. The protein kinase exerts its antiviral effect by phosphorylating the α subunit of eukaryotic initiation factor 2, with the utilization of ATP. This in turn inhibits the binding of initiator tRNA to ribosomes, resulting in diminished protein synthesis. The third enzyme associated with the development of the antiviral state is a phosphodiesterase that does not require ds RNA for activation. It degrades the C-C-A terminus of tRNA, preventing peptide chain elongation, and also degrades the 2-5A with consequent regulation of 2-5A synthetase activity (Baglioni, 1979; Schmidt et al., 1979).

IFN has also been reported to affect other steps in the viral infectious cycle. For instance, in some DNA viruses, IFN interrupts viral mRNA transcription rather than translation (Oxman, 1977). High-dose IFN has recently been shown to strongly inhibit receptor-mediated endocytosis of vesicular stomatitis virus, resulting in much less virus entering IFN-treated cells (Whitaker-Dowling et al., 1983).

The mechanisms by which IFNs inhibit herpesviruses are poorly understood. The complexity of these viruses has made such analysis difficult. Recent studies suggest that in a herpes simplex virus (HSV) in vitro infection of mouse L cells, virus replication is inhibited at an early step prior to DNA synthesis (Panet and Falk, 1983). HSV-1 DNA polymerase and thymidine kinase, both β proteins that precede viral DNA synthesis, were reduced in IFN-treated cells. These results are perhaps analogous to the inhibition of SV40 T antigen by IFNs. Both involve inhibition of a DNA virus at a very early step in replication. The existence of multiple IFN subtypes and multiple sites of action suggest that no simple conclusions on IFN mechanisms will soon be forthcoming. Moreover, evidence from murine models that host genes can influence IFN-mediated resistance to individual viruses suggests that an individual's IFN response may be quite specific. This could explain, in part, the differential susceptibility of individuals to viral infections, as well as differential responses to exogenously administered IFN (Hirsch and Hammer, 1982).

IV. IMMUNOMODULATORY EFFECTS OF INTERFERON

The antiviral action of IFN is due in part to its direct induction of an "antiviral state" in host cells. However, many other non-antiviral activities of IFN result indirectly in an enhanced ability of the host to combat viral infection. These other effects of IFNs on cellular structure and/or function can be divided into two general categories:

A. Immunoenhancement

Low-dose IFN-α and IFN-β in an in vivo mouse model have been shown to result in an enhanced antibody response (Braun and Levy, 1972).

Increased expressions of IgG Fc receptors (Itoh *et al.*, 1980), as well as β_2-microglobulin and HLA-A, B, and C antigens have been demonstrated on human peripheral blood lymphocytes exposed to purified IFN-α *in vitro* (Hokland *et al.*, 1981). Human IFN-γ has been reported to be more potent then IFN -α or IRN-β in inducing HLA and β_2-microglobulin mRNA and protein synthesis (Wallach *et al.*, 1982). All three species of IFN may stimulate NK cells, at least *in vitro* (Perussia *et al.*, 1980). Enhanced phagocytosis by macrophages exposed to IFN has been described, as well as enhanced T-cell-mediated and antibody-dependent cellular cytotoxicity (ADCC) (Epstein, 1979; Zarling *et al.*, 1978).

B. Immunosuppression

High-dose IFN can suppress both primary and secondary antibody responses, possibly by its antiproliferative effect on B cells (Epstein, 1979) or by the activation of suppressor T cells (Aune and Pierce, 1982). IFN can decrease the number of IgM Fc receptors on human lymphocytes (Itoh *et al.*, 1980). It has also been shown to depress delayed hypersensitivity reactions *in vitro* (DeMaeyer *et al.*, 1975), to decrease leukocyte migration inhibition (Szigeti *et al.*, 1980), and to inhibit monocyte maturation (Lee and Epstein, 1980). Chronic administration of IFN can result in marked depression of NK cell activity (Maluish *et al.*, 1982).

At first glance, many of the described effects of IFN appear contradictory. These effects can vary markedly, depending upon the type or dose of IFN, the timing and duration of exposure, the dose of antigen, the type of assay system, and the genetic makeup of the host (DeMaeyer and DeMaeyer-Guignard, 1980). Clinical trials of human IFNs have yet to define consistent stimulatory or suppressive effects on immune responsiveness.

V. INTERFERON AND THE HERPESVIRUSES

A. Cytomegalovirus

The inherited susceptibility or resistance to lethal CMV infection in congenic strains of adult mice appears linked to the *H-2* complex (Grundy *et al.*, 1981); the mechanisms of this resistance are unclear but are not related to antibody levels, nor to loss of ability by the cells to support viral replication. In some strains, resistance is controlled by non-*H-2* genes and the ability to produce IFN and to develop NK cell activity appear to be of critical importance (Grundy *et al.*, 1982; Bancroft *et al.*, 1981). In man, no clear-cut pattern of inherited resistance has been described for primary CMV infections, and no particular HLA types appear to predispose to CMV infections (Patel *et al.*, 1978). However, NK cells recovered from healthy young adults are capable of cytotoxic activity against both uninfected and CMV-infected fibroblasts, independent of donor immune status (Starr and Garrabrant, 1981). Preincubation of NK cells with IFN enhances the killing. These cells could play an important role as a front line of defense in primary CMV infection, prior to the generation of specific antibody and immune T lymphocytes.

Murine models have shown the importance of cytotoxic T cells in recovery from CMV infection (Quinnan et al., 1980; Ho, 1980). In humans, the importance of cytotoxic mechanisms in recovery has been evaluated in bone marrow transplant recipients (Quinnan et al., 1982). The outcome of CMV infections did not correlate with the nature of the underlying disease, the type of transplant, the pretransplantation CMV antibody status, nor the lymphocyte proliferative responses to CMV antigens or Con A. There was, however, a significant correlation between recovery from CMV infection and the development of a CMV-specific cytotoxic response by both HLA-restricted and nonrestricted T lymphocytes. The presence of depressed NK and antibody-dependent killer cells also correlated with increased mortality. As IFN can augment cytotoxicity in vitro, it may be critical to development of an effective response.

In acute CMV mononucleosis, decreased lymphocyte responsiveness has been documented, as manifest by decreased proliferative responses to mitogens and antigens (Rinaldo et al., 1980; Carney and Hirsch, 1981). In addition, lymphocytes from CMV mononucleosis patients show diminished levels of IFN production following challenge with mitogen or antigens (Levin et al., 1979; Rinaldo et al., 1980). During convalescence these responses normalize. In addition, there is a reversal of the normal ratio of peripheral blood lymphocytes with phenotypic markers of T-helper and T-suppressor cells (Carney et al., 1981). The increase in T-suppressor cells may help explain the immunosuppression, but a monocyte-macrophage population was also found to reduce the cellular response to Con A (Rinaldo et al., 1980; Carney and Hirsch, 1981). Whether exogenous IFN will help to reconstitute these deficient responses is under investigation.

CMV infection is believed by most investigators to be capable of inducing IFN in mice (Hamilton, 1982). The ability of CMV infection to elicit production of IFN in humans has been more difficult to demonstrate. Although some investigators have detected circulating IFN during active CMV infections (O'Malley et al., 1975; Rhodes-Feuillette et al., 1983), further studies are necessary to confirm these observations.

That CMV infection is susceptible to IFN is well established (Glasgow et al., 1967; Postic and Dowling, 1977). Several factors can influence the IFN sensitivity of CMV in vitro, such as the source of virus, the indicator cells employed, the size and timing of the inoculum, and whether the inoculum is cell-free virus or not. An infected cell inoculum is 300-fold less sensitive to IFN than its cell-free counterpart (Holmes et *al.*, 1978). Clinical isolates appear more susceptible to IFN *in vitro* than are laboratory strains.

B. Varicella–Zoster Virus

An association has been described in the immunocompromised host between recovery from varicella and a specific cell-mediated immune response, indicated by lymphocyte blastogenesis or lymphocyte-mediated cytotoxicity (Steele *et al.*, 1977), as well as the local production of IFN within the vesicles (Armstrong *et al.*, 1970). Absent or low levels of circulating IFN have been documented in varicella–zoster infections (Stevens and Merigan, 1972). Delayed local production of IFN has been described in patients with extradermatomal spread or prolonged disease associated with the use of immunosuppressive drugs such as cytosine arabinoside (Stevens *et al.*, 1973). During recovery from herpes zoster, peak intravesicular IFN titer correlated better than cell counts with cessation of dissemination (Stevens *et al.*, 1975). It was hypothesized that a few sensitized lymphocytes produce IFN, which subsequently helps augment an effective cellular response. IFN production occurred well before detection of any antibody response. VZV has been shown to be sensitive to the antiviral effects of IFN (Rasmussen *et al.*, 1977). Sensitivity is increased by low-input multiplicity of virus and by a cell-free inoculum in a similar fashion to CMV.

Various classes and types of antibodies have been described during varicella–zoster infections. IgG antibodies have been found to persist for life while IgA and IgM antibodies disappear after a few months (Brunell *et al.*, 1975). The ability of zoster immune globulin (ZIG) to completely prevent varicella in normal children (Brunell *et al.*, 1969) would support a protective role of humoral immunity. However, in immunocompromised patients, larger doses of ZIG may only modify the disease but do not prevent varicella (Brunell *et al.*, 1972; Gershon *et al.*, 1974), suggesting a major role for cell-mediated immune mechanisms, including IFN production, in these hosts. Moreover, varicella–zoster infections are major complications of disorders of cellular, but not humoral, immunity.

Transfer factor was found to be protective against varicella when administered to a group of leukemic children with no previous immunity to the virus (Steele *et al.*, 1980). Previously, the same transfer factor was shown to be capable of inducing VZV-specific lymphocytotoxicity, lymphocytic blastogenic response, and production of leukocyte inhibitory factor in individuals without prior exposure to VZV antigen (Steele, 1980). This would suggest a direct effect of transfer factor on lymphocytes, possibly through the production of lymphokines such as IFN.

C. Epstein–Barr Virus

The pathogenesis of human EBV infections has been extensively studied. The virus is thought to replicate initially in the oropharynx where it presumably gains access to local B lymphocytes. Infection of the lymphocyte population triggers them to proliferate, and results in the appearance of a series of specific antibodies to EBV as well as a more general polyclonal B-cell activation (Rosen *et al.*, 1977). Anti-early antigen (EA) antibodies and anti-viral capsid antigen (VCA) antibodies of the IgM class appear first. These are followed by anti-VCA antibodies of the IgG class and eventually by the appearance of anti-Epstein–Barr nuclear antigen (EBNA) antibodies. The increased number of EBV-infected B cells also results in the proliferation of different subclasses of T lymphocytes, which constitute the majority of the circulating atypical lymphocytes during infectious mononucleosis (Pattengale *et al.*, 1974). Among the T cells are found a population with suppressor activity, which have been shown to be capable of inhibiting the outgrowth of EBV-transformed B cells (Schooley *et al.*, 1981). The mechanisms of this suppression are not entirely clear, but T cells from normal donors are capable of suppressing EBV infection *in vitro* (Thorley-Lawson *et al.*, 1977). This suppressive effect has been shown to have a biphasic nature (Thorley-Lawson, 1980). An early phase is manifest within 24 hr of the EBV infection, prior to transformation of the B cells. A second phase of suppression then occurs, which is mediated through EBV-specific, HLA-restricted cytotoxic T cells (Misko *et al.*, 1980; Slovin *et al.*, 1983), as well as through non-specific cytotoxic T cells (Moss *et al.*, 1979). The early suppression of B-cell proliferation following infection can be blocked by the use of anti-IFN antibodies, and human leukocyte IFN, in the absence of T-suppressor cells, can produce similar suppression. These findings implicate IFN as the mediator of the early suppressive effect on the infected B cells (Thorley-Lawson, 1981). The exact cellular source of IFN in these *in vitro* experiments has not been established. Peripheral T lymphocytes can produce IFN in response to autologous EBV-infected lymphoblastoid cell lines (Trinchieri *et al.*, 1977). If the T cells were the only source, one would expect IFN-γ production, which would not be neutralized by anti-IFN-α antibodies. According to Thorley-Lawson (1981), neonatal B cells were resistant to the suppressive effect of IFN. Similar results have been reported by Menezes *et al.* (1976). However, other investigators (Lai *et al.*, 1977; Doetsch *et al.*, 1981; Garner *et al.*, 1984) have found cord blood mononuclear cells to be equally susceptible to the suppressive effect of IFN-α. These conflicting observations remain largely unexplained.

Whether or not the same mechanisms are operative during infectious mononucleosis is unclear at the present time. In support of these mechanisms, there is a report of a young girl who died of a severe persistent EBV infection and who was found to have normal humoral and cellular immunity except for a defect in IFN-γ production (Virelizier *et al.*, 1978). The second phase of suppression of an EBV infection *in vitro* also has a counterpart *in vivo* during infectious mononucleosis. In the convalescent phase of infectious mononucleosis, an EBV-specific memory T cell appears that is cytotoxic when tested *in vitro* against EBV-infected B cells (Rickinson *et al.*, 1980).

Non-EBV-specific cellular immunity has been found to be depressed during infectious mononucleosis, as expressed by cutaneous anergy and depressed T-cell responsiveness to mitogens, recall antigens, and allogeneic lymphocytes *in vitro* (Mangi *et al.*, 1974). The role, if any, of IFN in affecting these functions is not known.

EBV has been shown, like the other herpesviruses, to be able to induce IFN production in human cell lines (Adams *et al.*, 1975) and to be susceptible to the antiviral effects of IFN (Lvovsky *et al.*, 1981; Garner *et al.*, 1984).

D. Herpes Simplex Virus

Animal models of HSV infection have suggested a major role for the genetic background of the host in susceptibility to infection (reviewed by Lopez, this volume). Natural resistance (the ability of an organism to resist an infection never having encountered it before) to HSV-1 is governed in mice by two independently segregating dominant genes that are independent of the *H-2* locus (Lopez, 1980). The expression of this resistance is felt to be mediated through NK cells. HSV-2-induced hepatitis in young mice was found to be controlled by one X-linked dominant gene (Mogensen, 1977). In this model, macrophages were felt to be the main effector cells. It is unclear whether similar patterns of genetic resistance apply to man. The only support for this comes from reports of increased susceptibility of patients with HLA-DRw$_3$ to recurrent herpes keratitis and HLA-A$_1$ patients to herpes labialis infections (Meyers-Elliott *et al.*, 1980: Russell and Schlaut, 1975). Human neonates with disseminated HSV infection (Ching and Lopez, 1979) and certain AIDS patients with chronic ulcerative herpes lesions (Siegal *et al.*, 1981) have also been reported to have depressed NK activity, although it is uncertain what role genetic factors have in this diminished activity.

HSV spreads by causing fusion of cells and can escape neutralizing antibodies. It does, however, induce new antigens on infected cell surfaces, which may elicit a response from cytotoxic cells. Various cell types are capable of killing HSV-infected cells: nonadherent mononuclear cells (K cells), monocyte-macrophages, and granulocytes have been reported to participate in ADCC (Shore *et al.*, 1976; Kohl *et al.*, 1977; Oleske *et al.*, 1977). Macrophages are capable of restricting HSV infection within surrounding permissive Vero cells (Morahan *et al.*, 1980), and appear important in age-dependent resistance to this agent (Hirsch *et al.*, 1970). Adult stimulated mouse macrophages produce more IFN than neonatal macrophages and are better able to destroy intracellular HSV. The use of substances that are selectively cytotoxic to macrophages, e.g., silica and antimacrophage sera, results in increased mortality of HSV-1-infected mice (Zisman *et al.*, 1970). Substances stimulatory to macrophage activity, such as *C. parvum* and pyran blue, decrease mortality to HSV in mice (Starr *et al.*, 1976). HSV-infected human fibroblasts can be killed by macrophages and the killing is enhanced by IFN; this occurs in the absence of specific antibodies (Stanwick *et al.*, 1980).

Human IFN-α has been found in very high titer (mean 20,200 U) in vesicle fluid from lesions of recurrent herpes labialis (Overall *et al.*, 1981). In another study by the same group (Spruance *et al.*, 1982), IFN titers in vesicular fluid correlated with the age of the lesions and the virus titer within the fluid but not with the clinical severity of the illness. The titers observed in vesicle fluid are much higher than those required to inhibit HSV replication in human cell cultures (Overall *et al.*, 1980; Lazar *et al.*, 1980; Matsubara *et al.*, 1980). This implies that local production of IFN might play a role in localization of the infection and might contribute to recovery. IFN production has also been reported in the CSF of patients with HSV encephalitis (Hilfenhaus and Ackerman, 1981; Vezinet *et al.*, 1981). Recently, peripheral mononuclear cells from patients with recurrent herpes labialis have been shown to spontaneously secrete IFN-γ and small amounts of IFN-α when cultured *in vitro* (Cunningham and Merigan, 1983). No serum IFN could be detected. Peak supernatant IFN-γ levels were proportionate to the time interval to the next recurrence. This study confirms the report by Rasmussen *et al.* (1974) that IFN-γ titers generated by HSV antigen-stimulated macrophage–lymphocyte cultures correlated directly with the time elapsed to the next recurrence. No correlation has been demonstrated between titers of mononuclear cell IFN and severity or duration of the herpetic lesions.

VI. PROPHYLAXIS AND THERAPY WITH INTERFERON

The ideal antiherpes agent would be one that limits viral replication, prevents or cures latency, is relatively nontoxic, and promotes repair in virus-damaged tissue. No agent thus far fulfills such stringent requirements. IFN, being a natural substance, was once thought to be nontoxic. However, more extensive experience with both impure and cloned preparations suggests both immediate toxicity (fever and local pain), as well as dose-dependent adverse effects on bone marrow, CNS, and possibly other organs. Most of these side effects are reversible on discontinuation, and are relatively easily controllable. IFNs are active *in vitro* against each of the four human herpesviruses, possess the ability to contain herpes infections in animal models, and are potent immunomodulators. These attributes continue to render IFNs prime candidates for clinical trials. We shall briefly outline important animal and human trials on the potential use of IFN against herpesvirus.

A. Animal Studies

The species specificity of VZV, EBV, and human CMV has prevented the study of animal models with these agents. In contrast, a number of studies have addressed the role of IFN in murine HSV infections. The use of a variety of different routes of inoculation, varying doses of virus, and varying endpoints has made the interpretation of murine studies difficult.

DeClercq and Luczak (1976) studied a model of disseminated human newborn HSV infection and were able to show decreased mortality in young mice infected with intranasal HSV-1 when they were pretreated with high-dose mouse IFN. IFN failed to reduce the mortality if given after the viral challenge. IFN also could not reduce the mortality of immunosuppressed (antithymocyte serum) mice with disseminated HSV-1 infection (Worthington et al., 1977). Exogenous IFN has reduced the mortality of mice infected intraperitoneally with HSV-2 when administered as late as 24 hr after viral inoculation, but when HSV was inoculated intranasally, no reduction in mortality or increased survival could be shown (Olsen et al 1978).

Although human CMV is species specific, Cruz et al. (1981) were able to demonstrate reduced murine CMV replication in liver and spleen, and delayed seeding of salivary gland in newborn mice pretreated with low-dose IFN. However, no protection or delay in time of death could be documented by Kern et al. (1978) using high-dose IFN in another murine CMV model.

Arvin et al. (1983) administered high-dose human leukocyte IFN to *Erythrocebus patas* monkeys during the incubation phase of a laboratory outbreak of simian varicella and demonstrated a marked reduction in attack rate. This study supports the possible use of IFN as a postexposure prophylaxis among immunosuppressed humans exposed to VZV.

B. Human Studies

1. Herpes Simplex Virus

Prophylactic IFN-α has been evaluated in a double-blind study of systemic administration on the reactivation of HSV (Pazin et al., 1979). Thirty-seven patients with a past history of herpes labialis who had been suffering from trigeminal neuralgia and were undergoing microvascular decompression of the trigeminal sensory root were evaluated. None of the patients in this study was receiving radiation or chemotherapy; all were relatively healthy but had received high-dose corticosteroids before and after surgery. The administration of 7×10^4 U IFN/kg body wt per day for 5 days starting on the day before surgery resulted in a statistically significant reduction in reactivation of HSV, as determined by the number of herpetic lesions or positive throat cultures, taken together. Taken alone, the frequency of herpetic lesions and their size was decreased but not in a statistically significant manner (5 of 19 patients on IFN vs. 10 of 18 patients on placebo). A telephone survey published 1 year later (Haverkos et al., 1980) showed that of the 37 patients contacted, 9 in the placebo group and 9 in the IFN group recalled having had one or more lesions in the interval since the original study. The authors concluded that there is no evidence that the brief course of IFN had any effect on subsequent reactivation. Whether higher IFN dosage or alteration in its administration schedule could have improved results remains unresolved.

Prophylactic trials of IFN-α were also conducted in renal transplant recipients, using 3×10^6 U twice weekly for 6 weeks (Cheeseman et al., 1979). Eleven of sixteen IFN-treated seropositive patients excreted HSV, as compared to 13 of 14 placebo-treated patients. Virus shedding began 1–3 weeks after transplantation in both groups. Approximately half the seropositive patients and two-thirds of those who excreted HSV had lesions. No relation was observed between the presence, severity, or extent of lesions and IFN treatment.

Several trials of recombinant IFN-α, alone or in combination with acyclovir, are currently under way in the prophylaxis and treatment of human genital herpes infections. IFN-β and IFN-γ have not yet been studied in this disorder.

The earliest controlled trial of the use of IFN in human herpetic infection was in recurrent HSV keratitis, using topical therapy. A study comparing 49 patients treated topically with human leukocyte IFN (6.4 $\times 10^4$ U/ml, twice daily) to 46 placebo-treated patients showed an equal number of recurrences in both groups when evaluated at 18 months (Kaufman et al., 1976). In another study, 78 patients were treated with minimal debridement and then stratified to receive high-dose (33×10^6 U/ml, one drop per day) or moderate-dose (11×10^6 U/ml, one drop per day) topical IFN or placebo; a statistically significant difference in the number of recurrences was found between the high-dose treatment group and the placebo group (Coster et al., 1977). Other studies have concluded that there is no significant difference between human leukocyte IFN or human fibroblast IFN in this infection, and that as little as 10^6 U/ml can be of benefit as long as it is combined with debridement (Sundmacher et al., 1978a,c).

Treatment of acute herpes keratitis is amenable to many antiviral agents. However, the requirement for debridement with its attendant morbidity has led to a trial of trifluorothymidine (TFT) alone or in association with high- or low-dose IFN (Sundmacher et al., 1978b). No significant differences were found between TFT alone and TFT plus low-dose human leukocyte IFN (1×10^6 U/ml), but corneal healing time was significantly improved in the group receiving TFT plus high-dose IFN (30

$\times$ 10^6 U/ml). Topical IFN-α also potentiates the antiviral activity of topical acyclovir against dendritic herpes keratitis (Colin et al., 1983).

2. Varicella–Zoster Virus

In 1978, a well-designed therapeutic study of 90 patients with herpes zoster and cancer provided strong evidence for the ability of IFN-α to reduce cutaneous dissemination (Merigan et al., 1978). This was a randomized double-blind placebo-controlled study of three dosage schedules of IFN-α. The two higher doses, 1.7 or 5.1 $\times$ 10^5 U/kg per day, were found to be much more effective than the lower dose regimen of 4.2 $\times$ 10^4 U/kg per day in reducing cutaneous dissemination and the severity of postherpetic neuralgia. Visceral complications were six times less frequent in the IFN recipients compared to the placebo group.

A follow-up study of short course therapy with 2.55 $\times$ 10^5 U/kg of IFN-α intramuscularly every 12 hr for four doses demonstrated no effect on acute pain or disease progression in the primary dermatome. IFN-treated patients had a modest diminution in distal cutaneous spread and in both severity and duration of postherpetic neuralgia (Merigan et al., 1981).

A randomized double-blind controlled trial in children with varicella undergoing either chemotherapy or radiotherapy for malignancy has shown beneficial effects of IFN-α (Arvin et al., 1978, 1982). Nine patients received high-dose IFN (minimum of 4.2 $\times$ 10^4 U/kg body wt per day), until no vesicles had appeared for 24 hr (average of 6.4 days) and an equal number received placebo. Subsequently, 14 similar patients received an initial dose of 3.5 $\times$ 10^5 U/kg body wt per day for 48 hr, followed by 1.75 $\times$ 10^5 U/kg body wt per day for 72 hr, and 12 patients were given placebo. The IFN was administered intramuscularly at 12-hr intervals. All patients were enrolled within 72 hr of onset of the exanthem. The IFN group had significantly fewer new lesions, especially those on the higher dose regimen, and the number of patients with life-threatening dissemination was also reduced. Mortality consisted of two patients in the IFN group and three patients in the placebo group. IFN levels in the blood were 200–400 U/ml in the patients given the higher dose and 40–190 U/ml in the other IFN group, when measured 6–8 hr after intramuscular injection.

These studies indicate that in the immunocompromised host, relatively high-dose IFN-α can reduce the morbidity associated with VZV infections. Similar results have been obtained with vidarabine (Whitley et al., 1982) and acyclovir (Balfour et al,. 1983). Direct comparisons between IFN-α and these other agents have not been conducted.

3. Cytomegalovirus

The first placebo-controlled double-blind study of leukocyte IFN prophylaxis against CMV infections was reported in 1978 (Weimar et al.,

1978, 1979). Eight renal allograft recipients received 3×10^6 U of human IFN-β intramuscularly twice weekly for 3 months, starting a few hours prior to transplantation. Comparison was made with eight patients who received placebo. CMV infections occurred in four patients in the placebo group and in five of the IFN group as determined by antibody rises. Blood levels of IFN-β could not be detected after intramuscular injections, possibly because of local inactivation (Hanley *et al.*, 1979). Drawing any conclusion from this study is difficult because of the documented absence of circulating IFN, the small number of patients in each group, and the attempt at virological isolation only when viral infections were suspected on clinical grounds.

A larger randomized placebo-controlled double-blind study of renal allograft recipients who were studied with frequent virological and serological monitoring, was conducted by Cheeseman *et al.* (1979). Twenty-one patients received 3×10^6 U of IFN-α twice weekly for 6 weeks starting on the day of transplantation, and 20 patients received placebo. Twenty-four patients were seropositive for CMV prior to transplantation. Onset of CMV excretion was delayed in the IFN group to 7.2 weeks compared to 4.2 weeks in the placebo group. CMV viremia was more frequent in placebo (9/10) than in the IFN group (5/11), $p = 0.04$. Overall, the clinical outcome was not significantly altered.

Recently, a follow-up double-blind trial in renal transplant recipients has been reported (Hirsch *et al.*, 1983). Longer courses of prophylaxis were evaluated in seropositive patients susceptible to CMV reactivation. IFN-α (3×10^6 U) or placebo was administered three times a week for 6 weeks and then twice a week for 8 weeks (total of 102×10^6 U). Clinical signs of CMV infection were markedly reduced in IFN recipients. Seven of twenty-two placebo recipients and one of twenty IFN recipients developed CMV syndromes ($p = 0.03$). Opportunistic superinfections occurred only in placebo recipients, and CMV-associated glomerulopathy developed in one IFN and three placebo patients.

IFN therapy of ongoing CMV infections has not been encouraging. Uncontrolled trials in congenital infection, CMV pneumonia, or CMV retinitis have not suggested significant benefits (Chou *et al.*, 1984; Meyers *et al.*, 1980; Arvin *et al.*, 1976). These clinical trials indicate that while IFN may be of considerable use in the prophylaxis of CMV syndromes among high-risk patients, they are unlikely, as single agents to provide therapy for serious ongoing infections.

4. Epstein–Barr Virus

The only controlled study to evaluate IFN in EBV infections was conducted by Cheeseman *et al.* (1980) as part of a larger prophylactic trial in renal transplant recipients. Thirty-eight of the patients were seropositive for EBV. At the onset of the study, 8.6% of the patients excreted EBV from their oropharynx and this increased to 50% at the 5th month and beyond. Excretion of EBV was significantly more frequent in patients receiving antithymocyte globulin (65 versus 33%). Virus shedding could not be correlated with a fourfold or greater rise in titer of antibodies to VCA or EA of restricted type. IFN-α reduced the virus excretion from 65% in the placebo group to 38% (p = 0.08). The reduction was more impressive in the patients receiving antithymocyte globulin where excretion in the placebo group was 83% versus 45% in the IFN group (p = 0.07). IFN had no apparent effect on serological markers of EBC infection, or on clinical EBV syndromes.

EBV is implicated in a variety of lymphoproliferative neoplasms, as well as in nasopharyngeal carcinoma. Trials of recombinant and natural IFNS, either alone or in combination with other agents, are awaited in these conditions.

VI. FUTURE DIRECTIONS

Over the last 25 years, IFNs have gradually moved from the laboratory to the clinic. Much has been learned concerning the antiviral and immunomodulatory roles of IFNs, although information concerning their antiherpesvirus mechanisms remains meager.

The ultimate place of IFNs in the management of human herpesvirus infections is unclear. It is likely that prophylaxis in high-risk patients will be more useful than treatment of ongoing disease. Combination therapy with agents such as acyclovir holds promise against various herpesviruses (Levin and Leary, 1981; Starwick et al., 1981; Hammer et al., 1982; Smith et al., 1983), although such combinations may also prove more toxic than treatment with single agents. Combinations of different IFN types, such as α and γ, may also be more beneficial than either type alone. The availability of multiple subtypes of recombinant IFNs assures us that considerable additional information will be obtained in the years to come, not only on their mechanisms of action, but on their clinical utility in human herpesvirus infections.

ACKNOWLEDGMENTS. The authors' research was supported in part by Contract 43222 from the National Cancer Institute and by the Mashud A. Mezerhane B. Fund. F.L. is the recipient of a Research Fellowship from the Fonds de la Recherche en Santé du Québec.

REFERENCES

Adams, A., Lidin, B., Strander, H., and Cantell, K., 1975, Spontaneous interferon production and Epstein–Barr virus antigen expression in human lymphoid cell lines, *J. Gen. Virol.* **28**:219.

Armstrong, R. W., Gururth, M. J., Waddell, D., and Merigan, T. C., 1970, Cutaneous interferon production in patients with Hodgkin's disease and other cancers infected with varicella or vaccinia, *N. Engl. J. Med.* **283**:1182.

Arvin, A. M., Yeager, A. S., and Merigan, T. C., 1976, Effect of leukocyte interferon on urinary excretion of cytomegalovirus by infants, *J. Infect. Dis.* **133**(Suppl):A205.

Arvin, A. M., Feldman, S., and Merigan, T. C., 1978, Human leukocyte interferon in the treatment of varicella in children with cancer: A preliminary controlled trial, *Antimicrob. Agents Chemother.* **13**:605.

Arvin, A. M., Kushner, J. H., Feldman, S., Baehner, R. L., Hammond, D., and Merigan, T. C., 1982, Human leukocyte interferon for the treatment of varicella in children with cancer, *N. Engl. J. Med.* **306**:761.

Arvin, A. M., Martin, D. P., Gard, E. A., and Merigan, T. C., 1983, Interferon prophylaxis against simian varicella in *Erythrocebus patas* monkeys, *J. Infect. Dis.* **147**:149.

Aune, T. M., and Pierce, C. W., 1982, Activation of a suppressor T-cell pathway by interferon, *Proc. Natl. Acad. Sci. USA* **79**:3808.

Baglioni, C., 1979, Interferon-induced enzymatic activities and their role in the antiviral state, *Cell* **17**:255.

Balfour, H. H., Jr., Bean, B., Laskin, O. L., Ambrinder, R. F., Meyers, J. D., Wade, J. C., Zaia, J. A., Aeppli, D., Kirk, L. E., Segreti, A. C., Keeney, R. E., and the Burroughs–Wellcome Collaborative Acyclovir Study Group, 1983, Acyclovir halts progression of herpes zoster in immunocompromised patients, *N. Engl. J. Med.* **308**:1448.

Balkwill, F. R., Griffin, D. B., Band, H. A., and Beverley, P. C. L., 1983, Immune human lymphocytes produce an acid-labile α interferon, *J. Exp. Med.* **157**:1059.

Bancroft, G. J., Shellani, G. R., and Chalmer, J. E., 1981, Genetic influences on the augmentation of natural killer (NK) cells during murine cytomegalovirus infection: Correlation with patterns of resistance, *J. Immunol.* **126**:988.

Berg, K., and Heron, I., 1980, The complete purification of human leucocyte interferon, *Scand. J. Immunol.* **11**:489.

Branca, A. A., and Baglioni, C., 1981, Evidence that types I and II interferons have different receptors, *Nature (London)* **294**:768.

Braun, W., and Levy, H. B., 1972, Interferon preparations as modifiers of immune responses, *Proc. Soc. Exp. Biol. Med.* **141**:769.

Brunell, P. A., Ross, A., Miller, L. H., and Kuo, B., 1969, Prevention of varicella by zoster immune globulin, *N. Engl. J. Med* **280**:1191.

Brunell, P. A., Gershon, A. A., Hughes, W. T., Riley, H. D., and Smith, J., 1972, Prevention of varicella in high risk children: A collaborative study, *Pediatrics* **50**:718.

Brunell, P. A., Gershon, A. A., Uduman, S. A., and Steinberg, S., 1975, Varicella–zoster immunoglobulins during varicella, latency, and zoster, *J. Infect. Dis.* **132**:49.

Carney, W. P., and Hirsch, M. S., 1981, Mechanisms of immunosuppression in cytomegalovirus mononucleosis. II. Virus–monocyte interactions, *J. Infect. Dis.* **144**:47.

Carney, W., Rubin, R. H., Hoffman, R. A., Hansen, W. P., Healey, K., and Hirsch, M. S., 1981, Analysis of T lymphocyte subsets in cytomegalovirus mononucleosis, *J. Immunol.* **126**:2114.

Cheeseman, S. H., Rubin, R. H., Stewart, J. A., Tolkoff-Rubin, N. E., Cosimi, A. B., Cantell, K., Gilbert, J., Winkle, S., Herrin, J. T., Black, P. H., Russell, P. S., and Hirsch, M. S., 1979, Controlled clinical trial of prophylactic human-leukocyte interferon in renal transplantation, *N. Engl. J. Med.* **300**:1345.

Cheeseman, S. H., Henle, W., Rubin, R. H., Tolkoff-Rubin, N. E., Cosimi, A. B., Cantell, K., Winkle, S., Herrin, J. T., Black, P., Russell, P. S., and Hirsch, M. S., 1980, Epstein–Barr virus infection in renal transplant recipients, *Ann. Intern. Med* **93**:39.

Ching, C., and Lopez, C., 1979, Natural killing of HSV type 1 infected target cells: Normal human responses and influence of anti-viral antibody, *Infect. Immun.* **26**:49.

Chou, S., Dylewski, J. S., Gaynon, M. W., Eibert, P. R., and Merigan, T. C., 1984, Alpha interferon administration in cytomegalovirus retinitis, *Antimicrob. Agents Chemother.* **25**:25.

Colin, J., Chastel, C., Renard, G., and Cantell, K., 1983, Combination therapy for dendritic keratitis with human leukocyte interferon and acyclovir, Am. J. Ophthalmol. 95:346.

Coster, D. J., Falcon, M. G., Cantell, K., and Jones, B. R., 1977, Clinical experience of human leucocyte interferon in the management of herpetic keratitis, Trans. Ophthalmol. Soc. U.K. 97:327.

Cruz, J. R., Dammin, G. J., and Waner, J. L., 1981, Protective effects of low-dose interferon against neonatal murine cytomegalovirus infection, Infect. Immun. 32:332.

Cunningham, A. L., and Merigan, T. C., 1983, Gamma interferon production appears to predict time of recurrence of herpes labialis, J. Immunol. 130:2397.

DeClercg, E., and Luczak, M., 1976, Intranasal challenge of mice with herpes simplex virus: An experimental model for evaluation of the efficacy of antiviral durgs, J. Infect. Dis. 133 (Suppl):A226.

DeMaeyer, E., and DeMaeyer-Guignard, J., 1980, Host genotype influences immunomodulation by interferon, Nature 284:173.

DeMaeyer, E., DeMaeyer-Guignard, J., and Vandeputte, M., 1975, Inhibition by interferon of delayed-type hypersensitivity in the mouse, Proc. Natl. Acad. Sci. USA 72:1753.

DeStefano, E., Friedman, R. M., Friedman-Kien, A. E., Goedert, J. J., Henriksen, D., Preble, O. T., Sonnabend, J. A., and Vilcek, J., 1982, Acid-labile human leukocyte interferon in homosexual men with Kaposi's sarcoma and lymphadenopathy, J. Infect. Dis. 146:451.

Doetsch, P. W., Suhadolnik, R. J., Sawada, Y., Mosca, J. D., Flick, M. B., Reichenbach, N. L., Dang, A. Q., Wu, J. M., Charubala, R., Pfleiderer, W., and Henderson, E. E., 1981, Core (2'-5') oligoadenylate and the cordycepin analog: Inhibitors of Epstein–Barr virus-induced transformation of human lymphocytes in the absence of interferon, Proc. Natl. Acad. Sci. USA 78:6699.

Epstein, L. B., 1979, The comparative biology of immune and classical interferons, in: Biology of the Lymphokines (S. Cohen, E. Pick, and J. J. Oppenheim, eds.) pp. 443–514, Academic Press, New York.

Eyster, M. E., Goedert, J. J., Poon, M.-C., and Preble, O. T., 1983, Acid-labile alpha interferon: A possible preclinical marker of the acquired immunodeficiency syndrome in hemophilia, N. Engl. J. Med. 309:583.

Field, A. K., 1967, Inducer of interferon and host resistence. II. Multi-stranded synthetic polynucleotide complexes, Proc. Natl. Acad. Sci. USA 58:1004.

Garner, J. G., Hirsch, M. S., and Schooley, R. T., 1984, Interferon-alpha prevention of Epstein–Barr virus-induced B-cell outgrowth, Infect. Immun. 43:920.

Gershon, A. A., Steinberg, S., and Brunell, P. A., 1974, Zoster immune globulin, N. Engl. J. Med. 290:243.

Glasgow, L. A., Hanshaw, J. B., Merigan, T. C., and Petralli, J. K., 1967, Interferon and cytomegalovirus in vivo and in vitro, Proc. Soc. Exp. Biol. Med. 125:843.

Gray, P. W., Leung, D. W., Pennica, P., Yelverton, E., Najarian, R., Simonsen, C. C., Derynck, R., Sherwood, P. J., Wallace, D. M., Berger, S. L., Levinson, A. D., and Goeddel, D. V., 1982, Expression of human immune interferon cDNA in E. coli and monkey cells, Nature (London) 295:503.

Gresser, I., Bandu, M.-T., Brouty-Boye, D., and Tovey, M., 1974, Pronounced antiviral activity of human interferon on bovine and procine cells, Nature 251:543.

Grundy, J. E., Mackenzie, J. S., and Stanley, N. F., 1981, Influence of H-2 and non-H-2 genes on resistance to murine cytomegalovirus infection, Infect. Immun. 32:277.

Grundy, J. E., Trapman, J., Allan, J. E., Shellam, G. R., and Melief, C. J. M., 1982, Evidence for a protective role of interferon in resistance to murine cytomegalovirus and its control by non-H-2-linked genes, Infect. Immun. 37:143.

Hamilton, J. D., 1982, Cytomegalovirus and immunity, in: Monographs in Virology, Vol. 12 (J. L. Melnick, ed.), Karger, Basel.

Hammer, S. M., Kaplan, J. C., Lowe, B. R., and Hirsch, M. S., 1982, Alpha interferon and acyclovir treatment of herpes simplex virus in lymphoid cell cultures, Antimicrob. Agents Chemother. 21:634.

Hanley, D. F., Wiranowska-Stewart, M., and Stewart, W. E., II, 1979, Pharmacology of interferons. I. Pharmacologic distinctions between human leukocyte and fibroblast interferons, *Int. J. Immunopharmacol.* **1**:219.

Havell, E. A., 1975, Two antigenically distinct species of human interferon, *Proc. Natl. Acad. Sci. USA* **72**:2185.

Haverkos, H. W., Pazin, G. J., Armstrong, J. A., and Ho, M., 1980, Follow-up of interferon treatment of herpes simplex, *N. Engl. J. Med.* **303**:699.

Hayes, T. G., Yip, Y. K., and Vilcek, J., 1979, Interferon production by human fibroblasts, *Virology* **98**:351.

Heine, J. W., De Ley, M., Van Damme, J., Biliau, A., and De Somer, P., 1980, Human fibroblast interferon purified to homogeneity by a two-step procedure, *Ann. N.Y. Acad. Sci.* **350**:364.

Hilfenhaus, J., and Ackerman, R., 1981, Endogenous interferon in the cerebrospinal fluid of herpes encephalitis patients, *Proc. Soc. Exp. Biol. Med.* **166**:205.

Hirsch, M. S., and Hammer, S. M., 1982, Nucleoside derivatives and interferons as antiviral agents, in: *Current Clinical Topics in Infectious Diseases—3* (J. S. Remington, and M. N. Swartz, eds.), pp. 30–55, McGraw-Hill, New York.

Hirsch, M. S., Zisman, B., and Allison, A. C., 1970, Macrophages and age-dependent resistance to herpes simplex virus in mice, *J. Immunol.* **104**:1160.

Hirsch, M. S., Schooley, R. T., Cosimi, A. B., Russell, P. S., Delmonico, F. L., Tolkoff-Rubin, N. E., Herrin, J. T., Cantell, K., Farrell, M., Rota, T. R., and Rubin, R. H., 1983, Effects of interferon-alpha on cytomegalovirus reactivation syndromes in renal-transplant recipients, *N. Engl. J. Med.* **308**:1489.

Ho, M., 1980, Role of specific cytotoxic lymphocytes in cellular immunity against murine cytomegalovirus, *Infect. Immun.* **27**:767.

Hokland, M., Heron, I., and Berg, K., 1981, Increased expression of β_2-microglobulin and histocompatibility antigen on human lymphoid cells induced by interferon, *J. Interferon Res.* **1**:483.

Holmes, A. R., Rasmussen, L., and Merigan, T. C., 1978, Factors affecting the interferon sensitivity of human cytomegalovirus, *Intervirology* **9**:48.

Isaacs, A., and Lindenmann, J., 1957, Virus interference. I. The interferon, *Proc. R. Soc. London Ser. B* **147**:258.

Itoh, K., Inoue, M., Kataoka, S., and Kumagai, K., 1980, Differential effect of interferon on expression of IgG and IgM Fc receptors on human lymphocytes, *J. Immunol.* **124**:2589.

Kaufman, H. E., Meyer, R. F., Laibson, P. R., Waltam, S. R., Nesburn A. B., and Shuster, J. J., 1976, Human leukocyte interferon for the prevention of recurrences in herpetic keratitis, *J. Infect. Dis.* **133**(Suppl.):A165.

Kern, E. R., Olsen, G. A., Overall, J. C., Jr., and Glasgow, L. A., 1978, Treatment of a murine cytomegalovirus infection with exogenous interferon, polyinosinic-polycytidylic acid and polyinosinic-polycytidylic acid-poly-L-lysine complex, *Antimicrob. Agents Chemother.* **13**:344.

Kohl, S., Starr, S. E., Oleski, J. M., Shore, S. L., Ashman, R. B., and Nahmias, A. J., 1977, Human monocyte-macrophage-mediated antibody-dependent cytotoxicity to herpes simplex virus-infected cells, *J. Immunol.* **118**:729.

Lai, P. K., Alpers, M. P., and MacKay-Scollay, E. M., 1977, Epstein–Barr herpesvirus infection: Inhibition by immunologically induced mediators with interferon-like properties, *Int. J. Cancer* **20**:21.

Lazar, R., Breinig, M. K., Armstrong, J. A., and Ho., M., 1980, Response of cloned progeny of clinical isolates of herpes simplex virus to human leukocyte interferon, *Infect. Immun.* **28**:708.

Lebon, P., Moreau, M. C., Cohen, L., and Chang, C., 1975, Different effect of ouabain on interferon production and action, *Proc. Soc. Exp. Biol. Med.* **149**:108.

Lee, S. H. S., and Epstein, L. B., 1980, Reversible inhibition by interferon of the maturation of human peripheral blood monocytes to macrophages, *Cell Immunol.* **50**:177.

Levin, M. J., and Leary, P. L., 1981, Inhibition of human herpesviruses by combination of acyclovir and human leukocyte interferon, *Infect. Immun.* **32:**995.

Levin, M. J., Rinaldo, C. R., Jr., Leary, P. L., Zaia, J. A., and Hirsch, M. S., 1979, Immune response to herpesvirus antigens in adults with acute cytomegaloviral mononucleosis, *J. Infect. Dis.* **140:**851.

Lopez, C., 1980, Resistance to HSV-1 in the mouse is governed by two major, independently segregating, non-H-2 loci, *Immunogenetics* **11:**87.

Lvovsky, E., Levine, P. H., Fucillo, D., Ablashi, D. V., Bengalia, Z. H., Armstrong, G. R., and Levy, H. B., 1981, Epstein–Barr virus and *Herpesvirus saimiri:* Sensitivity to interferons and interferon inducers, *J. Natl. Cancer Inst.* **66:**1013.

Maluish, A. E., Conlon, J., Ortaldo, J. R., Sherwin, S. A., Leavitt, R., Fein, S., Weirnik, P., Oldham, R. K., and Herberman, R., 1982, Modulation of NK and monocyte acitivity in advanced cancer patients receiving interferon, in: *Interferons* (T. C. Merigan, and R. M. Friedman, eds.), pp. 357–386, Academic Press, New York.

Mangi, R. J., Niederman, J. C., and Kelleher, J. E., Jr., 1974, Depression of cell-mediated immunity during acute infectious mononucleosis, *N. Engl. J. Med.* **291:**1149.

Marcus, P. I., 1982, The interferon inducer moiety of viruses: A single moleculer of ds RNA, *Tex. Rep. Biol. Med.* **41:**70.

Matsubara, L. M., Imanishi, J., Yasuno, H., Kagami, K., Osaki, Y., Pak, C. B., and Kishida, T., 1980, Comparative study of the sensitivity of herpes simplex types 1 and 2 to human leukocyte interferon in vitro, *J. Dermatol. (Tokyo)* **7:**203.

Menezes, J., Patel, P., Dussault, H., and Joncas, J., 1976, Effect of interferon on lymphocyte transformation and nuclear antigen production by Epstein–Barr virus, *Nature* **260:**430.

Merigan, T. C., Rand, K. H., Pollard, R. B., Abdallah, P. S., Jordan, G. W., and Fried, R. P., 1978, Human leukocyte interferon for the treatment of herpes zoster in patients with cancer, *N. Engl. J. Med.* **298:**981.

Merigan, T. C., Gallagher, J. G., Pollard, R. B., and Arvin, A. M., 1981, Short-course human leukocyte interferon in treatment of herpes zoster in patients with cancer, *Antimicrob. Agents Chemother.* **19:**193.

Meyers, J. D., McGuffin, R. W., Neiman, P. E., Singer, J. W., and Thomas, E. D., 1980, Toxicity and efficacy of human leukocyte interferon for treatment of cytomegalovirus pneumonia after marrow transplantation, *J. Infect. Dis.* **141:**555.

Meyers-Elliott, R. H., Elliott, J. H., Maxwell, W. A., Pettit, T. H., O'Day, D. M., Terasaki, P. I., and Bernoco, D., 1980, HLA antigens in recurrent stromal herpes simplex virus keratitis, *Am. J. Ophthalmol.* **89:**54.

Misko, I. S., Moss, D. J., and Pope, J. H., 1980, HLA antigen related restriction of T lymphocyte cytotoxicity of Epstein–Barr virus, *Proc. Natl. Acad. Sci. USA* **77:**4247.

Mogensen, S. C., 1977, Genetics of macrophage-controlled resistance to hepatitis induced by herpes simplex virus type 2 in mice, *Infect. Immun.* **17:**268.

Morahan, P. S., Morse, S. S., and McGeorge, M. B., 1980, Macrophage extrinsic antiviral activity during herpes simplex virus infection, *J. Gen. Virol.* **46:**291.

Moss, D. J., Rickinson, A. B., and Pope, J. H., 1979, Long termed T-cell mediated immunity to Epstein–Barr virus in man. II. Activation of cytotoxic T cells in virus-infected leukocyte cultures, *Int. J. Cancer* **23:**618.

Nagata, S., Brack, C., Henco, K., Schamlock, A., and Weissman, C., 1981, Partial mapping of ten genes of human interferon-α family, *J. Interferon Res.* **1:**333.

National Institute of Allergy and Infectious Diseases, World Health Organization–U.S. National Centre on Interferon, 1980, Interferon nomenclature, *Nature* **286:**110.

Oleske, J. M., Ashman, R. B., Kohl, S., Shore, S. L., Starr, S. E., Wood, P., and Nahmias, A. J., 1977, Human polymorphonuclear leucocytes as mediators of antibody-dependent cellular cytotoxicity to herpes simplex virus-infected cells, *Clin. Exp. Immunol.* **27:**446.

Olsen, G. A., Kern, E. R., Overall, J. C., Jr., and Glasgow, L. A., 1978, Effect of treatment with exogenous interferon, polyriboinosinic-polyribocytidylic acid or polyriboinosinic-polyribocytidylic acid-poly-L-lysine complex on herpesvirus hominis infections in mice, *J. Infect. Dis.* **137**:428.

O'Malley, J. A., Al-Bussam, N., Beutner, K., Wallace, H. J., Gailani, A., Henderson, E. S., and Carter, W. A., 1975, Cytomegalovirus infection with acute myelocytic leukemia, *N.Y. State J. Med.* **75**:738.

Overall, J. C., Jr., Yeh, T., and Kern, E. R., 1980, Sensitivity of herpes simplex types 1 and 2 to three preparations of human interferon, *J. Infect. Dis.* **142**:943.

Overall, J. C., Jr., Spruance, S. L., and Green, J. A., 1981, Viral-induced leukocyte interferon in vesicle fluid from lesions of recurrent herpes labialis, *J. Infect. Dis.* **143**:543.

Owerbach, D., Rutter, W. J., Shows, T. B., Gray, P., Goeddel, D. V., and Lawn, R. M., 1981, Leukocyte and fibroblast interferon genes are located on human chromosome 9, *Proc. Natl. Acad. Sci. USA* **78**:3123.

Oxman, M. N., 1977, Molecular mechanisms of the antiviral action of interferon: Effects of interferons on the transcription of viral messenger RNA, *Tex. Rep. Biol. Med.* **35**:230.

Panet, A., and Falk, H., 1983, Inhibition by interferon of herpes simplex virus thymidine kinase and DNA polymerase in infected and biochemically transformed cells, *J. Gen. Virol.* **64**:1999.

Patel, R., Fiala, M., Berne, V., and Chatterjee, N., 1978, Cytomegalovirus infections in renal allograft recipients: Correlative studies with histocompatibility antigens, *N.Z. Med. J.* **87**:393.

Pattengale, P. K., Smith, R. W., and Perlin, E., 1974, Atypical lymphocytes in acute infectious mononucleosis, *N. Engl. J. Med.* **291**:1145.

Pazin, G. J., Armstrong, J. A., Lam, M. T., Tarr, G. C., Jannetta, P. J., and Ho, M., 1979, Prevention of reactivated herpes simplex infection by human leukocyte interferon after operation on the trigeminal root, *N. Engl. J. Med.* **301**:225.

Perussia, B., Santoli, D., and Trinchieri, G., 1980, Interferon modulation of natural killer cell activity, *Ann. N.Y. Acad. Sci.* **350**:55.

Postic, B., and Dowling, J. N., 1977, Susceptibility of clinical isolates of cytomegalovirus to human interferon, *Antimicrob. Agents Chemother.* **11**:656.

Preble, O. T., and Friedman, R. M., 1983, Characterization of acid-labile alpha interferon from patients with autoimmune diseases, in: *Antiviral Research*, p. 27, Abstr. 1, Elsevier, Amsterdam.

Quinnan, G. V., Manischewitz, J. E., and Ennis, F. A., 1980, Role of cytotoxic T lymphocytes in murine cytomegalovirus infection, *J. Gen. Virol.* **47**:503.

Quinnan, G. V., Kirmani, N., Rook, A. M., Manischewitz, J. E., Jackson, L., Moreschi, G., Santos, G. W., Saral, R., and Burns, W. H., 1982, Cytotoxic T cells in cytomegalovirus infection: HLA-restricted T-lymphocyte and non-T-lymphocyte cytotoxic responses correlate with recovery from cytomegalovirus infection in bone-marrow-transplant recipients, *N. Engl. J. Med.* **307**:7.

Rasmussen, L. E., Jordan, G. W., Stevens, D. A., and Merigan, T. C., 1974, Lymphocyte interferon production and transformation after herpes simplex infections in humans, *J. Immunol.* **112**:728.

Rasmussen, L., Holmes, A. B., Hofmeister, B., and Merigan, T. C., 1977, Multiplicity-dependent replication of varicella–zoster virus in interferon-treated cells, *J. Gen. Virol.* **35**:361.

Revel, M., Kimchi, A., Shulman, L., Fradin, A., Shuster, R., Yakobson, E., Chernajovsky, Y., Schmidt, A., Shure, A., and Bendori, R., 1980, Role of interferon-induced enzymes in the antiviral and antimitogenic effects of interferon, *Ann. N.Y. Acad. Sci.* **350**:459.

Rhodes-Feuillette, A., Canivet, M., Champsaur, H., Gluckman, E., Mazeron, M. C., and Peries, J., 1983, Circulating interferon in cytomegalovirus infected bone-marrow-transplant recipients and infants with congenital cytomegalovirus disease, *J. Interferon Res.* **3**:45.

Rickinson, A. B., Moss, D. J., Pope, J. H., and Ahlberg, N., 1980, Long-term T-cell-mediated immunity to Epstein–Barr virus in man. IV. Development of T-cell memory in convalescent infectious mononucleosis patients, *Int. J. Cancer* **25**:59.

Rinaldo, C. R., Carney, W. P., Richter, B. S., Black, P. H., and Hirsch, M. S., 1980, Mechanisms of immunosuppression in cytomegaloviral mononucleosis, *J. Infect. Dis.* **141:**488.

Rosen, A., Gergely, P., Jondal, M., and Klein, G., 1977, Polyclonal Ig production after Epstein–Barr virus infection of human lymphocytes in vitro, *Nature* **267:**52.

Rubin, B. Y., and Gupta, S. L., 1980a, Differential efficacies of human type I and type II interferons as antiviral and antiproliferative agents, *Proc. Natl. Acad. Sci. USA* **77:**5928.

Rubin, B. Y., and Gupta, S. L., 1980b, Interferon-induced proteins in human fibroblasts and development of the antiviral state, *J. Virol.* **34:**446.

Russell, A. S., and Schlaut, J., 1975, HLA transplantation antigens in subjects susceptible to recrudescent herpes labialis, *Tissue Antigens* **6:**257.

Schmidt, A., Chernajovsky, Y., Shulman, L., Federman, P., Berissi, J., and Revel, M., 1979, An interferon-induced phosphodiesterase degrading (2'-5') oligoisoadenylate and the C-C-A terminus of tRNA, *Proc. Natl. Acad. Sci. USA* **76:**4788.

Schooley, R. T., Haynes, B. F., Grouse, J., Payling-Wright, C., Fauci, A. S., and Dolin, R., 1981, Development of suppressor T lymphocytes for Epstein–Barr virus-induced B-lymphocyte outgrowth during acute infectious mononucleosis: Assessment by two quantitative systems, *Blood* **57:**510.

Sehgal, P. B., and Sagar, A. D., 1980, Heterogeneity of poly (I), poly (C)-induced human fibroblast interferon mRNA species, *Nature* **288:**95.

Shore, S. L., Black, C. M., Melewicz, F. M., Wood, P. A., and Nahmias, A. J., 1976, Antibody-dependent cell-mediated cytotoxicity to target cells infected with type 1 and type 2 herpes simplex virus, *J. Immunol.* **116:**194.

Siegal, F. P., Lopez, C., Hammer, G. S., Brown, A. E., Kornfeld, S. J., Gold, J., Hassett, J., Hirschman, S. Z., Cunningham-Rundles, C., Adelsberg, B. R., Parham, D. M., Siegal, M., Cunningham-Rundles, S., and Armstrong, D., 1981, Severe acquired immunodeficiency in male homosexuals, manifested by chronic perianal ulcerative herpes simplex lesions, *N. Engl. J. Med.* **305:**1439.

Slovin, S. F., Schooley, R. T., and Thorley-Lawson, D. A., 1983, Analysis of cellular immune response to EBV by using cloned T cell lines, *J. Immunol.* **130:**2127.

Smith, C. A., Wigdahl, B., and Rapp, F., 1983, Synergistic antiviral activity of acyclovir and interferon on human cytomegalovirus, *Antimicrob. Agents Chemother.* **24:**325.

Spruance, S. L., Green, J. A., Chiu, G., Yeh, T., Wenerstrom, G., and Overall, J. G., Jr., 1982, Pathogenesis of herpes simplex labialis: Correlation of vesicle fluid interferon with lesion age and virus titer, *Infect. Immun.* **36:**907.

Stanwick, T. L., Campbell, D. E., and Nahmias, A. J., 1980, Spontaneous cytotoxicity mediated by human monocyte-macrophage against human fibroblast infected with herpes simplex virus-augmentation by interferon, *Cell. Immunol.* **53:**413.

Stanwick, T. L., Schinazi, R. F., Campbell, D. E. and Nahmias, A. J., 1981, Combined antiviral effect of interferon and acyclovir on herpes simplex virus types 1 and 2, *Antimicrob. Agents Chemother.* **19:**672.

Starr, S. E., and Garrabrant, T., 1981, Natural killing of cytomegalovirus-infected fibroblasts by human mononuclear leucocytes, *Clin. Exp. Immunol.* **46:**493.

Starr, S. E., Visintine, A. M., Tomeh, M. O., and Nahmias, A. J., 1976, Effects of immunostimulants on resistance of newborn mice to herpes simplex type 2 infection, *Proc. Soc. Exp. Biol. Med.* **152:**57.

Steele, R. W., 1980, Transfer factor and cellular reactivity to varicella–zoster antigen in childhood leukemia, *Cell-Immunol.* **50:**282.

Steele, R. W., Keeney, R. E., Brown, A., III, and Young, E. J., 1977, Cellular immune response to herpesviruses during treatment with adenine arabinoside, *J. Infect. Dis.* **135:**593.

Steele, R. W., Myers, M. G., and Vincent, M. M., 1980, Transfer factor for the prevention of varicella–zoster infection in childhood leukemia, *N. Engl. J. Med.* **303:**355.

Stevens, D. A., and Merigan, T. C., 1972, Interferon, antibody, and other host factors in herpes zoster, *J. Clin. Invest.* **51:**1170.

Stevens, D. A., Jordan, G. W., Waddell, T. F., and Merigan, T. C., 1973, Adverse effect of cytosine arabinuside on disseminated zoster in a controlled trial, *N. Engl. J. Med.* **289:**873.

Stevens, D. A., Ferrington, R. A., Jordan, G. W., and Merigan, T. C., 1975, Cellular events in zoster vesicles: Relation to clinical course and immune parameters, *J. Infect. Dis.* **131:**509.

Stewart, W. E., II, and Desmyter, J., 1975, Molecular heterogeneity of human leukocyte interferon: Two populations differing in molecular weights, requirements for renaturation, and cross-species antiviral activity, *Virology* **67:**68.

Stewart, W. E., II, Lin, L. S., Wiranowska-Stewart, M., and Cantell, K., 1977, Elimination of size and change heterogeneities of human leukocyte interferon by chemical cleavage, *Proc. Natl. Acad. Sci. USA* **74:**4200.

Sundmacher, R., Cantell, K., Houg, P., and Neumann-Haefelin, D., 1978a, Role of debridement and interferon in the treatment of dendritic keratitis, *Albrecht von Graefes Arch. Klin. Exp. Ophthalmol.* **207:**77.

Sundmacher, R., Cantell, K., and Neumann-Haefelin, D., 1978b, Combination therapy of dendritic keratitis with trifluorothymidine and interferon, *Lancet* **2:**687.

Sundmacher, R., Cantell, K., Skoda, R., Hallermann, C., and Neumann-Haefelin, D., 1978c, Human leukocyte and fibroblast interferon in a combination therapy of dendritic keratitis, *Albrecht von Graefes Arch. Klin. Exp. Ophthalmol.* **208:**229.

Szigeti, R., Masucci, M. G., Masucci, G., Klein, E., and Klein, G., 1980, Interferon suppresses antigen and mitogen-induced leukocyte migration inhibition, *Nature* **288:**594.

Tan, Y. H., 1975, Chromosome-21-dosage effect on the inducibility of antiviral gene(s), *Nature* **253:**280.

Taniguchi, T., Ohno, S., Fuji-Kuriyama, Y., and Muramatsu, M., 1980, The nucleotide sequence of human fibroblast interferon cDNA, *Gene* **10:**11.

Thorley-Lawson, D. A., 1980, The suppression of Epstein–Barr virus infection in vitro occurs after infection but before transformation of the cells, *J. Immunol.* **124:**745.

Thorley-Lawson, D. A., 1981, The transformation of adult but not newborn human lymphocytes by Epstein–Barr virus and phytohemagglutinin is inhibited by interferon: The early suppression by T cells of Epstein–Barr infection in mediated by interferon, *J. Immunol.* **126:**829.

Thorley-Lawson, D. A., Chess, L., and Strominger, J. L. 1977, Suppression of *in vitro* Epstein–Barr virus infection, *J. Exp. Med.* **146:**495.

Trinchieri, G., Santoli, D., Dee, R. R., and Knowles, B. B., 1977, Anti-viral activity induced by culturing lymphocytes with tumor-derived or virus-transformed cells, *J. Exp. Med.* **147:**1299.

Vengris, V. E., Reynolds, F. H., Hollenberg, M. D., and Pitha, P. M., 1976, Interferon action: Role of membrane gangliosides, *Virology* **72:**486.

Vezinet, F., Lebon, P., Amoudry, C., and Gilbert, C., 1981, Synthese d'interferon au cours des encephalites herpetiques de l'adulte, *Nouv. Presse Med.* **10:**1135.

Virelizier, J. L., Lenoir, G., and Griscelli, C., 1978, Persistent Epstein–Barr virus infection in a child with hypergammaglobulinemia and immunoblastic proliferation associated with a selective defect in immune interferon secretion, *Lancet* **2:**231.

Wallach, D., Fellous, M., and Revel, M., 1982, Preferential effect of gamma interferon on the synthesis of HLA antigens and their mRNAs in human cells, *Nature* **299:**833.

Weck, P. K., Apperson, S., May, L., and Stebbing, N., 1981, Comparison of the antiviral activities of various cloned human interferon-α subtypes in mammalian cell cultures, *J. Gen. Virol.* **57:**233.

Weimar, W., Schellekens, H., Lameijer, L. D., Masurel, N., Edy, V. G., Billiau, A., and Desomer, P., 1978, Double-blind study of interferon administration in renal transplant recipients, *Eur. J. Clin. Invest.* **8:**255.

Weimar, W., Lameijer, L. D., Edy, V. G., and Schellekens, W., 1979, Prophylactic use of interferon in renal allograft recipients, *Transplant. Proc.* **11:**69.

Whitley, R. J., Soong, S., Dolin, R., Betts, R., Linneman, C., Alford, C. A., Jr., and the NIAID Collaborative Antiviral Study Group, 1982, Early vidarabine therapy to control the complications of herpes zoster in immunosuppressed patients, *N. Engl. J. Med.* **307**:971.

Whitaker-Dowling, P. A., Wilcox, D. K., Widnell, C., and Youngner, J. S., 1983, Interferon-mediated inhibition of virus penetration, *Proc. Natl. Acad. Sci. USA* **80**:1083.

Worthington, M. G., Conliffe, M., and Williams, J., 1977, Treatment of fatal disseminated herpes simplex virus, type 1, infection in immunosuppressed mice, *Proc. Soc. Exp. Biol. Med.* **156**:168.

Yip, Y. K., Barrowelough, B. S., Urban, C., and Vilcek, J., 1982, Purification of two subspecies of human gamma (immune) interferon, *Proc. Natl. Acad. Sci. USA* **79**:1820.

Zarling, J. M., Sosman, J., Eskra, L., Borden, E. C., Horoszewicz, J. S., and Carter, W. A., 1978, Enhancement of T cell cytotoxic responses by purified human fibroblast interferon, *J. Immunol.* **121**:2002.

Zisman, B., Hirsch, M. S., and Allison, A. C., 1970, Selective effects of anti-macrophage serum, silica and anti-lymphocyte serum on pathogenesis of herpes virus infection of young adult mice, *J. Immunol.* **104**:1155.

CHAPTER 19

Effects of Immunopotentiating and Immunomodulating Agents on Experimental and Clinical Herpesvirus Infections

GEORGES H. WERNER AND AURELIO ZERIAL

I. INTRODUCTION

As pointed out by Ching and Lopez (1979), patients with some types of immunodeficiency disorders, newborn infants, and patients treated with cytotoxic immunosuppressive drugs, either for cancer or for organ transplantation, are prone to unusually severe infections with herpesviruses (herpes simplex, cytomegalovirus, varicella–zoster). Herpes simplex virus (HSV) and cytomegalovirus (CMV) are a common cause of morbidity and mortality in bone marrow, renal, and cardiac allograft recipients. A defect in cell-mediated immunity to CMV has been reported both in children who excrete CMV in the first year of life and in their mothers (Starr et al., 1979). It appears reasonable therefore to assume that restoration of the immune capacity by appropriate immunopotentiating agents might be a way to control the severity of herpesvirus infections in immunodeficient patients and, by some stretch of the imagination, that such drugs might even be useful in the management of herpesvirus infections in immunologically normal subjects.

GEORGES H. WERNER AND AURELIO ZERIAL • Rhône-Poulenc Santé, Centre de Recherches de Vitry, 94407 Vitry-sur-Seine, France.

Indirect evidence for the beneficial effect of improving the immune capacity simply by providing an adequate diet is illustrated by events that occurred in 1960 in Capetown, South Africa (Tyrrell, 1982). Seriously ill children were admitted to the hospital suffering from a generalized, often fatal, infection due to HSV and involving the liver. Quite suddenly, however, the disease disappeared and, on investigation, it was found that the likely reason was that a voluntary group, with no thought of preventing infection, was supplying milk supplements to children living in underprivileged areas.

Within the last few years, a fairly large number of immunopotentiating (or immunomodulating) drugs have been described, some of which have already been used clinically. Our purpose is to review what has been achieved, experimentally in laboratory animals and clinically in humans, with these drugs in the treatment of primary or recurrent herpesvirus infections. In view of the fact that immunopotentiating agents may exert a variety of actions on the diverse effector mechanisms of the immune system, it is first necessary to delineate the immune mechanisms of resistance to and recovery from herpesvirus infections. Most studies have been performed in the mouse, in which a systemic, usually rapidly lethal, infection can be induced by intraperitoneal (i.p.) inoculation of human HSV-1 or HSV-2 or of murine CMV. This is also the system that has been most commonly used to test the activity of immunopotentiating agents on such infections.

When the viruses are introduced i.p., the first line of defense they encounter is the peritoneal macrophage. (Zisman *et al.*, 1970; Hirsch *et al.*, 1970; Lopez, this volume). Important roles for NK cells (Lopez, 1975, and this volume; Lopez *et al.*, 1980; Bancroft *et al.*, 1981, Morahan *et al.*, 1982) and interferons (IFN) as resistance mechanisms against HSV infections have also been well established (Gresser *et al.*, 1976; Engler *et al.*, 1982; Kirchner *et al.*, 1983). Evidence supporting these conclusions is reviewed by Lopez (this volume). Participation of T lymphocytes in the recovery mechanisms from HSV infections is suggested by studies from several laboratories (summarized by Nash *et al.*, this volume). Specifically, both cytotoxic T cells and delayed-type hypersensitivity are required for protection.

Other, still unknown, mechanisms may play a role in the resistance against HSV, as shown by studies performed with a strain of HSV-1 that does not cause lethal encephalitis in various inbred mouse strains, when injected by all routes except the intracerebral one (Kümel *et al.*, 1982). This strain replicates in various organs to an extent similar to that of lethal strains, but transport of this virus to the CNS or replication therein is efficiently restricted after peripheral inoculation. The blocking mechanism is not, however, a function of IFN induction or sensitivity nor is it due to enhanced NK cell activation; it appears to be independent of T-cell or macrophage functions but it is age-dependent, as newborn mice are fully susceptible to i.p. infection with this strain.

Thus, any substance capable of exerting stimulating or suppressive activities on macrophages and monocytes, on the NK cell system, on endogenous production of IFNs, on cytotoxic T lymphocytes, on polymorphonuclear leukocytes, or on antibody-producing cells will in theory be able to influence in a beneficial or harmful manner the outcome of a primary or recurrent infection with herpesviruses. Although this is likely the case when immunomodulating drugs are administered in the course of natural infections in humans or animals, it is doubtful that all the mechanisms described above come into play when a relatively simple system is used in testing these drugs, such as the i.p. inoculation of large doses of highly virulent HSV strains into fully susceptible mice. Yet, as will be seen, this is the way most natural or synthetic immunopotentiators have been studied thus far.

II. MICROBIAL AGENTS AND SUBSTANCES OF MICROBIAL ORIGIN

It has been known for many years that nonlethal infections of the mouse with bacteria or parasites can enhance the animal's resistance against antigenically unrelated infectious agents. Such a nonspecific immunostimulating effect has been examined by a number of authors against parenteral (usually i.p.) infection of the mouse with HSV-1 or HSV-2. BCG, a bovine strain of *Mycobacterium tuberculosis* of attenuated virulence, has been most extensively tested in this respect. It was shown (Starr *et al.*, 1976) that live BCG administered i.p. to newborn mice 6 days before i.p. challenge with HSV-2 increased the survival rate of the animals. Typhoid or brucella inactivated vaccines were ineffective in this system. In adult mice, intravenous (i.v.) infection with 10^6 live BCG bacilli increased their resistance against lethal challenge 15–30 days later with a strain of HSV-1, but it was without effect when challenge was performed with a strain of HSV-2 requiring immunosuppression of the animals with cyclophosphamide to be lethal (Floc'h and Werner, 1976). On the other hand, BCG infection by the i.v. route 7–10 days before intravaginal inoculation of HSV-2 in adult female mice did not protect against disease (vaginitis, posterior paralysis, encephalitis, and death) but rather tended to enhance its severity, whereas passive immunization with specific antiserum 4 hr before viral challenge afforded significant protection (Baker *et al.*, 1974). The same authors noted, however, that combination of BCG inoculation (-7 days) and antiserum treatment (-4 hr) provided the highest degree of protection.

As early as 1972, Larson *et al.*, showed that BCG immunization of rabbits increased their resistance to HSV-2 infection: whereas control rabbits died from encephalitis following vaginal or corneal inoculation of the virus, animals that had received 4×10^7 viable units of BCG i.v.

4 weeks before viral challenge also developed encephalitis but, in most cases, recovered.

All these experimental data suggest that in animals whose mononuclear phagocytic system has been stimulated through infection with mycobacteria, an enhanced resistance exists against dissemination of HSV to the brain. Attempts have been made to treat recurrent herpes genitalis in men and women by BCG immunotherapy. In one study (Anderson et al., 1974) 15 such patients were injected intradermally (i.d.) with BCG; all of them experienced a decrease in frequency and severity of the recurrences, the best response being found in the patients becoming and remaining tuberculin skin test-positive following BCG injection. These results were not confirmed in a later, more extensive trial, carried out in 100 patients (Corey et al., 1976): patients suffering from recurrent herpes genitalis were injected i.d. with BCG or a placebo (Candida antigen); the average number of recurrences per 100 days was similar in both groups, the mean duration of the lesions during the recurrences was also similar but, among women, the mean duration of pain was significantly shorter in the BCG recipients than in the placebo group.

Next to live BCG, heat- and formaldehyde-killed microorganisms of the anaerobic Corynebacterium genus (C. granulosum, C. parvum) have been extensively tested, in laboratory animals and in man, as nonspecific stimulants of the immune system and a number of studies have been devoted to the effects of such inactivated microbial preparations in mice infected with herpesviruses. Kirchner et al., (1977) showed that i.p. injection of C. parvum (10 mg/kg) 1 week before i.p. infection of adult mice with HSV-1 protected them from encephalitis and death, even when they had been rendered more susceptible to the virus by cyclophosphamide-induced immunosuppression. Treatments performed shortly before or after virus inoculation were ineffective. Similarly, mice treated 7–10 days before virus inoculation with C. parvum exhibited enhanced resistance against lethal infection with HSV-2 or with murine CMV (Glasgow et al., 1977). The increased resistance against HSV-2 could be transferred to recipient mice by peritoneal exudate cells from Corynebacterium-treated animals. The same authors (Morahan et al., 1977a) showed that inhibition of macrophage function by in vivo treatment of the mice with silica, while greatly increasing their susceptibility to HSV-2 infection, did not affect the antiviral activity of the immunostimulant. Contradictory results were reported more recently (Gabrielson et al., 1980): in BALB/c mice, silica treatment did not enhance susceptibility to HSV-2 infection but markedly reduced the protective activity of C. granulosum. Attempts to passively transfer protection to suckling mice with peritoneal cells from C. granulosum-treated adult mice were unsuccessful. In our laboratory (A. Zerial unpublished data), protection of adult mice against lethal i.p. challenge with HSV-1 was evident only when treatment with C. parvum was performed at least 5 days before virus inoculation; in vivo transfer of resistance was achieved with spleen or peritoneal cells from C.

parvum-treated mice. In the case of HSV-2 infection of adult mice, two treatments with C. parvum, 7 days and 1 day before virus inoculation, provided greater protection than a single treatment 7 days before infection.

The fact that in the hands of Morahan et al., (1977a) abrogation of macrophage function through silica injection did not significantly reduce the protective activity of Corynebacterium against HSV infection suggests that this protection may not be entirely mediated by the mononuclear phagocyte system—contrary to what is likely the case with BCG. Indeed, it has been shown (Ojo et al., 1978) that i.p. injection of C. parvum into adult mice causes a sharp increase of NK activity in peritoneal exudate cells, and a certain population of NK cells may thus be the most likely candidate as the target of Corynebacterium antiviral activity. NK cell activity is stimulated by IFN; it has been shown (Gabrielson et al., 1980) that injection of BALB/c mice with HSV-2 caused a threefold increase in splenic IFN level within 3 hr and that pretreatment with C. granulosum 1 week before HSV-2 inoculation increased the splenic IFN level an additional threefold. Whether NK cells are activated directly by C. parvum or indirectly through IFN induction remains to be elucidated. The important participation of the NK system in the protection against HSV infection by corynebacteria is also suggested by the observation that newborn mice, practically devoid of such a system, are not protected by C. parvum against HSV-2 infection.

As far as we known, in spite of the rather widespread use of Corynebacterium immunotherapy in cancer patients, and in contrast to what has been attempted with BCG in humans, no clinical study has been performed with Corynebacterium in patients suffering from herpesvirus infections. It would be worthwhile to investigate whether or not C. parvum-treated cancer patients are less prone to herpes simplex or zoster infections.

Besides live BCG and killed whole Corynebacterium microorganisms, a number of bacterial extracts and substances produced by bacteria have been tested for their capacity to stimulate the resistance of mice to HSV infection. Lipopolysaccharides (endotoxins) from gram-negative bacilli, given parenterally and at appropriate times before infection, do exert a protective effect against the infection of mice with bacteria and viruses, including HSV, but of all the viruses tested the latter appear to be the least sensitive to that effect (Rolly et al., 1974; A. Zerial unpublished data).

Extraction of live Brucella abortus microorganisms with aqueous ether yields an insoluble residue with low toxicity (BRU-PEL), which induces IFN (Youngner et al., 1974). Given shortly before intranasal infection with HSV-2 in 3-week-old mice, BRU-PEL decreased their mortality. In adult mice infected i.p. with HSV-2, protection by BRU-PEL was significant when it was injected by the same route 24 hr before virus challenge, nonexistent when treatment was performed 4 days before virus inoculation, but again significant when injection took place 7, 10, or 14 days before challenge (Kern *et al.*, 1976). When BRU-PEL was administered as an IFN inducer (i.e., 24 hr before infection), its antiviral activity was less marked than that of a synthetic IFN inducer such as poly I:C, but when it was given several days before infection, its antiviral activity was comparable to that of live *Brucella* organisms. In the latter case, this insoluble substance, which may persist in the macrophages having ingested it, protects against viral infection through stimulation of the phagocytic cells rather than through IFN induction.

Glucan, a water-insoluble 1,3-glucopyranose polysaccharide extracted from yeast cell walls, has been shown, when given i.v. to adult mice 2,4,6,8, or 19 days before i.p. inoculation of HSV-2, to decrease mortality in a highly significant way (DiLuzio *et al.*, 1982). When given by this route, glucan causes granuloma formation in the liver and the mechanism of its nonspecific antiviral activity may well be similar to that of BCG.

In conclusion, bacterial organisms or extracts that enhance the resistance of mice to HSV infection are those that exert a powerful and sustained stimulation of the mononuclear phagocyte system (e.g., BCG, glucan) or those that activate NK cells concomitantly with IFN induction (*C. parvum*).

Chemically defined substances have been isolated from bacterial cell walls and shown to possess most of the immunostimulating activities of whole bacteria or crude extracts. The best example is provided by muramyl dipeptide (MDP), a minimal structure of bacterial cell wall endowed with adjuvant activities on humoral and cellular immunity (Ellouz *et al.*, 1974). In spite of its marked enhancing activity on resistance of mice to a number of bacterial, parasitic, and fungal infections, MDP was not shown to protect mice against viral infections, notably against infection with HSV. Similar negative findings were made with a synthetic immunostimulating lipopeptide, lauroyl-L-alanyl-D-glutamyl-diaminopimelamic acid (Migliore-Samour *et al.*, 1980), first obtained by chemical coupling of a microbial tetrapeptide with lauric acid. This substance (LTP) protects mice against infection with *Listeria monocytogenes* or *Klebsiella pneumoniae* and, given subcutaneously (s.c.) at 1 or 3 mg/kg, markedly increases NK cell activity (tested against HYAC cells) in mouse peripheral blood. However, it was not found to enhance resistance of mice to HSV-1 infection, when administered at various intervals before and/or after viral challenge (A. Zerial unpublished data). On the other hand, a synthetic lipophilic distant relative of MDP, namely MTP-PE [*N*-acetyl-muramyl-L-alanyl-D-isoglutaminyl-L-alanine-(1′,2′-dipalmitoyl-*sn*-glycero-3′-hydroxyphosphoryloxy)-ethylamine] has been shown, when given intranasally 7 days before infection, to protect mice against mortality caused by intranasal infection with HSV-1 and guinea pigs against paralysis following intravaginal challenge with HSV-2 (Dietrich *et al.*, 1983).

III. TRANSFER FACTOR

This is not the place to speculate about the chemical nature and the possible mechanisms of immunomodulating activity of transfer factor (TF), a term used to describe a dialyzable extract of human or animal leukocytes, which can transfer specific cellular immunity from a skin test-positive donor to a skin test-negative recipient. For instance, marmosets were protected from fatal experimental infection with HSV-1 when they were treated with TF obtained from a human donor exhibiting a high degree of cellular immunity to that virus (Steele et al., 1976).

A double-blind placebo-controlled study was undertaken to evaluate the efficacy of TF (obtained and pooled from 8 donors with evidence of humoral and cellular immunity to HSV of either type but free from recurrences) in patients with severe recurrent herpes simplex of type 1 (6 patients) or type 2 (22 patients). Each patient received TF or saline s.c. every 2 months for a total of six doses. The results of this study (Oleske et al., 1978) were negative: TF therapy neither decreased the number or severity of recurrences nor modified in vitro responsiveness of the recipients' lymphocytes to HSV antigens. Subjective improvement was reported, however, by more than 50% of the patients both in the placebo and in the TF group: such a "placebo effect" must indeed be taken into consideration in all therapeutic trials involving patients with recurrent herpes.

Cellular immune reactivity to varicella–zoster virus (VZV) was transferred to 15 children with acute lymphocytic leukemia who were previously negative (lymphocyte cytotoxicity against VZV-infected cells, VZV-induced lymphocyte transformation), using TF obtained from adults who had recovered from chicken pox (Steele, 1980). No patients in relapse (3) converted immune response whereas 10 of 12 in remission developed positive reactivity in at least one of the assays; cytotoxicity was the most consistently positive test following TF administration. In a subsequent study (Steele et al., 1980), 61 children with acute lymphocytic leukemia and no immunity to varicella, were given TF (doses of 10^8 lymphocyte equivalents/7 kg body wt, s.c.) or placebo (saline) and followed for 12–30 months in a double-blind trial. Sixteen patients in the TF group and fifteen in the placebo group were exposed to varicella or zoster during this period and most of them (9 of 16 in the TF group, 8 of 15 in the placebo group) showed a serum antibody titer rise. Clinical chicken pox developed in 13 of the 15 exposed children in the placebo group versus only 1 of 16 in the TF group.

IV. THYMIC HORMONES (FACTORS)

Thymic hormones (or factors) is the name given to soluble mediators produced by the thymus, which act within this gland or at a distance to induce T-lymphocyte differentiation and maturation. They comprise polypeptides of various degrees of purity such as thymosin, thymopoietin, thymic humoral factor (THF), thymostimulin, serum thymic factor (FTS), and much work has been performed lately to characterize them better. Available now for clinical studies are: fraction 5 of thymosin, TP-1 (an extract prepared in a way similar to that of thymosin fraction 5), THF (prepared from calf thymus by ultracentrifugation and dialysis), thymostimulin (also prepared from calf thymus), and synthetic polypeptides prepared with the recently acquired knowledge of the amino acid sequence of some of the thymic hormones. The latter include: thymuline (FTS), a Zn-containing nonapeptide, thymosin alpha-1, and TP-5, a pentapeptide patterned after thymopoietin.

Thymic hormones are capable of restoring to normal levels some of the functions of T lymphocytes (such as E rosette formation or response to T-cell mitogens) in animals and humans exhibiting corresponding deficiencies. Thymuline can also stimulate NK cell activity *in vitro* and *in vivo* in the mouse and *in vitro* in human peripheral blood (Dokhelar *et al.*, 1983).

The first report of application of a thymic factor to the treatment of a herpesvirus infection in immunocompromised patients concerned THF therapy in four children who were treated with immunosuppressive cytostatic drugs for lymphoproliferative malignancy and who came down with varicella, a life-threatening infection in such patients. They received daily intramuscular (i.m.) injections of THF for 5–16 days, beginning on the first day of the varicella eruption. All four children recovered uneventfully; an increase in the count of peripheral blood lymphocytes and in rosette-forming cells was seen following THF treatment (Zaizov *et al.*, 1977). In a more recent study by the same group (Handzel *et al.*, 1981), 63 patients suffering from herpes zoster or varicella complicating various malignancies were treated with daily i.m. injections of THF: clinical and laboratory findings gave evidence of a positive effect of this treatment on the immunological mechanisms of recovery, including five cases of encephalitis (presumably herpetic) which showed rapid clinical improvement after a short course of THF therapy.

Clinical studies are under way with thymuline and encouraging results are being reported in some viral infections, such as generalized herpes in a thymectomized patient (Bach and Dardenne, 1982). Fiorelli *et al.*, (1983) treated with thymostimulin or a placebo 21 patients with primary or secondary immunodeficiencies (details of which were not given) suffering from recurrent herpes labialis. Thymostimulin-treated patients showed a significant reduction in the number and severity of recurrences together with higher levels of circulating T cells, of HSV-specific lymphocyte transformation, and of NK activity in peripheral blood. The same authors have initiated a double-blind study of thymostimulin treatment of recurrent HSV keratitis in immunologically normal patients. Therapy was started on the occasion of a recurrence of keratitis

(1 mg/kg i.m. per day for 1 week followed by the same dose twice a week for 3 months). Of the 22 patients who accomplished the 1-year period of observation required before breaking the code, 2 of the 12 thymostimulin-treated individuals and 6 of the 10 placebo recipients had recurrences. Thymostimulin treatment counteracted the decrease of circulating T cells that was associated with the infectious episode.

In conclusion, there appears to be an interesting potential for the use of thymic hormones in the treatment of HSV, VZV, and CMV infections, especially in immunocompromised patients but also in subjects in whom recurrence of HSV infection may be associated with transient impairment of T-cell functions.

V. SYNTHETIC INTERFERON INDUCERS

The possibility of using exogenous IFN for treating herpesvirus infections is discussed by Lebel and Hirsch in this volume. Mention must be made, however, of experimental work performed with synthetic IFN inducers, for these substances, in addition to (or in connection with) their IFN-inducing activity, exert *bona fide* immunomodulating effects. For instance, the double-stranded polyribonucleotide poly I : C is a good adjuvant of humoral immunity, whereas tilorone (a fluorenone derivative active orally) inhibits delayed-type hypersensitivity reactions. Figure 1 shows the protective activity in adult mice of poly I : C administered i.v. 6 hr before their i.p. infection with an HSV-2 strain. Figure 2 summarizes the results of an experiment in which tilorone was administered orally to adult mice 24 hr before their i.p. infection with an HSV-1 strain (A. Zerial unpublished data). In both cases, maximum efficacy was obtained when treatment with the inducer was performed in such a way that subsequent viral challenge coincided in time with peak IFN activity in the animals' plasma. One may therefore assume that protection of the mice was mediated through the antiviral activity of the IFN that was induced.

Poly I : C was shown to be a mediocre IFN inducer in humans but a poly I : C–polylysine complex has been described that is markedly more resistant than poly I : C to human serum nucleases and thus produces good plasma levels of IFN in man (Levine *et al.*, 1979). Tilorone does not induce IFN in man and, furthermore, exerts severe toxic effects. Other synthetic IFN inducers have been prepared more recently, such as pyrimidinones (Lotzova *et al.*, 1983), which are undergoing phase I trials in cancer patients. It might be of interest to study the efficacy of these novel IFN inducers in patients with severe herpesvirus infections.

Mention should also be made of pyran, a copolymer of divinyl ether and maleic anhydride, which induces IFN in mouse and man and enhances macrophage function. Administered i.v. or i.p. 24 hr before virus challenge, pyran increased resistance of adult mice to i.v. or intravaginal infection with HSV-2 (Morahan and McCord, 1975; Breinig *et al.*, 1978).

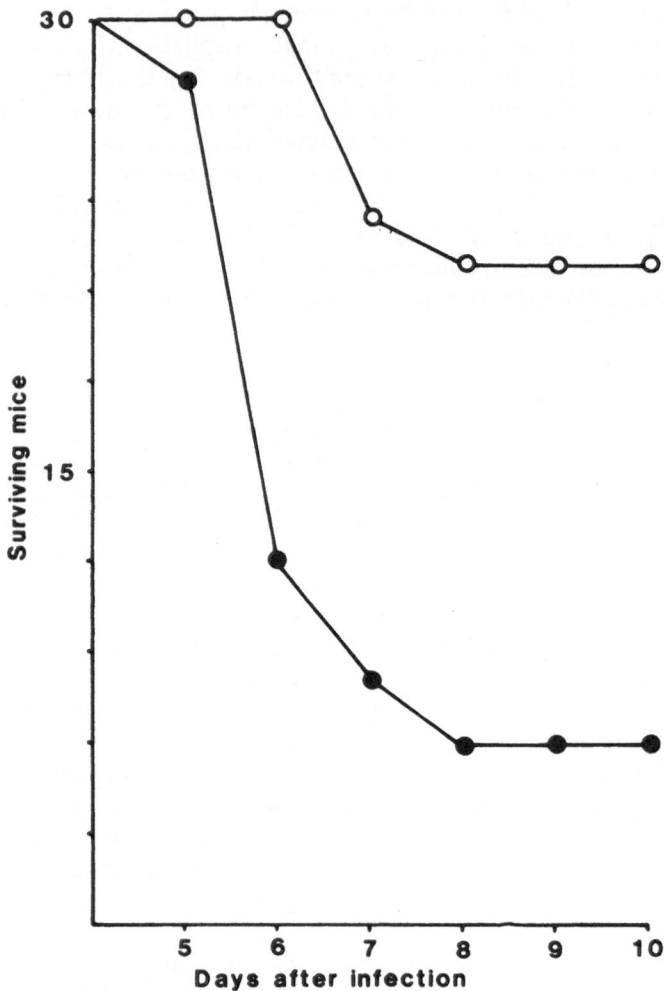

FIGURE 1. Activity of poly I:C (IFN inducer) on the infection of adult mice with a lethal strain of HSV-2. (O) Intravenous treatment (5 mg/kg body wt.) 6 hr before i.p. virus challenge; (●) HSV-2 controls.

Depression of macrophage function by *in vivo* treatment with silica did not inhibit this activity of pyran (Morahan *et al.*, 1977b).

Carbopol, a polymer of acrylic acid cross-linked with allylsucrose, administered i.p. to mice enhanced their resistance to intranasal challenge 1–4 days later with HSV-1, and a modest peak of IFN activity was found in the serum between 20 and 40 hr after treatment (DeClercq and Luczak, 1976). While in the case of pyran, IFN induction may be the chief mechanism of action, activation of the alveolar macrophages of the lung may better explain the effect of carbopol.

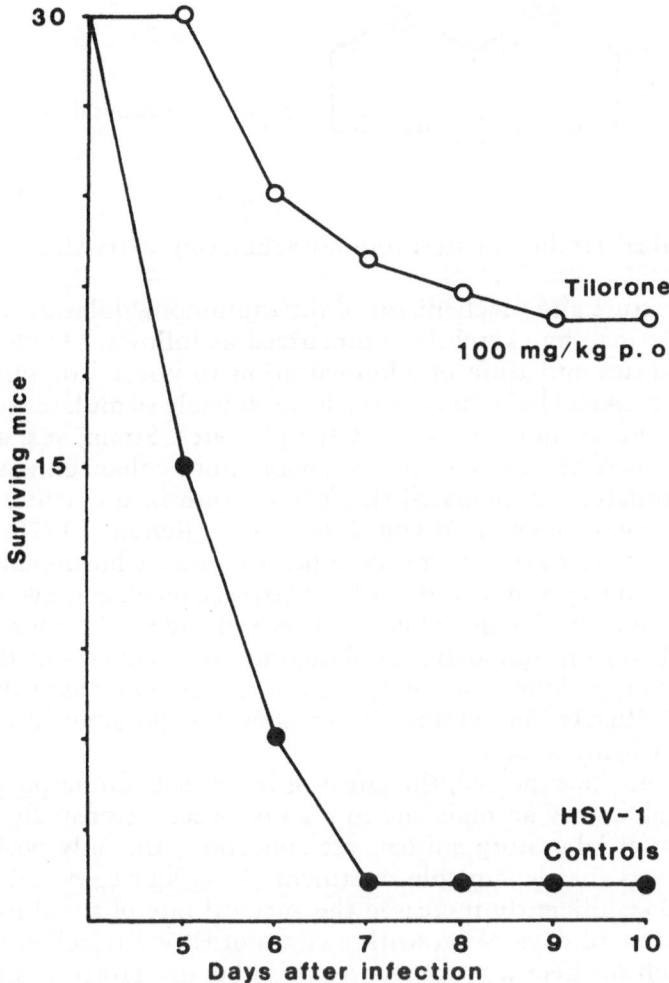

FIGURE 2. Activity of tilorone (IFN inducer) on the infection of adult mice with a lethal strain of HSV-1. (○) Oral treatment (100 mg/kg body weight) 24 hr before i.p. virus challenge; (●) HSV-1 controls.

VI. LEVAMISOLE

Tetramisole, an imidazothiazole compound, and its levorotatory isomer levamisole (formula in Fig. 3) have been used successfully for several years to treat nematode infestations in animals and man. In 1971, Renoux and Renoux found that treatment of mice with levamisole or tetramisole at the time of their immunization with a *Brucella* vaccine increased the protective efficacy of the latter. It was the first time that a simple synthetic compound had been shown to exert an immunoadjuvant effect, and because levamisole was already in use in humans as an anthelmintic

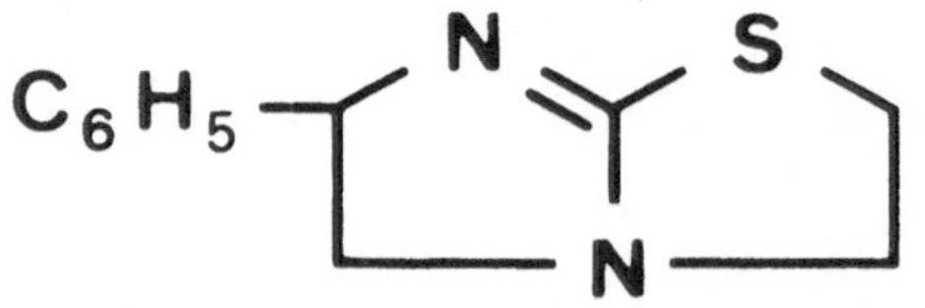

FIGURE 3. Chemical formula of levamisole.

drug, clinical studies of its immunomodulating activities were readily possible.

The nature and mechanisms of the immunomodulating activities of levamisole may be sketchily summarized as follows: "Under a narrow range of doses and time of administration to normal or immunocompromised men and laboratory animals, levamisole stimulates recruitment and functions of monocytes and T lymphocytes. Strain, sex, age and antigen modulate the activity of levamisole from enhancement to inhibition; stimulation is mediated through a serum factor, which is able to promote thymocytes in thymusless mice" (Renoux, 1978; Symoens, 1978). In several ways, levamisole appears to mimic hormonal regulation of the immune system and it has been stressed (Bach and Edelman, 1980) that there are similarities of action between thymic hormones, TF, and levamisole on immunocytes, as shown by the analogies of their effects on delayed-type hypersensitivity. These agents may share the capacity of nonspecifically "activating" T lymphocytes, possibly through an action on their membranes.

As could be expected, the effect of levamisole on herpesvirus infections in laboratory animals and in man has been extensively studied.

As far as laboratory animals are concerned, the only positive result reported was that levamisole treatment (3 mg/kg s.c., 4 and 24 hr after infection) significantly increased the survival rate of suckling rats inoculated i.p. at 10 days of age with a strain of HSV-2 (Fischer *et al.*, 1975, 1976). Such an effect was not reported by other investigators and we could not confirm it with the strain of HSV-2 that we used, which was markedly more virulent for suckling rats than that employed in the study summarized above (G.H. Werner, unpublished data). On the other hand, studies in mice have yielded uniformly negative results, with respect to protection by levamisole of newborn or adult animals infected i.p. or intravaginally with various strains of HSV-1 or HSV-2 (Starr *et al.*, 1976; Morahan *et al.*, 1977b).

Many clinical studies have been performed with levamisole in human patients afflicted with recurrent herpes labialis or genitalis. The protocols of treatment (dosage and frequency of administration of the drug) varied greatly from one study to another, making comparison of the conclusions difficult. Most early clinical studies were uncontrolled (i.e., did not include placebo-treated subjects): clinical improvement and decreased frequency of recurrences were claimed in several cases of facial, labial, corneal, or genital herpes (Kint and Verlinden, 1974; Kint *et al.*,

1974; Lods, 1976). Clinical improvement was associated with enhanced *in vitro* response of peripheral blood lymphocytes to HSV antigen in one study (O'Reilly *et al.*, 1977) and with a decreased response in another (Spitler *et al.*, 1975). Rabson *et al.* (1977) studied the random migration and chemotaxis of neutrophils of 10 patients suffering from recurrent mucocutaneous lesions due to HSV-1 or HSV-2. Chemotaxis to endo-toxin-activated serum (but not to hydrolyzed casein) was significantly lower than in healthy subjects; random migration was slightly reduced. After treatment with levamisole (150 mg/day orally for 3 days), 5 of the 10 patients showed improved chemotaxis of their neutrophils. In all patients, the defective chemotactic response could be corrected *in vitro* by addition of levamisole (10^{-3} M) to the culture medium.

More recently, several double-blind placebo-controlled studies of the efficacy of levamisole in recurrent herpes labialis or genitalis have been performed. Reviewing 10 such studies, Symoens *et al.*, (1979) found that overall, 107 of 172 patients (i.e., 62%) improved with levamisole versus 86 of 194 (44%) who improved while receiving a placebo. This marked placebo effect is, of course, a complicating factor. The beneficial effects of levamisole varied in nature from one study to another: decrease of recurrence frequency, decrease of severity of the recurrence episodes, more rapid healing of the lesions (which were almost never very obvious). Symoens *et al.* (1979) stress the point that the difference between placebo and levamisole appears to be most pronounced when individual respon-siveness is considered rather than mean differences between groups and they cautiously conclude that patients with recurrent herpes simplex are a very heterogeneous population, comprising subpopulations that respond to levamisole, placebo, both or neither.

Very few new studies have been published in recent years about the use of levamisole in treating recurrent herpes, this may be due to growing disenchantment with the drug for this application as well as to recog-nition of the potential risk of long-term levamisole therapy, namely the rare but severe occurrence of agranulocytosis.

VII. INOSIPLEX

At the present time, inosiplex is the sole immunomodulating agent on the pharmaceutical market in several countries (usually under the trade name Isoprinosine) with, as its main indication, the treatment of viral diseases, including herpes.

Inosiplex is a compound formed from inosine and the *para*-acetami-dobenzoate salt of *N,N*-dimethylamino-2-propanol, with a molar ratio of 1:3 (formula in Fig. 4). Its main characteristic is a very low level of tox-icity, allowing daily oral dosages of 4 g or even more in humans.

Early studies led to consideration of inosiplex as an antiviral com-pound (in the usual meaning of this word): for instance, it was shown

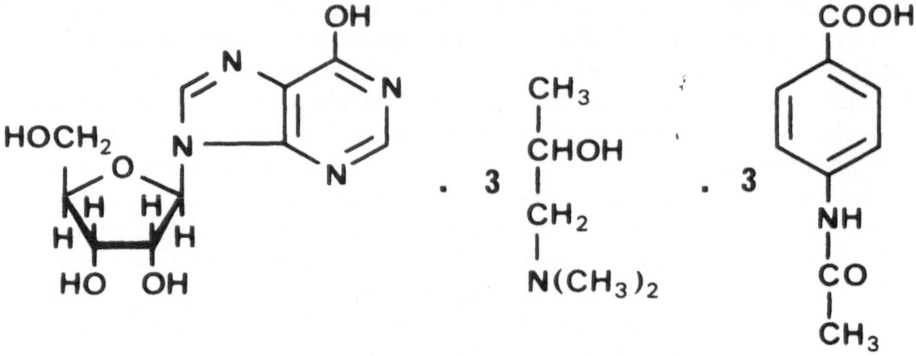

FIGURE 4. Chemical formula of inosiplex.

(Gordon *et al.*, 1974) that, under precise conditions of concentrations, timing of application, and methods of evaluation, inosiplex could inhibit the cytopathic effect of an HSV-2 strain (but not of an HSV-1 strain) in primary rabbit kidney cells. The same authors indicated that, given in the drinking water at large doses, inosiplex could protect hamsters from the encephalitis following corneal inoculation of HSV-1 or HSV-2. More recently, it became apparent that the *in vivo* antiviral effect of the compound is rather due to its immunomodulating properties than to a direct inhibitory activity on viral replication. These immunomodulating activities have been summarized as follows: "Inosiplex is a nontoxic inducer of lymphocyte differentiation and a potentiator of mitogen and/or virus-induced lymphocyte proliferation *in vitro* and *in vivo*; it modulates active T lymphocytes, cytotoxic T lymphocytes, lymphokine secretion and helper, as well as suppressor, functions" (Simon *et al.*, 1981). Taking all this into consideration, such a compound might be expected to exert a variety of effects on experimental and natural viral infections. Evidence for an *in vivo* activity of inosiplex on viral infections of laboratory animals has thus far been scanty. Negative results were reported when inosiplex was evaluated in a coordinated study in five different laboratories (Glasgow and Galasso, 1972): no protection was seen in mice and other animals experimentally inoculated with a variety of viruses, including HSV-2.

More recently, however, inosiplex was shown to exert some protective effects on experimental viral infections in laboratory animals when care was taken to administer the drug according to a therapeutic schedule (i.e., starting treatment *after* infecting the animals) and at an appropriate dosage (e.g., at a 1% concentration in the drinking water). A/J mice, normally highly susceptible to the lethal effect of an i.p. inoculation of HSV-1, were converted to moderate susceptibility when inosiplex was present in their drinking water from 24 hr after virus challenge (Hadden *et al.*, 1977). In our own unpublished studies, in which the compound was administered in the same way, increased survival was seen in mice inoculated i.p. with murine hepatitis virus or with HSV-1; this protective effect was slight, however, and not always reproducible. Inosiplex was tested in golden hamsters inoculated on their scarified cornea with a strain of HSV-2 (Ohnishi et al., 1983). Treatments were performed orally four times a day, starting 1 hr after virus inoculation, and for 21 days. Under these conditions, survival at 21 days was 20% in the control hamsters and 80% in the animals receiving a total daily dose of 100 mg/kg of inosiplex. The effect of lower doses was not significant statistically.

Altogether, inosiplex has been more extensively studied in humans than in laboratory animals. In addition to trials performed in volunteers (challenged with influenza or rhinoviruses), clinical studies of inosiplex have been conducted in several areas. According to an overview by Glasky et al. (1978), 165 such studies had been performed, involving more than 4000 patients with viral illnesses as diverse as influenza, hepatitis A, mumps, measles, recurrent herpes simplex, varicella, and herpes zoster. The general conclusion was that the drug exerted a favorable influence on the intensity and duration of the symptoms and accelerated recovery.

In a double-blind placebo-controlled study of the therapeutic activity of inosiplex on recurrent herpes labialis and herpes genitalis (Bradshaw and Sumner, 1977), the cellular immune status of the patients was evaluated prior to and after initiation of therapy. After a week of treatment, 13 patients on inosiplex showed an increase and 7 a decrease in the PHA responsiveness of their peripheral blood lymphocytes; in the group receiving the placebo, 9 patients showed an increased and 12 patients a decreased response. The ability of the lymphocytes to produce a lymphotoxin following PHA stimulation was more often augmented in the patients treated with the drug than in those receiving the placebo. Clinically, there was an indication that inosiplex therapy reduced the intensity of symptoms of recurrent herpes, reduced the frequency of new lesion formation, and increased the proportion of patients relieved of lesions during the first week of treatment. As in the case of levamisole, it appears that the response to treatment with inosiplex may vary from one individual to another.

In another double-blind placebo-controlled study (Corey et al., 1979), the immunomodulating and therapeutic effects of inosiplex were evaluated in patients affected by genital herpes. Thirty-nine patients were enrolled within 7 days after appearance of the lesions: 19 received 4 g daily of inosiplex orally and 20 were given a placebo. The responses of peripheral blood lymphocytes to PHA, Con A, and PWM were not significantly different between the two groups, but the lymphocytes from the drug-treated patients (especially those affected by a primary infection) displayed a higher response to HSV antigens than those from the placebo group. Clinical course of the primary infections was favorably affected by inosiplex: shorter duration of itching, shorter duration of virus shedding from genital lesions, and more rapid healing.

Beneficial effects of inosiplex against various forms of herpesvirus infections have been claimed in a number of other clinical studies, including open studies in herpetic keratitis (DiTizio *et al.*, 1979), herpetic encephalitis (Lesourd *et al.*, 1980), zoster (Torregrossa, 1978; Carco *et al.*, 1979), and a double-blind placebo-controlled study on recurrent mucocutaneous herpes (Bouffaut and Saurat, 1980).

VIII. MISCELLANEOUS IMMUNOMODULATING AGENTS

Over the last few years, several synthetic substance exhibiting immunomodulating activities have been described, some of which have reached the stage of phase I or phase II clinical studies (mainly for immunotherapy of cancer). These include azimexon, 1-[1-(2-cyano-1-aziridinyl-1-methylethyl]-2-aziridine carboxamide (Bicker *et al.*, 1979), bestatin, a *Streptomyces* metabolite of known chemical structure, [(2S,3R)-3-amino-2-hydroxy-4-phenylbutyryl]-L-leucine (Umezawa, 1981), and sodium diethyldithiocarbamate (Renoux and Renoux, 1979). In spite of the numerous studies performed *in vitro* and in laboratory animals with these substances, no report is available concerning their possible effects on viral infections.

Therafectin, a synthetic glucofuranose derivative [1,2-O-isopropylidene-3-O-3'-(N',N'-dimethylamino-propyl)-D-glucofuranose hydrochloride] first described by Majde and Gordon (1976), has been shown to stimulate *in vitro* and *in vivo* macrophage activities (Hadden *et al.*, 1979). *In vitro*, therafectin stimulates blood monocytes from healthy human subjects to secrete interleukin-1 (Horwitz *et al.*, 1983). Favorable effects of oral administration of low doses of the compound on experimental infections of laboratory animals with influenza, vaccinia, and herpes simplex viruses have been reported (P. Gordon and Greenwich Pharmaceuticals, personal communication); for instance, delayed and decreased mortality were observed in hamsters infected by the ocular route with HSV-1 or HSV-2 and treated with therafectin throughout the period following virus challenge.

NPT 15,392 is a synthetic immunomodulating substance [(erythro-9-(2-hydroxy,3-nonyl)hypoxanthine] currently under clinical evaluation in cancer patients. Like inosiplex, it induces *in vitro* T-cell markers on bone marrow prothymocytes and possesses stimulating properties on neutrophils, T lymphocytes, and NK cells (see review by Renoux and Wybran, 1983). We do not know of any study concerning the effects of this drug on experimental or clinical HSV infections.

Finally, it is interesting to note that an antiviral drug, with demonstrated clinical efficacy in HSV and VZV infections, has been shown to posses immunomodulating properties. Administered to mice at high but nontoxic doses, 9-β-D-arabinofuranosyladenine (vidarabine, Ara-A), an inhibitor of the DNA polymerases of herpesviruses, enhanced antibody production and delayed hypersensitivity reactions to various antigens (Hinrichs *et al.*, 1983). It is speculated that these immunostimulating activities may result from a selective inhibition by vidarabine of suppressor cell functions.

IX. CONCLUSION

At the present time, definite proof of the usefulness of immunopotentiating (or immunomodulating) agents in the nonspecific prevention and/or treatment of herpesvirus infections is lacking. Conflicting results have been reported with respect to the efficacy of such agents in experimental and in clinical infections. On the one hand, there are some powerful immunostimulants (such as BCG, *C. parvum*, glucan), which exhibit significant protective activities against HSV infections in the mouse but which are probably much too pleiotropic to be used in human medicine. On the other hand, at least two immunomodulating drugs (levamisole, inosiplex) are being used to treat HSV infections in man but they show only modest activity in experimental HSV infections of laboratory animals. IFN inducers show activity in murine systems but the efficacy against herpes-virus infections of the inducers that could be used in man is yet to be assessed. Encouraging results have been recorded with thymic hormones in immunocompromised patients but clinical and laboratory data are still preliminary. No finding has been reported as yet about the possible activities of monokines and lymphokines (other than some of the IFNs) in herpesvirus infections.

As things stand now, we see two essential requirements to make significant progress in this area. First, it is imperative that more sophisticated and relevant models of herpesvirus infections in laboratory animals be used to test immunomodulating agents, in place of the rapidly lethal systemic infection of the mouse with a virulent strain of HSV. Models of local and of recurrent HSV infections exist now in mice, guinea pigs, hamsters, and rabbits and they can be used to study pharmacological manipulation of the immune reactions that are involved in such infections. Because of the exacerbated pathogenicity of herpesviruses in immunocompromised patients, it will be important to design animal models of combined herpesvirus infection and immunodeficiency, in particular those in which virus reactivation would result from immunosuppressive therapy. Second, better knowledge must be gained of the mechanisms of action and precise cellular targets in the immune system of the various immunomodulating drugs, in order to define how and when they could be tested and used in experimental and clinical herpesvirus infections.

It must also be considered that a number of antiviral drugs, exhibiting high degrees of selectivity against herpesviruses, have been discovered in the recent past and are already used clinically, for the treatment of primary or recurrent HSV and VZV infections or for the prophylaxis of such infections in patients undergoing organ or bone marrow transplantation. These include, among others, acyclovir, bromovinyldeoxyuridine, and phosphonoformic acid. These drugs show unquestionable efficacy (by local, oral, or parenteral administration, depending on the case and on the agent) in treating primary infections caused by HSV-1, HSV-2, or VZV; they also tend to shorten the course of the recurrent episodes but they do not affect significantly the frequency of episodes in patients with recurrent herpes. The possible usefulness of immunomodulating drugs must therefore be viewed in connection with these novel chemotherapeutic approaches: one may, for instance, envisage the combination of an antiviral drug and an immunomodulating agent to achieve elimination of the virus and thereby prevention of recurrences. Again, a better knowledge of the immune mechanisms that should be activated to achieve this aim and of the modes of action of the immunomodulating drugs will be necessary for adequate therapeutic strategy.

Finally, if the use of herpesvirus vaccines becomes a reality in the near future, compounds such as muramylpeptides or lipopeptides could be useful as adjuvants of humoral and/or cell-mediated immunity, in conjunction with these vaccines.

REFERENCES

Anderson, F. D., Ushijima, R. N., and Larson, C. L., 1974, Recurrent herpes genitalis: Treatment with attenuated *Mycobacterium bovis* (BCG), *Obstet. Gynecol.* **43**:797.

Bach, J. F., and Dardenne, M., 1982, L'utilisation clinique des hormones thymiques, in: *Actualités de chimie Thérapeutique*, 9th Ser. (J. J. Panouse and J. F. Robert, eds.), pp. 91–100, Technique et Documentation-Lavoisier, Paris.

Bach, J. F., and Edelman, L., 1980, On the significance of the similarity of the immunological effects of transfer factor, levamisole and thymic hormones, in: *Thymus, Thymic Hormones and T Lymphocytes* (F. Aiuti and H. Wigzell, eds.), pp. 187–193, Academic Press, New York.

Baker, M. B., Larson, G. L., Ushijima, R. N., and Anderson, F. D., 1974, Resistance of female mice to vaginal infection induced by *Herpesvirus hominis* type 2: Effects of immunization with *Mycobacterium bovis*, intravenous injection of specific herpesvirus hominis type 2 antiserum and a combination of these procedures, *Infect. Immun.* **10**:1230.

Bancroft, G. J., Shellam, G. R., and Chalmer, J. E., 1981, Genetic influences on the augmentation of natural killer (NK) cells during murine cytomegalovirus infection: Correlation with patterns of resistance, *J. Immunol.* **126**:988.

Bicker, U., Ziegler, A. E., and Hebold, G., 1979, Investigations in mice on the potentiation of resistance to infections by a new immunostimulant compound, *J. Infect. Dis.* **139**:389.

Bouffaut, P., and Saurat, J. H., 1980, Isoprinosine as a therapeutic agent in recurrent mucocutaneous infections due to herpes virus, *J. Immunopharmacol.* **2**:193 (abstract).

Bradshaw, L. J., and Sumner, H. L., 1977, *In vitro* studies on cell-mediated immunity in patients treated with inosiplex for herpes virus infection, *Ann. N.Y. Acad. Sci.* **284**:190.

Breinig, M. C., Wright, L. L., McGeorge, M. B., and Morahan, P. S., 1978, Resistance to vaginal or systemic infection with herpes simplex virus type 2, *Arch. Virol.* **57**:25.

Carco, F. P., Fruttaldo, L., and Saliva, G., 1979, Studio preliminare sulla valutazione clinica dell'efficacia terapeutica dell'Isoprinosina nelle infezioni acute da herpes-virus, *Arch. Med. Intern.* **21**:589.

Ching, C., and Lopez, C., 1979, Natural killing of herpes simplex virus type 1-infected target cells: Normal human responses and influence of antiviral antibody, *Infect. Immun.* **26**:49.

Corey, L., Reeves, W. C., Vontver, L. A., Alexander, E. R., and Holmes, K. K., 1976, Trial of BCG vaccine for the prevention of recurrent genital herpes, Abstract No. 403, 16th Interscience Conference on Antimicrobial Agents and Chemotherapy, Chicago.

Corey, L., Chiang, W. T., Reeves, W. C., Stamm, W. E., Brewer, L., and Holmes, K. K., 1979, Effect of Isoprinosine on the cellular immune response in initial genital herpes virus infection, *Clin. Res.* **27**:41A.

DeClercq, E., and Luczak, M., 1976, Antiviral activity of carbopol, a cross-linked polycarboxylate, *Arch. Virol.* **52**:151.

Dietrich, F. M., Lukas, B., and Schmidt-Ruppin, K. H., 1983, MTP-PE (synthetic muramyl peptide): Prophylactic and therapeutic effects in experimental viral infections, Communication at 13th International Congress of Chemotherapy, Vienna.

DiLuzio, N. R., Williams, D. L., and Browder, W., 1982, Immunopharmacology of glucan: The modification of infectious diseases, in: *Advances in Pharmacology and Therapeutics II* (H. Yoshida, Y. Hagihara, and S. Ebashi, eds.), pp. 101–112, Pergamon Press, Elmsford, N.Y.

DiTizio, A., Mutolo, A., Glorialanza, G., Catone, E., and Romani, G. P., 1979, Methisoprinol, valutazione clinica della sua efficacia therapeutica nella cheratite da herpes virus hominis: Studio preliminare, *Ann. Oftalmol. Clin. Ocul.* **105**:341.

Dokhelar, M. C., Tursz, T., Dardenne, M., and Bach, J. F., 1983, Effect of synthetic thymic factor on natural killer cell activity in humans, *Int. J. Immunopharmacol.* **5**:277.

Ellouz, F., Adam, A., Ciorbaru, R., and Lederer, E., 1974, Minimal structural requirements for adjuvant activity of bacterial peptidoglycan derivatives, *Biochem. Biophys. Res. Comm.* **59**:1317.

Engler, H., Zawatzky, R., Kirchner, H., and Armerding, D., 1982, Experimental infection of inbred mice with herpes simplex virus. IV. Comparison of interferon production and natural killer cell activity in susceptible and resistant adult mice, *Arch. Virol.* **74**:239.

Fiorelli, M., Pivetti-Pezzi, P., Sirianni, M. C., De Liso, P., Testi, R., Russo, G., Pontesilli, O., Carbonari, M., and Luzi, G., 1983, Immunological aspects and thymic hormone therapy of human herpesvirus infections, in: *Thymic Factor Therapy* (N. A. Byrom and J. R. Hobbs, eds.), pp. 252–255, Academic Press, New York.

Fischer, G. W., Podgore, J. K., Bass, J. W., Kelley, J. L., and Kobayashi, G. Y., 1975, Enhanced host defense mechanisms with levamisole in suckling rats, *J. Infect. Dis.* **132**:578.

Fisher, G. W., Balk, N. W., Crumrine, M. H., and Bass, J. W., 1976, Immunopotentiation and antiviral chemotherapy in a suckling rat model of herpesvirus encephalitis, *J. Infect. Dis.* **133**:A217.

Floc'h, F., and Werner, G. H., 1976, Increased resistance to virus infections of mice inoculated with BCG (bacillus Calmette-Guérin), *Ann. Immunol. (Inst. Pasteur)* **127C**:173.

Gabrielson, D. A., Kelleher, J. J., and Varani, J., 1980, Effect of *Corynebacterium granulosum* immunopotentiation on the pathogenesis of herpes simplex virus type 2 in BALB/c mice, *Infect. Immun.* **30**:791.

Glasgow, L. A., and Galasso, G. J., 1972, Isoprinosine: Lack of antiviral activity in experimental model infections, *J. Infect. Dis.* **126**:162

Glasgow, L. A., Fischbach, J., Bryant, S. M., and Kern, E. R., 1977, Immunomodulation of host resistance to experimental virus infections in mice: Effects of *Corynebacterium acnes*, *Corynebacterium parvum* and bacille Calmette-Guerin, *J. Infect. Dis.* **135**:763.

Glasky, A. J., Kestelyn, J., and Romero, M., 1978, Isoprinosine ®: A clinical overview of an antiviral/immunomodulating agent, 7th International Congress of Pharmacology, Paris, Abstract No. 2267.

Gordon, P., Ronsen, B., and Brown, E. R., 1974, Anti-herpesvirus action of Isoprinosine, *Antimicrob. Agents Chemother.* **5**:153.

Gresser, I., Tovey, M. G., Maury, C., and Bandu, M. T., 1976, Role of interferon in the pathogenesis of virus disease in mice as demonstrated by the use of anti-interferon serum. II. Studies with herpes simplex, Moloney sarcoma, vesicular stomatitis, New-castle disease and influenza viruses, J. Exp. Med. **144:**1316.

Hadden, J. W., Lopez, C., O'Reilly, R. J., and Hadden, E. M., 1977, Levamisole and inosiplex: Antiviral agents with immunopotentiating action, Ann. N.Y. Acad. Sci. **284:**139.

Hadden, J. W., Englard, A., Sadlik, J. R., and Hadden, E. M., 1979, The comparative effects of isoprinosine, levamisole, muramyldipeptide and SM 1213 on lymphocyte and ma-crophage proliferation and activation in vitro, Int. J. Immunopharmacol. **1:**17.

Handzel, Z. T., Zaizov, R., Varsano, I., Levin, S., Pecht, M., and Trainin, N., 1981, The influence of thymic humoral factor on immunoproliferative disorders and viral infec-tions in humans, in: Advances in Immunopharmacology (J. Hadden, L. Chedid, P. Mullen, and F. Spreafico, eds.), pp. 83–88, Pergamon Press, Elmsford, N.Y.

Hinrichs, J., Kitz, D., Kobayashi, G., and Little, J. R., 1983, Immune enhancement in mice by Ara-A, J. Immunol. **130:**829.

Hirsch, M. S., Zisman, B., and Allison, A. C., 1970, Macrophages and age-dependent re-sistance to herpes simplex virus in mice, J. Immunol. **104:**1160.

Horwitz, D. A., Linker-Israeli, M., and Nishiya, K., 1983, Potentiation of interferon effects on natural cytotoxicity by therafectin, a new macrophage activator, Abstracts, 5th In-ternational Congress of Immunology, Kyoto.

Kern, E. R., Glasgow, L. A., and Overall, J. C., Jr., 1976, Antiviral activity of an extract of Brucella abortus: Induction of interferon and immunopotentiation of host resistance, Proc. Soc. Exp. Biol. Med. **152:**372.

Kint, A., and Verlinden, L., 1974, Levamisole for recurrent herpes labialis, N. Engl. J. Med. **291:**308.

Kint, A., Coucke, C., and Verlinden, L., 1974, The treatment of recurrent herpes infections with levamisole. Arch. Belg. Dermatol. **30:**167.

Kirchner, H., Hirt, H. M., and Munk, K., 1977, Protection against herpes simplex virus infection in mice by Corynebacterium parvum, Immunology **16:**9.

Kirchner, H., Engler, H., Schröder, C. H., Zawatzky, R., and Storch, E., 1983, Herpes simplex virus type 1-induced interferon production and activation of natural killer cells in mice, J. Gen. Virol. **64:**437.

Kümel, G., Kirchner, H., Zawatzky, R., Engler, H., Schröder, C. H., and Kaerner, H. C., 1982, Experimental infection of inbred mice with herpes simplex virus. V. Investigations with a virus strain non-lethal after peripheral infection, J. Gen. Virol. **63:**315.

Larson, C. L., Ushijima, R. N., Karim, R., Baker, M. B., and Baker, R. E., 1972, Herpesvirus hominis type 2 infection in rabbits: Effect of prior immunization with attenuated M. bovis (BCG) cells, Infect. Immun. **6:**465.

Lesourd, B., Rancourel, G., Huraux, J. M., Pompidou, A., Jacque, C., Denvil, D., Buge, A., and Moulias, R., 1980, Immunological restoration in vivo and in vitro, Isoprinosine therapy and prognosis of acute encephalitis, Int. J. Immunopharmacol. **2:**195 (ab-stract).

Levine, A. S., Sivulich, M., Wiernik, P. H., and Levy, H. B., 1979, Initial trials in cancer patients of polyriboinosinic-polyribocytidylic acid stabilized with poly-L-lysine, in car-boxymethylcellulose [poly (ICLC), a highly effective interferon inducer, Cancer Res. **39:**1645.

Lods, F., 1976, Traitement de l'herpès cornéen recidivant et des zonas ophtalmiques par le lévamisole, Nouv. Presse Med. **5:**148.

Lopez, C., 1975, Genetics of natural resistance to herpesvirus infections in mice, Nature (London) **258:**152.

Lopez, C., Ryshke, R., and Bennett, M., 1980, Marrow-dependent cells depleted by ^{89}Sr mediate genetic resistance to herpes simplex virus type 1 infection in mice, Infect. Immun. **28:**1028.

Lotzova, E., Savary, C. A., and Stringfellow, D. A., 1983, 5-Halo-6-phenyl-pyrimidinones: New molecules with cancer therapeutic potential and interferon-inducing capacity are strong inducers of murine natural killer cells, J. Immunol. **130:**965.

Majde, J. A., and Gordon, P., 1976, Immunomodulation by 1,2-O-isopropylidene-3-O-3'-(N'-N'-dimethylamino-N-propyl)-D-glucofuranose (SM 1213), a drug with antiviral activity, 16th Intersci. Conf. Antimicrob. Agents Chemother., Abstract No. 126.

Migliore-Samour, D., Bouchaudon, J., Floc'h, F., Zerial, A., Ninet, L., Werner, G. H., and Jollès, P., 1980, A short lipopeptide representative of a new family of immunological adjuvants devoid of sugar, Life Sci. 26:883.

Morahan, P. S., and McCord, R. S., 1975, Resistance to herpes simplex type 2 virus induced by an immunopotentiator (pyran) in immunosuppressed mice, J. Immunol. 115:311.

Morahan, P. S., Glasgow, L. A., Crane, J. L., Jr., and Kern, E. R., 1977a, Comparison of antiviral and antitumor activity of activated macrophages, Cell. Immunol. 28:404.

Morahan, P. S., Kern, E. R., and Glasgow, L. A., 1977b, Immunomodulator-induced resistance against herpes simplex virus, Proc. Soc. Exp. Biol. Med. 154:615.

Morahan, P. S., Coleman, P. H., Morse, S. S., and Volkman, A., 1982, Resistance of mice with defects in the activities of mononuclear phagocytes and natural killer cells: Effects of immunomodulators in beige mice and ^{89}Sr-treated mice, Infect. Immun. 37:1079.

Ohnishi, H., Kosuzume, H., Inaba, H., Ohkura, M., Shimada, S., and Suzuki, Y., 1983, The immunomodulatory action of inosiplex in relation to its effects in experimental viral infections, Int. J. Immunopharmacol. 5:181.

Ojo, E., Haller, O., Kimura, A., and Wigzell, H., 1978, An analysis of conditions allowing Corynebacterium parvum to cause either augmentation or inhibition of natural killer cell activity against tumor cells in mice, Int. J. Cancer 21:444.

Oleske, J. M., Starr, S., Kohl, S., Shaban, S., and Nahmias, A., 1978, Clinical evaluation of transfer factor in patients with recurrent herpes simplex virus infections, in: Current Chemotherapy, Vol. I (W. Siegenthaler and R. Lüthy, eds.), pp. 359–360, American Society for Microbiology, Washington, D.C.

O'Reilly, R. J., Chibbaro, A., Wilmot, R., and Lopez, C., 1977, Correlation of clinical and virus-specific immune responses following levamisole therapy of recurrent herpes progenitalis, Ann. N.Y. Acad. Sci. 284:161.

Rabson, A. R., Whiting, D. A., Anderson, R., Glover, A., and Koornhof, H. J., 1977, Depressed neutrophil motility in patients with recurrent herpes simplex virus infections: In vitro restoration with levamisole, J. Infect. Dis. 135:113.

Renoux, G., 1978, Modulation of immunity by levamisole, J. Pharmacol. Ther. 2:397.

Renoux, G., and Renoux, M., 1971, Effet immunostimulant d'un imidazothiazole dans l'immunisation des souris contre l'infection par Brucella abortus, C.R. Acad. Sci. Ser. D 272:349.

Renoux, G., and Renoux, M., 1979, Immunopotentiation and anabolism induced by sodium diethyldithiocarbamate, J. Immunopharmacol. 1:247.

Renoux, G., and Wybran, J., 1983, Immunopotentiators, in: Advances in immunopharmacology 2 (J. W. Hadden, L. Chedid, P. Dukor, F. Spreafico, and D. Willoughby, eds.), pp. 814–815, Pergamon Press, Elmsford, N.Y.

Rolly, H., Vertesy, L., and Neufarth, A., 1974, Studies on the chemotherapeutic actions of antiviral lipopolysaccharides, in: Progress in Chemotherapy, Vol. 2 (G. K. Daikos, ed.), pp. 1013–1018, Hellenic Society of Chemotherapy, Athens.

Simon, L. N., Maxwell, K., Ginsberg, T., and Glasky, A. J., 1981, Immunologic therapy of viral infections, in: Advances in Immunopharmacology (J. Hadden, L. Chedid, P. Mullen, and F. Spreafico, eds.), pp. 115–125, Pergamon Press, Elmsford, N.Y.

Spitler, L. E., Glogau, R. G., Nelms, D. C., Basch, C. M., Olson, J. A., Silverman, S., and Engleman, E. P., 1975, Levamisole and lymphocyte response in herpes simplex virus infections, Symposium on Antivirals with Clinical Potential, Stanford, Calif.

Starr, S. E., Visintine, A. M., Tomeh, M. O., and Nahmias, A. J., 1976, Effects of immunostimulants on resistance of newborn mice to herpes simplex type 2 infection, Proc. Soc. Exp. Biol. Med. 152:57.

Starr, S. E., Tolpin, M. D., Friedman, H. M., Pauker, K., and Plotkin, S. A., 1979, Impaired cellular immunity to cytomegalovirus in congenitally infected children and their mothers, J. Infect. Dis. 140:500.

Steele, R. W., 1980, Transfer factor reactivity to varicella–zoster antigen in childhood leukemia, *Cell. Immunol.* **50**:282.

Steele, R. W., Heberling, R. L., Eichberg, J. W., Eller, J. J., Kalter, S. S., and Kniker, W. T., 1976, Prevention of herpes simplex virus type 1 fatal dissemination in primates with human transfer factor, in: *Transfer Factor: Basic Properties and Clinical Applications* (M. S. Ascher, A. A. Gottlieb, and C. H. Kirkpatrick, eds.), pp. 381–386, Academic Press, New York.

Steele, R. W., Myers, M. G., and Vincent, M. M., 1980, Transfer factor for the prevention of varicella–zoster infection in childhood leukemia, *N. Engl. J. Med.* **303**:355.

Symoens, J., 1978, Treatment of the compromised host with levamisole, a synthetic immunotherapeutic agent, in: *Immune Modulations and Control of Neoplasia by Adjuvant Therapy* (M. A. Chirigos, ed.), pp. 1–9, Raven Press, New York.

Symoens, J., Decree, J., Van Bever, W. F. M., and Janssen, P. A. J., 1979, Levamisole, in: *Pharmacological and Biochemical Properties of Drug Substances* (M. E. Goldberg, ed.), pp. 407–464, American Pharmaceutical Association, Washington, D.C.

Torregrossa, F., 1978, Risultati terapeutici preliminari nel trattamento dell'herpes zoster con un nuovo agente antivirale, il Methisoprinol. *Acta Gerontol.* **28**:105.

Tyrrell, D. A. J., 1982, The abolition of infection: Hope or illusion? Rock Carling Fellowship Lecture, the Nuffield Provincial Hospitals Trust, p. 40.

Umezawa, H., 1981, *Small Molecular Immunomodifiers of Microbial Origin: Fundamental and Clinical Studies of Bestatin,* Pergamon Press, Elmsford, N.Y.

Youngner, J. S., Keleti, G., and Feingold, D. S., 1974, Antiviral activity of an ether-extracted nonviable preparation of *Brucella abortus, Infect. Immun.* **10**:1202.

Zaizov, R., Vogel, R., Cohen, I., Varsano, I., Shohat, B., Rotter, V., and Trainin, N., 1977, Thymic hormone (THF) therapy in immunosuppressed children with lymphoproliferative neoplasia and generalized varicella, *Biomedicine* **27**:105.

Zisman, B., Hirsch, M. S., and Allison, A. C., 1970, Selective effects of antimacrophage serum, silica and antilymphocyte serum on pathogenesis of herpesvirus infection of young adult mice, *J. Immunol.* **104**:1155.

Index

Acquired immune deficiency syndrome
(AIDS)
  CMV-related, 131, 137–139
  EBV-related, 184
  HSV-related
    interferon deficiency, 59
    natural killer cell activity, 379
  immunopathology, 114
  interferon production in, 59, 373
  simian, 156
  varicella-zoster-related, 373
Activated lymphocyte killers, 253
Acyclovir
  action mechanisms, 344–345
  cerebrospinal fluid concentrations, 349
  controlled trials of, 347–350
  for cutaneous HSV infections, 351
  as cytomegalovirus prophylaxis, 219
  development, 340
  drug delivery routes, 350
  as EBV therapy, 246
  as genital herpes therapy, 351
    intravenous, 353–354
    oral, 354
    topical, 352–353
  as herpes zoster therapy, 363–364
  as keratitis therapy, 383
  plasma concentrations, 348–349
  in renal failure, 349–350
  urinary clearance, 349
  as varicella–zoster therapy, 383
Adjuvants
  for EBV subunit vaccine, 331–332
  immune response regulation by, 116
AK cell: see Anomalous killer cell

Allografting, CMV infections and, 201–227
  antiviral agent prophylaxis, 219
  blood product transmission, 205, 206–
    207, 211–212, 220–221, 298, 299–
    300
  donor selection and, 210–211, 220, 221,
    298
  graft rejection, 205
  immunosuppression in, 140, 202–203,
    219–220
  latent virus reactivation, 208–210
  passive immunoprophylaxis, 212–215,
    299–300
  pathogenesis of, 202–205
  primary infection prevention, 210–216,
    297–310
  risk factors, 205–210
  severity, 203
  superinfection, 203, 205
  T cell activity, 126–140, 204, 205
  in transplanted tissue, 208, 298–299
  treatment, 220, 298–310
  vaccination against, 215–216, 221–222,
    297–310
  virus reactivation prevention, 216–219
Anemia, aplastic
  CMV-related pneumonia in, 203, 213
  natural killer cell response, 50
Anomalous killer cells, 244–245
Antibody(ies)
  anti-idiotype, 116
  CMV-specific, 122, 298–300
  EBV-specific, 378
    antibody-dependent cell-mediated
      cytoxicity and, 190

417

Antibody(ies) (cont.)
  EBV-specific (cont.)
    anti-early antigen, 378
    anti-Epstein–Barr nuclear antigen, 378
    anti-viral capsid antigen, 378
    in neoplastic disease, 184–192, 195
    neutralizing, 330–334
  HSV-specific
    cytolytic, 72
    in human convalescent sera, 70–71
    IgA and IgM classes, 77–78
    in vivo function, 79–81
    measurement, 71–74, 77–79
    neutralizing, 71–72
    precipitating, 72–73
  monoclonal: see Monoclonal antibodies
  varicella–zoster-specific, 320–342
  See also specific Immunoglobulins
Antibody-dependent cell-mediated
        cytotoxicity
  CMV-related, 123–124, 129–130
  EBV-related, 252, 253
    assay, 188–190
  HSV-related, 72, 379
    mediation, 48
  in Wiskott-Aldrich syndrome, 48
Antigen(s)
  EBV-specific
    cytotoxic lymphocyte detection, 175–
        176
    induced cell surface, 175–176
    late membrane, 252–253
    in marmosets, 149–150
    membrane, 330–332
    nuclear, 149–150, 171, 177, 192
    viral capsid, 185, 186, 187
  herpesvirus papio-specific, 153–154
  subunit: see Subunit vaccines
Antigen-presenting cell, 103–105, 115–116
Anti-idiotype immunization, 116
Antithymocyte globulin, 203
Antithymocyte sera, 39
Antiviral drugs
  clinical trials, 350–364
    cutaneous HSV infections, 351–354
    herpes simplex encephalitis, 355–359
    keratoconjunctivitis, 350–351
    neonatal HSV infections, 358–361
    VZV infections, 361–364
  controlled trials, 345–350
    acyclovir, 347–350
    vidarabine, 346–347
  as cytomegalovirus prophylaxis, 219

Antiviral drugs (cont.)
  development, 339–345
    for clinical evaluation, 340–342
    immunomodulating agent interaction,
        411–412
  See also names of specific drugs
9-β-D-Arabinofuranosyl adenine: see
        Vidarabine
1-β-D-Arabinofuranosyl cytosine: see
        Cytarabine
Arabinosyl nucleosides, 344
Arthus reaction, 107, 108
Asthma, 106
Ataxia telangiectasia, 179
Autoinoculation
  of genital herpes infection, 16
  of HSV infection, 266–267
Azathioprine, 203
Azimexon, 411

B cell
  CMV transport by, 300
  EBV-transformed, 113, 171–179
    in immunopathogenesis, 230, 234,
        235, 240, 244, 345–350, 378
    in vitro growth inhibition, 173–175
    proliferation, 171–172
    receptors, 252
    regression phenomenon, 255–256, 261
  T cell interaction, 173–176
  interferon inhibition of, 173–174
Baboon, lymphoproliferative disease in,
        152–156
Bacterial superinfection, 17
BCG
  as herpes genitalis therapy, 49
  HSV-2 resistance and, 397–398, 399
Bestatin, 411
Blast cell neoantigen, 244
Blastogenesis, CMV-induced, 141
Blood-borne infection
  CMV, 205, 206–207, 211–212, 220–221,
        298, 299–300
  genital herpes, 16–17
Bone marrow transplant patients
  CMV infections in, 126–140, 201–222,
        297–300, 395
    cytotoxic lymphocyte response, 126–
        129, 131, 132, 133–134
    donor selection and, 210, 298
    incidence, 202
    interferon prophylaxis, 218
    passive immunoprophylaxis, 212–215,
        298–300

Bone marrow transplant patients (*cont.*)
  CMV infections in (*cont.*)
    pathogenesis, 203
    steroid therapy, 135–137
    transfusion-acquired, 206–207, 298,
      299
    vaccination against, 216, 297–310
  HSV infections in, 395
Bromovinyl deoxyuridine, 345
  action site, 365
  development, 340
Brucella abortus, 399–400
Burkitt's lymphoma, EBV-related, 150,
    327, 328, 329, 334
  antibody titers, 187
  cellular immunity in, 193
  chromosomal translocations, 172–173
  virus isolation, 183

Cancer
  EBV-related, 183–199
    antibody-dependent cell-mediated
      cytotoxicity in, 188–190
    cellular immunity and, 171–179, 193–
      194
    clinical serology of, 184–190
    proteins in, 190–192
    seroepidemiological studies, 184
  varicella-zoster-related, 313–314, 319–
    320, 321, 402
Carbopol, 404
Cardiac transplant patients
  CMV infections in, 395
    blood transfusion transmission, 207,
      211–212
    donor selection and, 210
    incidence, 202
    transplanted tissue transmission, 208
    vaccination against, 216
  HSV infections in, 395
Cell-mediated immunity
  in CMV infections, 121–145
    in AIDS patients, 137–139
    antibody-dependent cell-mediated
      cytotoxicity and, 123–124, 129–130
    cytotoxic T cells in, 124–126, 134–139
    delayed-type hypersensitivity and,
      139–141
    effector cell function in, 122–126
    in vitro lymphocyte blastogenesis in,
      141
    natural killer cells in, 122–123, 132–
      139
    steroid therapy effects, 134–137

Cell-mediated immunity (*cont.*)
  in CMV infections (*cont.*)
    vaccine response, 141–142
    virus-specific cytotoxic lymphocyte
      response, 126–132
  in EBV infections, 171–182
    cell surface antigens in, 175–176
    immunological memory, 176–178
    mechanisms, 253–254
    natural killer cells in, 193–194
    in pathological conditions, 178–179
  T cell function in, 87–102
    in athymic mice, 88
    in B cell-suppressed mice, 89
    immune response regulation, 97–98
    immunological memory, 98–99
    in vivo responses, 93–97
    subsets, 89–93
  in varicella-zoster infections, 313, 401
Central nervous system, genital herpes
    infection in, 13–16
Cervical cancer, HSV-related, 70, 266
  vaccines and, 81, 273
Cervical infections, genital herpes-related,
    11, 20–22
Chediak–Higashi syndrome, 252, 261
Chicken pox: *see* Varicella-zoster virus
Children
  Burkitt's lymphoma in, 187
  HSV vaccination of, 275, 277, 284
  varicella-zoster in
    clinical illness, 313–315
    vaccination, 315–325
CMV: *see* Cytomegalovirus(es)
Corticosteroids
  CMV infection severity and, 203
  as EBV therapy, 246
Corynebacterium, 45, 398–399
Cutaneous HSV infection, 351–354
Cyclosporin, 178, 203, 220
Cytarabine, 340
Cytomegalovirus(es)
  blood-borne, 300
  dense bodies, 308, 309
  glycoproteins, 309
  noninfectious enveloped particles, 308–
    309
  passive antibodies, 298–300
  protein kinase, 309
Cytomegalovirus infections
  in bone marrow transplant patients, 298,
    201–222, 395
    cytotoxic lymphocyte response, 126–
      129, 131, 132, 133–134

Cytomegalovirus infections (*cont.*)
  in bone marrow transplant patients (*cont.*)
    donor selection and, 210, 298
    incidence, 202
    interferon prophylaxis, 218
    passive immunoprophylaxis, 212–215,
      299–300
    pathogenesis, 203
    steroid therapy, 135–137
    transfusion-acquired, 206–207, 298,
      299
    vaccination against, 216, 297–310
  in cardiac transplant patients, 395
    blood transfusion transmission, 207,
      211–212
    donor selection and, 210
    incidence, 202
    transplanted tissue transmission, 208
    vaccination against, 216
  cell-mediated immunity in, 121–145
    in AIDS patients, 137–139
    antibody-dependent cell-mediated
      cytotoxicity, 123–124, 129–130
    delayed-type hypersensitivity, 139–140
    effector cell function, 122–126
    immunosuppression, 140–141
    in vitro lymphocyte blastogenesis, 141
    natural killer cells, 122–123, 132–139
    steroid therapy effects, 134–137
    vaccine response, 141–142
    virus-specific cytotoxic lymphocyte
      response, 126–132
  cytotoxic T cells in, 124–126, 134–139,
    376
  infectious mononucleosis and, 376
  interferon and, 375–377
    production, 133
    as prophylaxis, 217–219, 309–310
    therapy, 383–384
  neonatal, 122
  organ allografting and, 201–227
    antiviral agent prophylaxis, 219
    blood transfusion transmission, 205,
      206–207, 211–212, 220–221, 298,
      299–300
    donor selection and, 210–211, 220,
      221, 298
    graft rejection, 205
    immunosupression and, 202–203,
      219–220
    latent virus reactivation, 208–210
    passive immunoprophylaxis, 212–215,
      229–300

Cytomegalovirus infections (*cont.*)
  organ allografting and (*cont.*)
    pathogenesis, 202–205
    primary infection prevention, 210–
      216, 297–300
    risk factors, 205–210
    severity of infection, 203
    superinfection, 203, 205
    T cell activity, 204, 205
    in transplanted tissue, 208, 298–299
    treatment, 220, 298–310
    vaccination effects, 215–216, 221–222,
      297–310
    virus reactivation prevention, 216–219
  prevention, 297–298
  in renal transplant patients, 395
    cytotoxic lymphocyte responses, 128,
      133–134
    donor selection and, 210–211
    graft rejection, 205
    immune complexes in, 107
    incidence, 202
    interferon effects, 218, 384
    pathogenesis, 202, 203, 205
    steroid therapy effects, 135–137
    transfusion-acquired, 207, 298
    vaccination against, 216, 302–305
  T cell cytoxicity, 128, 133–134, 204,
    205, 244
  thymic hormone therapy, 403
  transmission, 297, 298–300
    nosocomial, 205–206
    sexual, 205
    by transfusion, 205, 206–207, 211–
      212, 220–221, 298, 299–300
  vaccine, 297–312
    animal models, 300–301
    latency and, 307
    live vs. attenuated, 300
    in renal transplant patients, 302–303
    subunit, 307–309
    Towne strain, 141–142, 209–210, 301–
      307
  virion, 300, 308–309

Delayed-type hypersensitivity
  in CMV infections, 139–140
  in HSV infections, 91–92, 109–114, 396
  stimulation of, 115. 116
  T cell-mediated, 91–93, 97–98
Dense bodies, of CMV, 308, 309
1-(2-Deoxy-2-fluoro-β-D-arabinofuranosyl)-
    5-iodocytosine, 345
DHPG, 365

Dimethylsulfoxide, 351
DNA
    of EBV, 147
    of *herpesvirus papio*, 154–155
    of *herpesvirus saimiri*, 160
    of lymphotropic viruses, 147
    of varicella–zoster virus, 320

EBV: *see* Epstein–Barr virus infections
Effector cells
    in CMV infections, 122–126
    in HSV infections, 103–105
    origin, 121–122
    See also specific types of effector cells
Encephalitis, HSV-related, 15
    antibodies, 78
    incidence, 354–355
    therapy, 355–358, 411
Endotoxins, in HSV-infections, 40, 399
Epidemiologic studies, of genital herpes
        infections, 2–7
Epstein-Barr virus infections
    acyclovir bioactivation, 344–345
    antibodies, 378
        antibody-dependent cell-mediated
            cytotoxicity and, 190
        anti-early antigen, 378
        anti-Epstein–Barr nuclear antigen, 378
        anti-viral capsid antigen, 378
        in neoplastic disease, 184–192, 195
        neutralizing, 330–334
    antigens
        cytotoxic lymphocyte detection, 175–
            176
        induced cell surface, 175–176
        late membrane, 252–253
        in marmosets, 149–150
        membrane, 330–332
        nuclear, 149–150, 171, 177, 192
        viral capsid, 185, 186, 187
    associated diseases, 148
    B cells in, 113
        in immunopathogenesis, 230, 234,
            235, 240, 244, 345–350, 378
        in vitro growth inhibition, 173–175
        proliferation, 171–172
        receptors, 252
        regression phenomenon, 255–256, 261
        T cell interaction, 173–176
    Burkitt's lymphoma and, 324–325, 327,
        328–329, 333
        antibody titers, 187
    cancer and, 183–199
        antibody-dependent cell-mediated
            cytotoxicity in, 188–190

Epstein-Barr virus infections (*cont.*)
    cancer and (*cont.*)
        cellular immunity and, 171–179, 193–
            194
        clinical serology of, 184–190
        proteins in, 190–192
        seroepidemiological studies, 184
    cell-mediated immunity in, 171–182
        cell surface antigens, 175–176
        immunological memory, 176–178
        mechanisms, 253–254
        natural killer cell activity, 193–194
        in pathological conditions, 178–179
    DNA, 147
    as glomerulonephritis cause, 107
    immunological control, 171–182, 252–
        254, 327–328
    immunopathogenesis, 230–231
    infectious mononucleosis in, 150–151,
        229, 231, 259
        B cell response, 178
        immunopathology, 172, 327–328, 332
        T cell reactions, 113–114
        X-linked proliferative syndrome, 251
    interferon interactions in, 377–379, 384–
        385
        clinical applications, 257–259
        immunosuppression and, 257
        in vitro, 255–257
        infectious mononucleosis and, 259
        secretion defects, 257–259
    isolation, 183
    nasopharyngeal cancer and, 327, 329,
        385
        serological markers, 185–187
    pathogenesis, 377–379
    in primates, 147–152
        host range for disease induction, 148–
            150
        tumor clonality, 150–152
    proteins, 190–192
    serology
        diagnostic value, 184–187
        prognostic value, 187–190
    T cell activity in
        B cell interactions, 173–176
        in neoplastic disease, 193
        proliferation, 172
    treatment, 246–247
    vaccine, 329–337
        adjuvants, 331–332
        animal test species, 152
        membrane antigen, 330–332
        subunit, 247

Epstein-Barr virus infections (*cont.*)
  X-linked lymphoproliferative syndrome,
      178, 229, 230–250, 328
    B cell activity, 230, 234, 235, 240, 244
    in carrier females, 240–242
    clinical course, 233–234
    cytotoxicity studies, 236
    general characteristics, 231–232
    humoral immune response, 240, 243,
        244
    immunodeficiencies in, 178, 254
    immunologic studies, 233, 234–235,
        239–242
    in males, 232–240
    natural killer cell activity, 56, 230,
        235, 240, 245, 247
    pathogenesis, 242–245
    T cell activity, 230, 234, 235, 239,
        240, 241, 242, 244, 245
    virological studies, 235–240
Erythema multiforme, 107–108
[(Erythro-9-(2-hydroxy, 3-
        nonyl)hypoxanthine], 411
Esophagitis, CMV-related, 215
5-Ethyl-2'-deoxyuridine, 280

Fetus, CMV infections in, 297
Flagellin, 115
Fluoroarabinosylpyrimidine derivatives,
    365
Fluoro-iodo-arabinosyl-cytosine, 219, 341
Fungal superinfection, 17

Genetic engineering, of HSV vaccines, 279,
    281–89
  deletion mutants, 282–289
  intertypic recombinations, 282
Genetic resistance
  to HSV-1, 39–42
    interferon in, 57–58
  to HSV-2, 43, 47–48
Genital herpetic infections, 1–29
  BCG therapy, 49, 398
  clinical manifestations, 7–23
    cervicitis, 11, 20–22
    clinical course, 17–18
    first episodes, 7–8, 17–18
    future research, 18–19
    proctitis, 15, 23
    recurrent, 19–20
    ulceration, 10–11, 23, 24
    urethritis, 8, 22–23
  complications, 13–17
    of central nervous system, 13–16
    disseminated infections, 16–17

Genital herpetic infections (*cont.*)
  complications (*cont.*)
    extragenital lesions, 16
    local extension, 17
    superinfection, 17
  epidemiology, 2–7
    demographic factors, 2, 3, 4, 5–7
    HSV antibody prevalence, 4–6
    incidence, 6, 352
    prevalence, 2–4
  IgM sera levels, 78
  immunoprophylaxis, 28
  natural history of, 1–35
  in pregnant women, 16
  primary, 8–13
    complications, 13–17
    concomitant infections, 11
    HSV-1-related, 81
    pharyngeal infections, 11–13
    signs, 10–11
    symptoms, 8–10
  recurrent, 23–27, 352
    neutralizing antibodies, 81
    reinfection vs. recrudescence, 26–27
    "trigger" mechanisms, 26
  sexual transmission, 27–28
  therapy
    clinical trials, 351–354
    intravenous, 353–354
    levamisole, 406–407
    oral, 354
    phosphonoformate, 353
    topical, 352–353
    vidarabine, 347
  vaccines, 81, 267, 290
Gibbon, EBV infections in, 149
Gingivostomatitis, HSV-related, 351
Glomerulonephritis, 107
Glucan, 400
Glycoproteins, of CMV, 309
Guinea pig, CMV infections in, 301, 308

Hemophilus ducreyi, 23
Hemopoietic system, 41–42
Hepatitis, HSV-2-induced, 43, 379
  macrophage function in, 47–48
Hepatitis B, 347
Hepatitis B virus surface antigen, 116
Herpes ateles, 157
  lymphocyte transformation by, 161
  target cell, 158
Herpes genitalis, 49
Herpes labialis, 347, 351, 353, 402, 406–
    407

Herpes simplex virus infections
  in AIDS patients, 59, 379
  antibodies, 69–81
    cytolytic, 72
    in human convalescent sera, 70–71
    IgA and IgM classes, 77–78
    in vivo function, 79–81
    measurement, 71–74, 77–79
    neutralizing, 71–72
    precipitating, 72–73
  antibody-dependent cell-mediated
      cytotoxicity in, 48, 72, 379
  in bone marrow transplant patients, 395
  in cardiac transplant patients, 395
  cervical cancer and, 70, 81, 266, 273
  cutancous, 351–354
  encephalitis, 15, 78, 354–358, 411
  genital, 1–35
    antibody prevalence, 4–6
    BCG therapy, 49, 398
    clinical course, 17–18
    clinical manifestations, 7–23
    complications, 13–17
    concomitant vulvar and cervical
        infections, 11
    demographic factors, 2, 3, 4, 5–7
    epidemiology, 2–7
    first episodes, 7–8
    IgM sera levels, 78
    immunoprophylaxis, 28
    incidence, 6, 352
    levamisole therapy, 406–407
    natural history, 1–2
    pharyngeal infections and, 11–13
    in pregnant women, 16
    prevalence, 2–4
    primary, 8–13
    recurrent, 19–20, 23–27, 352
    sexual transmission, 27–28
    therapy, 347, 351–354, 406–407
    vaccine, 81
  humoral response, 69–86
    antibody measurement, 71–74, 77–79
    cytolytic antibodies, 72
    ELISA antibody measurement, 70, 73–
        74
    HSV-specific antibodies, 70–71
    IgA and IgM antibody classes, 77–78
    immunoblotting test, 74, 79, 80
    in vivo antibody function, 79–81
    neutralizing antibodies, 71–72
    precipitating antibodies, 72–73
    protein immunogenicity, 74–77
    radioimmunoassay, 70, 73–74

Herpes simplex virus infections (cont.)
  immunopathology, 97–98, 103–119
    antigen-presenting cells, 103–105
    erythema multiforme, 107–108
    IgE-mediated, 105–106
    immune complex, 106–109
    infectious mononucleosis, 107, 113–
        114
    keratitis, 108–109, 111–113
    management, 114–116
    mechanisms, 103–105
    in renal transplant patients, 107
    T cell-mediated, 109–114
  immunopotentiating agent effects, 395–
      416
    inosiplex, 407–411
    microbial, 397–400
    synthetic interferon inducers, 403–405
    thymic hormones, 401–403
    transfer factor, 401
  interferon interactions, 53, 57–60, 379–
      380, 382–383, 396
  latency
    cervical cancer and, 266
    reactivation and, 37–38
    vaccines and, 275, 281, 282, 284, 289
  natural killer cells in, 49–57, 379, 399
  natural resistance mechanisms, 37–68,
      87, 396
    endotoxins, 40
    genetic, 39–43
    hemopoietic system, 41–42
    interferon, 57–60
  neonatal
    maternal antibodies and, 80–81
    prevention, 358–359
    symptoms, 358
    therapeutic considerations, 359
    therapy, 360–361, 364
  pathogenesis, 37
  primary, 37, 38
  prophylaxis, 381–382
  protein immunoprecipitation, 72–73
  reactivation, 37–38, 69
  recurrent, 38
  in renal transplant patients, 395
  T-cell-mediated immune response, 87–
      102, 109–114,
    in athymic mice, 88
    in B-cell-suppressed mice, 89
    immune response regulation, 97–98,
        114–116
    in vivo responses, 93–97
    T cell memory, 98–99
    T cell subsets, 89–93

Herpes simplex virus infections (*cont.*)
  vaccines, 265–296
    autoinoculation, 267
    development, 268, 269, 273–274, 282–
      289
    developmental problems, 265–266
    efficacy, 267, 268–269, 270–271, 272,
      273, 276–278, 286, 289
    genetically engineered, 281–289
    inactivated virus, 267–273
    live, 279–289
    oncogenic, 273, 274, 281, 282
    placebo effect, 267–268, 271–273
    side effects, 271, 281
    standardization, 278–279
    subunit, 273–279
    as therapy, 267, 268–273
    virulent strains, 280
    wild-type virus, 266–267
  virulence, 280
  in Wiskott–Aldrich syndrome, 55–56
Herpes simplex virus 1
  cross-reactivity, 4–6
  genetic resistance to
    in animal models, 39–42
    interferon effects, 57–58
  genital infections, 81
  intertypic recombinations, 282, 283
  macrophage function and, 44–46
  mouse ear model of, 93, 94
  natural killer cell function and, 44, 45,
    53–55
  natural resistance to
    interferon in, 55
    macrophage in, 44–46
  neurovirulence, 38
  in newborn, 44, 45
  protein immunogenicity, 74–77
  recurrent infections, 73
  in trigeminal ganglia, 59–60
Herpes simplex virus 2
  cervical cancer and, 70
  cross-reactivity, 4–5
  genetic resistance to, 43
  genital infections, 1–29
    antibody response, 4–5
    recurrent, 81
  hepatitis and, 47–48
  host defense mechanisms, 38
  immunopotentiating agent effects, 397–
    398
  incidence, 280
  intertypic recombination, 282, 283

Herpes simplex virus 2 (*cont.*)
  natural resistance to, 379
    macrophage function in, 46–49
  in newborn, 48–49
  protein immunogenicity, 74–77
Herpes zoster virus
  interferon and, 377
  sacral, 14
  therapy, 362–364
Herpesvirus(es), heterologous, 281
Herpesvirus papio, 152–156
  antigens, 153–154
  DNA, 154–155
  experimental disease induction by, 155
  viral isolates, 153
Herpesvirus pongo, 156
Herpesvirus saimiri, 157–158, 330
  lymphocyte transformation by, 161–162
  target cell, 158–159
  tumor cell lines, 159–161
Histamine, interferon-induced release, 106
Hodgkin's disease, 179, 183
Homosexuals
  CMV infections in, 297–298
  HSV antibody rates in, 6
  HSV-related proctitis in, 23
Host defense mechanisms: *see* Natural
    resistance mechanisms
Human convalescent sera, HSV antibodies
    in, 70–71
Human T-cell leukemia virus, 156
Humoral immune response
  in EBV X-linked lymphoproliferative
    syndrome, 240, 243, 244
  in HSV infections, 69–86
    antibody measurement, 71–74, 77–79
    cytolytic antibodies, 72
    ELISA antibody measurement, 73–74
    HSV-specific antibodies, 70–71
    IgA and IgM antibody classes, 77–78
    immunoblotting test, 74, 79, 80
    in vivo antibody function in, 79–81
    neutralizing antibodies, 71–72
    precipitating antibodies, 72–73
    protein immunogenicity, 74–77
    radioimmunoassay test, 70, 73–74

Immune complex, 106–109
Immune globulin
  CMV, 212, 213–215, 221
  zoster, 212, 315, 362
Immune response modulation, 114–116,
    395–416

Immunoblotting, 74, 79, 80
Immunocompromised patients
  antiviral therapy in, 352
  varicella–zoster virus in, 314, 315, 318,
    319, 322, 323, 361–362
  See also Allografting
Immunocytolysis, of HSV-infected cells,
  70, 72
Immunodeficiency disorders
  HSV infections in, 395
  severe combined, 52
  See also X-linked lymphoproliferative
    syndrome
Immunoenhancement, by interferon, 374–
  375
Immunofluorescence detection
  of EBV antigen, 192
  of HSV antibodies, 70, 72
Immunoglobulin A antibodies
  assay, 185–187
  EBV-specific, 185–187, 190, 192
  HSV-specific, 77–78
  varicella–zoster-specific, 377
Immunoglobulin E antibodies, 105–106
Immunoglobulin G antibodies
  EBV-specific, 247
  HSV-specific, 77–78
  varicella-zoster-specific, 377
Immunoglobulin M antibodies
  CMV-specific, 209
  EBV-specific, 192
  varicella–zoster-specific, 377
Immunopathology
  of AIDS, 114
  of EBV infections, 230–231
  of herpesvirus infections, 97–98, 103–
    119
    antigen-presenting cells, 103–105
    erythema multiforme, 107–108
    IgE-mediated, 105–106
    immune complex, 106–109
    infectious mononucleosis, 107, 113–
      114
    keratitis, 108–109, 111–113
    management, 114–116
    mechanisms, 103–105
    in renal transplant patients 107
    T cell-mediated, 109–114
Immunopotentiating agents, 395–416
  antiviral drug interaction, 411–412
  inosiplex, 407–411
  levamisole, 405–408
  macrophage effects, 45
  microbial, 397–400

Immunopotentiating agents (cont.)
  synthetic interferon inducers, 403–405
  thymic hormones, 401–403
  transfer factor, 401
Immunoprecipitation, of HSV proteins, 70,
  72–73
Immunoprophylaxis
  active: see Vaccines
  passive
    CMV, 212–215, 298–300
    varicella–zoster, 212, 315, 362
Immunosuppression
  CMV-induced, 140–141, 202–203, 219–
    220
  interferon and, 257, 375
  manipulation of, 115
  T-cell-mediated immunity in, 88
  See also Allografting
Inactivated HSV vaccine, 267–268, 271–
  273
Infectious mononucleosis
  in Chediak-Higashi syndrome, 261
  CMV-related, 376
  definition, 251
  EBV-related, 150–151, 178, 229, 231,
    327–328, 333
    B cell response, 178
    immunopathology, 172
    interferon production, 259
    T cell response, 375, 378–379
  immunopathology, 107, 113–114, 172
  in Wiskott–Aldrich syndrome, 261
Influenza virus, 110, 224
Inhibition-passive hemagglutination test,
  70
Inosiplex, 407–411
Interferon
  in acquired immune deficiency
    syndrome, 59, 373
  in allograft recipients, 203
  α, 57, 58, 59, 255, 256, 257
  antigen-presenting cell interaction, 103,
    105
  antiviral mechanisms, 373–374
  B cell growth inhibition, 173–174
  clinical applications, 257–259
  immunosuppression in, 257
  in vitro, 255–257
  infectious mononucleosis and, 251–
    254, 259
  secretion defects, 257–259
  therapeutic applications, 246–247
  β, 57, 58, 59, 255, 256
  classification, 371–372

Interferon (cont.)
  CMV and, 133, 217–219, 309–310, 375–
    377
  definition, 371
  EBV and, 377–379
  future uses, 385
  γ, 57, 59, 255, 256, 257
  HSV and, 57–60, 379–380, 396
    effector cells, 58–59
    natural killer cell response, 53
  HSV-1 and, 57–58
    macrophage function, 46
  histamine release and, 106
  immunomodulating effects, 374–375
  in renal transplant patients, 309–310
  synthetic inducers of, 403–405
  system, 254–255
  therapeutic applications, 380–385
    animal studies, 381
    human studies 381–385
  varicella–zoster and, 377
Idoxuridine, 339, 343, 350

Kaposi's sarcoma, EBV-related, 184
Keratitis, HSV-related
  immunopathology, 108–109, 111–113
  therapy, 343, 350–351
    acyclovir, 383
    inosiplex, 411
    interferon, 382–383
    thymostimulin, 402–403
    trifluorothymidine, 382–383
    vidarabine, 347

Late membrane antigen, 252–253
Latency
  CMV vaccine effects, 307
  of HSV
    in cervical cancer, 266
    delayed-type hypersensitivity and, 111
    reactivation, 37–38
    vaccines and, 275, 281, 282, 284, 289
Lauroyl-L-alanyl-D-glutamyl-
    diaminopimelamic acid, 400
LCMV: see Lymphocytic choriomeningitis
    virus
Leukocyte migration inhibition factors,
    176–177
Levamisole, 405–408
Live vaccine, for HSV, 279–289
  advantages, 279–280
  antigen expression, 281
  genetically engineered, 281–289
  heterologous herpesvirus, 281
  mutant virulence, 280

Liver transplant patients, CMV infections
    in, 208
Lymphocyte(s)
  large granular, 122–123, 132–134
  virus-specific cytotoxic responses, 89–
    91, 109–114, 126–132
Lymphocyte-defined membrane antigen,
    175, 176
Lymphocytic choriomeningitis virus, 105,
    107
Lymphoid disease, EBV-related, 150: see
    also X-linked lymphoproliferative
    syndrome
Lymphokine(s)
  in CMV-infected cells, 125–126
  in EBV-infected cells, 176–178
  natural killer cell activity and, 125
Lymphokine production assay, 176–178
Lymphoma
  EBV-related, 183
    animal model, 157
    antibody profile, 179
    X-linked lymphoproliferative disease
      and, 231
  T-cell-tropic herpesviruses-related, 157–
    162, 164
Lymphoproliferative syndrome, 229–247
  EBV-related, 328
    in marmosets, 149–152
  herpesvirus papio-related, 152–156
  See also X-linked lymphoproliferative
    syndrome
Lymphotropic herpesviruses, 147–170
  associated diseases, 148
  characteristics, 147–148
  DNA, 147
  herpesvirus papio, 152–156
  in New World primates, 152–156
  in Old World primates, 156–157
  proteins, 147
  T-cell-tropic, 157–162
  Tupaia herpesviruses, 162–163
  See also Epstein-Barr virus infections
Lysosomal enzymes, 44

Macrophage
  CMV transport by, 300
  in delayed-type hypersensitivity, 109
  in HSV resistance, 379–380, 394
  in HSV-1 resistance, 44–46
    animal model, 44–46
    in man, 48–49
  in HSV-2 resistance, 46–49
    animal model, 46–48
    to hepatitis, 47–48
    in man, 48–49

Macrophage (*cont.*)
  mature cells, 43–44
  precursors, 43
Marek's disease, 329, 330
  vaccine, 281
Marmosets
  EBV infections in, 149–152
  herpesvirus papio in, 155
  herpesvirus saimiri in, 159–161
  T-cell-tropic herpesvirus in, 158, 159
Mast cell, 105, 106
Measles virus, 242
Meningitis, gentital herpes infection and,
    13–14, 15
Mice: *see* Mouse
Microbial immunopotentiating agents,
    397–400
Monkey
  EBV vaccination, 331–333
  HSV vaccination, 276, 287–289
Monoclonal antibodies
  in EBV-protein purification, 190–192
  HSV vaccine antigens and, 278–279
Monoclonal lymphoid disease, 150
Monocyte, 43, 124
Mouse
  CMV infections in
  cytotoxic lymphocyte response, 131–132
    natural killer cell activity, 132–133
    vaccines, 300–301
  HSV infections in, 80–81
    genetic resistance, 39–43
    vaccines, 276, 284–287
  HSV T-cell-mediated immune response,
      87–102
    in athymic mice, 88
    in B-cell-suppressed mice, 89
    immune response regulation, 97–98
    in vivo response, 93–97
    T cell memory, 98–99
    T cell subsets, 89–93
Mucocutaneous herpes infections, therapy
    for, 347, 411
Mumps virus, 244
Muramyl dipeptide, 400
  Mycobacterium tuberculosis, 397
Myelitis, HSV-related, 13, 15

Nasopharyngeal cancer, EBV-related, 183,
    327, 329, 385
  cellular immunity in 193
  serological markers, 185–187
Natural cytotoxic cell, 123

Natural killer cell
  in CMV infections, 122–123, 132–134
    depressed activity, 134–139
    lymphokine interaction, 125
  in EBV infections, 174–176, 178, 193–
      194, 230, 235, 240, 245, 247
  effector cells, 51–53
  heterogeneity of, 50–52
  in HSV infections, 49–57, 396–397, 379
    cell lineage, 52
    corynebacteria activator, 399
    interferon effects, 53
  in HSV-1 infections, 44, 45, 53–55
  interferon effects, 50, 53
  in Wiskott–Aldrich syndrome, 55–56
  in X-linked lymphoproliferative
      syndrome, 56
Natural resistance mechanisms, 37–68, 87,
    396
  endotoxins, 40
  genetic, 39–43
  hemopoietic system, 41–42
  interferon, 57–60
Neisseria gonorrhoeae, 3, 11
Neutrophil, 124
Newborn
  CMV infections in
    antibodies, 122
    transfusion-acquired, 206, 298, 299
  EBV infections in, 247
  HSV infections in
    maternal antibodies and, 80–81
    natural killer cell activity, 56, 379
    prevention, 358–359
    severity, 395
    symptoms, 358
    therapeutic considerations, 359
    therapy, 347, 360–361, 364
  HSV-1 susceptibility, 44, 45
  HSV-2 infections in, 48–49
  lymphocytic choriomeningitis virus in,
      107
  varicella in, 313
Noninfectious enveloped particles, 308–
    309
Nosocomial transmission, of CMV, 205–
    206
Nucleoside analogs, 246, 340
Null cell: *see* Natural killer cell

2′,5′-Oligoadenylate synthesis, 230–231
Oncogenic viruses, virus-host relationship,
    171
Orangutan, lymphotropic herpesvirses in,
    156

Organ transplants: *see* Allografting; specific types of transplants

Osteopetrosis, natural killer cell activity in, 49–50

Owl monkey, T-cell-tropic herpesvirses in, 157–158

Pharynx, HSV infections of, 11–13

Phosphonoformate
  clinical trials, 353
  development, 340
  HSV mutant resistance to, 280

Placebo effect, in vaccine development, 267–268, 271–273

Pneumonia, CMV-related, 203, 213, 214–215, 218

Poly I:C, 403, 404

Polyclonal lymphoid disease, 150

Polymorphonuclear leukocytes, 105, 106–107, 108–109

Polypeptides, of EBV-infected cells, 191–192

Prednisone, 134–135

Pregnancy, genital herpes infections in, 16

Primates
  EBV infections in, 147–152
    host range for disease induction, 149–150
    tumor clonality, 150–152
  lymphotropic viruses in, 147–170
    associated diseases, 148
    characteristics, 147–148
    DNA, 147
    herpesvirus papio, 152–156
    herpesvirus pongo, 156
    proteins, 147
    T-cell-tropic, 157–162
    Tupaia, 162–163

Proctitis, HSV-related, 15, 23

Promonocytes, 43

Protein(s)
  EBV-specific, 190–192
  HSV-specific
    immunogenicity, 74–77
    immunoprecipitation, 72–73
  lymphotropic virus-specific, 147

Protein kinase, CMV-associated, 309

Pyran, 403–404

Pyrimidinones, 403

Radioimmunoassay, of HSV antibodies, 70, 73–74

Recombination, intertypic, 282, 283

Reinfection, recrudescence, vs., 26–27

Renal failure, acyclovir clearance in, 349–350

Renal transplant patients
  CMV infections in, 131, 395
    cytotoxic lymphocyte response, 128, 133–134
    donor selection and, 210–211
    graft rejection, 205
    immune complexes, 107
    incidence, 202
    interferon interaction, 218, 384
    pathogenesis, 202, 203, 205
    steroid therapy effects, 135–137
    transfusion-acquired, 207, 298
    vaccination against, 216, 302–305
  HSV infections in, 395
    interferon prophylaxis, 218, 309–310

Reovirus, 116

Restriction enzyme analysis, of genital herpetic reinfection, 26–27

Rhesus monkey, HSV vaccination of, 276

RNA viruses, interferon induction by, 373–374

Sacral radiculopathy syndrome, 13

Salvia, CMV transmission in, 205

Secretions, HSV excretion in, 27–28

Sera, HSV-antibody typing in, 73

Seroepidemiology
  of EBV, 184
  of genital herpetic infections, 4–6

Serology, of EBV-related cancers, 184–190, 194–195

Serum thymic factor, 402

Severe combined immunodeficiency disease, 52

Sexual transmission
  of CMV infections, 205, 297
  of genital herpetic infections, 27–28

Sodium diethyldithiocarbamate, 411

Spider monkey, T-cell-tropic herpesvirus infections in, 157

Squirrel monkey, T-cell-tropic herpesviruses infections in, 157, 158

Steroid therapy
  for CMV infections, 134–137
  for EBV infections, 246

Subunit vaccines
  CMV, 307–309
  EBV, 247
  HSV, 273–279
    in animal models, 274–276
    clinical efficacy, 276–278
    future trends, 278–279
    preparation, 274

Superinfection
  in CMV infections, 203, 205
  in genital herpetic infections, 17

T cell
  accessory cells, 123
  in AIDS patients, 137–139
  in CMV-infections, 244
    in allograft patients, 204, 205
    depressed activity, 134–139
    infected fibroblasts, 124–126
    interferon and, 376
    virus-specific cytotoxicity, 128
    virus transport by, 300
  in EBV-infections, 171–179, 261
    antigens, 175–176
    B cell interactions, 173–176, 253–254, 261
    infectious mononucleosis, 378–379
    neoplastic disease, 193
    proliferation, 172
    X-linked lymphoproliferative syndrome, 230, 234, 235, 239, 240, 241, 242, 244, 245
  in HSV-cell-mediated immunity, 87–102, 109–114
    in athymic mice, 88
    in B-cell-suppressed mice, 89
    helper cells, 92
    as herpes saimiri target cell, 158–159
    immune response regulation, 97–98
    in influenza, 244
    in vivo response, 93–97
    in mice, 89–91
    T cell memory, 98–99
    T cell subsets, 89–93
  memory, 98–99
  precursors, 125
  suppressor cells
    in delayed-type hypersensitivity, 109–114
    in HSV infections, 92–93
T-cell-tropic herpesviruses, 157–162
  antigen modulation, 158–159
  herpes saimiri, 159–161
  host range, 157–158
  in vitro transformation, 161–162
  lymphocyte tropism, 158–159
  pathogenicity, 157–158
Tetramisole, 405–406
Therafectin, 411
Therapy, 339–369
  antiviral drugs
    clinical trials, 350–364
    controlled trials, 345–350
    development, 339–345

Therapy (cont.)
  interferon, 380–385
    animal studies, 381
    human studies, 381–385
  See also names of specific drugs
Thymic hormones, 401–403
Thymostimulin, 402
Thymuline, 402–403
Tilorone 403, 405
Towne strain, of CMV, 141–142, 209–210, 301–307
Transfer factor, 377, 401
Transfusion transmission, of CMV, 205, 206–207, 211–212, 220–221, 298, 299–300
Transplacental transmission, of CMV, 297, 299
Treponema pallidum, 23
Trifluorothymidine
  action mechanisms, 343
  as keratitis therapy, 350, 382–383
Trigeminal ganglia, HSV-infection in, 58–60
Tupaia glis belangeri, 148
Tupaia herpesviruses, 162–163

Ulceration, genital, 10–11, 23, 24
Urethritis, 8, 22–23
Urine, CMV transmission in, 205

Vaccines
  adjuvants, 116
  anti-idiotypic, 116
  CMV, 297–312
    in allograft patients, 215–216, 221–222
    animal models, 300–301
    latency, 307
    live vs. attenuated, 300
    in renal transplant patients, 302–305
    subunit, 307–309
    Towne strain, 141–142, 209–210, 301–307
  EBV, 329–337
    adjuvants 331–332
    animal model, 152
    membrane antigen, 330–332
    need for, 329–330
  genital herpes, 81, 267
  HSV, 265–296
    autoinoculation, 267
    cervical cancer and, 81, 273
    development, 268, 269, 273–274, 282–289
    developmental problems, 265–266

Vaccines (*cont.*)
  HSV (*cont.*)
    efficacy, 267, 268–269, 270–271, 272,
      273, 276–278, 286, 289
    genetically engineered, 281–289
    inactivated virus, 267–273
    latency, 275, 281, 282, 284, 289
    live, 279–289
    oncogenic, 273, 274, 281, 282
    placebo effect, 267–268, 271–273
    side effects, 271, 281
    standardization, 278–279
    subunit, 273–279
    as treatment, 267–273
    virulent strains, 280
    wild-type virus, 266–267
  Marek's disease, 281
  varicella–zoster virus, 315–324
    clinical trials, 316–320
    development, 315–317
    dosage, 321–322
    future use, 323–324
    malignancy and, 318–320, 321, 322–
      323
    problems with, 320–321
    reinfection and, 320–321
    safety, 322–323
    zoster as complication, 322, 323
Vaccinia virus, as adjuvant, 116, 281
Varicella–zoster virus, 313, 361–362
  acyclovir therapy, 348
  antibodies, 377
  cancer and, 313–314, 319–320, 321,
    322–323
    thymic factor therapy, 402
  cell-mediated immunity to, 313
    transfer factor in, 401
  DNA, 320
  immune globulin, 212
  interferon interactions, 377, 383
  therapy, 314, 347–348
    clinical trials, 361–364
  vaccines, 315–324
    clinical trials, 316–320
    development, 315–317
    dosage, 321–322
    future use, 323–324
    malignancy and, 318–320, 321, 322–
      323
    problems with, 320–321
    reinfection and, 320–321
    safety, 322–323
  zoster rash, 314–315, 322, 323

Vidarabine
  action mechanisms, 344
  controlled trials, 346–347
  as CMV prophylaxis, 219
  development, 340, 343
  immunomodulating properties, 411–412
  therapeutic applications
    encephalitis, 355–358
    genital herpes, 347
    keratitis, 350
    neonatal HSV infections, 360–361, 364
    varicella–zoster virus, 314, 361–363,
      383
  therapeutic index, 343
Viral shedding, in genital HSV infections,
    9, 11, 12, 22
Viremia, CMV-related, 215, 218
Virion, of CMV, 300, 307–309
Virulence, of HSV, 280, 281
Virus–host relationship, of oncogenic
    viruses, 171
Vulvar infections, genital herpes-related,
    11

Wiskott–Aldrich syndrome, 252, 261
  antibody-dependent cell-mediated
    cytotoxicity in, 48
  HSV susceptibility in, 48, 55–56
  natural killer cell function in, 55–56

X-linked lymphoproliferative syndrome,
    178, 229, 230–250, 328
  B cell activity, 230, 234, 235, 240, 244
  in carrier females, 240–242
  clinical course, 233–234
  cytotoxicity studies, 236
  general characteristics, 231–232
  humoral immune response, 240, 243,
    244
  immunodeficiencies in, 178, 254
  immunologic studies, 233, 234–235,
    239–242
  in males, 232–240
  natural killer cell activity, 56, 230, 235,
    240, 245, 247
  pathogenesis, 242–245
  T cell activity, 230, 234, 235, 239, 240,
    241, 242, 244, 245
  virological studies, 235–240

Zoster immune globulin, 212, 315, 362